Contemporary Ergonomics and Human Factors 2010

Contemporary Ergonomics and Human Factors 2010

Editor

Martin Anderson FIEHF, EurErg, CMIOSH

Health and Safety Executive, Bootle, UK

CRC Press
Taylor & Francis Group
Boca Raton London New York

CRC Press is an imprint of the
Taylor & Francis Group, an **informa** business
A BALKEMA BOOK

Institute of
Ergonomics &
Human Factors

CRC Press
Taylor & Francis Group
6000 Broken Sound Parkway NW, Suite 300
Boca Raton, FL 33487-2742

First issued in paperback 2019

ISBN 13: 978-1-138-43447-9 (hbk)
ISBN 13: 978-0-415-58446-3 (pbk)

Typeset by Macmillan Publishing Solutions, Chennai, India

Visit the Taylor & Francis Web site at
http://www.taylorandfrancis.com

and the CRC Press Web site at
http://www.crcpress.com

Contents

Preface

These are the proceedings of the Annual Conference of the Institute of Ergonomics & Human Factors held in April 2010 at Keele University, Staffordshire, UK. The conference is a major international event for ergonomists and human factors specialists, and attracts contributions from around the world.

On 20th October 2009, the Ergonomics Society changed its name to the Institute of Ergonomics & Human Factors. Previously known as *Contemporary Ergonomics*, the title of these conference proceedings has therefore been renamed *Contemporary Ergonomics and Human Factors* to reflect this change.

Papers are chosen by a selection panel from abstracts submitted in the autumn of the previous year and the selected papers are subject to a peer review process before they are published in *Contemporary Ergonomics and Human Factors*.

This year, the proceedings contain 'full' papers and 'open' papers. Both categories of papers have been peer-reviewed; however, open papers do not follow the traditional scientific or technical format. The first part of these proceedings contains the plenaries and full papers; the second part contains the open papers followed by posters. The list of contents includes all papers combined together, arranged by session (topic).

The Institute of Ergonomics & Human Factors is the professional body for ergonomists and human factors specialists based in the United Kingdom. It also attracts members throughout the world and is affiliated to the International Ergonomics Association. It provides recognition of competence of its members through its Professional Register. For further details contact:

Institute of Ergonomics & Human Factors
Elms Court
Elms Grove
Loughborough
Leicestershire
LE11 1RG
UK

Tel: (+44) 1509 234 904
Fax: (+44) 1509 235 666

Email: ergsoc@ergonomics.org.uk
Webpage: http://www.ergonomics.org.uk

DONALD BROADBENT LECTURE

GREEN ERGONOMICS: EMBRACING THE CHALLENGES OF CLIMATE CHANGE

Margaret Hanson

Principal Ergonomist, WorksOut

The challenges presented by climate change, and moving to a lower carbon and lower energy way of living will continue to impact on us all. Business, government, society and individuals all need to consider our responses to some of the most significant challenges of our day. So what might this mean for ergonomics? Might changes in the way we travel, communicate, work, generate energy, produce food, consume goods and deal with waste, produce opportunities for ergonomics? Do we have something to contribute to greening our economy?

This talk will present some of these issues, and consider how ergonomists might respond to the challenges and opportunities that are arising.

PLENARY SPEAKERS

KEEPING IT REAL: THE IMPORTANCE OF THE INTERFACE

Sarah Sharples

Associate Professor in Human Factor, Nottingham University

As ergonomists, it seems intuitive that the design of the interface has a critical impact on the success of a product, whether in a work, leisure or educational environment. However, it is sometimes difficult to demonstrate this importance. This talk will consider examples of work drawn from areas such as virtual reality, pedestrian navigation systems, social networking technologies and transport control interfaces to consider how the design of the interface does contribute to operator performance and satisfaction, and how this interacts with other aspects of cognitive ergonomics and design.

RAISING SAFETY STANDARDS IN HAZARDOUS INDUSTRIES: A REGULATOR'S PERSPECTIVE

John Wilkinson

Team Leader, HSE Hazardous Installations Inspectorate

After ten years of developing the inspection and industry approach to human factors in major hazards control and prevention for the onshore chemical and allied industries, this is a regulatory (and personal) reflection on achievements, on what worked and didn't work, current challenges, and a look forward at the next ten years. All of our achievements have substantially raised the profile of human factors in the onshore major hazard industries and also the degree of focused activity in addressing them.

There's still a way to go though, in how people think about hazard and risk control, and about accidents and incidents, and this means personal and cultural change on a large scale – so this is not a short-term project.

ERGONOMICS IN EXTREME ENVIRONMENTS; AN EXPLORER'S PERSPECTIVE

Phil Wickens

Remote mountain guide and photographer

Lightweight travel in remote locations presents numerous challenges if it is to be carried out both safely and efficiently. Conditions are often challenging, and in the worst cases potentially lethal. Those who operate in polar and mountain environments have to cope with low temperatures, strong winds, high altitudes, sustained damp and desert conditions, which present numerous ergonomic challenges to participants and equipment designers alike. This talk, based on Phil's own expeditions and work, will present the challenges faced by those operating in extreme expedition environments, and aims to facilitate further discussion about the ergonomics of such activity.

INSTITUTE LECTURE

THE PERFECT IS THE ENEMY OF THE GOOD: IMPLICATIONS FOR ERGONOMICS RESEARCH AND PRACTICE

Peter Buckle

Professor of Health Ergonomics, Robens Centre for Health Ergonomics

Research, by it's very nature, seeks to find 'the truth' or 'the perfect'. But how much perfection do we actually need as practitioners? This Lecture will provide a professional perspective, based on almost 30 years of research and application. It will explore why and where ergonomics and human factors research is undertaken and the impact of this research on application. It will consider what triggers our research, how we reach an understanding of a problem and how this in turn influences our interventions and the evaluation of their efficacy.

AGEING

HEALTH, SAFETY AND HEALTH PROMOTION FOR OLDER WORKERS

J.O. Crawford, R.A. Graveling, H. Cowie, K. Dixon & L. MacCalman

Institute of Occupational Medicine UK

Changing demographics within the UK associated with increased longevity and reduced birth rates do bring a number of challenges with an ageing population. The following paper reports on a systematic review carried out to gather the evidence on the health safety and health promotion needs of older worker. The process identified that there are a number of age related physical and mental changes that occur but these can be moderated by increased physical, increased intellectual activity and improvement of other lifestyle factors. In reviewing accident data, older workers are at a reduced risk of accidents but are more at risk of a fatal accident. Although there is an increased risk of developing disease with age, many chronic diseases can be controlled and workplace adjustments made. Research suggests that occupational health intervention can reduce the risk of early retirement and health promotion interventions are seen as positive by older workers. There are still a large number of research gaps in this area. Guidance from the current evidence base is presented in this paper.

Introduction

Demographic change including increased longevity and a reduction in the birth rate within the UK and comparable EU countries are set to bring a number of challenges to us. While it is positive people are living longer, the economically active population (those aged 16 to 64 years) is set to reduce from 65% to 60% of the total population. Within the 50 to retirement age group, only 50% of individuals are still working one year before retirement age (National Statistics, 2005). At the time of this report, of workers over 50 years, 25% were working part-time and 17% were self-employed. Furthermore at this time there were 2.7 million people not working, 49% dependent on benefits, 9% with occupational pensions but 18% had retired involuntarily with limited choices or distorted incentives. There have been a number of government initiatives to improve this as the cost estimates ranged from £19-£31 billion without costing the human and social costs of poverty, inequality and ill health. Difficult questions have also arisen with regard to pension provision within the UK. Thus there is likely to be continued efforts made to extend working life and return individuals currently unemployed to the workplace.

The following paper summarises a systematic evidence review that aimed to identify the health, safety and health promotion needs of older workers and reports on guidance developed from the available evidence to enable better health and safety management for older workers.

Methodology

The review was carried out using the method developed by the Centre for Reviews and Dissemination to address the following questions. (The University of York, 1996)

1. What are the health, safety and health promotion needs of older workers and does this differ from the rest of the workforce?
2. How are those needs being addressed – what is effective and ineffective?
3. What guidance can be developed to aid in the management of older workers in relation to health, safety and health promotion?

Initial scoping work had identified that to address these questions, an understanding of age related physical and mental change in relation to work should also be addressed by the review. However, the main component of the review was to identify occupational health, safety and health promotion interventions that were effective in maintaining the health of older workers. A search strategy was developed in conjunction with a steering group. The strategy included 40 key words or phrases and inclusion and exclusion criteria including defining older workers as those of 50 years and older. Seventeen databases and 4 websites were searched for documents from 1990 onwards. The full search strategy is documented in (Crawford *et al.*, 2009).

On conclusion of the searches, an initial screening process was carried out to evaluate titles and abstracts against the inclusion and exclusion criteria. Full publications were then obtained and data were extracted and a quality assessment made of each paper based on the criteria below.

Table 1. Quality Assessment Criteria.

Strength of Evidence	Description
***	Strong evidence, provided by consistent findings in multiple, high quality scientific studies
**	Moderate evidence, provided by generally consistent findings in fewer, smaller or lower quality scientific studies
*	Limited or contradictory evidence, produced by one scientific study or inconsistent findings in multiple scientific studies
–	No scientific evidence

Results

Initial searches identified 180 publications relevant to the review. On completion of screening and data extraction, 59 papers were included in the review, 119 were excluded from the review as containing no data or no relevant data and a further two books could not be obtained in time for the review.

Age Related Change – Physical and Psychological

Research identifies that there are reductions in aerobic capacity with age approximating to a 10% reduction for each decade of life (Shephard, 1999, Shephard, 2000a, Savinainen *et al.*, 2004c, Savinainen *et al.*, 2004b, Savinainen *et al.*, 2004a, Gall and Parkhouse, 2004). This reduction was less in individuals maintaining physical activity and those without disease.

Anthropometric studies have identified that there is an increase in BMI caused by a reduction in height and increase in weight in both cross-sectional and longitudinal studies (Savinainen *et al.*, 2004c, Savinainen *et al.*, 2004b, Savinainen *et al.*, 2004a, Gall and Parkhouse, 2004). Muscle strength was also found to reduce with age but again this can be improved upon by maintaining physical activity (Shephard, 1999, Savinainen *et al.*, 2004c, Savinainen *et al.*, 2004b, Savinainen *et al.*, 2004a, de Zwart *et al.*, 1996). Two studies suggested that there may be a training effect occurring in specific muscle groups for individuals involved in heavy physical work (Gall and Parkhouse, 2004, Schibye *et al.*, 2001). In relation to musculoskeletal disorders (MSDs), both incidence and prevalence have been found to increase in relation to age (Cassou *et al.*, 2002, de Zwart *et al.*, 1997). However, this relationship is often confounded by factors including the use of different diagnostic methodologies and exposure duration being greater in older workers (Woods and Buckle, 2002). Woods and Buckle (2002) suggest that age-related imbalance between physical workload and capacity is likely to increase the risk of MSDs

A further two studies examined the need for recovery in older workers (Kiss *et al.*, 2008, Devereux and Rydstedt, 2009). These studies identified that an increased need for recovery was associated with increasing age, high physical and high psychological demands, monotonous work and working for more than 42 hours per week.

Balance was also found to change with age. Postural and functional balance was evaluated in 4 occupational groups including construction workers, fire-fighters, nurses and home care workers (Punakallio, 2003). The results identified that a reduction in functional balance was increased with age. However, construction workers and fire-fighters had better functional balance compared to the other two groups. This suggests that a training effect exists for those who have demand placed on their balance abilities.

Tolerance to heat has also been addressed but it has been concluded that age does not have an effect on heat tolerance but older people are more susceptible to heat related problems (Pandolf, 1997, Pandolf, 1991). It appears that heat tolerance is decreases due to cardiorespiratory changes. Other risk factors identified with an ageing population include the reduced thermorgulatory ability of those with Type 2 diabetes (Wick *et al.*, 2006). It is vital that any health surveillance or risk assessments for individuals undertaking hot work should take into account these issues,

The reduction in reaction time with age has been well documented (Benjamin and Wilson, 2005). However, it is difficult to translate laboratory based studies into the workplace. Morgan (2004), reviewed psychological aspects of ageing and found that this is due to a slowing of central of central processing time which is accompanied by an increase in caution; thus a trade-off between speed and accuracy. Positively, ageing brings with increased accumulated knowledge and experience. There are large inter-individual differences within the research but any interventions for older people should include mental stimulation as well as physical activity (Morgan, 2004)

There was little research identified with regard to mental wellbeing in older workers. The two papers reviewed identified that social support, risk reduction strategies for stress and improving coping strategies were important (De Lange *et al.*, 2006, Gershon *et al.*, 2002)

Working time was also covered by two research papers. These identified that older workers involved in heavy manufacturing were more likely to have adverse work outcomes if they were hourly paid (this meant in more physically demanding jobs) and working more than 60 hours per week (Allen *et al.*, 2008) . This suggests that excessive overtime in physically demanding jobs will have a negative impact. Further research has also identified that reductions in the work ability index were also associated with shiftwork and found to reduce more quickly in females carrying out night work (Costa and Sartori, 2007). This was thought to be related to the dual roles of females and a healthy worker effect in men continuing to do shiftwork.

Accidents and Injuries in Older Workers

In reviewing accident and injury data from different countries the reader should be reminded that data is collected in a variety of ways which will impact on its quality. These data suggest that older workers were found to be less of an accident risk than younger workers (under 24) but females over 55 had the highest estimated incidence rate (HSE, 2008). It was suggested this was due to the higher number of females in blue collar occupations in this age group. Although the risk of serious non-fatal injuries was lower in the over 55 year plus group, injuries were more severe and recovery took longer (Rogers and Wiatrowksi, 2005, Laflamme and Menckel, 1995). Employer engagement was identified as being key in return to work after an injury but no significant changes in function were identified on return to work (Pransky *et al.*, 2005).

The number of fatalities in older workers has been found to be higher in comparison with other groups (HSE, 2008, Rogers and Wiatrowksi, 2005, Grandjean *et al.*, 2006). Larger numbers of incidents were found in agriculture, construction and transportation including highway incidents, being struck by an object and falls to a lower level (Personick and Windau, 1993).

Data on ill-health in older workers have identified that there is an increased risk of developing disease with age but this is not always a reason to exclude an individual from work. Adjustments can be made within the workplace to allow employment to continue. Although the largest source of absenteeism at work is short-term non-certified absence, older workers over 55 do take more days away from work (HSE, 2008). What is concerning is that for new cases of work-related illness including musculoskeletal and stress, those over 45 years have the highest estimated prevalence rate (HSE, 2008)

Safety, Health and Health Promotion Interventions for Older Workers

Although data has identified that older workers are more at risk of a fatal injury and take longer to recover from injury, no intervention studies were identified to reduce strategies in this groups of workers.

A number of occupational health interventions were reviewed but their quality was limited. One good quality study identified that an intervention involving occupational physicians and line managers reduced early retirement and improved work ability when assessed 6 months post intervention (de Boer *et al.*, 2004).

Health promotion studies were also limited in number and quality. Health checks, counselling and health condition tests were seen as positive by older workers (Naumanen, 2006a, Naumanen, 2006b). However, (Shephard, 2000b) suggests that worksite health promotion can improve both wellbeing but there is currently little evidence of long term maintenance of behaviour change. He stresses the importance of participation by all groups of workers.

Guidance

One of the aims of the review was to understand the impact of age-related change on ability to work and if relevant be able to apply interventional evidence to aid in the health and safety management of older workers. The lack of evidence in this area has been problematic in the development of guidance but based on age-related change and ergonomic principles, the following is a summary of the guidance.

Physical Capacity

The job demands for all workers must meet the aerobic and muscular capacity of the individual. Where those demands are too great, is this due to the job being too demanding and requires re-design or is it due to the abilities of the individual. Thus objective assessment of job demands is necessary to be able to aid this process.

Consideration of the whole workforce is also essential as jobs cannot be designed for the strongest and fittest employees. Using ergonomics design including workplace design; work-rest scheduling, assessment of repetitive work, risk assessment and risk reduction of manual handling activities and ensuring there is a good reporting system in place.

Shift Work

Working long hours (over 60 hours per week in physically demanding jobs) and night working are potential risk factors for older workers. When managing people who work night shifts the following recommendations by Costa and Sartori (2007) should be followed.

(i) consider limitation or absention from night work after 45-50 years of age
(ii) give priority to transfer to day work
(iii) where possible give the choice of the preferred shift
(iv) reduce work load
(v) shorten working hours and/or increase rest periods
(vi) arrange more frequent health checks
(vii) give proper counselling and training on coping strategies concerning sleep, diet, stress management and exercise.

As there are large differences in how well individuals cope with shiftwork, proper support and health surveillance are vital.

Heat Tolerance

Although heat tolerance is not directly related to age, individual health and fitness are going to impact on the ability to tolerate heat. Thus for individuals involved in hot work, regular health assessments should be made to ensure their fitness is maintained and their health status has not changed.

Working Environment

Although hearing and vision change with age, some of these changes can be ameliorated by the use of risk management with regard to noise exposure and noise reduction measures including hear protection throughout working life. In the same vein, some age-related changes in vision can also be corrected via the use of corrective lenses or improvements in lighting in the workplace.

High Risk Industries

In certain high risk industries there may be a need to consider more frequent health assessment of staff to ensure they are able to do their work safely. However, any assessments made need to be objective and occupationally related; that is, relevant to the occupation and the task being carried out.

Psychological and Psychosocial Factors

Although there are a number of changes with age there is no evidence to support the view that mental impairment is a problem for those over 50. Although changes do occur with regard to reaction time, this is compensated by greater experience and knowledge. Access to training and continued intellectual stimulation is important for all workers. Despite the many myths associated with learning in older people updating of skills is important to all workers but consideration should be made of the training media for different groups.

There is currently little research available regarding mental wellbeing in older workers. Lack of social support and the development of coping strategies for high stress environments are needed. The lack of research does not mean there are no problems for this group of people as self-reported data indicates that work-related stress, anxiety and depression are higher in the over 50s. Any measures taken in the workplace must also include those over 50.

Safety, Occupational Health and Health Promotion

Accident data has identified that older workers are more at risk of a fatal injury, take longer to recovery from non-fatal injuries and employer engagement is vital to aid successful return to work. The types of accidents identified in older workers include sprains, strains, fractures and dislocations. Thus accident prevention and accident analysis post incident are vital tools in reduction strategies; as is ensuring workers of all ages are involved in this process.

Ill health is not an inevitable outcome of the ageing process and although there is an increased likelihood of developing disease with age, many diseases can be treated and controlled and allow individuals to remain at work. Self-reported data including work-related illness has identified that musculoskeletal symptoms and stress anxiety and depression are reported more often by the over 55s. This data can allow prioritisation of action for workplace interventions in the over 50s.

Occupational health interventions involving physicians, line managers and the worker can reduce the risk of retirement and sickness absence in older workers. Health checks have also been seen to be an important resource by older workers.

Health promotion initiatives in relation to older workers were found to be limited in number. There is the potential to improve and maintain health through this route but a number of difficulties have been identified. These include ensuring access to activities is encouraged for all age groups of workers, identifying from older workers whether age specific groups would encourage attendance, for multi-component interventions, maintaining attendance at all components and as with other health promotion activities allowing time during the working day to attend.

Conclusion

At the current time there is a lack of high quality research interventions to aid in the management of older workers in the workplace although age related change may have an impact on ability to work. However, using evidence based on research and consensus guidance can be provided in the management of this group of workers. It would be hoped that any improvements in the workplace will benefit all workers not just those over 50.

Acknowledgement

The authors would like to thank IOSH for funding this research project

References

Allen, H., Woock, C., Barrington, L. and Bunn, W. 2008, Age, Overtime, and Employee Health, Safety and Productivity Outcomes: A Case Study. *Journal of Occupational & Environmental Medicine*, 50, 873

Benjamin, K. and Wilson, S. 2005, *Facts and misconceptions about age, health status and employability*, (Health and Safety Laboratory, Buxton)

Cassou, B., Derriennic, F., Monfort, C., Norton, J. and Touranchet, A. 2002, Chronic neck and shoulder pain, age, and working conditions: longitudinal results from a large random sample in France. *Occupational & Environmental Medicine*, 59, 537

Costa, G. and Sartori, S. 2007, Ageing, working hours and work ability, *Ergonomics*, 50, 1914–1930

Crawford, J.O., Graveling, R.A., Cowie, H., Dixon, K. and MacCalman, L. 2009, *The health, safety and health promotion needs of older workers: An evidence-based review and guidance*, Report Number 09.4 (IOSH, Wigston, UK)

de Boer, A., van Beek, J., Durinck, J., Verbeek, J. and van Dijk, F. 2004, An occupational health intervention programme for workers at risk for early retirement; a randomised controlled trial, *Occupational and environmental medicine*, 61, 924–929

De Lange, A., Taris, T., Jansen, P., Smulders, P., Houtman, I. and Kompier, M. 2006, Age as factor in the relation between work and mental health: results of the longitudinal TAS survey, *Occupational Health Psychology*, 1

de Zwart, B.C.H., Broersen, J.P.J., Frings-Dresen, M.H.W. and van Dijk, F.J.H. 1997, Musculoskeletal complaints in the Netherlands in relation to age, gender and physically demanding work, *International archives of occupational and environmental health*, 70, 352–360

de Zwart, B.C.H., Frings-Dresen, M.H.W. and van Dijk, F.J.H. 1996, Physical workload and the ageing worker: a review of the literature, *International archives of occupational and environmental health*, 68, 1–12

Devereux, J.J. and Rydstedt, L.W. 2009, Does the older workforce with high work demands need more recovery from work, *Contemporary Ergonomics 2009*, 189–196

Gall, B. and Parkhouse, W. 2004, Changes in physical capacity as a function of age in heavy manual work, *Ergonomics*, 47, 671–687

Gershon, R.R.M., Lin, S. and Li, X. 2002, Work Stress in Aging Police Officers. *Journal of Occupational and Environmental Medicine*, 44, 160

Grandjean, C.K., McMullen, P.C., Miller ,K.P., Howie, W.O., Ryan, K., Myers, A. and Dutton, R. 2006, Severe occupational injuries among older workers: Demographic factors, time of injury, place and mechanism of injury, length of stay, and cost data, *Nursing and Health Sciences*, 8, 103–107

HSE. 2008 *Self-reported work-related illness and workplace injuries in 2006/07: Results from the Labour Force Survey*, (Health and Safety Executive, Sudbury, Suffolk)

Kiss, P., De Meester, M. and Braeckman, L. 2008 Differences between younger and older workers in the need for recovery after work, *International archives of occupational and environmental health*, 81, 311–320

Laflamme, L. and Menckel, E. 1995 Aging and occupational accidents a review of the literature of the last three decades, *Safety Science*, 21, 145–161

Morgan, K. 2004, Psychological aspects of ageing, *Psychiatry*, 3, 8–10

National Statistics. 2005, *Focus on Older People*, (Palgrave MacMillan, London)

Naumanen, P. 2006a, The health promotion of aging workers from the perspective of occupational health professionals, *Public Health Nursing*, 23, 37–45

Naumanen, P. 2006b, Opinions of ageing workers on relative importance of health promotion. *International journal of nursing practice*, 12, 352

Pandolf, K.B. 1997, Aging and human heat tolerance, *Experimental aging research*, 23, 69–105

Pandolf, K.B. 1991, Aging and Heat Tolerance at Rest or During Work, *Experimental aging research*, 17, 189–204

Personick, M. and Windau, J. 1993, Characteristics of older workers' injuries, *Fatal Workplace Injuries in*

Pransky, G.S., Benjamin, K.L., Savageau, J.A., Currivan, D. and Fletcher, K. 2005, Outcomes in work-related injuries: A comparison of older and younger workers, *American Journal of Industrial Medicine*, 47, 104–112

Punakallio, A. 2003, Balance Abilities of Different-Aged Workers in Physically Demanding Jobs, *Journal of Occupational Rehabilitation*, 13, 33–43

Rogers, E. and Wiatrowksi, W.J. 2005, Injuries, Illnesses, and Fatalities among Older Workers, *Monthly Labor Review*, 128, 24

Savinainen, M., Nygård, C. and Ilmarinen, J. 2004a, A 16-year follow-up study of physical capacity in relation to perceived workload among ageing employees, *Ergonomics*, 47, 1087–1102

Savinainen, M., Nygård, C.H., Korhonen, O. and Ilmarinen, J. 2004b, Changes in Physical Capacity Among Middle-Aged Municipal Employees Over 16 Years, *Experimental aging research*, 30, 1–22

Savinainen, M., Nygard, C.H. and Ilmarinen, J. 2004c, Workload and physical capacity among ageing municipal employees—a 16-year follow-up study, *International Journal of Industrial Ergonomics*, 34, 519–533

Schibye, B., Hansen, A.F., Sřgaard, K. and Christensen, H. 2001, Aerobic power and muscle strength among young and elderly workers with and without physically demanding work tasks, *Applied Ergonomics*, 32, 425–431

Shephard, R.J. 2000a, Aging and productivity: some physiological issues, *International Journal of Industrial Ergonomics*, 25, 535–545

Shephard, R.J. 2000b, Worksite health promotion and the older worker, *International Journal of Industrial Ergonomics*, 25, 465–475

Shephard, R.J. 1999, Age and Physical Work Capacity, *Experimental aging research*, 25, 331–343

The University of York: Centre for Reviews and Dissemination. 1996, *Undertaking Systematic Reviews of Research on Effectiveness*, (York Publishing Services, York)

Wick, D.E., Roberts, S.K., Basu, A., Sandroni, P., Fealey, R.D., Sletten, D. and Charkoudian, N. 2006, Delayed threshold for active cutaneous vasodilation in patients with Type 2 diabetes mellitus, *Journal of applied physiology*, 100, 637–641

Woods, V. and Buckle, P. 2002, *Work, inequality and musculoskeletal health*, (HSE Books, Sudbury)

IS AGE/AGEING A RISK FACTOR FOR WORK-RELATED MUSCULOSKELETAL DISORDERS? A LITERATURE REVIEW

Olanrewaju O. Okunribido[1], Tony Wynn[1] & David Lewis[2]

[1] *Health and Safety Laboratory, Harpur Hill, Buxton, UK*
[2] *Health and Safety Executive, Liverpool, UK*

This work was undertaken to provide a review of the literature about ageing and work-related musculoskeletal disorders (MSD). Evidence suggests that attitudes towards ageing are changing; more employers now regard older workers as a valuable asset. Older workers are more susceptible to work-related MSD than younger workers due to decreased functional capacity. However, the propensity for injury is related to the difference between the demands of work and the worker's physical work capacity rather than their age. Once risk factors associated with work are adjusted for, older workers are no more likely to suffer an MSD than younger ones. An older workforce has implications for the health and safety responsibilities of employers. It is recommended that awareness campaigns be implemented to disseminate the benefits of ageing workers and raise awareness of the workplace barriers to their continued employment.

Introduction

Work related musculoskeletal disorders (MSD) are impairments of bodily structures, such as muscles, joints, tendons, ligaments and nerves, which are caused or aggravated primarily by the performance of work and by the effects of the immediate environment in which work is carried out (European Agency for Safety and Health at Work, 2008). They carry a high cost in terms of lost workdays in addition to medical treatment costs, making them an important issue for employers. For Great Britain, it has been reported that work-related MSDs affect around 1.0 million people a year (Jones *et al.*, 2006). Silverstein (2008) identified MSDs as one of the most frequently reported occupational illnesses among older workers and Peele *et al.* (2005) opined that MSDs in working populations might have a more pronounced effect on older workers than young workers since, physiologically, young adults have superior muscular performance peaking around age 33 and recovery time for musculoskeletal injuries lengthens with age.

The Health and Safety Executive's (HSE) policy concerning ageing and work, which is to fit the job to the worker, expects the risk assessment to take account of an employee's age and if necessary make appropriate workplace changes to accommodate this. HSE considers age in itself is not a risk factor for work-related MSD. Benjamin and Wilson (2005) showed that there are many misconceptions about the effect of age on employability and indicated that employers, workers and even health and safety professionals, do not always share HSE's current policy view. A review of the evidence regarding ageing and working with specific reference to work-related MSD was therefore considered necessary to ensure that important issues had not been overlooked, and that future advice was evidence based. The present work was undertaken to identify current scientific thinking about the subject, to evaluate whether individuals become more susceptible to MSD in the workplace as they age and to identify the implications of an ageing workforce for the health and safety responsibilities of employers.

Methodology

Studies included in this review were selected from searches of electronic databases (MEDLINE, OSHROM, Ergonomics Abstract), using key words (Ageing, Older, Worker, Injury, Disorders, Musculoskeletal, Pain, Interventions, Capability, Workability, etc.). The search was restricted to studies published from 2003 onwards. Studies were included if they were focused on older workers (aged 50 years and above), included older workers in the study population, compared outcomes between younger (25 years old or less) and older workers and examined the work-relatedness of musculoskeletal disorders (including symptoms and outcomes. The majority of the studies reviewed were in English and published in peer-reviewed journals.

Results

Demographic trends and functional capability

The literature regarding demographic trends, suggests the nature of the labour force in many countries, particularly developed countries, is changing, and that older workers are becoming more prevalent in the workplace. According to Silverstein (2008), increased life expectancy due to improvements in health and a decrease in birth rate over the years is leading to a progressive ageing of society. Whiting (2005) and Hotopp (2007) provided evidence that the rate at which companies are employing older workers has increased significantly since 1992, with a marked increase over the last decade and that, workers over the age of 50 are already becoming a defined part of the labour market. A growing number and proportion of older workers are also predicted for the next 25 years due to an anticipated shortage of younger workers (HSE, 2008). The data suggests the mean age of the UK population will rise from 39.6 years to 42.6 by 2031 and 44 by 2050, and it is projected that there will be an increase in the size of the population aged less than 45 years old (+2.7 million) by 2031, mainly due to the impact of migration. However, it is expected that the increase in the number of over 45 year olds during this period will be far greater (+7.8 million) than that of the less than 45 year olds. Thus, as older workers retire the flow of new workers required to replace them will be insufficient. Harris and Higgins (2006) identified a reasonable quantity of information relating to the retention of older workers and some useful interventions. However, while many employers did show increased appreciation of the value of older workers, only a few actually implemented measures, or increased their intake of older workers. This they attributed to insufficient information about laws governing workplace bias and equal opportunity, fear of being open to discrimination charges and a sparse evidence base for comprehensive programs and policies.

The studies on functional capability, e.g. Savinainen *et al.* (2004); Buchman *et al.* (2007); Attwood (2005); Welch *et al.* (2008), indicate three main ageing-related musculoskeletal changes: a reduction in joint mobility (range of movement), a decrease in muscular strength, and the slowing of reaction and movement times. These changes are influenced by individual genetics and lifestyle, as well as the environment in which individuals work and live (Buchman *et al.*, 2007). Therefore, physically active and fit older individuals may, in reality, be able to outperform those many years younger than them and the type of job that is done may have either a strengthening or wearing effect on physical capacity. The studies indicate that musculoskeletal functional capacities decline progressively with age, but several factors other than chronological age, such as level of physical activity and the demands of the work, contribute to determine an individual's decrease in capabilities.

Prevalence/incidence and consequences of injury

Generally, the studies (Kaila-Kangas *et al.*, 2006; Landau *et al.*, 2008; Hotopp, 2007) report higher prevalence/incidence values for older workers than younger workers, and higher values for those who leave work due to disease compared to those who continue in work till retirement.

Between the ages of 51 and 62 years, the prevalence of musculoskeletal disorders may increase as much as 15% among workers, with more pronounced increases occurring in physically demanding occupations, especially where such occupations do not maintain or improve strength (Savinainen *et al.*, 2004). Furthermore, aged workers suffer more serious but less frequent workplace injuries than younger workers. The MSDs they suffer are often the result of a failure to match the work-based requirements of a task to the functional capacity of workers (Silverstein, 2008). Thus, older workers in physically demanding occupations are more likely to report musculoskeletal injury complaints (back, neck, upper/lower extremities) than their younger counterparts.

Three main consequences of MSDs are discussed: sick leave/absence from work; functional impairment, and medical costs. Studies into these (Peele *et al.*, 2005; Gardner *et al.*, 2008) generally suggest the consequences are more severe for elderly workers than younger workers, in terms of longer recuperation, lost work time and costs, for the same conditions. However, a higher probability of sickness absence has been reported for younger workers in terms of higher total counts of recorded absence days, and a decreased number of one-day short-term absences has been reported for older workers (Leaker, 2008).

Risk factors for injury

The studies concerning the risk factors for injury, e.g., Breslin and Smith (2005), Hartman *et al.* (2003) and Landau *et al.* (2005), suggest various factors including age. However, the majority, including some that applied prospective study designs, reported no age effect or a higher relative risk for work factors independent of age compared with psychosocial factors and personal factors.

Peek-Asa *et al.* (2004) examined incidence rates of acute low-back injury by age, gender, job-specific lifting intensity and length of employment in a large prospective cohort study of manual handlers (n = 2152) and a multi-variate analysis of the data. The central hypothesis was that age-specific rates of acute low-back injury would be similar when stratified by the exploratory variables. The results showed that workers aged ≥ 55 years had a rate ratio (RR) of 0.84 (CI 0.69-0.92) compared to those aged <45 and a trend of decreasing rates with increasing age, which was more apparent for men than for women. Workers aged between 45–54 years were 0.69 times as likely to sustain an injury causing days away from work (CI 0.53-0.88), and those aged ≥ 55 were slightly but not significantly less likely to have missed workday injuries than workers aged <45 (RR 0.93, CI 0.70-1.23). For all age groups, the rate of injury decreased with length of employment and increased with increasing lifting intensity. The authors concluded that workers aged 55 and over were no more likely to suffer a low back injury at work than younger workers (aged 45 or less), irrespective of the lifting intensity in the work and their length of employment.

Werner *et al.* (2005) investigated the incidence rate and risk factors of upper extremity tendonitis (UET) with a longitudinal study design. 501 active workers were involved as participants and they were followed over 5.4 years. The factors found to have the highest predictive value for identifying persons likely to develop UET in the future included older age (>40 years), high BMI (>30), complaint at base line, history of injury and job with high effort requirement. The authors concluded that older, heavier workers who also had a history of discomfort or injury were at the highest risk for developing new injuries.

Breslin and Smith (2005) investigated age-related differences in work injuries, using a cross-sectional study design, with an emphasis on adjusting for the potential confounding effects of job characteristics. The results showed that, adolescents and young adult workers, especially males, showed elevated rates of work injury compared to adults who were ≥ 35 years. The injured young workers had proportionally more acute, traumatic injuries, and fewer musculoskeletal injuries than older workers. Multivariate analysis indicated that differences in the types of jobs young workers held played an important role in their elevated risk for injury. For men this was such that age related differences were reduced when job characteristics were controlled for, though even in the fully adjusted model, young males 15–34 years continued to exhibit a moderately elevated risk of injury. The age differences in injury rates for women were less marked than those for males.

Landau *et al.* (2008) carried out a cross-sectional workload-strain-symptoms study on 256 workers and workstations in a company of the automotive industry, to identify the independent role played by age. They used recursive partitioning to create regression trees to identify features or combinations of features exhibiting a strong association with the target variables and influences attributable to worker age. The analysis, like cluster analysis, makes it possible to segment a sample into sub-groups which are homogeneous, but also different from each other. In contrast to most cluster analysis procedures, it is also possible to identify the variables or combination of variables causing segmentation. The distribution for presence of musculoskeletal symptoms derived from the study revealed no correlations with worker age. Indeed, although multivariate analysis showed a strong association of age with workload, such that more unfavourable high workload jobs (based on expert rating) were occupied by younger workers and low workload jobs occupied by older workers, the association with physical health symptoms did not have any primary explanatory value. The authors opined from the results, that potential age-related performance deficits in the older workers were more or less completely compensated for by allocation to jobs less likely to cause them strain.

Discussion

This work was undertaken to examine more closely the role of age, specifically older age, as a risk factor for MSD injury in the workplace. The objective was to enable evidence based assessment of the impact that an increase in the number of older workers would have on duty holder responsibilities and HSE's responsibilities for provision of guidance. The literature reviewed showed that MSDs are a problem amongst both young and older workers in many workplaces but tend to be more severe for elderly workers than younger workers. Furthermore, there was evidence that older workers are already a significant component of the workplace as there is a shortage of younger people to replace them, should they retire, and case studies of organisational efforts to implement age friendly policies in the workplace were also identified (Buckle *et al*., 2008; Landau *et al*., 2008; Martin, 2005; Moyers and Coleman, 2004). These observations indicate that there is a general change in attitude towards ageing and increasing willingness of employers to retain/take on older workers.

Susceptibility to MSD

Benjamin and Wilson (2005), concluded that lifestyle, education, socio-economic status, genetics, stress, exercise, nutrition and healthcare needs have an equal if not greater importance than age as determinants of individual health. The current work identified that human functional capacity declines progressively with age, but that several factors, other than chronological age, such as an individual's level of physical activity and the physical demands of the work, made a greater contribution to susceptibility for MSD during work. Some of the studies reviewed, e.g. Peek-Asa *et al.* (2004), reported no significant association of age with prevalence/incidence of injury or they reported similar/higher risk levels for younger workers compared to older workers. Breslin and Smith (2005) found that when the potential confounding effects of work characteristics were adjusted for, young male workers (aged

35 years or less) showed higher rates of work-related injury than those aged over 35 years. Other studies, such as Hartman *et al.* (2003), and Landau *et al.* (2005), suggested that irrespective of their age, workers employed in physically demanding occupations, are more likely to report underlying health problems than those in sedentary occupations. Furthermore, it was identified that older workers are often forced to work closer to their individual maximum capacity than younger workers; typically, in most industries the demands of work do not change with the passage of time (Savinainen *et al.*, 2004; Kenny *et al.*, 2008). Thus, though a higher prevalence of MSD is often reported for older workers compared to younger workers, this may reflect the fact that many older workers are working closer to their physical capacity (Silverstein, 2008). These observations suggest the propensity for injury is related more closely to the difference between the demands of work and the worker's physical work capacity (or work ability) than their age. Physical work capacity (and/or work ability) rather than age should be the criterion used to determine if an individual is capable of performing a specific job and the likely level of risk of MSD it presents (Yassierli *et al.*, 2007).

Implications of ageing workforce

According to Alpass and Mortimer (2007), the main challenge an ageing workforce brings, is the need to deal with demographic change from a rapidly growing economically viable population to a slower growing and rapidly ageing population. These authors noted that the main working-age population (15–64 year old group) will probably contract after 2020, and opined that the effect of the trend on the workforce depends partly upon whether people decide to continue working into old age or whether they choose to retire. They identified a number of issues that employers would need to consider as a consequence of increased labour participation of older people, such as impact on working patterns, use of technology and corporate attitudes towards older workers. Silverstein (2008) identified the need for employers to encourage older employees to stay longer in work as they age by taking steps that supported their productive capacities and minimised their vulnerabilities. Arndt *et al.* (2005) identified two policy implications of ageing workforces: first, the need to sustain the health of workers generally and older workers particularly and, secondly, the need to develop prevention measures (including job alternatives in sufficient numbers). Both these require increased understanding of the needs of older workers. Yassierli *et al.* (2007) identified the need to consider individual factors and workers' physical capabilities when defining procedures for assigning tasks. Letvak (2005) identified the need to control job attributes which impact on health, especially high job demands, as well as the need for schemes that provide the workforce with increased support. These observations suggested two main groups of implications of an ageing workforce for health and safety in the workplace, those related to future requirements and those for workplace design and worker accommodations:

Future requirements

First, clear understanding of the needs of older workers is necessary if work, work equipment and work environments are to be designed accordingly.

Secondly, employers will need to be able to anticipate the physical and cognitive capabilities of their workers; there will be an increased challenge, i.e., to find the adjustments needed at work due to decreases of capability.

Workplace design and accommodation

The strategies applied for provision/adoption of workplaces will need to be such that they build upon workers' strengths and protect against their vulnerabilities, i.e., the physical work environments support the needs and capabilities of older workers, the general working environments are age friendly and preserve the capacity of employees to function safely and effectively as they age.

Conclusions

Attitudes towards ageing and work are changing. More employers regard older workers as a valuable asset and they are willing to keep current employees on for longer periods past the usual retirement age. However, while many do now appreciate the value of older workers, only a few implement measures to support and increase their retention of older workers. MSDs are a problem amongst both young and older workers in many workplaces but tend to be more severe for elderly workers than younger workers. Older workers are also more susceptible to work-related MSDs than younger workers because of decreased functional capacity. However, the propensity for injury is related more to the difference between the demands of work and the worker's physical work capacity (or work ability) than to age. An older workforce has implications for the health and safety responsibilities of employers. These include providing additional support for worker requirements, changing the workplace attitudes towards ageing, providing a positive knowledge base, adjusting the workplace design and improving worker/employer relationships (co-operation).

It is recommended that awareness campaigns are implemented to disseminate the benefits of ageing workers in the workplace and raise awareness of those elements of the workplace that are not well matched to their needs. The expectation is that this will change the attitudes of employers and employees towards ageing and aged workers.

References

Alpass, F. and Mortimer, R. 2007, *Ageing workforces and ageing occupations: A discussion paper*, (Department of Labour, Wellington, New Zealand)

Arndt, V., Rothenbacher, U., Daniel, U., Zscshenderlein, B., Schuberth, S. and Brenner, H. 2005, Construction work and risk of occupational disability: A ten-year follow-up of 14474 male corkers, *Occupational and Environmental Medicine*, 62, 559–566

Attwood, D.A. 2005, Ergonomics solutions for the ageing worker in the process industries. *Proceedings of the 2005 ASSE Professional Development Conference: Ernest N. Memorial Conference Centre, New Orleans, Louisiana, June 12–15,*

2005, (American Society of Safety Engineers, Des Plaines), Volume Session No. 501

Benjamin, K. and Wilson, S. 2005, *Facts and misconceptions about age, health status and employability*, (Health and Safety Laboratory, Buxton), HSL/2005/20

Breslin, F.C. and Smith, P. 2005, Age-related differences in work injuries: A multivariate, population-based study, *American Journal of Industrial Medicine*, 48, 50–56

Buchman, A.S., Boyle, P.A., Wilson, R.S., Bienias, J.L. and Bennet, D.A. 2007, Physical activity and motor decline in older persons, *Muscle Nerve*, 35, 354–362

Buckle, P., Woods, V., Oztug, O. and Stubbs, D. 2008, Understanding workplace design for older workers: A case study. In P.D. Bust (ed.) *Contemporary Ergonomics 2008*, (Taylor & Francis, London), 3–7

European Agency for Safety and Health at Work 2008, *Safety and health at work European good practice awards: Prevention of work-related MSDs in practice*, (OSHA, Bilbao: Spain), TE-81-08-190-EN-C

Gardner, B.T., Dale, A.M., Van Dillen, L., Franzblau, A. and Evanoff, B.A. 2008, Predictors of upper extremity symptoms and functional impairment among workers employed for 6 months in a new job, *American Journal of Industrial Medicine*, 51, 932–940

Harris, S. and Higgins, G. 2006, *A review of organisational interventions that can prevent early retirement among older workers*, (Health and Safety Laboratory, Buxton), RMS/06/03

Hartman, E., Oude Vrielink, H.H., Huirne, R.B. and Metz, J.H. 2003, Sick leave analysis among self-employed Dutch farmers, *Occupational Medicine*, 53, 461–468

Hotopp, U. 2007, The ageing workforce: A health issue?, *Economic & Labour Market Review*, 1, 30–35

HSE. 2008, *Health and Safety Statistics 2007/08*, (HSE Books, Sudbury, Suffolk)

Jones, J.R., Huxtable, C.S. and Hodgson, J.T. 2006, *Self-reported work-related illness in 2004/05: Results from the Labour Force Survey*, National Statistics Edition, (HSE Books, Sudbury, Suffolk)

Kaila-Kangas, L., Keskimaki, I., Notkola, V., Mutanen, P., Riihimaki, H. and Leino-Arjas, P. 2006, How consistently distributed are the socioeconomic differences in severe back morbidity by age and gender? A population based study of hospitalisation among Finnish employees, *Occupational and Environmental Medicine*, 63, 278–282

Kenny, G.P., Yardley, J.E., Martineau, L. and Jay, O. 2008, Physical work capacity in older adults: Implications for the ageing worker, *American Journal of Industrial Medicine*, 51, 610–625

Landau, K., Rademacher, H., Meschke, H., Winter, G., Schaub, K., Grasmueck, M., Moelbert, I., Sommer, M. and Schulze, J. 2008, Musculoskeletal disorders in assembly jobs in the automotive industry with special reference to age management aspects, *International Journal of Industrial Ergonomics*, 38, 561–576

Leaker, D. 2008, Sickness abscence from work in the UK, *Economic & Labour Market Review*, 2, 18–22

Letvak, S. 2005, Health and Safety of older nurses, *Nursing Outlook*, 53, 667–672

Martin, M.P. 2005, Holistic ergonomics: A case study from Chevron Texaco, *Professional Safety*, 50, 18–25

Moyers, P. and Coleman, S.D. 2004, Adaptation of the older worker to occupational challenges, *Work*, 22, 71–78

Peek-Asa, C., McArthur, D.L. and Kraus, J.F. 2004, Incidence of acute low-back injury among older workers in a cohort of material handlers, *Journal of Occupational and Environmental Hygiene*, 1, 551–557

Peele, P.B., Xu, Y. and Colombi, A. 2005, Medical care and lost workday costs in musculoskeletal disorders: Older versus younger workers, *International Congress Series*, 1280, 214–218

Savinainen, M., Nygard, C.H. and Ilmarinen, J. 2004, Workload and physical capacity among ageing municipal employees–a 16-year follow-up study, *International Journal of Industrial Ergonomics*, 34, 519–533

Silverstein, M. 2008, Meeting the challenges of an ageing workforce, *American Journal of Industrial Medicine*, 51, 269–280

Welch, L.S., Haile, E., Boden, L.I. and Hunting, K.L. 2008, Age, work limitations and physical functioning among construction roofers, *Work: A Journal of Prevention, Assessment and Rehabilitation*, 31, 377–385

Werner, R.A., Franzblau, A., Gell, N., Ulin, S.S. and Armstrong, T.J. 2005, A longitudinal study of industrial and clerical workers: predictors of upper extremity tendonitis, *Journal of Occupational Rehabilitation*, 15, 37–46

Whiting, E. 2005, The labour market participation of older people, *Labour Market Trends*, 113, 285–296

Yassierli, Y., Nussbaum, M.A., Iridiastadi, H. and Wojcik, L.A. 2007, The influence of age on isometric endurance and fatigue is muscle dependent: a study of shoulder abduction and torso extension, *Ergonomics*, 50, 26–45

APPLIED

USE OF HUMAN MODELLING SOFTWARE IN A COMMERCIAL DESIGN ENVIRONMENT: A SAMMIE CAD CASE STUDY

P.N. Day[1], C. Riley[2], C. Rohan[1], S.J. Summerskill[3] & R. Marshall[3]

[1]*Consumer Experience, NCR Financial Solutions Group Ltd, Dundee. DD2 4SW*
[2]*Consumer Experience, NCR Global Solutions Ltd, Swords, Co. Dublin, Ireland*
[3]*Department of Design & Technology, Loughborough University*

Communicating design recommendations in a large commercial organisation is a challenge. As a result of the number of competing perspectives that are part of modern product development, user-centred input needs to be promoted during the design process. The following paper describes recent work by the Consumer Experience [Cx] team at NCR in utilising the human modelling system SAMMIE during product design and evaluation. We have found this approach useful to communicate findings in a readily understandable, visual manner and thus convince stakeholders of the validity of the usability recommendations across a globally distributed design team. This case study describes the analysis of visibility and reach for both able-bodied and wheelchair users during the development of self-service terminals such as automated teller machines (ATMs) and kiosks. In order to adequately design for global markets, an approach is described using worldwide anthropometric models. This approach is presented as an alternative to the use of national percentile data. The authors have found the use of the SAMMIE human modelling tool to be effective and suggest that similar approaches may be of use to the wider ergonomics community when communicating with different stakeholders involved in the product development process.

Introduction

As subject matter experts within a large multinational corporation, the authors often have to communicate and justify the results from user-centred investigation to colleagues from different disciplines. These findings can be from a number of sources including empirical data from traditional usability studies, data from heuristic reviews or recommendations based on experience. It is a challenge to present this information in a form that is meaningful and compelling to colleagues from different disciplines. A considerable amount of time can be spent summarising data in multiple forms or using specially constructed physical models to communicate usability problems. For instance, when assessing visibility to a portion of a self-service terminal, such as an automated teller machine (ATM) fascia, we have had to resort to such labour intensive methods as constructing physical models of a concept, manually holding a camera at eye heights corresponding to people of varying stature and overlaying vision lines on annotated drawings (e.g. taken from computer aided design or CAD packages). In a similar manner, although reporting statistically significant results from studies has resonance with colleagues from related scientific disciplines, alternative presentation forms are often needed to make a more compelling case for design changes, which sometimes means that valuable input is only made later in the design process.

This paper describes recent work conducted using three dimensional human modelling software during the product development process at NCR. It is hoped this will be of benefit to both other practitioners in related industries and those considering the application of such tools in industrial environments.

Use of human modelling software

The authors have been using the human modelling software package SAMMIE CAD, for approximately twelve months. Human modelling systems have been shown to offer a valid approximation of real world fitting trials (Marshall et al., 2009; Summerskill et al., 2009) and the SAMMIE systems has successfully been used for rapid, early assessments of reach and visibility at NCR.

One particular advantage of using such a system is the ability to use existing three dimensional CAD models for usability analysis. Early concept work tends to move straight from sketches to high-level volume models with only the necessary features being defined. Refinement of these concepts into more detailed models continues in CAD. These CAD models can be imported into SAMMIE at any stage from very early concepts through to fully detailed models, meaning that quick assessments can readily be made without the need for more time-consuming and costly physical models. The authors are not advocating the complete replacement of physical prototyping, though a system like SAMMIE is useful for the quick evaluation of designs at an early stage in the development process.

Use of SAMMIE with worldwide anthropometric models

There is a significant challenge in designing products for a worldwide population, with a given percentile varying significantly between national populations. For instance, taking PeopleSize (2008) as a source, a 5th percentile female has stature of 1479, 1517 or 1559 mm for Chinese (urban), British and Dutch (1986) female populations respectively. Arm length (shoulder to fingertip) varies in a similar manner with a British female having an arm length of 665 mm and a Chinese urban female of 622 mm. For this reason, for quick assessments we tend to use seven worldwide anthropometric models for ambulant users and three models for the wheelchair user population collated from global anthropometric data to approximate the worldwide ambulant population, as described by Smith and Coventry (2002). These models have been designed to reflect the range of the world population, enclosing both the lower and higher ends of the national populations for which data exists. Each of these models can readily be created in SAMMIE, either by deriving from a corresponding percentile, or by defining key measures and defining based on semi-automated methods. More detailed analysis can be conducted using multivariate human models, or more specific analysis can make use individuals from the HADRIAN database.

Use of SAMMIE with individual anthropometry from HADRIAN

In addition to using SAMMIE with standard anthropometric models, we have been given access to a portion of the HADRIAN sample set (Marshall *et al.*, 2009). HADRIAN human models represent specific people from whom anthropometric and joint constraint data have been gathered. The HADRIAN sample contains 103 members over half of which had some form of disability. The HADRIAN dataset allows designs to be tested to ensure that they are accessible to a wide range of people, including those with disabilities such as arthritis and lower limb paralysis.

NCR has a long tradition of promoting inclusive design and our self-service terminals, including ATMs, have led the way in offering accessible features such as private audio, large, clearly legible keys, compact user interfaces and tactilely discernable card insertion and audio features. HADRIAN is useful for helping us go beyond the various accessibility guidelines, standards and regulations for self-service terminals and continue to promote universal design in this area.

As an aside, it is interesting to note that SAMMIE predictions of wheelchair users' reach are congruent with the standards and guidelines in this area, such as CAE (2002) and ADAAG (2004). In most cases the standards describe reach envelops that correspond with the comfortable reach of a smaller individual in a wheelchair. For example, Figure 1 shows the comfortable reach limit for a small (10th percentile UK female for stature) wheelchair user and the maximum highs for interface elements. The lowest line in this case is that corresponding to the upper limit of the ADAAG guidelines; namely 1220mm. It can clearly be seen that this corresponds closely with SAMMIE/ HADRIAN predictions for comfortable reach by a small wheelchair user at this style of terminal.

The Use of SAMMIE for predicting reach

Quick assessments of reach to individual interface components can readily be achieved, in practice this is often by using the aforementioned standard anthropometric models. Particular focus is paid to people at either end of the physical spectrum, such as small users in a wheelchair and very tall standing users. For example, Figure 2 shows an early concept for a self-service terminal assessed for reach by a small wheelchair user in a typical position, at approximately 45° (i.e. a compromise between a front and side approach to the terminal). An additional example is given of a concept with a large portrait touch screen. Reach envelopes for a small wheelchair user (Figure 2) and large ambulant user (Figure 3), are shown

Figure 1. Reach from wheelchair, lowest line at 1220 mm (product detail obscured).

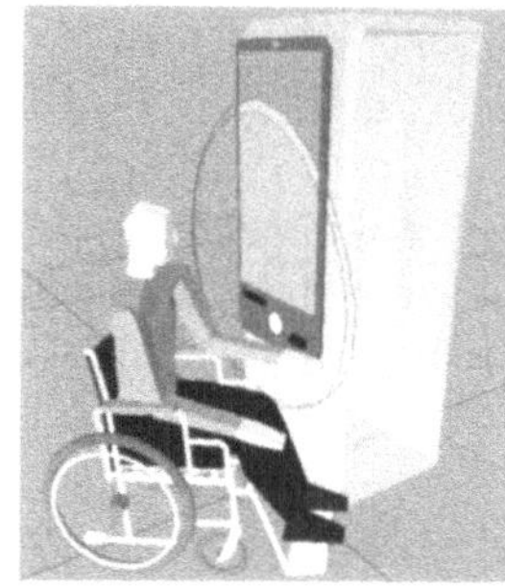

Figure 2. Reach from wheelchair showing reach contours.

Figure 3. Reach for tall ambulant user.

Figure 4. Reach for multiple users.

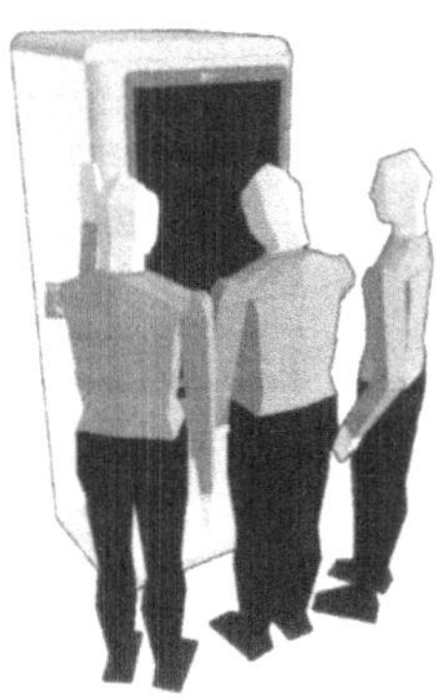

Figure 5. Reach for group of users at public terminal.

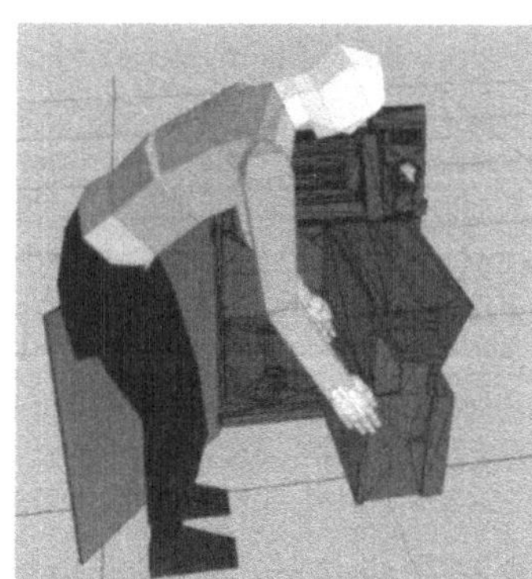

Figure 6. Reach for tall service/ maintenance personnel.

below as well as for a group of users, Figure 4 & Figure 5. It should be noted that the images included in this paper are for illustration only; often multiple iterations of posture and limb position are undertaken for detailed analysis.

Figure 6 to Figure 8 show how SAMMIE has been used to assess reach for service personal that replenish, maintain and service terminals. Based on analysis using

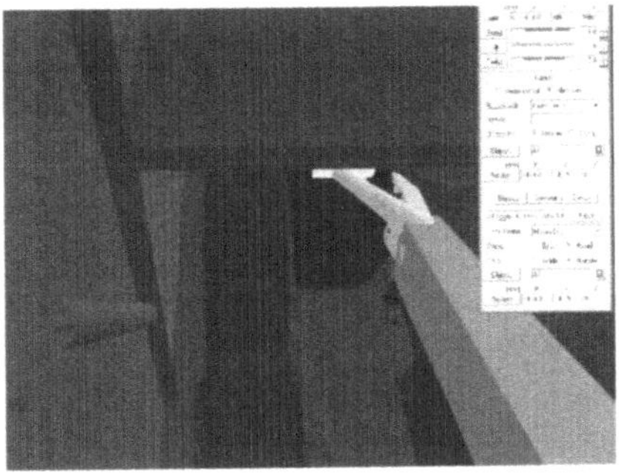

Figure 7. Reach & visibility, human view.

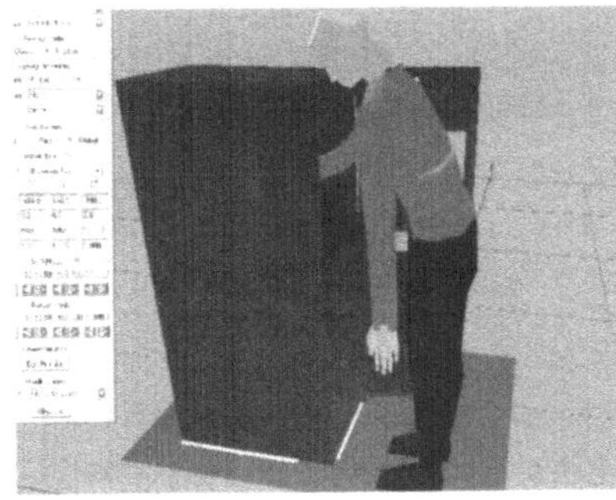

Figure 8. Reach & visibility, side view.

Figure 9. Short retraction – clearance view.

Figure 10. Long retraction – clearance.

SAMMIE a number of changes have been made to the service and replenishment mechanisms of newer designs.

Further initial work conducted using SAMMIE in a drive-up simulation demonstrated that this could be of benefit, particularly when investigating tasks that require higher levels of manual dexterity than just touching a screen. However this type of evaluation would require access to different vehicle models and further investigation of postures adopted and model fidelity.

The use of SAMMIE for predicting space volumes

In addition to assessing reach SAMMIE has been used to assess clearance and access requirements for different tasks. It has been useful to conduct quick evaluations of the types of postures people would need to be adopted for specific designs and subsequently define the space volumes that would allow comfortable postures. This helps us to define the space volume required to service a machine, i.e. the service footprint. Quick assessments can be made of how far modules should be retracted to provide sufficient space for replenishment, Figure 9 and Figure 10 with particular attention being paid to areas that require postures that takes up more space, Figure 11 and Figure 12.

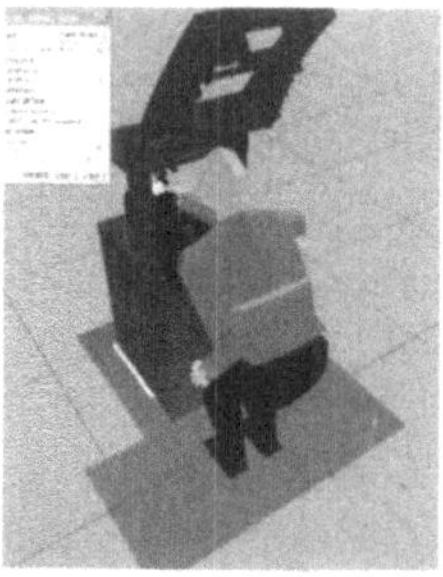

Figure 11. Crouching posture.

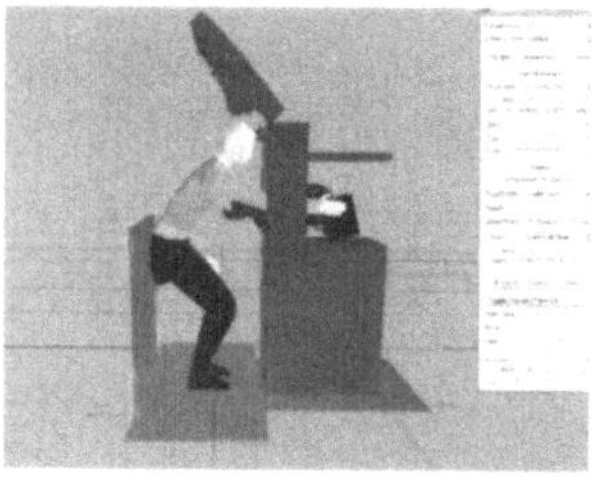

Figure 12. Crouching posture, side view.

Figure 13. Gripping media, concept 1.

Figure 14. Gripping media, concept 2.

As well as broad assessment of posture and clearance, more detailed analysis can be conducted using hand models representing a number of predefined postures or grips. This is particularly useful when considering designs involving hand clearances for both consumer and service personal. An example of this in use is shown in Figure 13 and Figure 14.

Use of SAMMIE for assessing visibility

A further quick analysis technique that we have been found useful is the assessment of visibility. Worst case scenarios can be assessed, such as a very tall user with a product mounted at its lowest installation height and posture can be set to a reasonable representation of that which a user may adopt. SAMMIE allows concepts to be 'viewed' through the eyes of the human (Figure 7, Figure 15 & Figure 16). Again, this offers an easy way of illustrating potential usability problems to other project stakeholders.

Unlike other packages that the authors have used, SAMMIE appears to offer a more accurate representation of what would actually be seen. Other systems can have an over-constrained view as if looking through a pair of binoculars. Furthermore, anecdotal tests with physical models and volunteers of differing statures suggest that the SAMMIE tool offer a realistic approximation of real life visibility.

Figure 15. Visibility assessment of moulding.

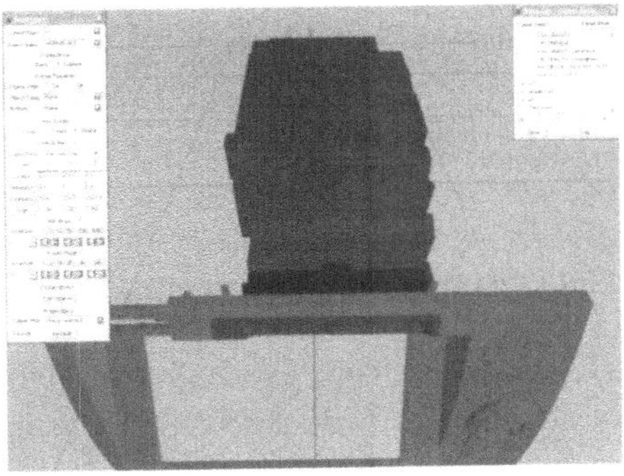

Figure 16. Visibility assessment of moulding – human view.

SAMMIE for communication

SAMMIE has proven useful for early evaluations of reach and visibility and we have also found significant benefit in using human modelling software as a communication tool. In addition to talking about the percentage of people accommodated by a design, the authors have found value in taking an image from SAMMIE and showing a product concept together with an a person using it. Previously usability issues have been raised and communicated within the organisation through a combination of verbal discussion and reports. Depending on the audience, it has been necessary in the past to *convince* people of the importance of a usability issue or the ramifications of a particular design decision. In our experience images from SAMMIE have helped to bridge the gap between ergonomists and other professionals, as we can use such images to show people the problem at hand. Being able to *see* the problem directly can be more influential than summarising information through tables or statistical analysis. In fact, a few screen grabs from a human modelling tool have can do more to convince others of the importance of user-centred design than any number of tables, statistical analysis or reports.

By contrast, prior to using a tool like SAMMIE, usability findings tended to be reported textually, or in the case of empirical data, by means of percentages and results of analysis. Illustrative photographs were sometimes used, and often a physical mock up was required to either investigate the issue concerned, or to communicate the particular concern with this issue. This has been of particular benefit in discussions with industrial designers, mechanical engineers, technical writers, product managers, technical project managers and marketing specialists. All groups have seemed to grasp problem areas more quickly when shown a couple of screen shots from SAMMIE. In some cases, this has resulted in changes being made after seeing SAMMIE images, whereas before using SAMMIE issues would have been judged to be insufficiently severe to warrant any action being taken despite recommendations being made to the contrary.

In communicating with others within a large development community, our experience has truly been that a picture is worth a thousand words, particularly as a thousand words are often contained in a report that may not be read!

Conclusions: Our Experience with SAMMIE

The Cx group at NCR has used SAMMIE to evaluate the physical design of numerous product concepts. The chief advantages of the tool have been the ability to evaluate concepts early in the design phase and to quickly provide feedback on subsequent design iterations. The ability to use existing CAD data for analysis and the availability of the HADRIAN dataset are further advantages. Human modelling systems are not a one stop shop for usability assessment. Such systems do not simulate textures, different lighting conditions or a host of other factors. Human modelling systems cannot be used to replace user-testing or real world evaluations; rather SAMMIE has proved to be a useful complement to these methods.

Perhaps the main advantage of using SAMMIE though has been its use as a communication tool. Instead of conveying usability input through reports, percentiles or tables of anthropometric data, a language perhaps best understood by usability professionals, SAMMIE allows information to be summarized in a readily understandable, pictorial way. Designers and engineers tend to be more familiar with CAD designs and wireframe models than statistical significance, making human modelling systems closer to a shared language across product development process. The use of SAMMIE has helped us bridge the gap between usability professionals and other stakeholders in the product development process.

References

Centre for Accessible Environments (CAE), 2002. Access to ATMs: UK Design Guidelines.

Federal Register Part VI Architectural and Transportation Barriers Compliance Board, 2004. 36 CFR Part 1191 *Americans with Disabilities Act Accessibility Guidelines: Accessible Automated Teller Machines (ADAAG).*

Marshall, R., Gyi, D.E., Case, K., Porter, J.M., Sims, R.E., Summerskill, S.J. and Davis, P.M., "A design ergonomics approach to accessibility and user needs in transport", *Contemporary Ergonomics 2009*, Bust, P.D. (ed), Taylor & Francis, The Ergonomics Society Annual Conference 2009, London, UK, April 2009, pp. 51–60.

Smith, K.T. and Coventry, L. 2002, Designing for a world population. In *Contemporary Ergonomics 2002,* (Taylor and Francis, London), 340–345.

Summerskill, S.J., Marshall, R., Case, K., Gyi, D. E., Sims, R.E. & Davis, P., 2009. Validation of the HADRIAN System Using an ATM Evaluation Case Study. *Proceedings of the 2nd International Conference on Digital Human Modeling: Held as Part of HCI International 2009*, Springer, 727–736.

REACH MODELLING FOR DRIVE-UP SELF-SERVICE

P.N. Day[1], C. Rohan[1], L. Coventry[4], G. Johnson[3] & C. Riley[2]

[1]*Consumer Experience, NCR Financial Solutions Group Ltd, Dundee, DD2 4SW*
[2]*Consumer Experience, NCR Global Solutions Ltd, Swords, Co. Dublin, Ireland*
[3]*Advanced Technology & Research, NCR Financial Solutions Group Ltd*
[4]*PaCT Lab, Department of Psychology, University of Northumbria*

People using a self-service terminal such as an automated teller machine (ATM) tend to adjust their physical position throughout a transaction. This is particularly apparent with terminals that are designed to be used from a vehicle (i.e. drive-up automated teller machines or ATMs). Existing predictive tools tend to focus on static reach and provide limited predictions for how far people are willing to stretch to complete a task. Drive-up self-service products have 3 main challenges; the variability of vehicles, people and driver behaviour. Such conventional tools are therefore of limited use in understanding how much people are willing to move to use a self-service terminal.

Work is described to build in-house predictive models based on 2 large empirical studies of reach in a drive-up installation. These 2 studies assessed comfortable and extended reach from 10 vehicle categories. Extended reach was defined as stretching/leaning as far as participants would normally be willing to in order to complete a drive-up transaction. Findings from these studies indicated that participants are prepared to adopt more extreme postures at drive-up than in other situations with extended reach at drive-up being significantly different to what might be seen at a walk-up kiosk.

Introduction

Self-service terminals (SSTs) such as automated teller machines (ATMs) or check-in kiosks are commonly used in a walk-up environment. Traditional models for workstation design and the models published in Smith and Coventry (2002) provide guidance and relevant reference material about product design for walk-up terminals. However, in some regions, these self-service terminals are also offered in a drive-up configuration with the consumer remaining in their vehicle to use the terminal. Typically these are found in North America and the Middle East (e.g. Saudi Arabia, Kuwait).

Using a terminal from a vehicle imposes significant pressures on the consumer and if traditional predictions of static reach were to be used it would be impossible to design a product to fit within the reach envelope predicted by these tools. However we know that these drive-up products are used and gaining in popularity.

For this reason further understanding is required of where a person can and will reach to from a vehicle for an infrequent and fast interaction such as collecting a ticket for car parking, ordering fast food, or getting cash from an ATM. Three key challenges in predicting reach for drive-up self-service are the variability of vehicles, people and driver behaviour. Our approach has therefore been to carry out studies to understand more about this environment. These studies have included the following:

- Studies of vehicle sizes
- Observations of people using live drive up ATMs
- Study of what drive-up users consider to be comfortable and extended reach in a drive-up environment
- Two studies using different prototype ATM layouts, with measurements of actual driver behaviour and user feedback on the acceptability of interface layout
- The development of predictive tools to be used for future design work.

Constraints for drive-up self service

Vehicle dimensions

Making a single solution that works for all shapes and sizes of consumers is difficult enough, but the variability increases dramatically when vehicles are involved. Not only is the size of the car important, the shape of the window and the ability of the driver to park close to the object they must reach must also be considered.

The first challenge to overcome is that vehicle dimensions vary significantly. In fact, the variation between key vehicle dimensions, such as seat pan height or height to window top and height to window bottom, is greater than the differences between human anthropometric variation. This is a particular challenge in markets such as North America as they have vehicles at both ends of the spectrum (e.g. very small, low sports cars at one end, and very tall pickup trucks at the other). For instance, when considering seat pan as an example, a small sports car, such as a Chevrolet Corvette, may have a seat pan height of 290 mm from the ground whereas a large pickup truck, such as a Dodge Ram, may have a seat pan height of 970 mm.

In addition to this variation in vehicle height, other key constraining factors include the vehicle window (height of bottom & top of window), position of the A and B pillars (where A pillar is the material that supports the windscreen at either side and B pillar is the pillar to the rear of the driver's window), proximity to the steering wheel, and the presence of seat belts.

Vehicle categories

In our research we have identified 10 vehicle categories as follows:

A. Sports car
B. Compact car

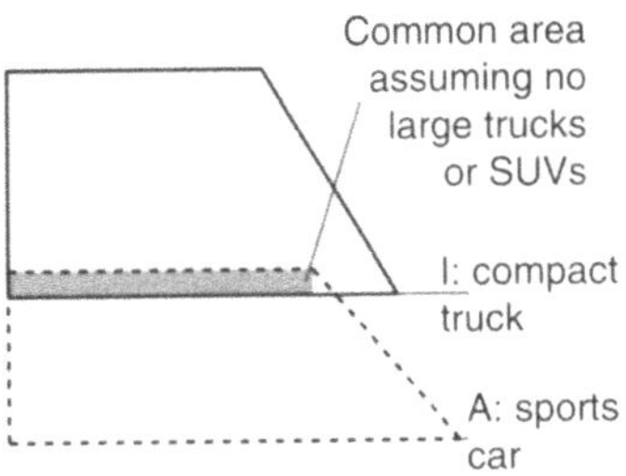

Figure 1. Compact truck to sports car windows.

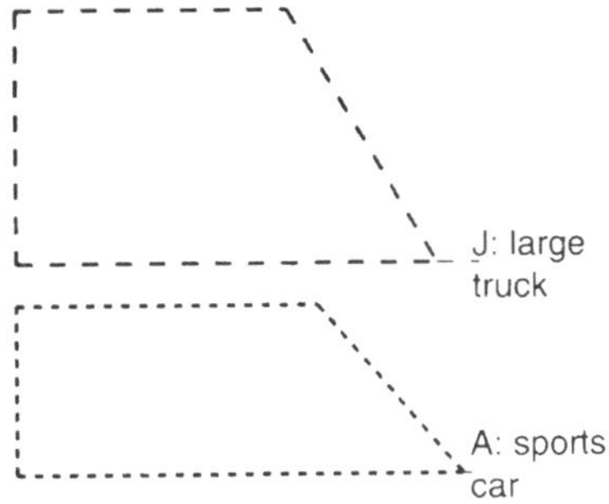

Figure 2. Large truck to sports car windows.

Table 1. US vehicle ownership 2004–2005.

Vehicle Description	2005 (%)	2004 (%)
a: sports	4	3
b: compact car	16	15
c: midsize car	21	21
d: fullsize (large) car	5	5
e: minivan (MPV)	6	7
f: compact SUV	7	7
g: midsize SUV	16	16
h: large SUV	4	5
i: compact truck	4	4
j: medium & large truck	15	16
k: motorbike	0	0
l: large commercial vehicles	0	0
m: full size van	2	2

C. Midsize car
D. Fullsize (large) car
E. Minivan (Multi-purpose vehicle, or MPV)
F. Compact sports utility vehicle (SUV)
G. Midsize SUV
H. Fullsize (large) SUV
I. Compact pickup truck
J. Midsize & large pickup truck

Vehicle categories were derived primarily from US vehicle categorisations, and then harmonised across other national classification schemes (e.g. Euro NCAP, ACRISS car classification codes). When considering window height, the B pillar, window top and window bottom can be summarised graphically giving a clear view of the distribution of vehicle dimensions across the vehicle population. All of these factors combine to ensure that there is little overlap between the vehicle categories Figure 1 (or no overlap at all when considering sports cars and large SUVs or pickup trucks, Figure 2) and an extremely limited zone of reach for all consumers in all vehicles.

Vehicle ownership

Data was gathered in 2006 regarding ownership of vehicles by NCR from available data for the previous two years (US Bureau of Transportation Statistics, 2005, US Automotive parts industry annual assessment, 2005 and the Road Ahead for the US Auto Industry, 2005). The dominant vehicle categories are compact and midsize cars, midsize SUVs & medium/large trucks for both 2004 and 2005.

Drive-up evaluations

Two studies were conducted at an NCR facility in Duluth, Georgia. The first study with 104 participants (with at least seven participants for each of the previously mentioned ten vehicle categories) evaluated how far people were willing to reach to use a drive-up ATM, with measures being taken for comfortable and extended reach. In addition, this study also assessed two prototype ATM layouts, with measurements of actual driver behaviour and user feedback on whether or not aspects of the layout were acceptable. The second study, with 74 participants validated driver behaviour and user feedback on a revised ATM layout (which had been changed as a result of the first study).

Variability in parking

Our research has found that consumers vary significantly in their positioning of vehicles when using a drive-up facility. Parking position varied significantly both in terms of horizontal alignment on the drive-up terminal (i.e. how far along the interface they parked), and distance from the ATM to the vehicle. In one study (104 participants), it was found that distance from ATM to shoulder of participant (when sitting at rest) varied from 575 to 1084 mm. This is a problem as this was beyond arm length for many participants. For example, a 5th percentile US female arm length (shoulder to fingertip) is 666 mm (PeopleSize 2008). In the same study, horizontal alignment (position of shoulder along the interface) varied from 32 to 685 mm (relative to the left-hand edge of the interface) as illustrated in Figure 4.

Challenges of designing for a worldwide population

In addition to this variability in vehicle demographics there is the challenge of designing for a worldwide population, with a given percentile varying significantly between national populations. For instance, taking PeopleSize (2008) as a source,

Figure 3. Examples of reach from a range of vehicle types.

a 5th percentile female has stature of 1479, 1517 or 1559 mm for Chinese (urban), British and Dutch (1986) female populations respectively. Arm length (shoulder to fingertip) varies in a similar manner with a British female having an arm length of 665 mm and a Chinese female of 622 mm. For this reason we use 7 worldwide anthropometric models for ambulant users as published by Smith and Coventry (2002).

Resultant postures

As can be seen in Figure 5, consumers at a drive-up terminal are willing to reach far beyond what would be acceptable in other environments. In particular, the difference between postures that are commonplace in a walk-up environment (such as an ATM in a town centre, or a self-checkout kiosk in a supermarket) and what is commonplace in drive-up are significant. For instance, in a drive-up environment some consumers are willing not only to lean beyond what would be considered comfortable, but are also willing to twist their torso, pivot about 1 arm, and sometimes even opening their vehicle door in order to reach the terminal. These findings were validated by field observations (before and after these studies) of actual behaviour at real drive-up ATMs, with similar postures being adopted by consumers at drive-up ATMs. For example, Figure 6 shows a consumer in a large pickup truck with their door open, pivoting on their left arm while reaching with the right to reach an actual drive-up ATM, Figure 7 shows a consumer in a midsize SUV with their door open stretching with their left hand.

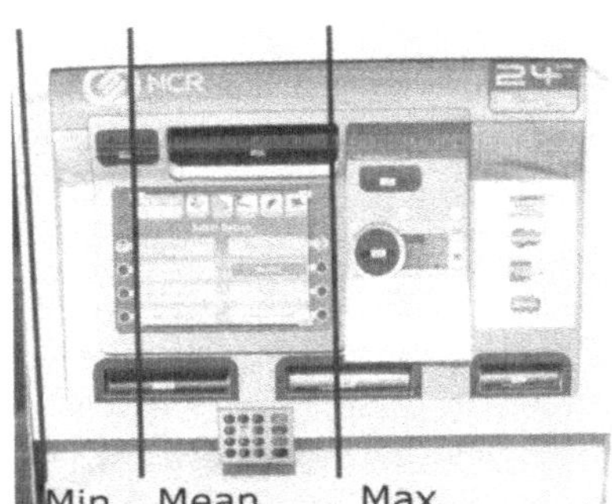

Figure 4. Horizontal parking position.

Figure 5. Examples of extreme postures at drive-up.

Figure 6. Large pickup truck.

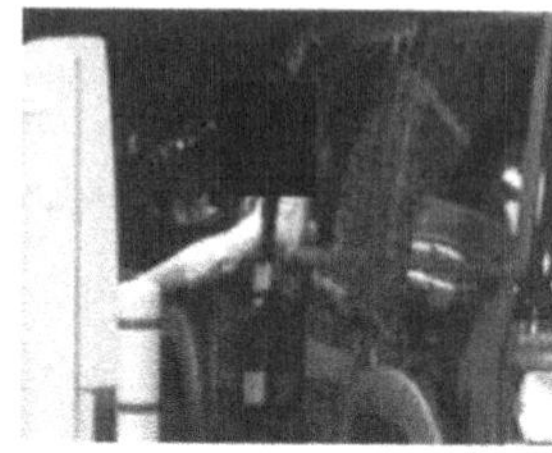

Figure 7. Midsize SUV.

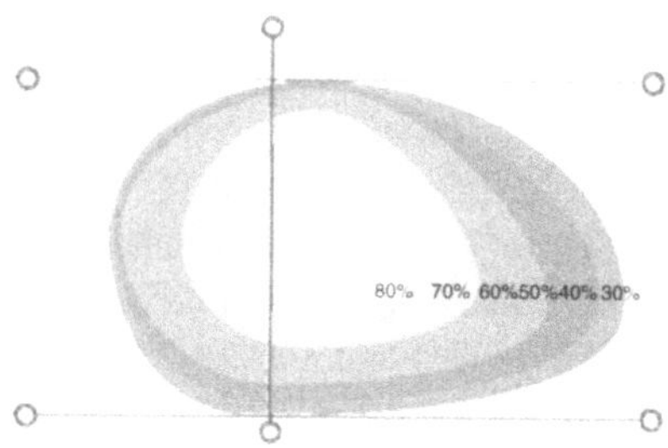

Figure 8. Comfortable reach, Psihogios and Swope (1997).

Main findings

The main findings from these two studies included recommendations and revisions of the ATM designs that were investigated and information that could then be used to inform the development of a predictive reach tool for drive-up. A key finding was the importance of minimising depth in the interface.

Predicting reach with existing tools

The challenge of accurately predicting reach in an applied area from static anthropometry tables is known (see for example Pheasant, 1995, Robinette and Hudson, 2006) as is the need for subsequent fitting trials. Application areas such as drive-up, that tend to involve more extreme stretching postures are more challenging as is modelling these constraints effectively with existing tools. For this reason the authors have built in-house predictive models of reach for drive-up.

The traditional approach for predicting reach assumes a single parking position for one or two vehicle types, and then makes predictions based on static anthropometry for a number of percentiles, for example 5th percentile US female and 95th percentile US male (Psihogios and Swope, 1997). This assumes that people would park in the optimal position horizontally along the interface. An example of this can be seen in Figure 8. This approach does not cover the range of vehicles currently in common usage, and only gives predictions of what might be considered comfortable reach. In particular, the extreme postural adjustments that are commonly seen in drive-up are not readily predicted by such tools.

Table 2. Number of participants in each vehicle type.

Sports	Compact car	Midsize car	Large car	Mini van	Compact SUV	Midsize SUV	Large SUV	Compact truck	Large truck	Bike
10	11	11	11	10	10	15	7	10	8	1

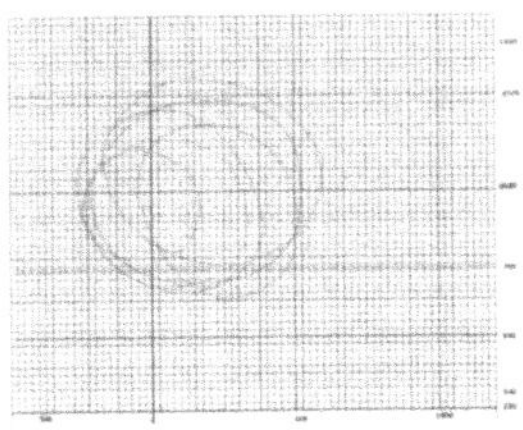

Figure 9. Comfortable reach – compact cars (raw).

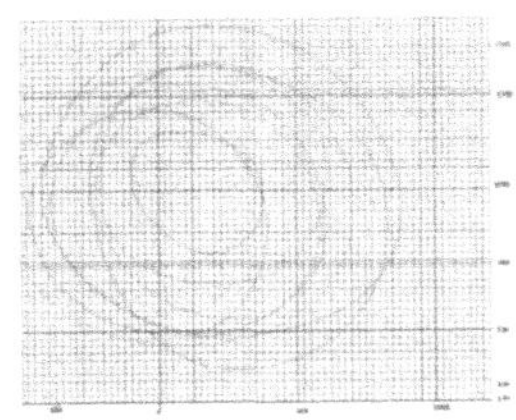

Figure 10. Extended reach – compact cars (raw).

Predicting reach – our approach

As a result of the challenges previously discussed, a two-stage study was conducted with 104 and 74 participants respectively. This study was a self-reported comfort zone approach, rather than a classic marker-based dynamic anthropometric study, as one of the key challenges was that consumers appeared to reach beyond what would normally be acceptable. A new predictive tool for reach was derived from empirical data. Participants were recruited such that all 10 vehicle categories (from sports cars to large pickup trucks) were adequately represented (Table 2).

Comfortable and extended reach were measured separately. Comfortable reach was defined as involving minimal movement, just moving the arm; i.e. static reach. This was expected to give an indication of an optimal zone, which is useful in optimising the design and provides a best case for comparisons. Extended reach was defined as "as far as you would be willing to reach when using a drive-up ATM"; i.e. dynamic reach. This included leaning and lifting from the seat, and gives an indication of possible reach.

Key vehicle dimensions were measured including height to window top & bottom, width of window bottom, and parking distance and horizontal alignment. Parking position was also self-reported with participants tending to report that they aligned on the centre of the display, whereas their actual measured shoulder position was just inside the left-hand edge of the screen. Reach was measured by each participant parking next to a large sheet of paper, being asked to mark how far they could comfortably reach (in a circular manner about the shoulder), and then how far they would be willing to reach to use a drive-up terminal (see Figure 9 and Figure 10 for an example of some of the raw data obtained for compact cars).

The results were then combined for each vehicle category and normalised (so each vehicle category was weighted equally irrespective of how many vehicles were

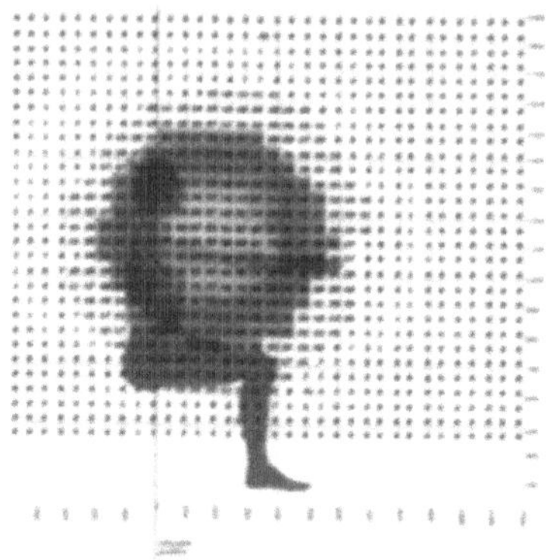

Figure 11. Comfortable reach (US).

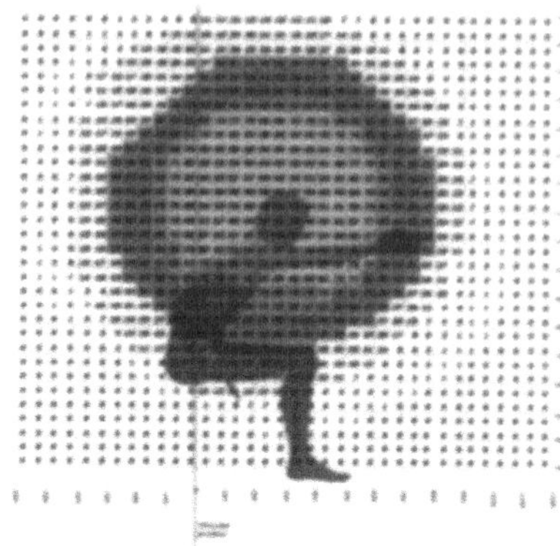

Figure 12. Extended reach (US).

actually sampled). Predicted reach can then be combined across all 10 vehicle categories (e.g. for US and Canada) or for a subset of vehicle categories (e.g. for Europe) thus giving predictions of comfortable and extended reach expressed as a percentage for these populations (Figure 11 and Figure 12). Colour coding was added to give a quick graphical indication of predictions of reach and to help in readily communicating the findings.

Limitations

Although the study had good ecological validity as it was conducted in a car park, with drive-up ATM models laid out as if within a drive-up lane and participants in their own vehicle, there may be slight differences between what people reported they were willing to do and what they actually do in practice. In addition, as this study was conducted in the US with US vehicles, there may be limitations when generalizing these results to other populations as there may be differences in culture, behaviour, driving position and willingness to accept extreme postures between the populations.

Conclusions and future work

Use of drive-up terminals often involves unusual and extreme postures. However, people find these acceptable (participants cited convenience and perceived security as two factors for this). Predicting what would be acceptable in terms of reach in these environments would not be possible with existing tools. New predictions of comfortable and extended reach were therefore derived from empirical data. Another key finding was to move from assuming a fixed parking position to gaining an appreciation of depth as a very important factor in a drive-up terminal interface. As a result we now suggest that drive-up interfaces should minimize reach in all three dimensions, recognizing the significant additional benefits that reductions in depth of the user interface have in improving reach for all users.

These predictions have subsequently been applied to the development of a new drive-up ATM. Initial feedback suggests that this drive-up ATM offers significant benefit over other designs. Predictions have been demonstrated to confirm previously gathered knowledge in terms of the challenges with existing products, and are expected to provide a powerful tool in the development of future drive-up solutions.

An extension of this work is to use more-detailed human modeling tools such as SAMMIE (Porter *et al.*,2004) and apply commonly occurring postures in order to provide more-detailed 3-dimensional predictions of reach. Initial work demonstrated that this would be of particular benefit when considering tasks that require high levels of manual dexterity (e.g. use of a biometric fingerprint reader). However, this has not been explored further, as it requires significant investment in terms of obtaining appropriate 3-dimensional models of the relevant vehicles.

Acknowledgements

The authors acknowledge the funding and support of NCR Corporation.

References

PeopleSize 2008 Professional software, Open Ergonomics Ltd.

Pheasant, S.T., 1995. Anthropometry and the design of workspaces,. In J.R. Wilson and E.N. Corlett (eds.) *Evaluation of Human Work (2nd edition)*, (Taylor and Francis, London), 557–573.

Porter, J.M., Marshall, R., Freer, M. and Case, K., 2004. SAMMIE: a computer aided ergonomics design tool. In: N.J. Delleman, C.M. Haslegrave, and D.B. Chaffin (eds.) *Working Postures and Movements – tools for evaluation and engineering*, (Boca Raton: CRC Press LLC), 454–462.

Psihogios, J. and Swope, G., 1997. Drive-up ATM Design recommendations based on user anthropometry and automobile type. NCR Human Factors Consulting internal report.

Robinette, K.M. and Hudson, J.A., 2006. Anthropometry. In G. Salvendy (Ed.) *Handbook of human factors and ergonomics (3rd edition)*. (John Wiley & Sons, New Jersey), 322–339.

Office of Aerospace and Automotive Industries Manufacturing and Services, International Trade Administration. June 2005, *The Road Ahead for the U.S. Auto Industry (RA2005)*, U.S. Department of Commerce.

Office of Aerospace and Automotive Industries Manufacturing and Services, International Trade Administration. June 2005, *U.S. Automotive Parts Industry 2005 Annual Assessment*, U.S. Department of Commerce.

Smith, K.T. and Coventry, L. 2002, Designing for a world population. In *Contemporary Ergonomics 2002*, (Taylor and Francis, London), 340–345.

U.S. Bureau of Transportation Statistics. 2005, *State Transportation Statistics*, U.S. Department of Transportation, Research and Innovative Technology Administration, Bureau of Transportation Statistics, December 2005.

EYE FATIGUE WHEN VIEWING BINOCULAR DISPLAYS: EFFECTS OF MISMATCH BETWEEN ACCOMMODATION AND VERGENCE

K.M. Chang[1], R.H.Y. So[1*], H.C. Huang[2], A.K.C. Lam[3] & P. Ting[3]

[1]*Department of Industrial Engineering and Logistics Management, The Hong Kong University of Science and Technology, Hong Kong SAR*
[2]*Department of Electronics Engineering, The Hong Kong University of Science and Technology, Hong Kong SAR*
[3]*School of Optometry, The Hong Kong Polytechnic University, Hong Kong SAR*

This paper reports an experiment conducted to investigate the levels of eye fatigue caused by viewing stereo images on a binocular display with compatible and incompatible lens focuses. Twelve participants watched four 30-minute stereoscopic slide shows under four viewing conditions that were exhaustive combinations of two levels of lens focus (40 cm and 200 cm) and stereo depths (40 cm ± 0.3 dioptres and 200 cm ± 0.3 dioptres). Five Landolt C rings would appear at a pre-determined depth on each slide and participants would locate the opening of the centre ring. Rated levels of eye fatigue were taken every two minutes. Results show that when using a binocular display of 200 cm lens focus, viewing stereo images with incompatible stereo depths (40 cm) as compared to compatible depths of 200 cm significantly increased the rated levels of eye fatigue ($p < 0.001$, $F_{1,383} = 33$, ANOVA). This carries serious implications to the manufacturers of binocular displays with fixed lens focus of 200 cm. Contrary to previous assumption, adjusting the lens focus of the displays to 40 cm to match images with a stereo depth of 40 cm did not result in significant reduction in eye fatigue ($p > 0.5$, $F_{1,383} = 0.36$, ANOVA). The lack of benefits had been due to an opposite confounding effects of accommodation on eye fatigue. Data also indicate that when the confounding effect of accommodation was in agreement with the compatibility effects between lens focus and stereo depth (i.e., a change of lens focus from 40 cm to 200 cm to match images of stereo depths of 200 cm), very significant reduction in rated levels of eye fatigue were reported ($p < 0.001$, $F_{1,383} = 55.5$, ANOVA).

Introduction

When viewing a real object, our eyes accommodate for clear focus and verge so that the visual axis of both eyes will point towards that object. In other words, the vergence of eyes will change in accordance to the accommodation of the eyes. However

*The corresponding author: rhyso@ust.hk

this relationship can be disturbed when viewing stereo images with changing stereo depths presented on a binocular display with fixed lens focus. The incompatibility between the varying stereo depth and the fixed lens focus can post an unnatural demand on human vergence eye movements. This may cause eye fatigue.

Yano *et al.* (2004) reported a significant increase in the level of eye fatigue when viewing stereoscopic high-definition TV images. Yano concluded that the eye fatigue among his participants was caused by viewing the stereo images whose stereo depths were outside the depth of focus. In their study, only one focal length was studied and there was no comparison between matching and non-matching effects of lens focus and stereo depth.

Miyao *et al.* (1996) reported that viewing of stereoscopic images with varying stereo depth between 0.8 m and 2 m presented on a CRT display located at 1 m from viewers caused loss of accommodation. It has also been reported that most viewers complained of stiffness in neck and shoulders, eye pain, and dry eyes after the experiment. However, only the mismatched condition was studied in their experiment. Again, there was no comparison between matching and non-matching effects of lens focus and stereo depth.

Emoto *et al.* (2005) reported the closest study to our experiment. Participants were required to view stereoscopic images with changing stereo depth cues displayed on a high-definition TV. A pair of variable prisms was installed in front of participants' eyes, the prism power was varied to produce varying vergence load on eyes. In the mismatched condition, no lens was used. In the matched condition, compensation lenses were used to match vergence with accommodation. It was found that in the mismatched condition, the chance of seeing double image was higher than in the matched condition. There is no significant difference on the subjective visual fatigue ratings between the two conditions. Although, both match and mismatch conditions were explored but the viewing distance was fixed at 30 cm. In other words, matching effects between vergence and accommodation at far viewing distance (say 200 cm) was not investigated.

Kuze and Ukai (2008) reported that viewing stereoscopic moving images projected on a 90 inches back-projected screen produced significantly higher eye strain ($p < 0.001$), focus difficulty ($p < 0.05$) and general discomfort ($p < 0.05$) in viewers when compared to viewing non-stereoscopic moving images projected on the same screen. Again, their study did not investigate the condition in which stereo depths and lens focus were matching to each other.

Mon-Williams and Wann (1998) studied the deficits in binocular vision and eye fatigue symptoms after watching a cross sign oscillating between 40 cm to infinity (frequency 0.3 Hz) for ten minutes. They reported significant changes in visual acuity, symptoms, distant heterophoria, distant-associated and near-associated heterophoria. However, the distance between the viewers and the display was fixed at 40 cm and all virtual objects had stereo depth behind the display (i.e., stereo depths of greater than 40 cm). In other words, only the condition in which the mismatched lens focus was shorter than the stereo depth was studied. Again, there was no comparison between matching and non-matching effects of lens focus and stereo depth.

In summary, previous studies have reported conflicting results in viewing stereoscopic images at a depth in appropriate to the lens focus. While Emoto *et al.* (2005) reported no significant changes in rated levels of visual fatigue; others have reported significant increases in eye fatigue (Kuze and Ukai, 2008; Miyao *et al.*, 2006; Mon-Williams and Wann, 1998; and Yano *et al.*, 2004). In addition, none of the previous studies have examined the problem of eye fatigue with a factorial combination of lens focus and stereo depth.

Objectives and hypotheses

As explained in the last section, a study investigating a full factorial combination of lens focus (accommodation) and vergence (stereo depth) on levels of eye fatigue among viewers of a binocular display has not been found. This paper reports a two-factor, (lens focus and stereo depth) two-level (40 cm and 200 cm) within-subject experiment conducted to study the levels of eye fatigue after viewing moving images displayed on a binocular display. It was hypothesized that (i) viewing duration will significantly increase levels of eye fatigue in all conditions (H1), (ii) mismatch between lens focus and stereo depth will significantly increase levels of eye fatigue (H2), and (iii) reduction in lens focus (i.e., increases in accommodation) will significantly increase levels of eye fatigue (H3).

Method and Design

Experimental Design

Nine male and three female Chinese participated all four experimental conditions that were the factorial combinations of two lens focuses (40 cm and 200 cm) and two stereo depths (40 cm $\pm$ 0.3 dioptres and 200 cm $\pm$ 0.3 dioptres). Order of presentation was balanced using three 4×4 Latin squares. Pre-experiment vision screening tests confirmed that all participants had normal or corrected vision with far and near visual acuities better than 20/20, stereo acuity better than or equal to 40 arc seconds and normal colour discrimination ability. The participants were naive with respect to the nature and aims of the experiment. They were instructed to search and locate the openings of Landolt C rings as soon as they appeared on the display. Participants received a stipend for their participation. The experimental protocol was approved by the Human Subject Committee of the Hong Kong University of Science and Technology.

Apparatus

A custom-built head-mounted display (HMD) was used in the experiment. The two micro-display LCD panels (resolution: 688 pixels (Horizontal) $\times$ 480 pixels (Vertical)) were supplied by Integrated Microdisplays Limited (iMD), Hong Kong (model: iSDTV704C). The diagonal field of view was 30 degrees for each eye. The lens focus could be adjusted from 40 cm to 200 cm. A chin rest was used to restrain the viewers' head positions. The study was conducted in an acoustic room supplied by Industrial Acoustics Company (H.K.) Ltd (model no: IAC 400A-CT) under complete darkness.

Figure 1. A … eoscopic slide with Landolt … (adopted from Chang *et al.*, 2009).

Stimuli and tasks

The inter-pupillary distance of each participant was measured and used to customize a set of stereoscopic slides showing a group of five Landolt C rings at various pre-determined depths. A snapshot of a pair of slides is shown in Figure 1. The slides were prepared using 3D Studio Max™ Version 3.0. The outer radius, inner radius, and the thickness of each Landolt C ring was 125 mm, 75 mm and 50 mm, respectively. The openings of each Landolt C and the distance between each ring were 50 mm. The five rings were positioned as a cross as shown in Figure 1 with their faces perpendicular to the central axial along the depth of a virtual room (6000 mm (depth) × 3000 mm (width) × 2000 mm (height)). The stereoscopic views were captured by placing two virtual cameras on the wall of the virtual room at a depth of 0 mm and facing into the room. The cameras were placed at a height of 1000 mm and at the positions along the width of 1500 mm ± half of the inter-pupillary distance (IPD) of that participant. The captured stereo images were projected to the binocular display via a multiple screen display card (Inno3D NVIDIA GeForce6200). During the experiment, slides with rings shown at different stereo depths would be displayed and the variations of the depth were set to within ± 0.3 dioptres of the stereo depth appropriate to the selected experimental condition. To the participants, the rings would appear to be moving back and forth randomly. The range of depth variation is within the depth of focus and was consistent with previous study (±0.3 dioptres, Yano *et al.*, 2004). Participants were required to read out the direction of the opening of the Landolt C ring in the middle as possible as it appeared as a single image. The directions of the openings (up, down, left or right) of the Landolt C rings were randomized.

Procedure

Before the start of a testing session, the optical axes of the binocular lens were adjusted to match the participant's IPD and the lens focus of the display was adjusted to the appropriate level (40 cm or 200 cm) according to the selected condition. During a session, the stereo depth of the Landolt C rings and the direction of openings of the centre C ring would change from slide to slide and participants were instructed to verbally report the direction of the openings of the centre Landolt C as soon as a single stereoscopic image was formed. Each session lasted for 30 minutes and participants were required to rate his/her level of eye fatigue according to a 7-point eye fatigue rating: 0 = no symptom; 1 = any unpleasant symptom related to my eyes, however slight; 2 = mild unpleasant symptom related to my eyes; 3 = mild fatigue related to my eyes; 4 = mild to moderate fatigue related to my eyes; 5 = moderate fatigue related to my eyes but can continue; 6 = moderate fatigue related to my eyes, and want to stop. Rated level of eye fatigue was taken every two minutes resulting in sixteen data points per participant per session. At least 48 hours separation was given before the same participant was allow to return for another session.

Results

Seven hundred and sixty-eight eye fatigue level data points were collected (16 data points per viewer × 12 viewers × 4 conditions × 768 data points). Analysis of Variance (ANOVA) was conducted to study the effect of lens focus and stereo depth on the eye fatigue level. The ANOVA result indicated that: lens focus ($F_{1,767} = 22.71$, $p < 0.0001$), stereo depth ($F_{1,767} = 4.87$, $p < 0.05$), duration of exposure ($F_{15,767} = 99.25$, $p < 0.0001$) and the interaction between lens focus and stereo depth ($F_{1,767} = 31.87$, $p < 0.0001$) all have significant effects on the eye fatigue levels. Results supported hypotheses H1 and H3. Since the effects of matching lens focus with stereo depth is essentially the interaction between the two effects, subsequent analyses will focus on the significant two-way interaction between the effects of lens focus and the effects of stereo depth. As the data within each condition did not have a normal distribution, non-parametric statistical tests were used alongside with ANOVAs.

Results of further statistical analyses indicated that when lens focus was set at 200 cm, rated levels of eye fatigue resulted from viewing stereo images at the matched stereo depth of 200 cm ± 0.3 dioptres was significantly lower than that resulted from viewing images at a mismatched stereo depth of 40 cm ± 0.3 dioptres ($F_{1,383} = 32.99$, $p < 0.0001$, ANOVA; $p = 0.041$, Wilcoxon signed-rank test). This supports hypothesis H2.

When viewing stereo images with depth of 200 cm ± 0.3 dioptres, rated levels of eye fatigue level resulted from using a display with a matched lens focus of 200 cm was significantly lower than that resulted from using a mismatched lens focus of 40 cm ($F_{1,383} = 55.46$, $p < 0.0001$, ANOVA; $p = 0.0205$, Wilcoxon signed-rank test). However, when lens focus was set at 40 cm, rated eye fatigue levels resulted from viewing images at a matched stereo depth of 40 cm ± 0.3 dioptres was not significantly different from that resulted from viewing images with a mismatched stereo depth of 200 cm ± 0.3 dioptres. ($F_{1,383} = 5.38$, $p = 0.021$, ANOVA; $p = 0.125$, Wilcoxon signed-rank test). Likewise, when viewing stereo images with depth of 40 cm ± 0.3 dioptres, the rated levels of eye fatigue resulted from using a matched lens focus of 40 cm was not significantly different from that resulted from using a mismatched lens focus of 200 cm ($F_{1,383} = 0.36$, $p = 0.548$, ANOVA; $p = 0.9868$, Wilcoxon signed-rank test). This rejects hypothesis H2.

Discussion and Concluding Remarks

The effects of matching lens focus to images with changing stereo depths on rated eye fatigue level have been studied at near (40 cm) and distant (200 cm) accommodation. When using a head-mounted display (HMD) of fixed lens focus of 200 cm, viewing stereo images at depths around 200 cm ± 0.3D will generate significantly less rated levels of eye fatigue than viewing similar images at depths around 40 cm ± 0.3D (ANOVA: $p < 0.0001$; Wilcoxon: $p < 0.05$). This is an important finding that carries serious implications to the manufacturers of HMDs with fixed lens focus of 200 cm. The results also suggest that game developers should avoid using images of stereo depths of 40 cm ± 0.3D if their potential customers were to view the games on a HMD of fixed lens focus of 200 cm. The authors acknowledge that the reported significant effects of matching lens focus with stereo depth were confounded with the effects of changing stereo depths (NB: this was inevitable as matching lens focus with stereo depth require either lens focus or stereo depth to be changed). To isolate the effects of changing stereo depth from the effects of matching, the analyses were repeated by controlling the stereo depth to be 200 cm ± 0.3D. A similar result was found: rated levels of eye fatigue were significantly less when images were viewed using HMDs of matched lens focus (i.e., 200 cm) instead of mismatched lens focus (i.e., 40 cm) (ANOVA: $p < 0.0001$; Wilcoxon: $p < 0.05$). In this case, the effects of matching lens focus with stereo depth have been isolated from the effects of changing stereo depths but the authors acknowledge that the finding is now, inevitably, confounded with the effects of accommodation. An examination of the confounding effects in the above analyses suggest that both the confounding effects of changing stereo depth or lens focus tended to reduce the levels of eye fatigue. In other words, the confounding effects were complimenting the main effects of matching lens focus with stereo depth.

Further analyses were conducted when the confounding effects of accommodation and convergence were in opposite to the effects of matching lens focus and stereo depth. Viewing images with stereo depths mismatched with lens focus resulted in non-significant effects of matching. In particular, rated levels of eye fatigue when viewing images at stereo depths of 40 cm ± 0.3D with HMDs were not affected by changing lens focus from 40 cm to 200 cm. Also, when the lens focus was set at 40 cm, viewing images with stereo depths at 40 cm ± 0.3D and 200 cm ± 0.3D results in similar levels of eye fatigue. These findings suggest that adjusting the lens focus to match with the changing stereo depths may not be beneficial – a finding that is opposed to Shibata *et al.* (2005) who reports a significant benefit of adjusting lens focus. This result also rejects hypothesis H2. This is not an encouraging result for those who are developing new technology to dynamically adjust the lens focus of HMDs (e.g. Sugihara *et al.*, 1999, Emoto *et al.*, 2005). This finding also disagree with the general predictions made in Kuze and Ukai (2008), Miyao *et al.* (2006), Mon-Williams and Wann (1998) and Yano *et al.*, (2004).

An examination of all the findings from this study suggest that the effects of matching lens focus with stereo depth can significantly reduce levels of eye fatigue when the confounding effects of changing stereo depth or accommodation also reduce levels of eye fatigue. When the effects of matching are in opposition to the effects of accommodation or stereo depth, the overall changes in eye fatigue become not significant.

Survival analyses were conducted to study the effects of stereo depth (i.e., vergence), lens focus (i.e., accommodation), and their interactions (i.e., matching) on the time required for viewers to report a 'mild to moderate' level of eye fatigue (i.e., a rating of 4 on the 7-point scale). Results indicate that changing lens focus (i.e., changing accommodation) was a significant ($p < 0.05$) predicting factor and matching lens focus with stereo depth was a marginal significant ($p < 0.1$) predicting factor.

Acknowledgement

This study was partially funded by Hong Kong Research Grants Council through the Earmarked Grant HKUST6128/03E and a research grant (A-PA0F) from the Hong Kong Polytechnic University.

References

Emoto, M., Niida, T. and Okano, F. (2005) Repeated vergence adaptation causes the decline of visual functions in watching stereoscopic television. Journal of Display Technology, 1(2), 328–340

Chang, K.M., So, R.H.Y., Huang, H.C., Lam, A.K.C. and Tang, P. (2009) Effects of matching lens focus with stereo depths on visual fatigue when viewing binocular displays. Awaiting publication

Kuze, J. and Ukai, K. (2008) Subjective evaluation of visual fatigue caused by motion images. Displays, 29, 159–166

Miyao, M., Ishihara, S.Y., Saito, S., Kondo, T.A., Sakakibara, H. and Toyoshima, H. (1996) Visual accommodation and subject performance during a stereographic object task using liquid crystal shutters. Ergonomics, 39(11), 1294–1309

Mon-Williams, M. and Wann, J.P. (1998) Binocular virtual reality displays: When problems do and don't occur. Human Factors, 40(1), 42–49

Shibata, T., Kawai, T., Ohta, K., Otsuki, M., Miyake, N., Yoshihara, Y. and Iwasaki, T. (2005) Stereoscopic 3-D display with optical correction for the reduction of the discrepancy between accommodation and convergence. Journal of the SID, 13(8), 665–671

Sugihara, T., Miyasato, T. and Nakatsu, R. (1999) An evaluation of visual fatigue in 3-D displays: Focusing on the mismatching of convergence and accommodation. IEICE transactions on electronics, Vol. E82-C, No. 10, 1814–1822, October

Yano, S., Emoto, M. and Mitsuhashi, T. (2004) Two factors in visual fatigue caused by stereoscopic HDTV images. Displays, 25, 141–150

MIND THE GAP: RESEARCH AND PRACTICE IN ERGONOMICS AND HUMAN FACTORS

S.T. Shorrock & A.Z.Q. Chung

Department of Aviation, The University of New South Wales, Sydney, NSW 2052, Australia

Significant discord has been aired regarding the widening scientist-practitioner gap in several disciplines (e.g. psychology, healthcare), especially with reference to research published in academic journals. Very little research on this important issue has been identified in ergonomics and human factors (E/HF). Using an online questionnaire, practitioners were asked about their application of scientific research findings published in peer-reviewed journals. A total of 587 usable responses were collected, spanning 46 countries. This article describes some of the key differences and correlations concerning reading, usefulness, and barriers to application among respondents who varied in terms of organisational type, percentage of work time devoted to application versus research, society membership, and experience. Several solutions proposed by respondents are summarised.

Introduction

Wilson (2000) defines ergonomics as the theoretical and fundamental understanding of human behaviour and performance in purposeful interacting sociotechnical systems, and the application of that understanding to the design of interactions in the context of real systems. This definition captures the relationship between research and application. Indeed, Wilson argues that any acceptable definition of ergonomics must emphasise the need for, and the complementarity between, research and practice. However, for ergonomists with an applied role, particularly those providing internal or external consulting services to organisations, research is often not so complementary to practice (Meister, 1999; Kirwan, 2000).

A widening gap between researchers and practitioners has been observed in many disciplines. This phenomenon has been variously termed the 'research-practice gap', 'theory-practice gap', 'academic-practitioner divide' or 'researcher-practitioner divide'. This gap especially concerns research published in academic journals, and impacts research utilisation, defined as, "the direct implementation of research findings into practice", characterised by "processes that involve reading, evaluating, and implementing research findings to improve understanding, solve problems, and make decisions in practice" (Kim, 2005, p. 439). Many disciplines such as industrial, work and organisational (IWO) psychology, library and information science, and healthcare have identified the common problem that practitioners often fail to implement research findings and researchers often do not address questions relevant to practitioners (e.g. Gelade, 2006).

There is very little research on this important issue in E/HF, and yet the history of E/HF is replete with references from the early 1960s to a schism between researchers and practitioners (Meister, 1999; Waterson and Sell, 2008). Meister's (1999) original research represents a rare attempt to examine the issue (among many others) in any depth. He conducted a survey of 46 of the most experienced members of the Human Factors and Ergonomics Society (HFES). The respondents recognised the need for research and practice but did not understand how they relate. He found that "there is tension between researchers and practitioners who see little value in the products of each other's activities" (p. 223). There was a feeling among practitioner respondents that research published in E/HF journals was largely irrelevant to practitioner problems.

Meister (1999) also performed a content analysis of over 621 empirical and non-empirical papers published in *Human Factors* and the Proceedings of the Annual Meetings of the HFES. Meister asserted that the effect of research on HFE practice is extremely tenuous, and that there is an astonishing lack of relationship between theory and practice. In examining research adequacy, he argued that "HFE research is not useful to what should be two of its primary consumers: the practitioner and designer working in system development… Simply phrased, the discipline has done a rotten job of applying its research to real-world problems" (p 264).

The reasons for the problems in research utilisation are complex, but empirical research suggests that the issues pertain to both research and practice. The current study aims to examine E/HF practitioners' application of research findings published in scientific journals via an international survey of practitioners. We hypothesised that those working in academic/research institutions, and those devoting a high proportion of work time to research, would read more research, find research more useful to practice, and perceive different barriers to research application. We also hypothesised that the reading frequency would be higher for members of E/HF Societies and Associations (hereafter 'Societies'). We speculated that the perceived journal usefulness may differ between Society members and non-members, and that reading frequency and perceived journal usefulness may correlate with experience.

Method

Development of the Questionnaire

A questionnaire was designed according to Dillman's (2007) 'tailored design method'. The items for the questionnaire were drawn from the psychology, library and information science, and healthcare literature. Following initial development, the questionnaire was pre-tested in three stages:

1. Review by knowledgeable colleagues. The questions, response sets and format were reviewed by the two authors plus a Lecturer and a Professor at the University

of New South Wales. Three members of the review group were members of Human Factors and Ergonomics Societies and qualified to PhD level. All were active E/HF researchers with experience of survey research. The review group discussed the number and type of questions, response categories and scales, question flow and questionnaire format.

2. Interviews with potential respondents. The amended questionnaire was presented in person to five E/HF practitioners. A 'think aloud' method was used with three of the respondents, and a retrospective interview was used with the remaining two respondents. Further amendments were made to the questionnaire based on these interviews.
3. Small pilot study. A small pilot study was performed to emulate the procedures proposed for the study, using the draft web-based questionnaire. This was intended to examine further question wording, non-responses, open-ended responses, and response distributions. Ten E/HF practitioners based in Australia, UK, Canada, France and Singapore completed the questionnaire. Final minor amendments were made based on the responses.
4. Final check. A final check of the questionnaire was performed independently by two HF researchers at the University of New South Wales.

Questionnaire Format

The questionnaire was hosted on www.surveymonkey.com, and comprised a welcome and information page followed by 20 questions in three sections: Section 1. About you; Section 2. About your access and attitude to published research; Section 3. About possible barriers and facilitators to using research findings in practice.

Procedure and Respondents

No universal list of E/HF practitioners is available for this type of research. While E/HF Societies do maintain lists of members, these are confidential, so it was not possible to determine a total population size. Additionally, it was difficult to define the attributes of an E/HF practitioner, partly because the profession is not regulated and practitioners are not required to join a Society. Our preferred definition was a person eligible to join an IEA Federated Society as an ordinary member. However, this was impractical because non-members may not know their eligibility and membership categories vary between Societies. Our operational definition used in the invitation to participate was therefore simply "practitioners who apply an understanding of E/HF to the design of interactions in real settings" (adapted from Wilson, 2000). The following recruitment methods were used to reach as many practitioners as possible:

1. E/HF Societies. The Presidents and Secretariats of each of the 46 International Ergonomics Association (IEA) Federated Societies were contacted in May 2009 to request that their Societies invite their members to complete the survey. Following a reminder email sent out four weeks later, a total of 30 Societies responded that they would distribute the invitation (it is not known if other Societies did the same).

2. LinkedIn Groups. A notice entitled "Invitation to participate: Human Factors/Ergonomics practitioner survey" was posted to members of 16 groups associated with E/HF on the professional networking website www.linkedin.com. Additionally, 471 members of these groups were sent the invitation individually via LinkedIn email (i.e. all of the members that could be contacted by email and indicated that they practiced E/HF).
3. Professional contacts. Eighty-two professional contacts were contacted by regular email and asked to complete the survey, and to send on the link to their own E/HF practitioner colleagues. Respondents could choose to complete the questionnaire anonymously, or leave their email address (on the questionnaire or separately by email) to receive the results. There was no material incentive to complete the questionnaire.

A total of 587 usable questionnaires were received. The key demographic categories and top five responses are shown in Table 1 (rows do not add up to 100% because only the top five responses are listed and some questions allowed multiple responses). In addition, respondents' estimated work time over their most recent 12 months as an active E/HF professional was split on average as follows: research (mean 28%), application (mean 31%), education (mean 26%) and other (mean 15%).

Results

Access and attitude to published research

Respondents were asked how often they read scientific journal articles concerning E/HF or related fields over their most recent 12 months as an active E/HF

Table 1. Key demographic categories from 1st (most frequent) to 5th.

	1st (most frequent)	2nd	3rd	4th	5th
Organisation types	Private sector org. – not consultancy 27%	Private sector consultancy 21%	Academic/Research institution 16%	Self-employed/Independent 14%	Government org./agency/dept. 14%.
Countries of work	US 33%	UK 17%	Australia 10%	Netherlands 5%	Canada 5%
University-level qualifications	E/HF 56%	Psychology 32%	Engineering/Computing 24%	Biol./Anat./Physiol./Med. 10%	Physio./Occ. Therapy 8%
Society memberships	US 28%	UK 18%	Australia 6%	Canada 4%	Netherlands 4%
Industries	Manufacturing 36%	Transport & storage 35%	Human health 22%	Defence 21%	Info. & comm. 21%
Application areas	Tasks, jobs & work processes 48%	Displays, controls & HCIs 46%	Safety management 38%	Workplaces & furniture 26%	Training 25%

Table 2. Responses to IEA-endorsed journals.

IEA-endorsed Journal	Mean usefulness out of 4 (Q13)	Top 3 most useful to me (Q12)	I've not read this (Q13)
Accident Analysis & Prevention	2.7	3%	62%
Activites	2.7	1%	85%
Applied Ergonomics*	3.1	29%	20%
Ergonomics*	3.0	28%	19%
Human Factors*	3.1	22%	20%
Human Factors and Ergonomics in Manuf.*	2.7	2%	60%
Int. J. of Human-Computer Interaction	2.7	2%	42%
Int. J. of Industrial Ergonomics*	2.8	9%	50%
Int. J. of Occ. Safety and Ergonomics	2.7	2%	54%
Journal of Safety Research*	2.5	1%	63%
Safety Science*	2.6	6%	64%
Theoretical Issues in Ergonomics Science*	2.4	2%	67%
WORK	2.7	2%	74%

professional. Seven ordinal categories were provided, from "Never during the 12-month period" to "11+ articles per month". The median response was 1 article per month. In terms of how respondents accessed these articles, the mode response was from a library/library website (university/other) (52%). The next most frequently chosen categories were via a personal or Society journal subscription (46%) and freely (no cost) from non-library websites (45%). Purchasing articles or issues was the least common means of access (15%).

Respondents were also asked which scientific journals concerning E/HF or related fields were most useful to them as an E/HF practitioner (if any). Respondents could name up to three journal titles as free text. A total of 419 respondents entered one or more journal(s). *Applied Ergonomics*, *Ergonomics* and *Human Factors* were the named the three most useful journals. Separately, respondents were asked to rate the usefulness of 13 IEA-endorsed scientific journals to them as an E/HF practitioner. The journals were listed with a four-point rating scale from "Not useful to me" (1) to "Very useful to me" (4), with options for "I've not read this journal" and "I've read this journal but I'm not sure how useful it is to me". Only three journals (*Applied Ergonomics*, *Ergonomics* and *Human Factors*) were read by 60% or more of the respondents. These journals were also rated most useful. Table 2 details usefulness and reading data for the IEA-endorsed journals. (The mean usefulness scores exclude "not read" and "not sure" ratings.) Respondents working primarily in academic/research institutions rated 8 journals significantly more useful to practice than those in other organisations (marked * in Table 2; $p < .01$, t-tests, one tailed).

Next, respondents were asked to rate the usefulness of E/HF or related scientific journals as a whole to them as an E/HF practitioner. The mean and median responses aligned with "Moderately useful to me" (3 out of 4). Statistical analyses showed that reading frequency and overall rated journal usefulness was significantly higher for the following (all one tailed unless stated otherwise):

Table 3. Top 5 barriers to research application for respondents working in academic/research institutions vs. other organisations.

Academic/Research Institutions ($n = 92$)	Other Organisations ($n = 492$)
1. Decision makers will not allow application of research	1. The relevant literature is not compiled in one place
2. The research has methodological inadequacies	2. Implications for practice are not made clear in the article
3. Key stakeholders will not cooperate with application of research findings	3. I do not have time to read research
4. I do not feel I have enough authority to change current organizational processes	4. The research is not relevant to my practice
5. There is insufficient time on the job to apply new ideas	5. Journal articles are not readily available/accessible

- respondents who worked in academic/research institutions vs. other organisations (reading frequency $U = 12079.500$, $p < .001$; usefulness $t(151) = 6.398$, $p < .001$);
- Society members vs. non-members (reading frequency $U = 27118.000$, $p < .01$; usefulness $t(537) = 2.705$, $p < .01$, two-tailed);
- respondents who spent a higher proportion of time (>50%) engaged in research vs. a higher proportion of time engaged in application (>50%) (reading frequency $U = 3191.500$, $p < .001$; usefulness $t(305) = 4.553$, $p < .001$).
- respondents who spent a higher proportion of time (>50%) engaged in research vs. a lower proportion of time engaged in research (≤50%) (reading frequency $U = 6786.500$, $p < .001$; usefulness $t(537) = -3.677$, $p < .001$);
- respondents who spent a lower proportion of time (≤50%) applying E/HF vs. a higher proportion of time applying E/HF (>50%) (reading frequency $U = 33142.500$, $p < .001$; usefulness $t(537) = 5.128$, $p < .001$).

There was a statistically significant but very weak negative correlation between years of professional experience in E/HF and overall journal usefulness ($r = -.087$, $n = 539$, $p < .05$). There was no significant correlation between experience and reading frequency ($\rho = .047$, $n = 587$, $p > .05$).

Possible barriers and facilitators to using research findings in practice

In this section, respondents were asked to rate 29 statements concerning current barriers to their application of E/HF research findings, as a practitioner. The barriers were adapted from Funk *et al.* (1991) to apply more closely to E/HF. Tables 3 and 4 shows the top 5 rated barriers for different types of respondent. For the sample as a whole, the barriers related primarily to research communication and secondarily to the organisation.

Finally, respondents were asked how practitioners could improve the application of E/HF research findings to their own practice, and how researchers could help to improve the application of E/HF research findings to practice. The findings were

Table 4. Top 5 barriers to research application for respondents engaged in 51–100% research vs. 51–100% application.

51–100% Research ($n = 42$)	51–100% Application ($n = 290$)
1. The relevant literature is not compiled in one place	1. The relevant literature is not compiled in one place
2. I do not feel I have enough authority to change current organizational processes	2. Implications for practice are not made clear in the article
3. The research has methodological inadequacies	3. I do not have time to read research
4. Decision makers will not allow application of research	4. The research is not relevant to my practice
5. Key stakeholders will not cooperate with application of research findings	5. Journal articles are not readily available/ accessible
6. Implications for practice are not made clear in the article	

analysed thematically and each suggestion was placed into one or more categories relating to clusters of 'barrier' items. The top five suggestions for practitioners were (in rank order): 1. Increase networking and collaboration among researchers and practitioners; 2. Increase reading and awareness of the research; 3. Improve awareness, support, cooperation, and authority from others; 4. More application or more appropriate application of research findings/recommendations; and 5. More research and publication by practitioners.

The top five suggestions for researchers were (in rank order): 1. Ensure research focus and methodology are more applied, relevant, practical and generalisable; 2. Increase networking and collaboration among researchers and practitioners; 3. Form more definitive conclusions, and clearer implications or applications; 4. Report research findings in different (non-journal) formats (e.g. blogs, Society publications); and 5. Report in a more understandable, clear and readable manner.

Discussion

The findings of the survey provide valuable insights into E/HF practitioners' reported access to, and attitudes toward, research published in E/HF-related journals, and factors that affect their research application. On average, respondents read around one article per month. *Applied Ergonomics*, *Ergonomics* and *Human Factors* were perceived as the most useful journals and respondents rated E/HF related scientific journals overall as moderately useful to them as practitioners.

A number of research-practice 'gaps' were evident. Reading frequency and overall rated journal usefulness were significantly higher for respondents who worked in academic/research institutions, respondents who spent a higher proportion of time engaged in research, respondents who spent a lower proportion of time engaged in application, and Society members. Those working primarily in academic/research institutions rated the majority of the IEA-endorsed journals as significantly more useful to practice than those working in other organisations.

The top five barriers for those working in academia/research versus other organisations were different, with no common ground. Four of the five top barriers for 'academics' related to characteristics of the organisation; only one related to research. For 'non-academics' the situation is reversed: four of the five top barriers related to research communication; only one related to the organisation. ('Practitioner-centredness' is, not surprisingly, the dominant theme for practitioners.) This pattern extended to suggestions to improve research application aimed at researchers: four of the five top suggestions aimed at researchers involved characteristics of the research. The top five suggestions aimed at practitioners were a mixture of practitioner, organisational and research factors. One suggestion ranked in the top five for both practitioners and researchers: increased networking, communication and collaboration between researchers and practitioners.

The dominance of barriers related to research may parallel the drift from pragmatic to pedantic science in IWO psychology journals, where the vast majority of research has been generated by academics, with a decline in the number of articles by practitioners and academic-practitioner collaborations (Anderson *et al.*, 2001). When academic researchers become distanced from practical concerns, external validity is undermined, and the quality and practical relevance of research is reduced (Bartunek, 2007).

The secondary organisational barriers may reflect the pressures in organisations to produce rapid results, contributing to a shift from pragmatic to popularist science in practice (Wall, 2006). These barriers may also reflect the limited opportunity for longer-term research-based solutions and the relatively low value placed on journal publications (Kirwan, 2000).

This research has highlighted a number of gaps between research and practice associated with organisational affiliation, research/application involvement, and Society membership. Those practitioners with the most power to effect change in organisations may be benefiting least from E/HF research. The research-practice gaps could be reduced by addressing a number of factors associated with the research, the organisation, and individual practitioners and researchers. More liberal access to research and better synthesis of research are issues that need to be addressed by Societies and publishers. However, better communication and collaboration among researchers and practitioners; more relevant and applicable research focus, methodology and reporting; different formats for reporting research findings; clearer writing; a greater commitment to reading and application: these are factors over which ergonomists have a degree of control, and should take seriously if we are ever to close the gap between research and application in our discipline.

References

Anderson, N., Herriot, P. and Hodgkinson, G.P. 2001, The practitioner–researcher divide in industrial, work and organizational (IWO) psychology: Where are we now, and where do we go from here? *Journal of Occupational and Organizational Psychology*, 74, 391–411

Bartunek, J.M. 2007, Academic-practitioner collaboration need not require joint or relevant research: Toward a relational scholarship of integration, *Academy of Management Journal*, 50 (6), 1323–1333

Dillman, D.A. 2007, *Mail and Internet Surveys: The Tailored Design Method, 2nd Ed*, (John Wiley & Sons, Hoboken, New Jersey)

Funk, S.G., Champagne, M.T., Weiss, R.A. and Tornquist, E.M. 1991, Barriers: The barriers to research utilisation scale, *Applied Nursing Research*, 4 (1), 39–45

Gelade, G.A. 2006, But what does it mean in practice? The Journal of Occupational and Organizational Psychology from a practitioner perspective, *Journal of Occupational and Organizational Psychology*, 79, 153–160

Kim, K. 2005, Perceived barriers to research utilization by Korean university librarians, *The Journal of Academic Librarianship*, 31 (5), 438–448

Kirwan, B. 2000, Soft systems, hard lessons, *Applied Ergonomics*, 31, 663–678

Meister, D. 1999, *The History of Human Factors and Ergonomics*, (Laurence Erlbaum, Mahwah, NJ)

Wall, T. 2006, Is JOOP of only academic interest? *Journal of Occupational and Organizational Psychology*, 79, 161–165

Waterson, P. and Sell, R. 2006, Recurrent themes and developments in the history of the Ergonomics Society, *Ergonomics*, 49 (8), 743–799

Wilson, J.R. 2000, Fundamentals of ergonomics in theory and practice, *Applied Ergonomics*, 31, 557–567

REDUCING THE LOAD – DEVICES TO REDUCE THE PHYSICAL DEMANDS OF FIRE-FIGHTERS DURING A SIMULATED RESCUE

Karen Henley[1], **Miles Richardson**[2] **& Shaun Baker**[3]

[1]*Canada Life Limited, UK*
[2]*University of Derby, UK*
[3]*Shropshire Fire and Rescue Service, UK*

Fire-fighters are routinely required to enter hot, smoke filled environments in order to search for and rescue casualties using a physically demanding 'flat drag' rescue. Two new prototype devices to aid casualty rescue were developed and tested along side the traditional flat drag method. Twenty-one serving fire-fighters completed a simulated rescue using each of the three methods in a randomised order. The results demonstrated that the two prototype methods out performed the traditional flat drag method of casualty rescue, in terms of time taken, perceived effort and physiological demand.

Introduction

Fire-fighting requires hard physical work whilst wearing heavy equipment in demanding environments (Holmer and Gavehed, 2006) and musculoskeletal injuries are common (Conrad *et al.*, 2008). When working in such adverse conditions, it is the primary objective to save human lives which means finding and removing casualties from the building as quickly as possible with the least risk to the fire-fighter and casualty. In order to ensure that the casualty has the best chance of survival it is imperative that they are kept close to the floor. If the casualty is elevated in any way then they are at an increased risk of smoke inhalation and burns. In order to keep the casualty close to the floor the fire-fighters are taught to perform a flat drag rescue (Figure 1). This requires the fire-fighter to kneel on the floor, grasp the casualty under both arms with the casualty's neck and shoulders resting on the fire-fighters knees in order to offer some support for the head and neck. The fire-fighter then 'shuffles' backwards on their knees dragging the casualty along, keeping them low to the floor. There have been concerns raised that this method is slow and increases the risk of injury to the fire-fighter.

Research in this area has mainly focused on the physiological aspects that affect the fire-fighters in rescue situations (e.g. Heimburg *et al.*, 2006; Holmer and Gavehed, 2006), very little research has focused on the unique manual handling needs that are required in the fire service, with no literature apparent on flat drag rescue. Therefore, the following study evaluates the flat drag method alongside the use of two new prototype devices by taking relevant objective and subjective measures during a simulated rescue.

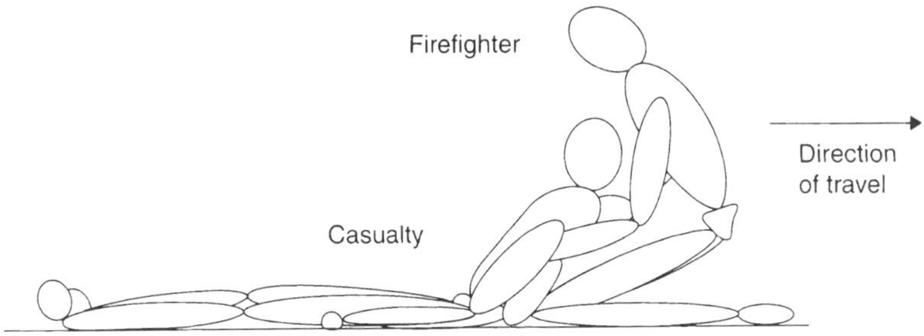

Figure 1. Flat drag rescue.

Prototype Device Design Process

The traditional flat drag rescue currently taught to fire-fighters undertaking rescues in hot, smoke filled environments appears to be flawed. A UK Fire Service station manager felt the method is slow and places enormous strain on the lower back and knees of the fire-fighter. Any alternative method which is quick, safe and reduces the load could potentially reduce the risk of fire-fighters getting injured and also increase the chances of survival for the casualty. Desk research indicated that although there are rescue devices available, there appeared to be nothing specifically for rescuing casualties in a hot, smoke filled environment. It appeared that using devices to assist with a flat drag rescue that allowed firefighters to keep an upright posture could reduce risk of injuries. As the casualty must be kept low, a device with handles that attaches to the casualty is required. The new devices also needed to be easy to use in limited visibility; allow quick application; be compact, small and lightweight; protect the casualties head and neck and fit a variety or body shapes and sizes.

Two pieces of equipment were designed. The first an adjustable belt with handles extending off it with two straps across the handles offering some head and neck support (Figure 2). The belt is clipped around the casualty's waist, enabling the fire-fighter to stand up and drag the casualty out of the fire zone, keeping them low to the floor (Figure 3). Whilst the figure indicates backwards movement, there is more flexibility for the fire-fighter to turn towards their direction of travel.

The second piece of equipment was a sheet with a fully adjustable belt around it and handles at the top offering protection to the head and neck whilst the casualty was being dragged (Figure 4). The belt is clipped around the casualty's waist thus 'wrapping' the sheet around the casualty, supporting the head and neck while enabling the fire-fighter to drag them out (Figure 5). Again whilst the figure indicates backwards movement, there is more flexibility for the fire-fighter to turn towards their direction of travel.

Method

Following a pilot study to test the two new prototype devices, they were evaluated alongside the traditional flat drag method during a simulated rescue. The study

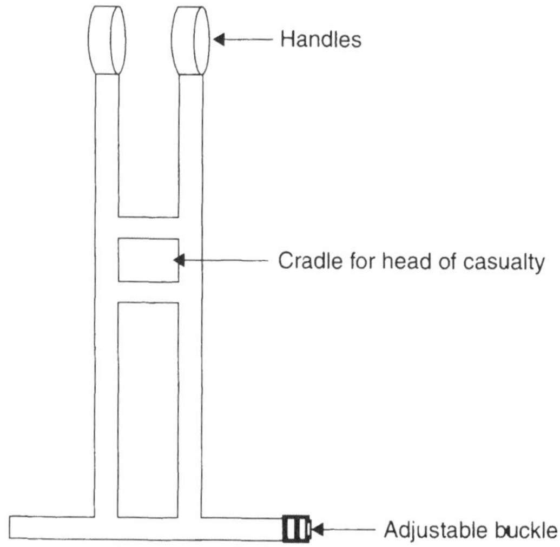

Figure 2. Belt design.

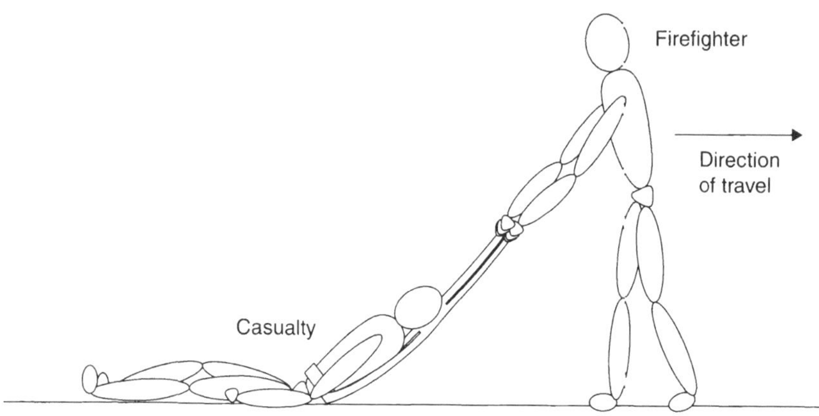

Figure 3. Belt method.

took place over two days in a fire service training centre and was overseen by the instructors to ensure that participants were properly briefed and safety was not compromised.

Design

The independent variable was the rescue method used, either the traditional flat drag method, sheet method or belt method. These were tested using a repeated measures design with the order of use randomized across the participants. The dependant variables measured were time taken, pre to post trial increase in heart rate and a post trial Borg Rating of Perceived Exertion. Following the use of each

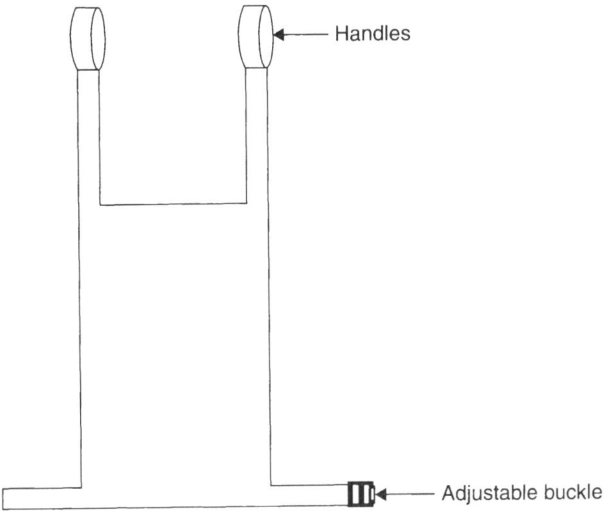

Figure 4. Sheet design.

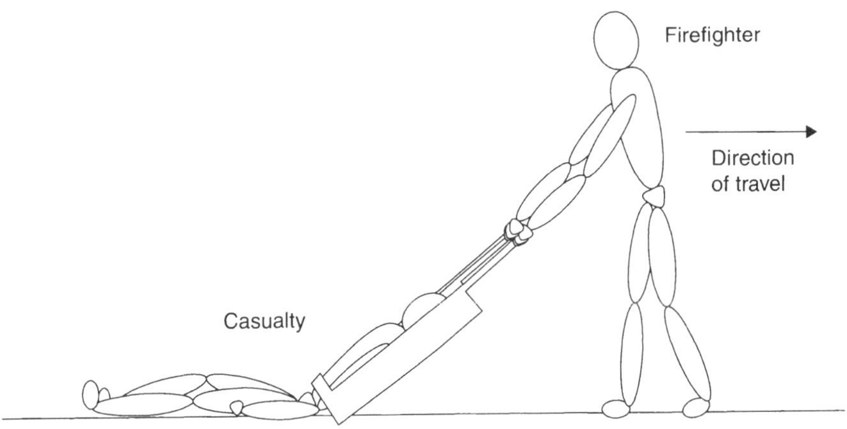

Figure 5. Sheet method.

method participants also completed a short questionnaire asking what they liked and disliked about each method.

Participants

Twenty-one serving fire-fighters with a mean age of 38 years took part, one participant was female. The mean length of service was 11 years. Of the 21 participants taking part five had undertaken a rescue in a hot, smoke filled environment in the last year, 12 had undertaken this specific rescue at some point in their career, but not in the last year, and four had never undertaken this rescue outside of the training centre.

Table 1. Time taken, perceived exertion and heart rate change for the three methods.

	Time taken (seconds)		Perceived Exertion (Borg)		Increase in Heat Rate (beats per minute)	
Method	Mean	SD	Mean	SD	Mean	SD
Flat drag	146.7	51.2	17.6	2.1	76.5	12.0
Sheet	92.9	35.8	11.0	1.8	64.7	14.1
Belt	90.4	40.0	11.4	1.7	65.1	10.5

Procedure

A fire service training building was filled with thick cosmetic smoke to impair visibility, but heat was not used inconsideration of the fire-fighters safety when using prototype devices. 80 kg dummies were used and placed in a downstairs room of the firehouse, the fire-fighters all started and finished at the same point and used the same route. The route was a 15 meter dash in and then a 15 meter return trip with the dummy being kept low.

The participants were briefed by the fire service training instructors and given a short training session on each aspect of the trial and the chance to practice all methods. All participants were familiar with the layout of the firehouse and they were given the location of the dummy along with the route. All participants were in full personal protective clothing, including breathing apparatus and kneepads.

On instruction the fire-fighters entered the building and timing commenced. After exiting the building their heart rate and time taken was recorded. The participant then completed the Borg Rating of Perceived Exertion and questionnaire asking for their opinions on the method used.

Results

Time Taken

The traditional method took significantly longer to undertake with a mean time of 147 seconds compared to a mean of 93 seconds for the Sheet and 90 seconds for the Belt. A repeated measures ANOVA test found the method used had a significant affect on time taken, $F\ (2, 40) = 14.47$, $p < 0.01$. Bonferroni post hoc analysis showed a significant difference between the flat drag and both the sheet method ($p < 0.01$) and the belt method ($p < 0.01$). There was no significant difference between the sheet and the belt methods ($p = 1.00$).

Borg Rating of Perceived Exertion

The Borg rating of perceived exertion for the flat drag method had a mean rating of 18 (Very Hard), whilst both the sheet and belt methods were rated as 11 (Light).

A repeated measures ANOVA test found the method used had a significant affect on perceived exertion, F (1.525, 30.496) = 138.65, $p < 0.01$. Bonferroni post hoc analysis showed a significant difference between the flat drag and both the sheet ($p < 0.001$) and the belt method ($p < 0.001$). There was no significant difference between the sheet and the belt methods ($p = 0.641$).

Increase in Heat Rate

The mean heart rate increased by 77 beats per minute for the flat drag method, compared to 65 beats per minute for the sheet and the belt methods. A repeated measures ANOVA test found the method used had a significant affect on heart rate increase, F (1.557, 31.140) = 8.83, $p < 0.01$. Bonferroni post hoc analysis showed a significant difference between the flat drag and both the sheet method ($p < 0.05$) and the belt method ($p < 0.01$). There was no significant difference between the sheet and the belt methods ($p = 1.00$).

Questionnaire Results

Reoccurring comments from the qualitative data collected following the use of each method were identified. The flat drag method was liked because it kept the casualty low, there was no 'messing about just grab the casualty and go' and no added equipment was necessary. It was disliked because it was felt to be hard work, painful on hamstrings and lower back, covered a small distance for a lot of effort, very energy sapping and had the potential for injury. The participants' would only use as there is no other method taught.

The sheet method was liked as it was quick and easy to use, improved maneuverability, allowed standing up and minimal exertion. It was disliked because of the time taken to attach the device to the casualty. Participants stated they would use it again.

The belt method was liked as it was quick and allowed standing. It was disliked because it was difficult to put on the casualty, the straps were confusing, it was fiddly and offered poor head support. Participants stated they would use it again.

The participants were not specifically asked to comment on which method they preferred, however in the comments on the questionnaires six stated they liked the sheet method best, five commented the sheet and the belt were both comparable and the remaining ten participants did not comment on their preferred method. It was also apparent that although the two alternative methods provided minimal exertion and allowed an upright posture, participants' remained skeptical that they would ever be used in an emergency situation as the temptation is to just grab the casualty and get out.

Discussion

It is clear that using the two prototype devices was quicker and required less exertion than the traditional flat drag technique. Further, serving fire-fighters without exception preferred the prototype methods as opposed to the traditional flat drag rescue, which is routinely taught as the preferred method of casualty evacuation. However, it was also clear from comments that although they acknowledged that the two alternative methods provided minimal exertion and allowed them to stand, they remained skeptical that they would ever be used in an emergency situation as the temptation is to just grab the casualty and get out, even though the time taken in the simulated rescue included the time taken to deploy the prototype devices and was significantly lower than the time taken for the flat drag rescue.

The new devices appeared to address the issues put forward regarding the flat drag rescue and from the data collected the prototype methods appear both quicker and provide a reduction to the physiological load. However further research needs to be undertaken to develop and evaluate the new devices and associated work practices further and address the concerns of the serving fire-fighters regarding the addition of yet more equipment and the perception that time spent attaching the devices will slow them down in a real emergency.

References

Conrad, K.M., Reichelt, P.A., Lavender, S.A., Gacki-Smith, J. and Hattle, S. 2008, Designing ergonomic interventions for EMS workers: Concept generation of patient-handling devices, *Applied Ergonomics*, 39, 792–802.

Heimburg, E.D., Rasmussen, A.K. and Medbo, J.I. 2006, Physiological responses of firefighters and performance predictors during a simulated rescue of hospital patients, *Ergonomics*, 49, 111–126.

Holmer, I. and Gavhed, D. 2007, Classification of metabolic and respiratory demands in fire fighting activity with extreme workloads, *Applied Ergonomics*, 38, 45–52.

PERCEPTION OF HAND-ARM VIBRATION, COMPARISON OF TWO TOOL TYPES AND THE EFFECT OF USING A TYPICAL ANTI-ABRASION WORK GLOVE

Erik Lewry[1], Miles Richardson[1] & Neil J. Mansfield[2]

[1] *University of Derby, UK*
[2] *Loughborough University, UK*

This study investigates the relationship between hand-arm vibration (HAV), perception and exposure magnitude for two tool types. It also examines the effect on perception of tool type (rotary and percussive) and the use of anti-abrasion gloves. 20 male operatives were tested to determine their perception of HAV across 5 magnitudes of vibration, for a hand held hammer drill (1–7 m/s^2) and random orbital sander (2–9 m/s^2), in both glove-on and glove-off conditions. Results showed that the vibration produced by the rotary power tool was perceived as higher than the percussive tool across the three vibration magnitudes (2, 4 and 5 m/s^2). Use of the anti-abrasion gloves resulted in a significant decrease in perception of HAV for both the sander and the drill. Analysis of the relationship between perception and actual hand-arm vibration for both tools under the no-glove condition showed adherence to Stevens' Power Law, however this was not the case for the glove condition.

Introduction

Hand-arm vibration syndrome is a potentially debilitating set of symptoms that was believed to affect 288,000 people in the UK in 1998 (Palmer *et al.*, 1999), with around 510 new cases reported in 2007, (HSE, 2009). Ideally perception of HAV could be used as a preliminary risk filter for tasks that may be exposing individuals to hazardous levels, in preference to using potentially flawed manufacturers' data (Rimell *et al.*, 2008), or carrying out complex monitoring exercises. Additionally operatives' perception could be used to identify where the risk assessment is not valid, for example in situations where the tool is not functioning correctly, is being used in the wrong way or has a worn consumable. However this all relies on operatives' perception being reliable for the majority of industrial situations.

Anti-abrasion gloves are frequently worn whilst handling power tools due to other risk factors such as cuts and skin puncture. It has been suggested that the use of gloves often results in a reduction in the grip force and feed force applied to the tool, two factors that significantly affect the transmission of, and response to, vibration. This is due to improved friction between the glove and the tool compared to the hand and the tool. Mital (as cited in Sampson & Neikerk, 2004), reported a drop of

20% in the grip force and 30% in the feed force when wearing simple gloves; this is supported by work by Chang *et al.* (1999) where nylon gloves were seen to reduce the grip force by 18.2%. Further, given that the international standard weighting system is based on vibration perception and is weighted based on equal sensation, this would suggest that where vibration magnitudes are similar the perception of vibration will be the same regardless of tool type. Finally, Stevens' Power Law shows that the relationship between stimulus magnitude and perception is a power function and as long as conditions are identical the exponent is constant for the type of stimulus. It is important that if the perception of HAV is to be a feasible control measure that the relationship of source vibration magnitude to perceived vibration level is understood. Therefore this study asks whether perception of HAV is affected by tool type (rotary and percussive) or by the use of commonly used anti-abrasion gloves, and whether the relationship between HAV magnitude and perception for the two tool types follows Stevens' Power Law.

Method

Participants

20 male volunteer personnel from the local fire department were used for this study, neither of the power tools used in the testing were commonly used by the personnel during their work.

Design

There were three Independent Variables: Tool Speed (equated to measured vibration magnitude), Tool Type (Rotary or Percussive) and Gloves (no gloves or gloves). To avoid the influence of confounding variables, particularly practice, fatigue and in part temporary threshold shift, a balanced Latin square design was used. The two tool types were balanced, but separated in the Latin square design, this was considered necessary to prevent the tester from having to remove and replace the accelerometer and risk a slightly different placement or tightness of the unit and thus change the applied vibration.

A powered pistol grip hammer drill was chosen to represent the percussion type tooling and a random orbital sander was used to represent the rotary tool type. A tri-axial accelerometer with a frequency range 6.3–1250 Hz, was used in both experiments. This accelerometer was connected to a 01dB Maestro vibration meter able to record the vibration magnitude individually for the X, Y and Z axes and combine the three to give an overall weighted vibration magnitude in accordance with ISO 5439:1-2001.

For 50% of the trials participants did not wear gloves; for the other 50% of the trials the operative wore a pair of correctly sized (BS EN 420) Nitrile coated Kevlar gloves (Showa Kevlar); a commonly used general protection glove for resistance to cuts, puncture and for general hand injury protection. These gloves are not sold as anti-vibration gloves.

Table 1. Tool speeds and vibration magnitudes for the two tool types.

Orbital sander		Hammer drill	
Speed (on tool)	Approximate vibration magnitude (m/s^2)	Speed (on tool)	Approximate vibration magnitude (m/s^2)
1	2	2	1
2	4	3	2
3	5	4	4
4	8	5	5
5	9	6	7

The dependent variable, individuals' perception of the magnitude of vibration, was recorded using the Borg CR10 category rating scale. The validity and reliability of this scale has been demonstrated for psychophysical tests (Hager, 2003) and in particular hand-arm vibration perception (Ajovalasit and Giacomin, 2008; Neely & Burstrom, 2005). Reliability of the scale has been proven for numerous uses, but in particular for hand-arm vibration by Wos, Marek, Noworol and Borg (1988) where reliability coefficients were around .90.

Other factors were also considered, namely type of tool used normally, experience of tool use and knowledge of vibration. Posture, grip strength, grip surface, temperature, time exposed, tool weight, age, sex, smoking, hobbies (with potential vibration exposure), pre-existing medical conditions, previous exposure on day of test, chemical use, humidity, push force and noise level were all controlled during the testing.

Procedure

Each participant was informed of the aim of the experiment and pre and post monitoring questionnaire was used to collect relevant information on the participants. Each tool was held by the operative in one hand, the dominant hand, with the index finger on the trigger of the tool and the hand resting on the main handle; both tools were held in the same orientation with the arm at 90° with the feet in a stable position. The grip strength applied to the tool was approximately 7 kg force. Both tools were operated in air with no consumable.

In each test the A_{hv} reading (vibration total value of frequency weighted root mean squared acceleration) for the period of use was recorded along with the A_{hw} reading (root mean squared single axis acceleration value of the frequency weighted hand transmitted vibration, BS 5439:1-2001) for each of the three axis of vibration over a 10 second period. Immediately after each test the participant was asked to give his perceived vibration rating on the Borg CR10 scale.

The participant rested for 3 minutes between tests to control the effect of temporary threshold shift. During this time the speed of the tool was adjusted by the experimenter in accordance with the experimental design. The tests were carried out for both tools in the glove and no glove condition.

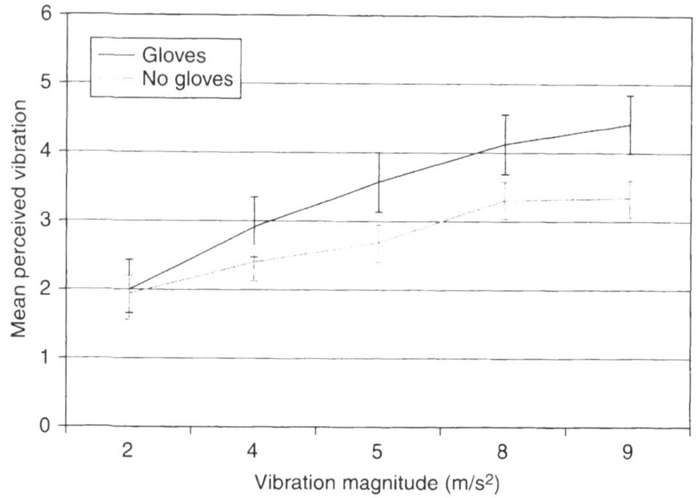

Figure 1. showing mean perceived vibration vs. vibration magnitude for the random orbital sander in the glove and no glove condition.

Results

Effect of glove use

The data for the two tool types showed that the no-glove condition was perceived as higher vibration than during the glove condition as can be seen from Figure 1 below. This difference was examined using a two way repeated measures ANOVA, the results of which showed the perception of vibration was significantly lower in the gloved condition compared to the no glove condition, $F(1,19) = 7.23$, $p = 0.015$, $r = 0.49$ and $F(1,19) = 9.75$, $P = 0.06$, $r = 0.55$ for the sander and drill respectively.

Effect of tool type

The tools were compared using a two way repeated measures ANOVA. The analysis shows that the perception of vibration was significantly lower for the drill than for the sander, $F(1,19) = 14.95$, $p = 0.001$, $r = 0.64$. The random orbital sander was shown to be perceived as having a higher vibration magnitude than the hammer drill at all vibration magnitudes monitored, see Figure 2.

Relationship of vibration perception and actual vibration magnitude

Exponents were calculated for each individual participant under the four conditions (sander glove, sander no glove and drill glove, drill no glove) using regression analysis (vibration magnitude as the IV and perception as the DV) and the power relationship proposed by Stevens. In each case an exponent and the R^2 value were recorded to allow analysis of the best fit nature of these exponents. Analysis of this data using repeated measures ANOVA (for the 4 conditions) reveals that for the

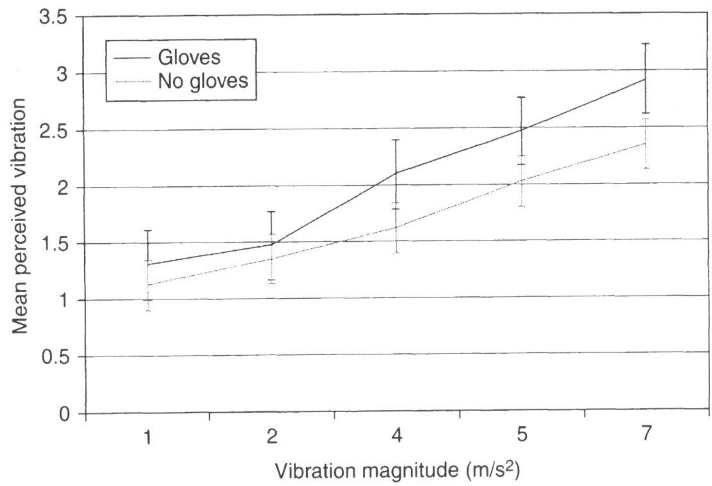

Figure 2. showing mean perceived vibration vs. vibration magnitude for the hammer drill in the glove and no glove condition.

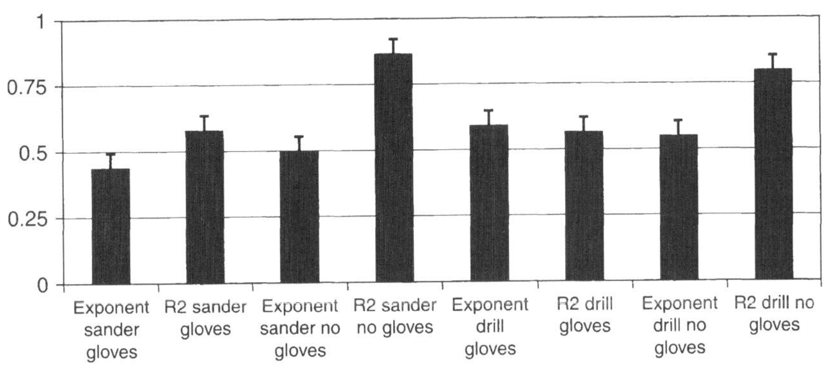

Figure 3. Mean exponents and R^2 values for the participants' perception of vibration magnitude vs. actual vibration magnitude for each experimental condition.

exponents there are no significant effects for either the tool or the glove conditions ($p > 0.05$). However the exponent is not the full information, the R^2 value identifies how well the data agrees with the power function line. Analysis of the R^2 values for the participants under the 4 conditions using repeated measure ANOVA shows that there is a significant effect (with a large effect size) on the fit of the line from the glove condition $F(1,15) = 37.62$, $p < 0.001$, $r = 0.84$. This can be clearly seen in Figure 3 below as the no glove conditions have a higher mean R^2 value than the glove condition. This demonstrates that the power function fits better for the no glove condition than for the glove condition and that this is true for both tool types.

The R^2 factors for the two tool types (under the no glove condition) were compared using a paired t-test, this identified no significant difference ($p > 0.05$) between the

R^2 factors for the tool types, suggesting that the fit of the power law was statistically similar for both the drill and the sander. As a result the exponents for the no glove conditions were compared using a paired t-test, this showed that there was not a significant difference ($p > 0.05$) in the relationships between the perceived and the actual vibration magnitude, between the two tool types. The mean exponents for the no glove conditions are 0.55 and 0.51 for the drill and sander respectively, in contrast to the 0.95 identified by Stevens for a frequency of 60 hz on the finger.

Effect of other factors

The effects of previous tool use, familiarity with the tool types, smoking and hobbies (that have some vibration exposure) were all found to not be significant ($p > 0.05$).

Discussion

Effect of tool type

Analysis of the results show that the hammer drill is perceived as having less hand arm vibration than the orbital sander, at the same actual vibration magnitude. This is in contrast to the hypothesis made that the two tool types would have the same perceived vibration at the same vibration magnitude due to the use of the Wh weighting scale (BS EN ISO 5439:1, 2001) which takes into account frequency and is based on equal sensation data.

Effect of glove use

The use of anti-abrasion work gloves had the hypothesised effect of decreasing the perception of vibration at the hand and arm of the participant despite having minimal impact on the actual vibration level. It is suggested that this due to 3 factors, firstly the reduced palm and finger contact force with the tool handle associated with better friction between the glove and the handle, than the hand and the handle (resulting in a lower grip strength); which leads to a reduction in the transmission of vibration (Dong *et al.*, 2008). Secondly different points of contact of the hand with the tool handle, which whilst not necessarily reducing the transmitted vibration, has the effect of changing the perception of the vibration. Thirdly it has been suggested (Mital, 1999 as cited in Sampson and Neikerk, 2004) that in addition to decreasing grip force the use of gloves can result in a decreased strength throughout the hand-arm system, this again affects how vibration is transmitted into and through the hand-arm system.

Relationship between perception of vibration and actual vibration magnitude

Analysis of the results shows that there is a power relationship between the perceived vibration and the actual vibration in this research. However the average exponent in this case is much lower than that identified by Stevens (1986) (0.51 and 0.55 compared with 0.95 predicted by Stevens). This is due to the research approach, original work with vibration perception carried out by Stevens and others looked at perception at the fingertips only, with the vibration exciter being simply a small probe applied to a small area of the fingertip.

Previous research on the perception of hand arm vibration and comparison with the Stevens Power Law found power function exponents differed across vibration frequencies, for magnitudes from 0.16 to 5 m/s^2 (Neely and Burstrom, 2005). The results at 63 and 125 Hz (0.5 and 0.45) show close agreement with the average exponents recorded in this study (0.51 for the sander and 0.55 for the drill), suggesting that it is higher frequency vibration that is dominating the perception in these tools.

In conclusion this study has shown that different tool types (rotary and percussive) despite having similar power relationships, are perceived differently by operatives, with the rotary tool type being perceived as higher vibration than the percussive tool type. In industry this means that operatives' ability to identify vibration levels in different tool types is likely to be ineffective as a control measure or as a risk assessment tool.

In addition the use of a simple pair of commonly used anti-abrasion gloves further confounds the ability of the operatives to perceive the vibration magnitude. This all suggests that relying on operatives to identify when they are at higher risk hand-arm vibration magnitudes would not be feasible, although using operatives to report when the same tool is perceived as increasing in vibration so that factors such as the maintenance of the tool, the condition of the consumable, the work practices or the task itself, can be examined to identify areas of potential risk, is certainly to be encouraged as an additional control layer to the original assessment of risk.

References

Ajovalasit, M. and Giacomin, J., 2008, Linear dependency of the Subjective Perceived Intensity of Steering Wheel Rotational Vibration, *International Journal of Industrial Ergonomics*, 39, 58–67

Chang, C.H., Wang, M.J., and Lin, S.C., 1999, Evaluating the Effects of Wearing Gloves and Wrist Support on the Hand-arm Response While Operating an In-line Pneumatic Screwdriver, *International Journal of Industrial Ergonomics*, 24, 473–481

Dong, J.H., Dong, R.G., Rakheja, S., Welcome, D.E., McDowell, T.W. and Wu, J.Z., 2008, A Method for Analyzing Absorbed Power Distribution in the Hand and Arm Substructures when Operating Vibrating Tools, *Journal of Sound and Vibration*, 311, 1286–1304

Hager, K.M.R., 2003, Reliability of Fatigue Measures in an Overhead Work Task: A study of Shoulder Muscle Electromyography and Perceived Discomfort, scholar.lib.vt.edu/theses/available/etd-01042004-132729/unrestricted/Kris_Hager_Thesis.pdf

Health and Safety Executive, 2009, Table IIDB02: Prescribed industrial diseases; non – lung diseases in England, Wales and Scotland by disease (a) 1995-2007, www.hse.gov.uk/statistics/tables/iidb02.htm

Neely, G. and Burstrom, L., 2005, Gender Differences in Subjective Responses to Hand-arm Vibration, *International Journal of Industrial Ergonomics*, 36, 135–140

Palmer, K.T., Coggon, D., Bendall, H.E., Pannett, B., Griffin, M.J. and Haward, B.M., 1999, *Hand-transmitted Vibration: Occupational Exposures and their Health Effects in Great Britain*, (HSE Books, Sudbury)

Rimell A.N., Notini L., Mansfield N.J. and Edwards D.J., 2008, Variation between manufacturers' declared vibration emission values and those measured under simulated workplace conditions for a range of hand-held power tools typically found in the construction industry, *International Journal of Industrial Ergonomics*, 38, 661–675

Sampson, E., and Van Niekerk, J.L., 2003, *Literature Survey on Anti-Vibration Gloves*, (JL van Niekerk and Associates)

Stevens, S.S., 1986, *Psychophysics: Introduction to its Perceptual, Neural and Social Prospects*, (Transaction Publishers, Edison)

Wos, H., Marek, T., Noworol, C. and Borg, G., 1988, The Reliability of Self-Ratings based on Borg's Scale for Hand-arm Vibrations of Short Duration: Part II, *International Journal of Industrial Ergonomics*, 2, 151–156

THE IMPACT OF PERSONAL PROTECTION EQUIPMENT (PPE) ON THE PERFORMANCE OF UK FIREFIGHTERS

Panagiotis C. Sotiriadis[1] & Sarah R. Fletcher[2]

[1]*Hellenic Air Force Academy*
[2]*Cranfield University, UK*

The impact of Personal Protective Equipment (PPE) on the performance of firefighters during operations has not been extensively investigated. The aim of this study was to elicit firefighters' perceptions of the impact that PPE has on their performance during operational tasks. The results indicate that the PPE provided is appreciated as essential to perform their tasks without compromising safety considerations. However, mobility and sensory impairments due to the restrictions of items of PPE are detrimental to performance. The use of the same basic PPE equipment in every operational task proved to be the main performance drawback due to physical and body temperature issues. Participants of the study made recommendations for the alleviation of such impairments caused by PPE on their operational performance.

Introduction

Task accomplishment in human activity often involves exposure to hazards to life and health. Any kind of clothing or equipment that is worn or used to protect people from those potential hazards is called Personal Protective Equipment (PPE) (Bensel and Santee, 2006). The term is applicable to a wide range of items, used in all manner of tasks, from the most simple, such as cultivating a garden, to the most complex, such as mine disarming in a biochemically polluted environment. The most commonly used PPE items include gloves, body suits, helmets, eye and face protection, respirators and earplugs.

PPE items, and especially those that are designed to protect their wearers from extremely hazardous environments, generally do not look like comfortable equipment to wear and carry. It is also likely that highly protective PPE items will limit mobility and performance capabilities due to a range of factors, such as manufacturing materials, shape, size, weight and interrelation with the environment. PPE could, therefore, become a mechanical burden on the user (Bensel, 1997). It seems that there is a potential trade-off between the level of protection and level of performance.

Firefighting is generally considered an occupation strongly associated with the use of PPE. PPE is an issue of paramount importance in the operational life of fire fighters, and is vital for their safety. Fire fighters have to face a wide range of operational tasks in all weather conditions and in every possible terrain (rural,

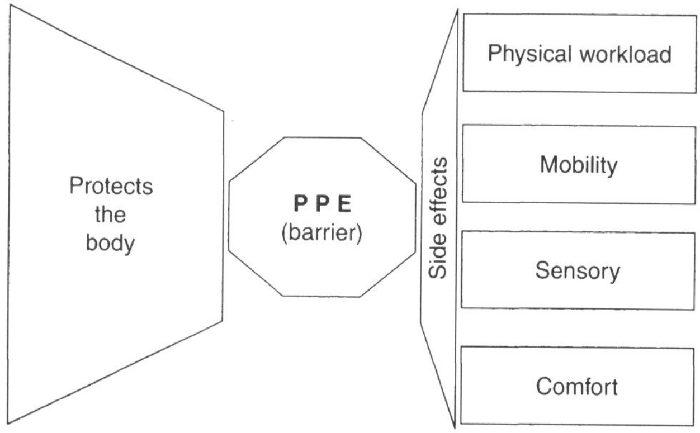

Figure 1. PPE concept schematic.

urban, exterior, interior), and in each case their PPE has to ensure their safety, as well as the safety of their colleagues and everyone involved in the operation, while enabling them to achieve a certain level of performance in order to fulfill the task in hand. PPE provides the necessary protection needed for a firefighter to remain safe, and to remain in the operational field for a longer period of time. The improved protection that PPE now provides to firefighters could be the reason for the significant reduction in injuries that has been evident in fire grounds (Khanzadeh *et al.*, 1995). Moreover, it is not only the design of the PPE that is important; the appropriate use of such equipment is of equal importance (Steffens, 2000). Thus, it is vital for a fire fighter to be equipped properly, and to make use of the equipment provided. Figure 1 summarises the concepts involved in PPE, showing both positive and negative aspects.

Qualitative research focusing on the effects of PPE on firefighting is limited. The aim of this study was to elicit firefighters' perceptions of the impact that PPE has on their performance during operational tasks. To do this, participants' opinions of specific PPE items were investigated, as well as their recommendations for improvements.

Methodology

Design

Qualitative interviewing was considered the most suitable method to achieve the aim of the study as this would provide in-depth subjective data. The interviews were semi-structured, to address issues and questions relating to the objectives of the study, combined with a number of exploratory unstructured interviews also focusing on the themes of the study. The interview framework was designed to address the following issues:

- What kind of PPE is used
- How PPE affects operational performance

Table 1. Study sample composition.

	FFT Type	Participants Number	Percentage on Type Population (%)
Male	WLT	32	11.85
	RTN	17	12.14
	Total	49	11.95
Female	WLT	2	25
	RTN	3	47.85
	Total	5	33.33
	TOTAL	54	12.7

- Proposals and recommendations
- The importance of using PPE
- How PPE affects decision making ability in operations

Recruitment of participants and sampling

Firefighters from one UK county Fire and Rescue Service took part in the study, from each of the service's 14 strategically positioned fire stations. Firefighters participated from both fully operational 'wholetime' (WLT) stations, and from 'retained' (RTN) stations where personnel generally work part-time in response to incident calls, both male and female. In total, 54 firefighters took part, constituting 12.7% of the total firefighter population for this regional service. A summary of the sample composition is presented in Table 1.

Data Analysis

All participants were interviewed at the fire stations they served during their normal shifts. All interviews were recorded. They were then fully transcribed for detailed analysis using the 'Template Analysis' method (King, 1998; Crabtree and Miller, 1999). The Template Analysis technique, which exists somewhere in between content analysis (Weber, 1985) and grounded theory (Strauss and Corbin, 1998), is used to interpret and analyse textual data by clustering themes (King, 2004). The substance of this technique is the creation of a list of 'codes' (template) representing issues of interest within the texts. In the study, the 'initial template' (King, 2004) contained a numbered of pre-determined codes derived mostly from the interview framework, clustered hierarchically in levels. Then the transcribed text was allocated to the defined codes. Through this process weaknesses in the initial template were identified and remedial adaptations were made, leading to the 'revised template', in which new codes were inserted and other were rearranged. The revision process was repeated two more times throughout the whole text. The 'final template', which was formulated at the point where no meaningful text was left uncoded, consisted of codes up to a fourth level.

A 'coding matrix' consisted of 776 coded statements was created. Triangulation of the coding (Robson, 2002) to affirm its accuracy was conducted by testing the

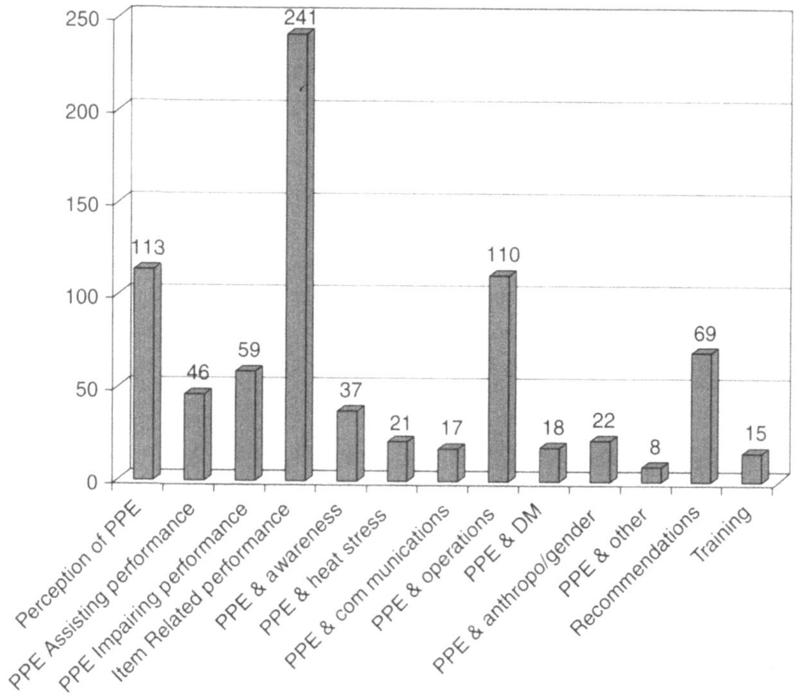

Figure 2. General distribution of coded statements.

inter-rater reliability of around one third of the codes (255) that were randomly selected. Statistical analysis produced a Kappa coefficient of 0.959, confirming ample inter-rater reliability.

Results

The general distribution of coded statements per defined code is presented in Figure 2. Most of the statements concerned the implications of specific items of PPE for the firefighters' performance. Furthermore, comments on the relationship between PPE and operational demands and of firefighters' individual perceptions of PPE were also of increased concern. Finally, a number of recommendations and suggestions were made.

Further analysis of the results regarding the way in which specific PPE items related to performance is shown in Table 2, where firefighting gloves, breathing apparatus and chemical protection suits were the most commonly mentioned issues. Figure 3 presents the distribution of statements regarding different types of firefighting operations.

69 proposal and recommendation statements were made, almost half of which (31) concerned specific PPE items. The remaining comments were about conceptual firefighting issues (31) and generic (7) improvement proposals.

Table 2. Analytical results per PPE item.

PPE Item	Statements			Remarks
Tunic and leggings	12	Positive	4	Excellent quality, protective
		Negative	8	Physically restrictive, heavy, bulky, increasing body temperature
Helmet	23	Positive	8	Protection from falling objects
		Negative	15	Restrictive in confined spaces, heavy, cumbersome
Boots	6	Positive	4	Reliable
		Negative	2	
FF gloves	59	Positive	13	Protect from burns
		Negative	46	Seriously impairs dexterity and sense of touch
RTC gloves	17	Positive	8	Increased dexterity and sense of touch
		Negative	9	Improper fit
Flash hood	11	Positive	4	Protects neck and ears
		Negative	7	Impairs hearing and situational awareness, uncomfortable when deformed
Breathing apparatus	51	Positive	14	Allows user to enter every fire incident and operate in hostile environments for a long time
		Negative	24	Functional problems, fit, comfort, size, weight, sensory impairment
		Remaining	13	commented on new and older set
Chemical protection suit	39	Positive	12	Protection from chemical hazards
		Negative	27	Mobility and hearing impairment, limited vision due to misting vision panels
Water rescue equipment	12	Positive	9	Protection, buoyancy
		Negative	3	Time to suit up
Eye protection	11	Positive	2	Eye protection
		Negative	9	Vision impairment due to mist and scratches

Discussion

PPE, as defined by the UK Health and Safety Executive Work Regulations (1992), is meant to provide protection against hazards to the wearer in the work environment. It has become obvious from participants' comments that the PPE equipment provided by the fire service has been totally successful in this regard. Firefighters recognised the high quality of their equipment and appreciate the protection it gives them. Moreover, it has been shown that this level of protection gives them the confidence to enter operations calmly, without having to worry about issues of equipment safety. The firefighters believed that this psychological status assists their performance, enabling them to concentrate on the task to be accomplished. The psychological state of the firefighters was also enhanced by the collective belief that their PPE equipment was better than that provided by other UK regional fire services. This belief makes the firefighters feel fortunate and also creates enhanced confidence in the brigade. Although it could be inferred that issues related to psychological

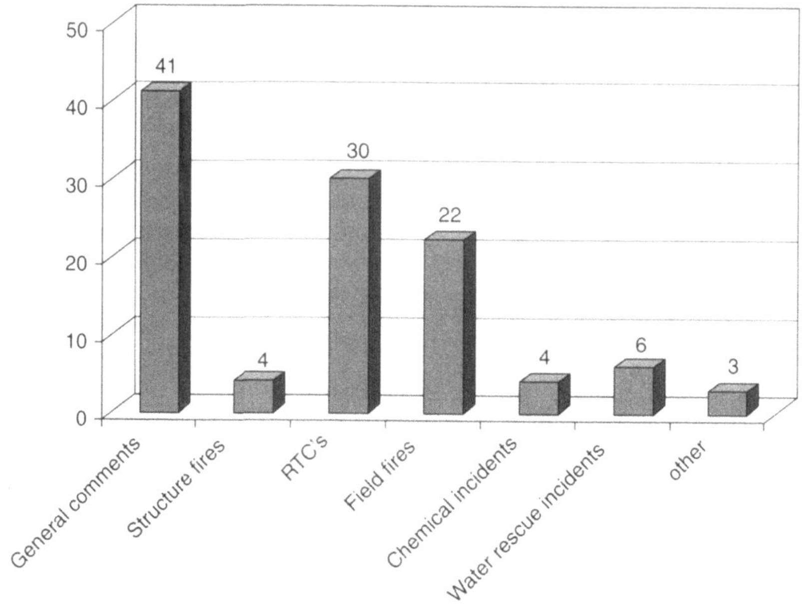

Figure 3. Coded statements distribution per operation.

stress (Hytten and Hasle, 1989; Kalimo *et al.*, 1980) were implicit in the comments concerning heat stress and discomfort, psychological disorders related to the use of PPE (Carter and Cammermeyer, 1985; Brooks *et al.*, 1983) or generally due to operational tasks were not directly mentioned by the participants. Thus, in terms of protection alone, the participants in this study overall indicated confidence and satisfaction. However, as set out earlier in the introduction, there is a potential trade-off between the levels of protection and performance.

Most participant statements concerned characteristics of specific PPE items or groups of items that are commonly used. There were no statements about uncommonly used items, such as PPE for working at height or other equipment (e.g. electrical safety gauntlets). As described, positive statements referred mainly to the protection that each item provides. Negative statements concerned issues related to the other side of the potential trade-off: performance. These were mainly in relation to mobility and sensory impairments due to the characteristics of particular items, such as the material from which they are manufactured, shape, size and weight. In addition, the ability to move unimpeded was found to be impaired by a number of items of PPE, such as the helmet or the breathing apparatus, especially in confined spaces. Despite that, those restrictions were felt to be minor to moderate factors in terms of their impact on the firefighters' performance. Comfort seemed to be a matter of personal preference and individual differences as what some firefighters considered comfortable was found uncomfortable by others.

Breathing apparatus seemed to be the PPE item of most concern, attracting the largest number of comments. The positive aspect of breathing apparatus is that it allows firefighters to enter environments that would otherwise be impossible, but it brings a number of functional, fitting and comfort problems leading to performance impairment during operations. Contrary to research which suggested that lightweight breathing apparatus sets seemed to be preferable to conventional ones due to reduced energy cost and comfort levels (Hooper *et al.*, 2001), participants in the study showed a preference for the conventional set, which they had used previously.

Flash hoods appear to impair firefighters' situational awareness during firefighting operations. This phenomenon was mentioned mainly in the context of structure firefighting. Before the use of flash hoods, the ear lobe served as a 'temperature sensor', informing firefighters of extensive environmental heat. Now, with the use of the flash hoods, this sense has been removed and firefighters say they remain unaware of the exact temperature conditions. Clearly, this could be become a factor in misjudging the environmental situation, resulting in wrong decisions being made in the operational field, and leading to degraded performance and increased risk taking behaviour. Despite that research conducted by the Institute of Occupational Medicine (IOM) concluded that task performance and ability to estimate temperature is not significantly impaired by the use of fire hoods (Johnstone *et al.*, 1995), it is noteworthy that very experienced firefighters who have worked with flash hoods for thirteen years still express the opposite opinion.

Heat stress (Parsons, 2003; Beshir and Ramsey, 1988) and related physical strain and exhaustion, changes in respiratory and heart rate, and intense fluid loss through sweating, were felt to be a major impairing factor in firefighters' performance during operations. Heat stress is an impairing factor in the decision making process, whilst also affecting firefighters' cognitive performance during operations (Bensel and Santee, 2006).

Anthropometric suitability in terms of the proper fit of PPE was mentioned by the participants as an influencing factor in operational performance. Female firefighters and males of smaller size in particular complained about the improper fit of equipment, which comes in pre-determined sizes and not individually tailored.

Impaired communication in fire fighting operations leads to reduced performance and safety problems (Varone, 2003). The issue of communications was highlighted by participants in this study, mainly with regard to situations involving breathing apparatus and chemical incidents. In cases where breathing apparatus is worn, noise and environmental conditions, along with increased respiration rate and the barrier that the face mask presents, were noted as the main factors in impaired communication. In chemical incidents, specifically those involving fire situations (in which breathing apparatus is worn), the chemical protection suit constitutes one more layer to impair communication in addition to the above, resulting in further reduced performance.

Firefighters have to meet a number of different operational challenges; chemical incidents, structure fires, field fires and various other specific situations. In all of these operational tasks, firefighters are kitted out with the same equipment, which is designed to be protective in severely hot environments and flashovers. This turned out to be the factor which impaired the firefighters' performance the most, according to the participants of this study. Working in non-fire related incidents whilst wearing firefighting attire increases the body temperature, resulting in heat stress with all the previously noted consequences. In addition, mobility constraints are inevitable due to the size of the equipment, which does not allow movement in confined spaces, such as inside a crashed car in road traffic collision. A dress down procedure would seem to be a valuable strategy to address this problem.

Conclusions and suggestions

A number of very interesting recommendations and proposals were made by participants, relating to improvement to specific characteristics of PPE items which, if acted upon, could directly resolve impaired operational performance due to sensory and mobility issues. While retaining the use of high specification fire fighting materials, new technologies and inventions could also improve and resolve such issues. The commonality of concerns regarding breathing apparatus indicates that this is an area worthy of further investigation. Proper communication during fire fighting operations, and its implications for operational effectiveness and performance could also be an issue for further research, particularly to address the lack of such studies in the literature. Similarly, the use of the flash hood is an issue that should be re-addressed in order to clarify whether or not firefighters' performance and safety is affected due to a lack of situational awareness, given the disparity between the data from this study and that produced by the IOM some years ago.

Overall, the fact that the same basic PPE equipment (fire kit) is used in any operation was considered to be the main operational performance drawback related to PPE, especially in road traffic collisions. The need to use different clothing and equipment according to the operational task in hand could be considered to be the main finding of this study.

Acknowledgements

We would like to thank all the participating firefighters whose contribution made this study possible.

References

Bensel, C.K. and Santee, W.R. 2006, Use of Personal Protective Equipment in the Workplace. In G. Salvendy (ed.) *Handbook of Human Factors and Ergonomics*, (John Wiley&Sons, England), 912–928

Bensel, C.K. 1997, Soldier Performance and Functionality: Impact of Chemical Protective Clothing, *Military Psychology*, 9, 287–300

Beshir, M.Y. and Ramsey, J.D. 1998, Heat Stress Indices: A Review Paper, *International Journal of Industrial Ergonomics*, 3, 89–102

Brooks, F.R., Xenakis, S.N., Ebner, D.G. and Balson, P.M. 1983, Psychological Reactions During Chemical Warfare Training, *Military Medicine*, 148, 232–235

Carter, B.J. and Cammermeyer, M. 1985, Biopsychological Responses of Medical Unit Personnel Wearing Chemical Defence Ensemble in a Simulated Chemical Warfare Environment, *Military Medicine*, 150, 239–248

Crabtree, B.J. and Miller, W.L. 1999, Using Codes and Code Manual: A Template Organising Style of Interpretation. In B.F. Crabtree and W.L. Miller (ed.) Doing Qualitative Research (Sage, London), 163–177

Hooper, A.J., Crawford, J.O. and Thomas, D. 2001, An Evaluation of Physiological Demands and Comfort between the Use of Conventional and Light Weight Self-Contained Breathing Apparatus, *Applied Ergonomics*, 32, 399–406

HSE, 1992, *A Short Guide to the Personal Protective Equipment at Work Regulations 1992*, (HSE, UK)

Hytten, K. and Hasle, A. 1989, Fire Fighters: A study of Stress and Coping, *Acta Psychiatrica Scandinavica*, 80/355, 50–55

Johnstone, J.B.G., Graveling, R.A., Butler, D.O., Butler, M.P., Cowie, H.A. and Hanson, M.A., 1995, *IOM Report TM/96/04 The Effectiveness of Fire Hood*, (IOM, Edinbrough)

Kalimo, R., Lehtonen, A., Daleva, M. and Kuorinka, I. 1980, Psychological and Biochemical Strain in Firemen's Work, *Scandinavian Journal of Work, Environment and Health*, 6, 179–187

Khanzadeh, F.A., Bisesi, M.S. and Rivas, R.D. 1995, Comfort of Personal Protective Equipment, *Applied Ergonomics*, 26/3, 195–198

King, N. 1998, Templates Analysis. In G. Symon, and C. Cassell, (ed.) *Qualitative Methods and Analysis in Organizational Research: a Practical Guide*, (Sage, London), 118–134

King, N. 2004, Using Templates in the Thematic Analysis of Text. In G. Symon, and C. Cassell, (ed.), *Essential Guide to Qualitative Methods in Organizational Research*, (Sage, London), 256–270

McQueen, R.A. and Knussen, C. 2006, *Introduction to Research Methods and Statistics in Psychology*, (Prentice Hall, Harlow)

Parsons, K.C. 2003, Second Edition, *Human Thermal Environments*, (Taylor & Francis, London)

Robson, C. 2002, Second Edition, *Real World Research*, (Blackwell Publications, Australia)

Steffens, J.T. 2000, *Fire fighter Safety: It's Your Responsibility – Participant Manual*, (VFIS, York)

Strauss, A. and Corbin, J. 1998, *Basics of Qualitative Research*, (Sage, U.S)

Varone, C. 2003, Firefighter Safety and Radio Communication, *Fire Engineering*, March 2003

Weber, R.P. 1985, *Basic Content Analysis*, (Sage, Beverley Hills)

ACADEMIC TRAINING AND PRACTICING ERGONOMISTS – MEETING THE NEEDS

C.A. Williams

School of Science, University of Derby, DE22 1GB

Ergonomics is described by the International Ergonomics Association (IEA) as a scientific discipline. Ergonomists work in both academia and practice; with academic research creating the evidence base which practitioners apply. The majority of both types of professional currently undergo the same initial professional development, often via MSc study. MSc level education has specific requirements outlined by The Quality Assurance Agency for Higher Education (QAA). The QAA requirements are better aligned with the needs of academics than practitioners, raising the question whether it is too much to ask for an academic course in a practical subject to meet the needs of both sets of potential professionals. Brief findings from some focus group research carried out with ergonomics academics and practitioners suggests this may be the case and lends some support to the idea that different training routes could be considered for the two groups. Potential reasons why splitting the training of academics and practitioners would be to the detriment of the field are put forward with some suggestions for gaining further information to examine these issues.

Introduction

Ergonomics is a scientific discipline according to the International Ergonomics Association (IEA) (IEA, 2009a). This status is born out by the existence in the UK of ergonomics and human factors researchers and academics in a number of Universities, whose role is to generate the scientific evidence base for our discipline, as well as teach.

In an effort to maintain standards across the profession, the IEA's code of conduct asserts that these ergonomics researchers have an obligation to undertake accurate and original research which is then communicated to the wider scientific community (IEA 2009b). Furthermore, they 'should pay special attention to communicating these findings so as to facilitate their practical application' (IEA 2009b).

The practical application of the evidence base is often undertaken by a different group of people (practitioners) about whom the IEA's code of conduct also has requirements for professionalism. These include integrity and ethical behaviour standards, in particular requiring practitioners to 'refrain from making misleading, exaggerated or unjustified claims for the effectiveness of their methods' and 'not advertise services in a way likely to encourage unrealistic expectations about the effectiveness and results of those services'(IEA 2009b).

Ergonomics is not unique as a discipline which requires some of its members (researchers) to generate a scientific evidence base which others (practitioners) then apply. A brief scan of relevant literature reveals that academic-practitioner debates occur in many other fields including, for example, information technology (Lee *et al.*, 2002), management (Rynes *et al.*, 2001) health services (Bartunek *et al.*, 2003) and marketing (Brennan, 2004). Though clearly not uncommon, the academic-practitioner split does raise interesting training dilemmas.

Current syllabi

These dilemmas are particularly marked for a discipline like Ergonomics whose main entry route in the UK is via a Masters Degree. This is in contrast to other professions where a Masters degree would not be entry into a subject but would instead be a specialisation for an individual already qualified in that area. In addition, other professions who have debated the academic-practitioner divide have professional training often not even linked with Universities, let alone at the level of a Masters Degree. This allows more freedom for the training to be applied and practical, without the constraining standards of academia.

These standards are specifically defined by The Quality Assurance Agency for Higher Education (QAA) in their framework for higher education qualifications (QAA, 2008). In this they explain that

'Master's degrees are awarded to students who have demonstrated:

- a systematic understanding of knowledge, and a critical awareness of current problems and/or new insights, much of which is at, or informed by, the forefront of their academic discipline, field of study or area of professional practice
- a comprehensive understanding of techniques applicable to their own research or advanced scholarship
- originality in the application of knowledge, together with a practical understanding of how established techniques of research and enquiry are used to create and interpret knowledge in the discipline
- conceptual understanding that enables the student:
  - to evaluate critically current research and advanced scholarship in the discipline
  - to evaluate methodologies and develop critiques of them and, where appropriate, to propose new hypotheses.'

(QAA, 2008)

This outline explains that Masters students should demonstrate they have current knowledge of the field of ergonomics and have the skills to evaluate and undertake research. There is no mention here of the practitioner skills – the only techniques referred to are those which are 'applicable to . . . research or advanced scholarship'.

It is therefore no surprise that a brief scan of a few available MSc. ergonomics syllabi in the UK demonstrates that research methods form a mandatory part of the offering. For example, the University of Derby's syllabus of 8 taught modules includes '*Investigations & Analysis in Ergonomics*'; a module covering qualitative and quantitative research methods (University of Derby, 2009). The Robens Centre for Public Health at the University of Surrey also has 8 taught modules on the MSc., 2 of which are compulsory and cover qualitative and quantitative research methods (The Robens Centre for Public Health, 2009). The University of Nottingham's Distance Learning MSc has 6 modules, one of which '*Ergonomics Methods*' deals with research methods (The University of Nottingham, 2009). Just this brief course check demonstrates that between an eighth and a quarter of the taught part of these Ergonomics Society recognised initial professional development (IPD) courses (which follow the Harmonising European Training Programmes for the Ergonomics Profession (HETPEP) guidance) are taken up with research methods training. This, in spite of the fact that many practitioners are unlikely to need these methods in the future.

Later on in the QAA framework reference is made to skills required for employment. Here it explains that 'typically, holders of the qualification will be able to:

- deal with complex issues both systematically and creatively, make sound judgements in the absence of complete data, and communicate their conclusions clearly to specialist and non-specialist audiences
- demonstrate self-direction and originality in tackling and solving problems, and act autonomously in planning and implementing tasks at a professional or equivalent level
- continue to advance their knowledge and understanding, and to develop new skills to a high level.
- 'holders will have:
- the qualities and transferable skills necessary for employment requiring:
  - the exercise of initiative and personal responsibility
  - decision-making in complex and unpredictable situations
  - the independent learning ability required for continuing professional development.'

(QAA, 2008)

It is noteworthy that, unlike the specific research skills outlined in the first excerpt from the QAA framework, the employment skills outlined in this second excerpt are generically applicable to both research and practice, rather than being specifically practitioner skills. In line with the mention of employment skills, however, the MSc courses referred to above also cover some aspects of professional practice too, via taught modules (e.g. Robens' optional module, Ergonomics, Practice & Management) and/or throughout all the taught materials (e.g. Derby's MSc. Course). In addition, all the courses include a dissertation element which will involve use of the learning from the research methods modules and potentially some professional practice learning, depending on the nature of the study.

The aim of these MSc courses, therefore, is to prepare the students for careers in ergonomics research or practice (or both). It is interesting, however, that in other disciplines such as psychology or more broadly the social sciences, it is possible to study for a Masters in research methods alone. Overall, there is a potential danger in Ergonomics that the current Masters route is insufficiently involved to develop the academic skills required to be a researcher, but is too academic to provide the skills that practitioners will need. Is it asking too much of an academic course in a practical subject to meet the needs of both sets of potential professionals? The following section outlines some research findings which suggest the answer might be 'yes'.

What do Ergonomists say. . . .

The conflict in training needs and work outputs between academics and consultants were raised during a focus group study examining expertise amongst ergonomists. Whilst a full account of the findings from that study will be presented elsewhere, the issue of the conflict between the requirements for academic work and those of consultancy came to the fore during 3 focus groups; 2 with ergonomics consultants and one with ergonomics academics who engage intermittently with consultancy.

Eight male and eighteen female ergonomists took part ($n = 26$) with a mean age of 35 ± 10.1 years. Their declared specialisms were Occupational Health/ Health and safety ($n = 14$), Physical ergonomics ($n = 3$), Human Factors ($n = 3$), HCI ($n = 2$), Usability ($n = 1$) and General ergonomics ($n = 3$). Their self determined job roles were; ergonomist consultant ($n = 10$), ergonomist consultant/trainer ($n = 9$), researcher ($n = 5$), ergonomist consultant/researcher ($n = 2$); all at various levels of seniority.

All 3 groups raised the need for Ergonomics knowledge to be practical, not just theoretical or academic, as exemplified by the following quote;

> ***Gp1- P1*** 'Well ideally I'd like people in a University setting – professors- to have some practical experience as well. . . . I think practical experience is important.'

A further participant equated the term 'expert' (the subject of the discussion) with 'academic'; when asked if he felt that that meant someone who was deeply knowledgeable about ergonomics he explained that he felt they were 'pointlessly knowledgeable . . . from an applied point of view. Fabulous from a research point of view'. . . ***(Gp 2- P7)***

Furthermore, all the groups identified dangers of academic approaches which generate knowledge which is not easily transferable to the field:

> ***Gp 3 – P1*** – 'Focussing on the nitty-gritty and losing the bigger picture – so for example all those studies which have got endless statistical analyses.. and you think well you're just completely off the mark in the first place.

However, the dangers of too pragmatic an approach to ergonomics issues was also raised as undermining expertise;

> ***Gp 3 – P1*** – 'But too practical to me spells consultant rather than human factors specialist because to me an expert is someone who can adapt and change and be insightful, as opposed to someone who goes along and does the same thing over and over'

In addition, research which was insufficiently rigorous or made inflated claims was highlighted as problematic;

> ***Gp 3 – P2*** 'And the flip side of that is the research where it kind of makes out to be quite a big thing when in fact there's five students that have done something.'

Training Ergonomists – do academics and consultants require different training?

Stanton (2005) explains that as a scientist, the ergonomist should, amongst other activities, develop hypotheses; use rigorous data collection and analysis techniques; and test theories. Conversely, as a practitioner the ergonomist will address real world problems; offer the most cost effective solutions and be required to seek compromises in difficult circumstances. He argues that most ergonomists will work somewhere between these two poles.

The findings outlined in section 3 above suggest that there are issues with ensuring that the training ergonomists receive enables them to function well at the poles of scientist or practitioner, with criticism of practitioner ergonomists who do not add to the ergonomics knowledge base and academics with insufficient practical experience who undertake unhelpfully unapplied research. The undertaking of scientifically poor research was also criticised.

Quite aside from the issue of the broad content of ergonomics courses (in terms of whether they should include research methods and ergonomics practice aspects for all students) these quotes indicate there may be finer detail issues with what is taught. For example, those studies where the 'nitty gritty' is focussed on without the understanding of what those applying it will need may be scientifically robust, but they are not obviously useful. Issues like this, however, whilst indicating a difference in the requirements for practitioners and researchers, are in fact a good argument for keeping the training of the two groups together.

In defence of keeping the two streams together – why have research methods for practitioners?

- So practitioners can read and learn from ergonomics research in a critical manner – It is entirely possible that practitioners will neither generate hypotheses to test theories, nor abstract theories to describe the world around them; broadly the aims of scientific endeavour (Holton, 1986). However, they do need to be able to read, understand and be critical of what others have done, to inform their opinion in the practitioner forum. For example, the evidence base for much of what we have accepted as the whole story in the physical ergonomics arena (such as the importance of biomechanical risk factors) is in fact proving to be much more limited than we first thought (Szabo & King, 2000) and of less importance in terms of return to work outcomes than psychosocial factors (Burton *et al.*, 2008). Practitioners need to be capable of being critical of the research done in their area and keep up to date with the latest findings.

- So that, given the right opportunity, practitioners can design and undertake sound research – On occasion, consultancy projects can expand into projects requiring research approaches to generate new knowledge and add to the knowledge base (Denis *et al.*, 2008). It is important that practitioners can design studies and analyse data to answer important questions when this opportunity arises.
- So practitioners can know how far they can or can't extrapolate from their work 'in the field' – The IEA's professional standards outlined in the introduction to this paper explained that practitioners should 'refrain from making misleading, exaggerated or unjustified claims for the effectiveness of their methods' and 'not advertise services in a way likely to encourage unrealistic expectations about the effectiveness and results of those services'(IEA 2009b). Ensuring a solid understanding of scientific and research methods should support practitioners in delimiting the extent to which their customers can generalise from their work, and the degree of certainty they can have. Knowing the boundaries of their work is also an important skill for researchers to have (Xiao & Vincente, 2000).

In defence of keeping the two streams together – what's the point of practitioner training for researchers?

- So researchers understand for what and for whom they're generating their research – In the past, Meister (1992) warned against too broad a split between research and practice explaining that 'much academic writing explains everything but at the same time very little, because one can do nothing with it.' He describes the difference between *explanatory* knowledge and *instrumental* knowledge, explaining it is the latter which is needed by the practitioner, and it is often not the latter which is generated by academia. This is perhaps echoed by the comment made by one practitioner that academics are 'pointlessly knowledgeable about' specific areas.
- So researchers ask useful questions – Understanding how the research will be used should encourage researchers to ask the kind of questions practitioners need answering. The questions should drive towards the production of *instrumental* knowledge as described by Meister (1992). Additionally, attention to the boundary conditions of the research produced is important (Xiao & Vincente, 2000) – how far can we generalise from the research done. This is the same issue as described for practitioners above, but practitioners will be reliant on the researchers explaining the boundaries of their work, so that the practitioners know how far they can apply it in the field.

- So researchers communicate their findings well – Understanding the type of situation practitioners will find themselves in should help researchers in 'communicating their findings so as to facilitate their practical application' (IEA 2009b). Again, this will involve the researchers having a good understanding of the limitations of their work but will also involve them ensuring they analyse their work to the point that practitioners can use it (Dekker & Nyce, 2004).

Next steps

A number of issues have been raised in this paper which touch on the quality and usefulness of some aspects of the current training for ergonomics researchers and practitioners. Whilst there are obvious differences in the training requirements of both groups, there are clear benefits in them covering the same content and studying together. Further investigation of the professional progression of both groups, post IPD qualification, and their perceptions of what is required would be a useful next step (Lee *et al.*, 2002), in order to find out formally if their needs are being met. As a new lecturer whose teaching load includes Research Methods and Ergonomics Methods, the coming year will shed more light on how best to provide the initial professional development in these areas to furnish students with the skills required to function well as academics or practitioners.

References

Bartunek, J., Trullen, J., Bonet1, E., & Sauquet, A., 2003, Sharing and expanding academic and practitioner knowledge in health care. *Journal of Health Services Research & Policy*, 8 , 62–68

Brennan, R. , 2004, Should we worry about an "academic-practitioner divide" in marketing? *Marketing Intelligence & Planning*, Volume 22, Number 5, 492–500

Burton, A.K., Kendall, N.A.S., Pearce, B.G., Birrell, L.N. & Bainbridge L.C., 2008, Management of Work-relevant Upper Limb disorders. *Contemporary Ergonomics 2008*, (Taylor and Francis, London), 425—430

Cornford, I. & Athanasou, J.,1995, Developing expertise through training. *Industrial and Commercial Training*, 27, 10–18

Dekker, S. & Nyce, J., 2004, How can ergonomics influence design? Moving from research findings to future systems. *Ergonomics*, 47, 1624–1639

Denis, D., St-Vincent, M., Imbeau, D., Jetté, C. & Nastasia, I. , 2008, Intervention practices in musculoskeletal disorder prevention: A critical literature review. *Applied Ergonomics*, 39, 1–14

Holton, G.J., 1986, *The advancement of science and its burdens*. Cambridge: Cambridge University Press

IEA 2009a (http://www.iea.cc/browse.php?contID=what_is_ergonomics accessed September 11th, 2009)

IEA 2009b http://www.iea.cc/browse.php?contID=edu_codeconduct accessed September 11th, 2009

Lee, S., Koh, S., Yen, D. & Tang, H., 2002, Perception gaps between IS academics and IS practitioners: an exploratory study. *Information & Management*, 40, 51–61

QAA, 2008, The framework for higher education qualifications in England, Wales and Northern Ireland August 2008 accessed September 16th, 2009 http://www.qaa.ac.uk/academicinfrastructure/FHEQ/EWNI08/FHEQ08.pdf

Meister, D., 1992, Some comments on the future of ergonomics. *International journal of Industrial Ergonomics*, 10, 257–260

Rynes, S.L., Bartunek, J.M., & Daft, R.L. (2001) *Across the great divide: Knowledge creation and transfer between practitioners and academics*. Academy of management journal, 44, 340–355

Stanton, N., 2005, Handbook of human factors and ergonomics methods. CRC Press: Boca Raton

Szabo, R. & King, K., 2000, Repetitive Stress Injury: Diagnosis or Self-Fulfilling Prophecy? *The Journal of Bone & Joint Surgery*, 82, 1314–1322

The Robens Centre for Public Health, 2009, http://www.surreyergonomics.org.uk/index.php?option=com_content&task=view&id=39&Itemid=22 accessed September 10th, 2009-09-11

University of Derby, 2009, http://psychology.derby.ac.uk/ergonomics/ accessed September 10th 2009

University of Nottingham, 2009, http://www.virart.nott.ac.uk/apperg/course Structure.html accessed September 10th 2009

Xiao, Y & Vincente, K.J., 2000, A Framework for epistemological analysis in empirical (laboratory and field) studies. *Human Factors*, 42, 87–101

CONSTRUCTION

HOW CAN WE PREVENT CONSTRUCTION ACCIDENTS? OUTCOMES FROM A STAKEHOLDER CONSULTATION: SOCIETAL AND INDUSTRY-WIDE INFLUENCES

C.H. Brace[1], A.G. Gibbs[2], M. Pendlebury[2] & P.D. Bust[2]

[1]*Scott Wilson Ltd., Manchester M1 6HS*
[2]*Loughborough University, LEICS LE11 3TU*

This paper presents some of the findings from a time-limited, independent research project that reviewed evidence about underlying causes of construction accidents and levers within and beyond health & safety systems to make further improvements. The triangulated method incorporated national and international expert consultations with a total of 72 participants from across the industry, plus a literature review. An initial 'sounding out' exercise was also undertaken to gather feedback on the practical strategies developed during the consultation process. Perspectives on the underlying causes identified: broader societal and industry-wide influences (macro); project and process factors (mezzo); and worker/supervisor/workplace causes (micro). A number of improvement strategies were suggested, some of which have been summarised in this paper.

Introduction

The UK has a large and exciting construction industry, which contributes over £100 billion annually to the gross domestic product (HSE, 2009). The construction industry has been recognised internationally as one of the most dangerous industries in which to work, with the statistics often explained in terms of the industry's inherently hazardous nature (Lingard & Rowlinson, 2005). There have been considerable efforts over many years directed at understanding accident causes (e.g. Haslam *et al.*, 2005) reducing the numbers of accidents and injuries through prevention. However, there is recognition that construction still has one of the highest rates of fatal and major injuries of all industries in the UK (HSE, 2009) and further improvement needs to be sought. In late 2008, the Secretary of State (SoS) at the Department for Work and Pensions (DWP) commissioned an inquiry into the underlying causes of construction fatal accidents. (The full findings of the inquiry can be found at: www.dwp.gov.uk/publications/policy-publications/fatal-accidents-inquiry.shtml and the details of the work presented in this paper can be found in Brace *et al.*, 2009). This paper presents findings from a time-limited, independent research project reviewing non-HSE (Health and Safety Executive) evidence about underlying causes of construction accidents and levers within and beyond health & safety systems that could be applied to make further improvements. The aims of this research were to explore the extent of complementary evidence about underlying or root causes of construction (fatal) accidents generated by parties outside HSE and to determine some practical strategies (levers) that might be adopted for accident reduction.

Approach

The triangulated method incorporated a literature review plus national and international expert consultations with a total of 72 participants from across the industry to gather: stakeholder opinion on underlying causes of accidents and strategies for prevention based on their experience of fatal and serious construction accidents; detail about existing accident investigation processes and approaches to learning from accidents; and case study examples of good practice regarding accident investigation and prevention from the UK as well as a range of different countries. The stakeholders consulted included: 27 UK senior construction industry expert stakeholders; 15 practitioners from the UK construction industry; 15 workers representing the UK's very small, hard to reach organisations/sole-traders, and; 15 overseas construction industry expert stakeholders. The consultation combined face to face and telephone interviews, plus email correspondence. Consequently, the research team created a prioritized 'list' of possible strategies that arose during the preceding research process. These were organised into main topic groupings and feedback on the potential improvement strategies was gathered from 14 industry stakeholders (who were separate individuals to those already involved in the research) via two focus groups. Upon completion of data gathering and feedback, based on researcher/stakeholder perceptions, the potential improvement strategies were given a rating of high, low or medium priority based on perceived impact, implementation and dependency. This paper focuses on summarizing some of the proposed intervention strategies that resulted from the research process.

Findings – Causes of Construction (Fatal) Accidents

There have been very few comprehensive studies reported in the literature on why or how construction accidents happen. Most studies on construction accidents focus on immediate causes, characteristics of accident victims or accident sequence. It is clear from the review that of the few studies examining the underlying causes and contributing factors to construction accidents, those that do have limitations. The underlying factors that have been documented in these studies (e.g. Arboleda & Abraham 2004; Behm 2005; Chi *et al.*, 2005) include: workplace management and culture; worker training and competency; worker attitude and behaviour; equipment factors; inappropriate/non-compliant procedures; lack of safety regulations and legislation; environmental factors; industry structure and set up. It is not possible to quantify the extent to which each factor plays a role in accidents due to the discrepancies in reporting and in data collection itself. As would be expected,

Table 1. A summary of the underlying causes under each of the classification groupings.

Macro Factors	Mezzo Factors	Micro Factors*
Immature corporate systems; Inappropriate enforcement; Lack of proper accident data; Lack of leadership from 'Government' as a key client; Lack of influence of trades unions in practice on most sites, especially for smaller projects.	Immature project systems and processes; Inappropriate procurement and supply chain arrangements; Lack of understanding and engagement by some of the design community; Lack of proper accident investigation / data and consequently, a lack of organisational learning.	A shortage of competent supervisors; A lack of individual competency and understanding of workers and supervisors; The ineffectiveness or lack of training and certification of competence; A lack of ownership, engagement and empowerment of, communication with and responsibility for workers and supervisors.

* These factors were also exacerbated by poor behaviour, cost pressures; poor equipment or misuse of equipment, including personal protective equipment; site hazards; poor employment practices; an itinerant workforce and inadequate management of and provision for vulnerable workers such as younger, older or migrant workers.

based on the accident hierarchy, there do not appear to be any major differences between the causal/underlying factors reported to be contributing to fatal accidents, compared to those for all accident injury outcomes. The underlying factors reported were complex and inter-related and for the purposes of this work they were grouped into the themes of macro, mezzo, and micro (it is important not to confuse this use of 'micro' to refer to worker, workplace and supervisor issues with the use of the term 'micro' to refer to very small organisations). Perspectives on the underlying causes gathered via the stakeholder consultation exercises identified similar and broader: societal and industry-wide influences (macro factors); project and process factors (mezzo factors); and worker/supervisor/workplace causes (micro factors). These influences can be viewed as potential defences against accidents, in line with approaches by Reason (1990) and applications to construction accidents by Gibb and colleagues (Gibb *et al.*, 2006). When active or latent failures at each level create holes in the defensive plates, accidents can occur, with the 'chance' element being represented by the chance of the holes in the various plates lining up to provide an opportunity for 'accident trajectory'. Amongst the stakeholder cohort, underlying causes of construction fatal accidents are documented in Table 1.

Findings – Strategies for Accident Prevention

Within the literature, a number of strategies have been proposed for preventing construction fatal accidents, including those focused on the technical, organisational and human factors, with many prevention strategies focused on particular accident mechanisms, such as falls from height. The expert stakeholder cohort were surveyed to find out what interventions they had found successful in reducing accident rates and other more 'futuristic', ideal world methods that they felt could be adopted to reduce accidents and increase safety within the industry based on their experience and understanding. Due to the relatively small number of fatal accidents that occur in the industry compared to serious and slight injury outcome accidents, stakeholders were asked to consider methods for preventing all accident outcomes. After reviewing the findings from the literature and stakeholder consultation exercises, a number of potential improvement strategies for reducing fatal and serious accidents at the macro, mezzo and micro level were proposed. Table 2 presents the full set of potential improvement strategies proposed but this paper focuses solely on the potential macro level improvement strategies. Another paper discusses the mezzo and micro level improvement strategies (Gibb *et al.*, forthcoming 2010).

These potential improvement strategies were themed under three categories: enforcement and compliance; competency and training; and culture and mindset, and are summarised in the following sub-sections, along with a rating of high, low or medium priority based on perceived impact, implementation and dependency (based on researcher/stakeholder perceptions). It should be noted that it is very unusual for (existing or former) interventions to be evaluated for their impact, which causes difficulty in determining their efficacy for reducing accidents improving safety.

Theme 1: Enforcement and compliance

Enhance enforcement activities (High Priority) It was widely suggested that enforcement by the regulator is the key improvement facilitator for the 'rogue' element of the industry. Based on finer analysis of the accident data, it was suggested that enforcement activities amongst target groups should be enhanced to raise awareness and priority of safety amongst the 'problem' areas of the industry (fairly but visibly). It was suggested that increasing the number of enforcement notices and subsequent prosecutions would create a higher profile and priority for safety, particularly amongst target groups, although we should be aware of promulgating a blanket 'stick' approach. However, as well as the 'stick' approach being appropriate on many occasions, the consensus was that 'one size does not fit all', and subsequently the 'carrot' is seen as being very important for many other circumstances, i.e. HSE to enhance positive enforcement.

Ensure Government is an exemplar client (High Priority) Government works encompass a significant proportion of the activities of the UK construction industry. Therefore, it was proposed that all UK Government departments follow the lead of some current departments in being exemplar clients in terms of health and safety.

Certify all construction organisations (Medium Priority) Within the industry, there are a large number of 'unengaged' small and medium enterprises (SMEs) resulting in unsafe behaviour and poor culture. To address the challenges of the small organisation sector and to engage SMEs especially those working on domestic projects, it was suggested by stakeholders that we (the UK construction

Table 2. A summary of the strategies under each of the classification groupings and cross cutting themes).

Macro Factors	Mezzo Factors	Micro Factors
Theme 1: Enforcement and Compliance		
Enhance enforcement activities; Ensure Government is an exemplar client; Certify all construction organisations; Link Building Control approval to health and safety; Focus on overall effective management systems; Divide HSE; Consider the creation of a Government construction body.	Enhance any connection between employment type and safety.	Increase the number of supervisors; Organisations to implement licensing requirements.
Theme 2: Competency and Training		
Provide free advice and training Conduct finer analysis of accident data; Consider advice/legislation for accident investigation and learning; Conduct evaluations of interventions; Implement competency standards; Develop training and competency in the design community.	Develop strong organisational competency and maturity; Conduct enhanced examination of accident and high potential incident data; Conduct in-house evaluations of interventions.	Develop individual competency and understanding.
Theme 3: Culture and Mindset		
Tackle safety from a consumer protection perspective; Outlaw inappropriate tools and equipment; Insurers to demand greater safety provision by organisations; Change HSE approach; Consider implementing an independent accident investigation board; Reduce poor employment practices.	Create a minimum time period before site work starts.	Make subcontractors part of the team; Remove the 'bonus payment'; Encourage and manage diversity in the industry.

industry and stakeholders) should consider requiring certification of all construction organisations.

Link Building Control approval to health and safety (Medium Priority) Stakeholders suggested that there are many bad practices being followed in industry, partly because there is a limit as to the amount of enforcement and 'checks' that can be conducted by the regulator. Therefore, the idea of making Building Control approval conditional on health and safety planning should be considered, e.g. organisations have to talk through their H&S plan with an H&S expert to ensure they are fully anticipating risk (similar to CDM: Construction (Design & Management) Regulations). The main function of Building Control is to ensure compliance with the Building Regulations under the Building Act 1984. The purpose of which is to ensure that building standards meet the requirements necessary to protect the health, safety of people in and around buildings. In the first instance, it was proposed that a review of how this could be practically implemented should be conducted.

Focus on overall effective management systems (Medium Priority) Good safety systems depend on good overall management systems. Therefore, it was considered that more focus by the regulator should be given to the overall effective management systems rather than solely on safety systems, hence consider appropriate training for HSE Inspectors to provide them with a better understanding of the commercial contractual complexities of the industry.

Divide HSE (potentially Medium Priority) Numerous issues were raised about the HSE and the challenges HSE faces in being both the regulator and the advisor. It was therefore suggested that the UK should consider dividing HSE in two (similar structure to OSHA and NIOSH or OSHA enforcement and OSHA consultation in the USA) to enable an arm for enforcement and another for assistance and to consider the outworking of the 'assistance' arm being assigned to the local authorities and building control at a local level.

Consider the creation of a Government construction body (potentially Medium Priority) The creation of a single Government body to manage the construction industry might be beneficial for safety. This body would be able to run a high level panel to which CEOs of companies who have had a fatality must give account (similar to the model adopted in Hong Kong). It was proposed that a review should be conducted on this suggestion.

Theme 2: Competency and training

Provide free advice and training (High Priority) It was frequently noted that health and safety awareness and competency across the industry are low, particularly within smaller organisations. Therefore, free information and training should be provided to the industry (focusing on SMEs) on key construction management issues (incorporating OHS).

Conduct finer analysis of accident data (High Priority) A better, more detailed understanding of the 'problem' (target) organisation types or sectors of the industry is required in order to target interventions better. Therefore, a finer analysis of accident data and investigations are required and findings publicised amongst the industry to raise awareness and understanding.

Consider advice/legislation for accident investigation and learning (High Priority) There can be scenarios arising within construction organisations when there is 'data overload' due to the large volume of accident data that is being collected. Therefore it was proposed that HSE/industry bodies provide guidance materials on best practice approaches to data collection and review to avoid 'data overload'. In order to place pressure on management to lead strong safety processes, legislation could be implemented for accident investigation/learning to ensure good practice is followed, led by the leaders of an organisation (e.g. as per Sweden, Hong Kong).

Conduct evaluations of interventions (High Priority) Typically, when interventions have been implemented in industry, there is limited evaluation of their efficacy. Therefore, (simple) evaluations of interventions should be arranged to determine whether or not changes are having the desired effect and to better inform the organisation and the industry. Additionally, it was noted that assistance should be provided to industry organisations on how to conduct worthwhile evaluations, e.g. including field workers to work with industry to ensure uptake and evaluation.

Implement competency standards (High Priority) It was suggested that if we're serious about making the construction industry effective, safe and well run, we need (competency and therefore) a licensing system. Therefore, the industry and Government should consider the implementation of competency standards for individuals' task and safety knowledge to standardise competency and aid awareness of what criteria need to be met.

Develop training and competency in the design community (High Priority) A key issue for improving safety in the build process was the need to improve competency and understanding amongst the design community. Examples of what could be developed included: following through on previous initiatives to regulate and accredit higher education courses to ensure there is enhanced and appropriate coverage of health and safety; put improved continuous professional development (CPD) and professional practice strategies in place so that there is improved understanding and engagement; place more emphasis on health and safety issues during initial professional development (IPD) and stress the essential role that designers and pre-construction planners can play in reducing health and safety risks.

Theme 3: Culture and mindset

Tackle safety from a consumer protection perspective (High Priority) It was suggested that, if awareness was raised amongst consumers, they would be more likely to only employ licensed professionals. Therefore, safety could be tackled from a consumer protection perspective, i.e. similar to systems operating in other industry sectors, e.g. CORGI (now the Gas Safe Register). Aimed at householders, emotive campaigns could be used to raise awareness about construction accidents and safety to make small-scale clients consider what is at stake by employing someone who is not competent.

Outlaw inappropriate tools and equipment (Low Priority) Strategies could be developed to encourage/force equipment suppliers and builders' merchants to outlaw inappropriate tools and equipment and to ensure that adequate safety directions and training is provided for all users, including do-it-yourself (DIY) enthusiasts and SMEs.

Insurers to demand greater safety provision by organisations (potentially High Priority) Insurers could demand greater safety provision by organisations in order to increase the emphasis on health and safety quality systems within an organisation. In the first instance it was suggested that the value of this idea is explored through consultation with the insurance sector.

Change HSE approach (Medium Priority) The challenge for HSE resources to pick up the rogue element and at the same time encourage the 'better end' of industry was acknowledged. However, a more positive and proactive approach by the HSE would assist industry in moving forwards. Therefore, it was proposed that the mindset of the regulator needs to change to enable them to more readily give credit for those organisations who are doing 'more right than wrong', in order to remove obstructive fear from the system.

Consider implementing an independent accident investigation board (Low Priority) It may be useful to consider the implementation of an independent accident investigation body. It was felt that such a body (similar to those currently operating for rail, aviation and maritime environments), which has the objective of learning and preventing accidents rather than apportioning blame, would enable more detail on the underlying causes of accidents to be investigated.

Reduce poor employment practices (Low Priority) It was recognised that many employment practices impact on health and safety in the workplace. Therefore, it was proposed that the Government should consider amending legislation (and subsequently enhancing enforcement) and systems to reduce the black market in undocumented workers and to eliminate/reduce the workers operating on a false self employed status.

Conclusions

This work has provided an insight into current industry practices and thinking about the causes of accidents and levers and strategies that could be adopted to improve health and safety and reduced accidents in the construction industry. A number of strategies have been proposed in this paper and these have been attributed a priority rating based on perceived impact, implementation and dependency. It is important to note that the responsibility of industry leaders and corporate managers extends beyond the macro to the mezzo and micro areas – actions of such leaders have a very significant impact on projects and on workers. There is an urgent need to evaluate interventions to establish their value in making the construction industry safer. The rich data gathered during this work was largely subjective opinion as there are only very limited numbers of 'hard facts' concerning the underlying causes of accidents and the efficacy of interventions in the construction industry, in the UK or internationally. The tight time constraints (6 weeks for the data collection phase caused by the SoS deadline) meant that the work was very intense but focused and the research team were unable to explore in depth all avenues that were captured during the research process. Despite its limitations, this work provides a picture of the current practice of construction health and safety and evaluates opinion on the challenges and opportunities facing the construction industry in this regard.

Acknowledgements

The work described in this paper was funded by the Health and Safety Executive (HSE). Its content, including any opinions and/or conclusions expressed are those of the authors alone and do not necessarily reflect HSE policy or the views of the Inquiry Chair, Rita Donaghy, CBE FRSA. The final report from the Inquiry including Rita Donaghy's recommendations can be found at: www.dwp.gov.uk/publications/policy-publications/fatal-accidents-inquiry.shtml

References

Arboleda, C.A. and Abraham, D.M., 2004. Fatalities in Trenching Operations-Analysis Using Models of Accident Causation. *Journal of Construction Engineering and Management,* 130(2), pp. 273

Behm, M., 2005. Linking construction fatalities to the design for construction safety concept. *Safety Science,* 43(8), pp. 589–611

Brace, C.L., Gibb, A.G.F., Pendlebury, M.C., and Bust, P.D., 2009. Secretary of State for Work and Pensions Inquiry into the underlying causes of construction fatal accidents. *Phase 2 Report: Health and safety in the construction industry: Underlying causes of construction fatal accidents – External research.* Available at: www.hse.gov.uk/construction/inquiry.htm

Chi, C., Chang, T. and Ting, H., 2005. Accident patterns and prevention measures for fatal occupational falls in the construction industry. *Applied Ergonomics,* 36(4), pp. 391–400

Gibb, A.G.F, Brace, C.L., Pendlebury, M., and Bust, P.D. (forthcoming 2010). How can we prevent construction accidents? Outcomes from a stakeholder consultation: project and workplace influences. In: *Proceedings of the cib Conference,* May 2010, Manchester

Gibb, A.G.F., Haslam, R.A., Hide, S., Gyi, D.E. Duff, A.R., 2006. Why accidents happen, *Civil Engineering, Proceedings of the Institution of Civil Engineers,* Vol. 159, November 2006, pp. 46–50, ISSN 0965 089 X

Haslam, R.A., Hide, S.A., Gibb, A.G.F., Gyi, D.E., Pavitt, T., Atkinson, S. and Duff, A.R., 2005. Contributing factors in construction accidents. *Applied Ergonomics,* 36(4), pp. 401–415

Health and Safety Executive (HSE), 2009. *Construction Division. The Construction Intelligence Report.* http://www.hse.gov.uk/construction/pdf/conintrep0405.pdf

Lingard, H. and Rowlinson, S., 2005. *Occupational Health and Safety in Construction Project Management.* Spon Press, ISBN 0 419 26210 5

Reason, J., 1990. *Human Error,* Cambridge University Prcss, ISBN 0 521 31419 4

COMMUNICATING HEALTH AND SAFETY IN CONSTRUCTION: THE USE OF PICTURES ON CONSTRUCTION SITE POSTERS

P.D. Bust, A.G.F. Gibb & K. Akwiwu

Department of Civil and Building Engineering, Loughborough University, Leicestershire, UK

Prior to the recent economic downturn there had been an increase in the migrant workforce in the UK with a large number of workers from many different nationalities working on projects such as the Channel Tunnel Rail Link, Heathrow Terminal 5 and the London 2012 Olympics. Health and safety managers are constantly looking for ways of overcoming language and literacy problems associated with a migrant workforce, to deliver training and reinforce health and safety messages. The use of posters without words can be used to overcome barriers created by language, low levels of literacy and visual problems (Wogalter *et al.*, 1997). This paper reports on work to examine the use of pictorials in posters to convey health and safety information. An appropriate information processing model was identified and the results suggest that well illustrated pictorials are easily comprehended by diverse users and colour and location of posters increase their ability to be noticed.

Introduction

Prior to the current recession the construction industry in the United Kingdom (UK) enjoyed one of its busiest periods of the past 20 years. In terms of employment, the total labour requirement was expected to rise from just over 2.4 million in 2005 to over 2.8 million by 2011. Which would have led to an increasing demand for migrant workers, particularly in Greater London and the South East (Dainty *et al.*, 2007).

The UK is not alone in Europe in experiencing an influx of migrant labour in construction. The highest estimates suggest that the number of illegal migrant workers on construction sites in Finland, has jumped from a handful to several thousand over the last few years. To reverse this unhealthy trend, the construction industry labour market organisations have jointly published a guidebook on the regulations governing the use of foreign labour (Artto 2004).

In the United States, the rise in the Hispanic population has not only affected the size of the construction labour force but also increased the burden on the management of health and safety with 40 percent of work-related deaths in the Houston, Texas area involving individuals with Hispanic surnames (Flory 2001).

The inability to immediately communicate via the spoken word on construction sites represents one of the major barriers to successful management of health and safety. When multinational consortia work alongside workforces derived from many countries, the opportunity for the message to be lost in translation is increased (Bust & Gibb 2006). In research to investigate health and safety on construction sites in developing countries it was recorded that large engineering companies were using pictorial and safety installations to overcome language and low literacy levels.

In recent research commissioned by the Institution of Civil Engineers (Dainty *et al.*, 2007), poor language and communication skill was the most cited employer reason for not using migrant labour. Clearly, where migrants lacked functional English, they were putting themselves and their fellow workers at risk, and many of those with poor English skills expressed a desire to learn the language (Dainty *et al.*, 2007).

The authors are currently working to understand the potentials and limitations of the current uses of (audio) visual media to communicate about health and safety to construction workers of increasingly diverse cultures and nationalities. In doing so some insights are taken from academics working in the subdiscipline of visual anthropology who have grappled with similar questions about how to communicate about the experiences of one group of people's to another (different) group for many years (Bust *et al.*, 2007). This paper reports on research carried out into the use of one particular area of visual communication of health and safety on construction sites: the use of health and safety posters.

Background to visual communications project

Ensuring accident free operations is a central goal and a major challenge for safety officials, managers of construction projects and indeed the organisation as a whole. Organisations spend huge amounts of money in ensuring that there is adequate and effective communication of health and safety messages to workers.

In an attempt by organisations to comply with regulations and with the increasing demand on contractors by clients to demonstrate increased commitment to health and safety, there has been a proliferation of the use of safety signs on construction sites. The proliferation trend is such that a commonly encountered problem with regards to the perceptual impact of safety signage is the tendency among employers to post a plethora of signage at a single or limited number of locations rather than at the point of hazard (Weyman, 2006). Also on the increase is the use of diverse methods of communicating health and safety messages to a diverse workforce, which sometimes differ from organisation to organisation. The question that ought to be asked therefore is; how effective are these modes of communication in communicating to the target audience. The very common and increasing challenge for UK construction companies is how to effectively communicate to an increasing number of non-English speaking workers.

In 2004, NCC Construction Sverige AB, Sweden, received the European Good Practice Awards in Safety and Health at Work from the European Work Environment

Agency. The company won the award for its ingenuity in developing the "Tysta boken" (The Silent Book). NCC has been using The Silent Book at construction sites since 2001 to communicate risks to workers. Using cartoons, the book illustrates what to do and what not to do in various hazardous situations based on NCC's work-related accident statistics. With The Silent Book, NCC was able to disseminate safety information across all its operations worldwide thereby overcoming potential translation problems.

This study focused on pictorial safety posters and the factors that determine their effectiveness. This is timely as safety poster campaigns within the construction industry appear to be on the increase.

Rationale for the Research

Construction work requires minimal reading and writing skills. It therefore attracts workers with low literacy skills and migrant workers. It has been suggested that language and literacy problems have contributed to increasing accident rates in both agriculture and construction (Trajkovski and Loosemore, 2006).

Greater safety awareness can be achieved through the effective communication of health and safety messages. However, with a considerable number of workers now having low literacy skills and to which English is not their first language this poses a serious challenge to employers in the UK. Available information suggests that pictorials are useful in conveying information to an audience of diverse lingual disposition (Young and Wogalter, 1990). Pictorials have the advantage of speed of appreciation in that they can convey instructions and messages at a glance. While there has been some research in other areas, such as, marketing, social science, and healthcare , there remains a paucity of research in the use of visual means of communication within the UK construction industry. Research in the area of warning communication is often left or assumed to be in the domain of human psychology and ergonomics. It is addressed in general terms and not relevant to the construction industry. Work needs to be carried out to understand how construction workers process and use the information they receive about construction related health and safety warnings.

The research objectives were to understand the human processing of warning information relating to the ability of safety posters to grab workers attention. This was investigated by the use of pictorials, colour, borders and identifying source on information on the posters in the context of construction site activities and low literacy levels of workers.

Methodology

Little research has been carried out in the domain of visual communication of safety messages in the construction industry. Little is known about the subject matter i.e. attitude of workers in construction towards visual methods of communication. The research investigated site workers' responses to warning messages on site.

A comprehensive and critical review of literature from previous research conducted in other disciplines provided a theoretical background. The literature review drew on similarities between research findings in different disciplines.

Interviews with operatives (21 No.) and site management staff (5 No.) were used to determine workers perception of the use of pictorials in communicating health and safety message on construction sites. Dependent variables (attention-grabbing, comprehension and source perception) which determine message effectiveness were measured through a non-experimental investigation.

There were 42 responses to a follow-up Questionnaire based on emerging issues from the interviews. Electronic questionnaires were targeted at safety managers, safety advisors and project managers in top 70 UK construction firms.

Sampling

The judgemental or purpose sampling method (Kumar, 2005) was used for this study. This type of sampling is used when it is the judgement of the researcher as to who can provide the best information to achieve the objectives of the study. Only those who, in the opinion of the researcher, have the required information and are willing to share were selected to be part of the study. The aim was to interview equal numbers of migrant (20 No.) and English speaking (20 No.) workers over four construction projects. The participating companies were selected from a list of top 100 construction companies in United Kingdom obtained from Construction News website.

Questionnaire Construction

The question areas were derived from the Communication Information Process Model (C-HIP), as shown in Figure 1. Categories of attitude to message source, attention and comprehension of illustrative warning messages were used. The closed-ended and open-ended questions were used to guide interviews with site operatives.

The participating companies were reluctant to release workers for long periods so the questions were short and succinct. Some dichotomous questions were also included for the same reason. Due to time constraints the questionnaires were designed to ensure that each interview would last between 10 to 15 minutes. Likert, ranking and semantic scales were used. The questionnaire was structured to cover questions relating to attitude, to source of message, attention-getting qualities and comprehension of warning messages.

Poster Selection and Manipulation

Four posters (Figure 2 below) were created by adopting two black and white images from the NCC Silent book (2000) and two colour images from the Trojan horse construction site safety messages report (Chinien and Cheyne, 2006). The selection was to address common health and safety issues faced in construction. Consideration was also given to activities most likely to be undertaken on an average construction site. For example the "Avoid Back Problems: Lift Correctly" could be considered to be an activity relevant to most sites. Two of the posters were related to safety issues and two to health issues.

Systematic variations of the pictorial warnings were created. Poster "A" had the company name added and poster "B" had the company name added and a border was introduced to give the poster a defined edge. Studies in the area of warnings in general have shown that the use of borders around warnings can improve the ability

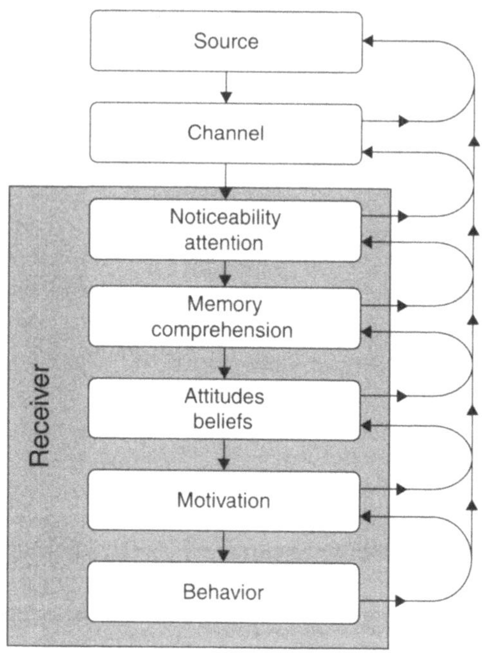

Figure 1. Communication human information process (C-HIP) Model with feedback loops illustrating that later stages influence earlier stages (Wogalter *et al.*, 1999).

Figure 2. Posters "A" – colour, "B" – b/w, "C" – b/w and "D" – colour.

of the poster to be noticed (Davies *et al.*, 1998). The poster "C" which is in black and white was left as was originally obtained. The poster "D" was also retained in its original form with no manipulations.

Results

In all, a total of 22 site operatives were interviewed with an fairly even age spread from workers in their 20 s to workers in their 50 s. Seventeen were English speaking with 5 non-English speaking. Nine of the workers had between zero and 5 years experience and 7 had more than 21 years experience. The sample was made up of the trades ground worker; labourer; carpenter; bricklayer; banksman and general foreman. For the mid-management, 6 interviews were conducted.

The questions for the operatives were structured under the main variables of investigation namely; source, attention and comprehension/memory. Most of the workers noticed two of the posters with colour. One of the respondents said he noticed all of the posters but added that it may have been the ones with colours that attracted his attention.

From Figure 3 below it can be seen that Images A and D which had coloured backgrounds were the most likely to attract attention. For images B and C the numbers of responses for likelihood of attracting attention are almost equal, but the negative response for Image C is higher than that of Image B. This could be due to the black lined frame around image B in addition to the yellow strip forming the background to the message "this message is from your safety manager".

QE2. How likely are the images to attract your attention?

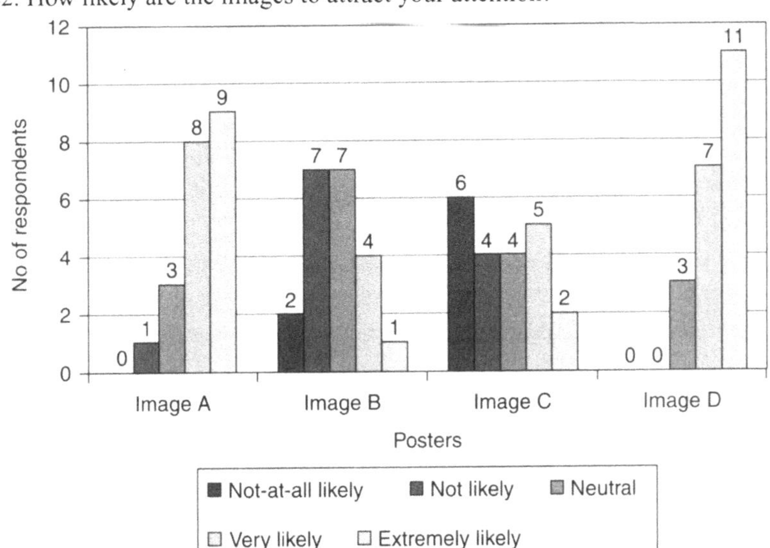

Figure 3. Responses to question E2.

The results showed that only 7 participants would go out of their way to check for new signs when they are on site as against the 12 who said they would not. The high number may be due to perceived low risk and familiarity with the environment.

Conclusions

This research looked at the use of visual means of communication, posters on construction sites, used by health and safety managers to overcome difficulty communicating health and safety information to workers with low literacy and language skills. From the literature review a model, C-HIP model, was identified which was appropriate for research into the use of warning pictorial posters in the construction industry. The model, see Figure 1, provides a framework for describing the human processing of warning information by dividing the process into separate stages that must be successfully completed for compliance behaviour to occur.

The investigation focused on the attention/noticeability stage of the model but also used the source of the poster to see if this affected the noticeability of the poster and looked at the affect of changes to the appearance of the posters on the comprehension of the health and safety information. The colour, pictorials, symbols and location of the posters were all manipulated in the work. The colour posters proved more noticeable than those in black and white. The attention of the operatives was easily drawn to the posters, in some cases, simply because of the tick (√) and the cross (x) on the poster, signifying what-to-do and what-not-to-do. This is in line with work by Sojourner and Wogalter (1997).

The illustrative pictorials were understood by all of the operatives regardless of language background. It is important to note that even in the presence of limited text or complete absence of the text, the pictorials where easily understandable. Simple line drawings have been shown, Moll (1986), to maximize comprehension of health information, especially for people with low literacy skills. Locating the posters at the point of hazard was most effective when compared to other locations (canteen, toilets and notice board) within the construction site. This is in agreement with other researchers' findings that warnings are most effective when located at the point of hazard, (Frantz & Rhodes, 1993; Wogalter, Barlow & Murphy, 1995).

These findings indicate that knowledge of message source did not have any significant effect on the perceived importance of the message by the operatives. However, the expert survey results suggested that knowledge of message source will have significant effect on target audience's perception regarding the seriousness of the message. The relevance of the posters to the task being carried out was found to affect the attention-grabbing capability of the poster. According to Wogalter and Vigilante (2006), stimuli that are personally relevant and interesting tend to attract more attention.

It was found that most of the companies had embarked on poster campaigns but only one organisation had assessed the effectiveness of the posters used in communicating health and safety messages. There is a need for target audience evaluation and feedback from poster campaigns in order to understand the effectiveness of the posters in gaining attention, changing attitudes and eventual compliance. Information obtained will provide valuable input into the creation and management of posters.

Limitations

The sample used for this work was small and so statistical analysis carried out has not been reported. It is hoped that this work will prove of use in preparation for future research work into this topic. The location of suitable sites was a limitation for this research. One of the intentions was to interview equal numbers of English and non-English speaking workers. However, it is not possible to predict the composition of the workforce available on site at any one time.

Indirect effects such as word-of-mouth among co-workers may also have influenced the results of the studies. For example, increased noticeability rates of the installed posters. It is not possible from this work to say if the posters were directly noticed by all workers or noticed by a few workers who informed the rest of the workers.

The answers given in the test for noticeability may not be as reliable as expected. Since the answer was being given in retrospect, there is a reliance on the memory of the respondent as to what caused the attraction to the poster. Another source of interference could be the individual's feeling of the need to provide a positive answer thereby claiming the poster drew attention even when it did not.

Further research

Migrant labour in construction and the resulting effects on health and safety create several highly complex socio-cultural problems. Any research therefore goes beyond simple instrumental solutions. The migrant workforce cannot be investigated as though it were an homogenous mass responding in the same way to all efforts in communication of safety messages.

Ideally a more nuanced understanding of the way in which health and safety can be seen to work in the many different contexts present in the construction industry and with the many different groups who now occupy the industry is required. If this route were followed a sophisticated understanding of the challenges confronting employers and migrant workers could be established. The result would encapsulate the way in which organisations and workers have already responded to locally constituted issues in a contingent manner.

References

Artto, J. (2004) Trade Union News from Finland. www.artto.kaapeli.fi/unions

Bust, P.D. and Gibb, A.G.F. (2006). Construction Health and Safety in Developing Countries. European Construction Institute. UK. ISBN 1873844 63 8

Bust, P D., Gibb, A.G.F. and Pink, S. (2007). Managing construction health and safety: Migrant workers and communicating safety messages. Safety Science, In Press, Corrected Proof, Available online 24 August 2007

Chinien, V. and Cheyne, A. (2006). Trojan horse health and safety messaging: An assessment of the long-term and behavioural impact on construction site operatives. Health and Safety Executive Research Report 505, Her Majesty's Stationery Office, Norwich, UK

Dainty, A.R.J. Gibb, A.G.F., Bust, P.D. and Goodier, C.I. (2007). Health, Safety and Welfare of Migrant Construction Workers in the South East of England. Report for the Institute of Civil Engineers

Davies, S., Haines, H., Norris, B. and Wilson, J.R. (1998). Safety pictograms: are they getting the message across? *Applied Ergonomics,* 29(1), pp. 15–23

Flory M., (2001). Solving the language barrier. Occupational Health & Safety. Waco: Jan 2001.Vol.70, Iss. 1; pg. 37, 2 pgs

Frantz, J.P. and Rhoades, T.P. (1993). A task analytical approach to the temporal placement of product warnings. Human Factors, 35, pp. 719–730

Kumar, R. (2005). Research methodology: a step-by-step guide for beginners. 2nd edn. Sage. London

Moll, J.M (1986), Doctor-patient communication in rheumatology: studies of visual and verbal perception using educational booklets and other graphic material. Annals of the Rheumatic Diseases, 45, pp.198–209

Sojourner, R.J. and Wogalter, M.S., (1997). The influence of pictorials on evaluation of prescription medication instructions. *Drug Information Journal,* 31, pp. 963–972

Trajkovski, S. and Loosemore, M. (2006). Safety implications of low-English proficiency among migrant construction site operatives. *International Journal of Project Management,* 24(5), pp. 446–452

Weyman, A.K. (2006). U.K. Safety Warning and Signage. In: M.S. Wogalter, ed, *Handbook of Warnings.* 1st edn. 2006: Lawrence Erlbaum Associates, pp. 565–580

Wogalter, M.S. and Leonard, S.D. (1999). Attention capture and maintenance. In: M.S. Wogalter, D.M. Dejoy and K.R. Laughery, eds, *Warnings and Risk Communication.* 1st edn. London: Taylor & Francis Ltd, pp. 123–148

Wogalter, M.S., Sojourner, R.J. and Brelsford, J.W. (1997). Comprehension and retention of safety pictorials. *Ergonomics,* 40, pp. 531–542

Wogalter, M.S., Barlow, T. and Murphy, S. (1995). Compliance to owner's manual warnings: Influence of familiarity of task-relevant placement of a supplemental directive. *Ergonomics,* 38, pp. 1081–1091

Wogalter, M.S. and Vigilante, W.J. (2006). Attention switch and maintenance. In: M.S. Wogalter, ed, *Handbook of warnings.* New Jersey: Lawrence Erlbaum Associates Inc, pp. 245–265

Young, S.L. and Wogalter, M.S. (1990), Comprehension and memory of instruction manual warnings: Conspicuous print and pictorial icons, *Human Factors,* 32, 637–649

DEFENCE

LIFEJACKETS IN SHOPPING MALLS: ERGONOMICS & PRODUCT DESIGN IN COUNTER TERRORISM

Alex W. Stedmon[1], Simon Harrison[2], David Ireland[2], Peter Fussey[3] & Richard Tew[2]

[1]*Human Factors Research Group University of Nottingham Nottingham, UK*
[2]*Product Design & Manufacturing Group University of Nottingham Nottingham, UK*
[3]*School of Law University of East London London, UK*

In response to the growing threat of international terrorism, the UK Government has developed its Counter Terrorism Strategy (CONTEST) which has four key elements: pursue, prevent, protect and prepare. The Royal Society for the encouragement of Arts, Manufactures and Commerce (RSA) conducts an annual 'Design Directions' competition with the aim of engaging the design community in socially relevant problems. In 2009 the focus of this competition was 'designing in counter terrorism' and this provided a unique opportunity to combine aspects of ergonomics and product design. This paper illustrates initial development of a lifejacket suitable for installation in crowded public spaces and focuses on a range of ergonomic factors used in developing the design solution (such as capturing user requirements, detailing the context of use, incorporating anthropometric data). The lifejacket concept was short-listed by the RSA for a presentation to the Home Office. The design was commended by several of the judges representing major counter terrorism agencies for its use of ergonomics in arriving at a practical and potentially usable solution.

International terrorism: protect and prepare

Recent terrorist activities have shown that the modus operandi has shifted towards indiscriminate, mass impact activities, often with coordinated and targeted secondary attacks aimed at inflicting maximum casualties. Such attacks have often been focussed on crowded public spaces (e.g. the Bali & Mumbai attacks). In these situations it can be very difficult to conduct counter terrorist initiatives but it is relatively easy for terrorists to plot their attacks. In the UK this was illustrated when the '7/7' bombers easily gained access to the London Underground network carrying rucksacks packed with explosives (Pape, 2005).

The success of counter terrorism is often measured against attacks that do not happen (because they have been foiled or other tactics have diverted the attack in some way). However, after the Provisional IRA's bombing of the Grand Hotel in 1984, the IRA issued an iconic statement to the security services, that is still as relevant today: 'you have to be lucky every time, we only have to be lucky once'. Under this pressure to counter every potential attack, it is realistic to assume that there will be occasions when terrorists will be successful and the first warning might be the first attack itself. If this is then part of a coordinated follow up initiative, what happens in the confused moments after that initial attack could define the overall impact and severity as well as having a major influence on perceptions of public safety in general.

Since the '9/11' attack on the World Trade Centre and the focus on Al Qa'ida as a major international terrorist threat, counter terrorism has been a high priority in the Western World. The impact has been far-reaching with even sporting events such as the 2008 Dakar Rally being cancelled, less than 24 hours before the start, due to a very real terrorist threat. Within the UK, since 2003, the Government has developed a counter terrorism strategy involving police, intelligence officers, emergency services, local authorities and community organisations (HM Govt, 2009). The Counter Terrorism Strategy (CONTEST), run by the Office of Security and Counter Terrorism (OSCT) has four key elements:

- **Pursue** - to stop terrorist attacks
- **Prevent** - to stop people becoming terrorists
- **Protect** - to strengthen defences against attacks
- **Prepare** - to mitigate against an attack if it cannot be stopped

CONTEST provides a framework for other agencies such as the National Counter Terrorism Security Office (NaCTSO) which is co-located with the Centre for the Protection of the National Infrastructure (CPNI) to focus on protect and prepare initiatives concerned with reducing the vulnerability of the UK, and UK interests overseas, to terrorist attacks. A major concern is crowded public spaces and protecting people going about their daily lives, free from fear. As a result, the Government has called for initiatives to 'design out' the terrorist threat to crowded places such as shopping centres and sporting venues (Building Design, 2007).

Design directions: capturing user requirements

The Royal Society for the encouragement of Arts, Manufactures and Commerce (RSA) has developed the annual 'Design Directions' competition which presents challenging design briefs to students and recent graduates to advance 'socially responsible' design (Gribben, 2006). The 2009 competition 'public spaces, safer places: designing in counter terrorism' was targeted at students from a range of disciplines and aimed to draw attention to the issues surrounding security and counter terrorism. In a novel twist, rather than 'designing out' terrorism, the competition focussed on 'designing in' counter terrorism. This is more than a semantic argument and underlines the belief that on some occasions terrorists might be successful and that it is not possible to design public spaces that deter every possible terrorist threat (Building Design, 2007). The brief, developed in conjunction with the Home Office, NaCTSO and the Royal Institute of British Architects, invited a design response to a public space in the aftermath of a fictitious terrorist attack scenario in the heart of a generic major European city.

Table 1. Key requirements capture (adapted from AOF, 2008).

Key areas	Requirements capture	Example
Operational processes	existing/proposed operational process	identify the operational effects and benefits that the users need to gain from exercising the capability
Human factors integration	interaction requirements between users and the capability	determine the human factors requirements relevant to the interactions between the users and the capability (not just the operators and the equipment)
Safety, environmental and sustainable development	safety and environmental requirements/constraints	threats to the safety of person, property and environment
Security	security requirements/ constraints	what security services the capability must deliver, both preventative and corrective
Non-operational requirements and constraints	miscellaneous/non-operational requirements and constraints	capture other miscellaneous non-operational requirements and constraints

Designing a solution for use in such circumstances presents unique problems in detailing user requirements. It is impractical and unethical to capture user needs in a real world setting as events unfold, and it is not always possible to interview victims after the event. As a result, an early strategy analysed video footage of previous attacks in order to profile potential users. However, this, in itself, did not provide a framework for capturing user requirements and so reference was made to the User Requirements Capture section of the Acquisition Operating Framework website (AOF, 2008) which sets out five key areas, as illustrated in Table 1.

From this basis it was important to consider the ergonomics underpinning the design solution (such as capturing user requirements, detailing the context of use, incorporating anthropometric data) and the importance of arriving at an inclusive design that enables users rather than excludes user (Gyi, Porter & Case, 2000) as well as incorporating ergonomics into the design process to arrive at a fully integrated solution and develop an awareness of ergonomics issues in product design (Ulrich & Eppinger, 2007).

Designing in counter terrorism: profiling users and developing concepts

Initial research illustrated that many recent terrorist attacks have involved sequenced blasts intended to inflict maximum casualties and fatalities. Typically, these attacks used a number of smaller devices to initially scare, disorientate and manipulate crowds with the intention of inflicting mass damage in secondary attacks detonated within crowds of civilians and rescue workers, or indiscriminate firearms attacks on the gathered crowds. Early in the design process it was clear that no single solution would suit all scenarios of use and so the solution thinking focussed on secondary blast scenarios. The rationale behind this was that it is virtually impossible to stop a dedicated terrorist attack. If obstacles are designed and put in place, a terrorist will go to extraordinary lengths to overcome them, as the '7/7' bombings illustrated, with terrorists taking their own lives in order to achieve their goal.

Using the elements of the CONTEST strategy as a basis, initial conceptual design work focussed on how to protect and prepare against terrorist attacks. The design thinking was based around minimising and reducing the effect of sequenced explosions in coordinated attacks, and managing the protection of members of the public within crowded spaces. Using ergonomics expertise and observing opportunistic video data (e.g. CCTV and mobile phone footage from real incidents, available over the internet through popular video sites) it was possible to begin profiling user behaviour and user requirements in terrorist attacks. In the moments following an initial attack, people do not seem to panic. If people are not directly in a blast zone they might be curious and seek out the source of the incident without realising they could be putting themselves in danger. Underpinning this is a requirement to get people to safety quickly, balanced against the knowledge that most casualties are caused by flying debris and glass. Protective clothing offers a solution for people caught 'in the open' but new initiatives are also emerging based on the notion of 'invacuation' where the aim is to stay within a place of safety rather than congregating in open public spaces. Recently this was illustrated in the Mumbai attacks where people survived in the hotel when they were able to lock themselves in their rooms.

Two concepts emerged from the initial design stage, both using standard conventions of emergency procedures from other applications: a floor lighting system to guide people to the nearest exit points and a protective lifejacket to protect from secondary attacks. The floor lighting concept was not developed further due to the complexity of retro-fitting it into existing public spaces, however it was incorporated in a novel manner into the lifejacket design. Lifejackets are familiar artefacts from air and sea travel and the premise was that people are familiar with them. In arriving at this solution, it was necessary to consider how the lifejacket might be used as well as the physical construction (i.e. properties of materials) designed to withstand various types of attack.

A lifejacket for shopping malls: key ergonomic features

The lifejacket was designed specifically for use in crowded public spaces (e.g. shopping malls, cinemas, mass transit terminals, sports stadia, etc) with a number of counter terrorism ideas factored in. It was designed to provide a high level of personal protection, a flexible means of installation in public spaces and to be used primarily in pre-blast situations (where a terrorist attack warning may have been issued) or in post-blast situations where a secondary attack might be likely. The lifejacket design is illustrated in Figure 1, a number of ergonomic issues are then explored that underpin the design.

Size & Weight

The lifejacket had to be a 'one size fits all' item. Within ergonomics this statement is rarely true but in terms of alternative design strategies, it was considered that

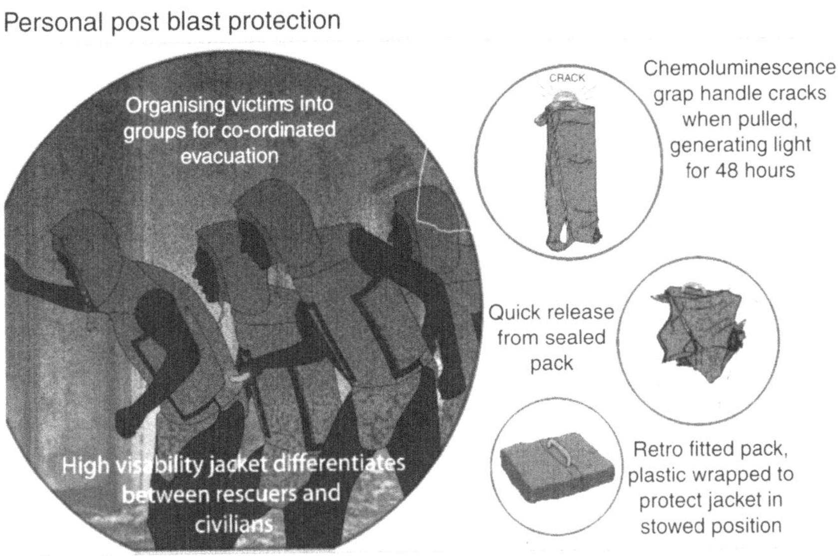

Figure 1. Personal lifejacket for post blast protection.

Table 2. Anthropometric data used in the design.

Anthropometric measurement	Minimum (mm)	Maximum (mm)	Application of data
Trunk Height*	500	650	length of lifejacket
Shoulder length*	121	173	width of shoulder sections for lifejacket
Waist circumference*	724	1093	guidance for adjustability of Velcro straps
Shoulder breadth (biacromial)**	325	430	overall width of lifejacket
Head length**	165	205	head length and breadth were used to
Head breadth**	135	165	determine the head aperture
Hand length**	160	205	size of packed lifejacket for ease of handling

designing for 'common fit' was the best approach in order to satisfy most of the users, most of the time. This was enhanced with adjustable Velcro pads for a more user specific fit. As the RSA was a UK based competition, key UK anthropometric data were used as guidance for a range of measurements for the lifejacket detailed in Table 2.

Minimum sizes were based on UK female 5th percentile data and maximum sizes were based on UK male 95th percentile data (from Peebles & Norris, 1998*; Pheasant & Haslegrave, 2006**). The weight of materials used in the construction of the lifejacket was an important factor. As well as providing the necessary protection, the construction of the lifejacket needed to be light and flexible. For this reason, lightweight 'smart' materials (e.g. D_3O and Honeywell Goldshield) were incorporated into the design so that the lifejacket would be light enough to unpack quickly and put on without being cumbersome to wear. Based on minimum lifting zone data for seated users, the required total maximum weight of the lifejacket was between 3 kg to 5 kg (Pheasant & Haslegrave, 2006).

Ease of use

The lifejacket was designed to be simple and intuitive to use in what would be very difficult circumstances when users might be disorientated and confused. Users would extract the lifejacket from under a seat or from a holding container and due to its flexible and practical design, suitable for most of the UK population (and, to an extent, other users around Europe), it would be placed over the user's head and adjusted to specific size using the large Velcro pads on either side. Under simple testing, it was anticipated that this would need to be achieved within approximately 10 seconds.

Installation

The lifejacket was designed to be installed in any public area, or places of work. It would be vacuum packed to reduce bulk, make it easier to handle and locate under seating or in bespoke holding containers located in public spaces (possibly designed as street furniture in their own right). The lifejacket design had the potential to be easily retro-fitted into current furniture designs and public spaces so that installation costs and user access to the product would not be compromised through bespoke or costly installations.

Protection

Ultimately the lifejacket needed to provide a suitable level of protection. Rather like an anti-mine vest, the lifejacket was designed to protect the vital organs in the user's torso. Protecting the limbs was not considered as part of the final design as this would make the lifejacket harder to put on in the first instance (and thus compromise the initial usability of the design), limit user movement, be more susceptible to anthropometric differences and ultimately would increase the cost of production. The lifejacket was designed to be made from a combination of new 'smart' materials (e.g. D_3O and Honeywell Goldshield) which provide excellent impact and puncture protection whilst remaining flexible when not under impact.

Implicit safety features

The handle, designed as a single use break-tube containing a chemoluminescence agent, would immediately activate as a 'glow-stick' as it was pulled from its location. This would provide a light source for up to 48 hours which might aid evacuation from smoke filled situations and/or could provide an aid to identifying people trapped in buildings. This idea exploited two novel analogies: floor based evacuation lighting systems (which in this design was implicit in the item rather than an artefact of the environment); and that of baby elephants holding on to their mother's tail so as not to get lost (as illustrated in Figure 1). A further safety feature was that it would be difficult for a terrorist to use the lifejacket and conceal any weapons or explosives

Table 3. Addressing user requirements in the design of the lifejacket.

Key areas	Requirements capture	Features of the lifejacket
Operational processes	existing/proposed operational process	simple lightweight protective device for pre- or post-blast scenarios
Human factors integration	interaction requirements between users and the capability	ease of use in situations of compromised personal safety where users may be confused and/or disorientated
Safety, environmental and sustainable development	safety and environmental requirements/constraints	timely protection from immediate danger, enhanced safety through implicit safety features
Security	security requirements/ constraints	protect vital organs of torso from flying debris
Non-operational requirements and constraints	Miscellaneous/non-operational requirements and constraints	must be freely available but protected against potential misuse

on their person (and if they did, then any explosive effect would be compromised with the blast occurring under the lifejacket).

Designed for potential misuse

As well as designing for normal use of the lifejacket, the design process also considered potential misuse of the lifejacket. The final design appeared to be a simple and practical solution however a number of factors proved to be problematic: could the lifejacket be safeguarded against being stolen from public spaces? Might the product become an attractive street fashion item or used for criminal activities? One strategy for reducing the theft of the lifejacket was to make it unappealing and impractical for all but its intended use. This approach caused some design dilemmas. Untraceable, stolen body armour with hoods could be an attractive proposition for criminal activities. Having full head protection, covering an individual's facial features, would have made these garments even more attractive to criminals wishing to hide their identity. This was one of the main reasons why there is no protection to the face with this product. It was also a concern that the inherent impact and puncture protection would be a very desirable attribute for street attire of gang members. One solution to this problem was to make the jacket in an unattractive colour or easily identifiable as an emergency garment so that its use 'out of context' would arouse suspicion.

Reviewing the user requirements

Returning to the key requirements earlier in the paper, it is possible to illustrate how the lifejacket design presents a viable solution as illustrated in Table 3.

Conclusion

The lifejacket design was developed as a concept and so no formal prototype was developed to conduct detailed user testing (this would be the next logical step in a formal design process). As such, it presents only a partial development of a potential final solution but still illustrates the importance of incorporating ergonomics into the design process at an early stage. The lifejacket design was short-listed by the RSA for a presentation to the Home Office and was commended by several of the judges representing major counter terrorism agencies for being a practical and potentially usable design.

Acknowledgement

The authors would like to thank the RSA judges in short-listing the design and for the positive feedback provided by the counter terrorism representatives. The authors are also indebted to the anonymous reviewers of the initial draft of this paper and their guidance in refining it for publication.

References

AOF (2008) Requirements and Acceptance, Policy, Information and Guidance on the Requirements and Acceptance aspects of UK MOD Defence Acquisition. (version 1.0.1, July 2008). Retrieved 6 November 2009 from http://www.aof.mod.uk/aofcontent/tactical/randa/content/urareastoconsider.htm #startcontent

Building Design (2007) *Ministers Call for Anti-Terror Design* (16 Nov 2007). Retrieved 20 Sept 2009 from http://www.bdonline.co.uk/story.asp?storycode=3100178

Gribben, R. (2006) *Designers Feel Way to Success* (26 May 2006). Retrieved 20 Sept 2009 from http://www.telegraph.co.uk/finance/2939662/Designers-feel-way-to-success.html

Gyi, D., Porter, M., & Case, K. (2000) Design Practice and 'Designing for All' *Proceedings of the International Ergonomics Association Conference*, San Diego, USA

HM Government (2009) *Pursue, Prevent, Protect, Prepare: The United Kingdom's Strategy for Countering International Terrorism*. Her Majesty's Stationary Office, ID 6072968 03/09 (Crown Copyright)

Pape, R.A. (2005) *Dying to Win. The Strategic Logic of Suicide Terrorism*. Random House Inc.

Peebles, L. & Norris, B. (1998) *ADULTDATA. The Handbook of Adult Anthropometric and Strength Measurements: Data for Design Safety*. London, Department of Trade and Industry

Pheasant, S.T., & Haslegrave, C. M. (2006) *Bodyspace: Anthropometry, Ergonomics and the Design of Work*. Taylor & Francis Ltd. London

Ulrich, K. & Eppinger, S. (2007) *Product Design and Development* (4th Edition). McGraw-Hill: Columbus, USA

RELIABILITY OF THE HFACS-ADF SAFETY CLASSIFICATION SYSTEM

N.S. Olsen & S.T. Shorrock

Department of Aviation, The University of New South Wales, Sydney, NSW, 2052, Australia

This article evaluates an adaptation of the human factors analysis and classification system (HFACS) adopted by the Australian Defence Force (ADF) to classify factors that contribute to safety incidents. Three field studies were undertaken to assess inter-coder consensus and intra-coder consistency in the use of HFACS-ADF within the context air traffic control (ATC). For all studies, agreement was low at the level of both fine-level HFACS-ADF descriptors and high-level HFACS-type categories. The studies suggest that the ADF adaptation of HFACS may be unreliable for incident analysis at the ATC unit level. Several reasons for the results are proposed, associated with the underlying HFACS model and categories, the HFACS-ADF adaptations, the context of use, and the conduct of the studies.

Introduction

The Human Factors Analysis and Classification System (HFACS; Shappell and Wiegmann, 2001) has been used extensively to identify the significant factors leading to an incident or accident. HFACS has an aviation heritage, and is applied most frequently in this domain, but it is a generic classification system. HFACS, and its derivatives, has been adapted, applied and promoted in several domains (and countries) in addition to commercial and private flying, including road, emergency medical transport, railroad, maritime, healthcare, surgery, maintenance, air traffic control, and military.

HFACS comprises four taxonomies. The overall structure is based largely on Reason's (1990) 'Mark 1' Swiss cheese model. The taxonomies contain 17 categories, each with a set of non-exhaustive, exemplar descriptors. The lack of specific descriptors within HFACS has led to criticism that HFACS' categories are underspecified (Beaubien and Baker, 2002).

The Australian Defence Force (ADF) previously developed a variant of HFACS, which we refer to as HFACS-ADF, to help identify and classify contributing factors in ADF Aviation Safety Occurrence Reports (ASORs) relating to events, incidents and accidents as defined in Annex 13 of the 1944 Chicago Convention. Similar to HFACS, HFACS-ADF contains 22 categories but adds a set of specific 'descriptors' beneath each category. In HFACS-ADF, these specific descriptors are terms available for classification. Some other adaptations of HFACS have similarly developed descriptors (e.g. O'Connor, 2008), while some evaluation studies have used the non-exhaustive set included within HFACS. The HFACS-ADF descriptors are stored in a database, which is used to help manage safety hazards.

Experience with HFACS-ADF and critical examination of its characteristics has raised a number of questions regarding its reliability and validity. The most critical question concerns whether HFACS-ADF can be used with adequate consensus between coders. This is important, because classification systems, such as HFACS-ADF, generally require input from multiple users – 100s in some cases – and therefore must achieve social consensus on discriminatory decisions in order to identify contributory factors consistently, which inform decisions on safety improvement countermeasures (Wallace and Ross, 2006). Classification systems are also used over a long time period, and so consistent coding over time is necessary. The studies reported in this article focus on the reliability of HFACS-ADF, as this is a necessary (but not sufficient) criterion for validity; if coders are unable to reach consensus and make consistent classification decisions over time, then the data produced are invalid.

HFACS (and its derivatives) has been evaluated in several analyses, predominantly by HFACS' developers (see Shappell *et al.*, 2007; Shappell and Wiegmann, 2001; Wiegmann and Shappell, 2001). However, independent reliability studies using metrics that can be compared across studies, are often lacking. Furthermore, adaptations of HFACS are often not evaluated at all. This article aims to assess the reliability of HFACS-ADF's categories and fine level descriptors, as used by 'non-expert' coders, to help determine the system's suitability as an incident analysis tool at the military air traffic control (ATC) section level of the ADF. Three field studies were performed within an ATC unit comprised of a tower and approach.

The HFACS-ADF taxonomies

HFACS-ADF comprises four main taxonomies; the names of these four closely correspond to those of the underlying original HFACS taxonomies (Shappell and Wiegmann, 2001), and the 'Mark 1' Swiss Cheese Model. Each taxonomy comprises a number of super-categories and 22 categories. The taxonomy names and the categories (without super-categories) are shown in Table 1. Each category is further broken down into several detailed descriptors. For instance, descriptors under 'adverse mental state' include 'fixation' and 'stress', and descriptors under 'resource management' include 'selection' and 'staffing/manning'.

While HFACS-ADF is substantially based on HFACS, there are several differences between the 2008 version of HFACS-ADF and the 2001 and 2007 versions of HFACS (Wiegmann and Shappell, 2001; Shappell *et al.*, 2007). Overall, HFACS (2007) and HFACS-ADF share 78% identical or similar categories. Many of the detailed descriptors in HFACS-ADF are identical to the examples suggested in the HFACS (non-exhaustive) list, though there are numerous differences in context and meaning. However, again note that HFACS only classifies at the level of categories.

Table 1. HFACS-ADF contributing factors categories.

Unsafe acts or conditions	Preconditions for unsafe acts	Deficient supervision	Organisational influences
Decision errors[1,2]	Adverse mental state[1,2]	Inadequate supervision[1,2]	Resource management[1,2]
Skill-based errors[1,2]	Adverse physiological state[1,2]	Planned inappropriate operations[1,2]	Organisational climate[1,2]
Perceptual errors[1,2]	Physical/mental limitations[1,2]	Failed to correct problems[1,2]	Organisational processes[1,2]
Knowledge information[3]	Equipment[2]	Supervisory problems[1,2]	
Routine[1,2]	Workspace[3]		
Exceptional[1,2]	Environment[2]		
	Crew Resource Management[2]		
	Training[3]		
	Qualifications authorisation[3]		

[1]HFACS-ADF category is associated with an identical, near identical or similar HFACS (2001) category within the same level of the corresponding taxonomy.
[2]HFACS-ADF category is associated with an identical, near identical or similar HFACS (2007) category within the same level of the corresponding taxonomy.
[3]HFACS-ADF category is not associated with an identical, near identical or similar HFACS (2001 or 2007) category within the same level of the corresponding taxonomy.

Study One

Method

This study tested inter-coder consensus within a single group of participants. A convenience sample of eleven air traffic control officers (ATCOs) from a Royal Australian Air Force (RAAF) ATC section took part in the study. The participants were representative of the unit, which comprised eighteen controllers, in gender, age, HFACS-ADF experience and endorsements held. Since the participants were already HFACS-ADF users, they had varied degrees of training and experience with HFACS-ADF.

Two incident reports were provided, chosen randomly from all reports available. Each report was one page in length (approximately 500 words), and typical of those classified in the ATC section. The reports had spaces marked to record codes from HFACS-ADF. The standard HFACS-ADF definitions within the ADF's Defence Aviation Safety Manual were provided.

Each participant completed the questionnaire individually over a period of one week. Each participant read the incident reports, and coded HFACS-ADF descriptors for the findings described in the report using the HFACS-ADF taxonomies. Due to errors in following the study instructions, only ten of the 11 participants presented classification results that were usable for the first incident, and nine participants presented results that were usable for the second incident. Participants were instructed to use the definitions of the taxonomies, categories and descriptors provided, and to refrain from collaborating.

The Index of Concordance was used to provide a percentage agreement (see Wallace and Ross, 2006). This involves the determination of agreement or disagreement for each pair of codes assigned by each pair of coders for each event. The number of pairs in a group of coders is calculated as follows: N (N – 1)/2, where N is the number of coders. The proportion of agreeing pairs of codes out of all the possible pairs of codes is calculated as follows: agreements / agreements + disagreements. This is converted to a percentage. So for instance, if four coders rate the same event using codes A, A, A and B, then this actually represents 50% agreement (three agreements out of six pairs) – not 75%. To calculate percentage agreement, differences in the number of codes assigned between coders are counted as disagreements.

The Kappa procedure (Cohen, 1960) is an extension of raw agreement that corrects for the number of agreements that would be expected entirely by chance. This was rejected because it is considered flawed for such taxonomic coding studies (see Ross *et al.*, 2004; Wallace and Ross, 2006). Percentage agreement is considered the 'industry standard' measure of agreement for safety data (Wallace and Ross, 2006). A criterion of 70% agreement between coders was adopted as a reasonable minimum, in accordance with Wallace and Ross (2006).

Results

Results are tabulated in Table 2 for both incidents at category level (e.g. 'Skill-based error') and descriptor level (e.g. 'Poor technique').

Table 2 shows that percentage agreement at descriptor level was very low and well below the 70% criterion. Percentage agreement improves at the higher 'category' level, but still does not approach the 70% criterion. On the basis of these results, intra-coder consensus was very low at the descriptor and low at the category level.

Study Two

Study one found low consensus between many coders using a small number of incident reports. It was considered plausible that the result could be due to features of the reports, and the number of coders used. A second study was performed with a

Table 2. Percentage agreement between coders for incident reports 1 and 2.

	Category level	Descriptor level
Incident 1	43.0%	20.7%
Incident 2	36.1%	18.5%
Overall	39.9%	19.8%

large number of incident reports, and comparisons between only one pair of coders for each – the original analyst and the study participant.

Method

This study tested inter-coder consensus between a single participant and one of several previous analysts, using the same incident reports. A convenience sample of one tower controller who was trained and experienced in HFACS-ADF participated in the study. A sample of 63 investigated incident reports tower and approach incident reports were used in the study. This sample of reports comprised all reports investigated in 2007.

The participant read each incident report and classified all factors that were considered to have contributed to the incident. The previously investigated reports each had two, three or four factors assigned, and had each been investigated by one of four separate analysts over the course of 2007. Each of the descriptors and categories assigned by the participant was compared with those previously assigned and percentage agreement was calculated. For this study, the percentage agreement referred to consensus between the participant and the original incident analyst, for each code. Differences in the number of codes assigned by the participant and the original incident analyst were counted as disagreements.

Results

The percentage agreement between the descriptors assigned by the participant and those assigned by the original analysts was 29.7% (total number of agreements for all incidents / [total number of agreements for all incidents + total number of disagreements for all incidents]). At descriptor level, only three of the 63 reports returned a percentage agreement of 100%, and 23 of the 63 reports returned a percentage agreement of 0%.

At category level, the percentage agreement between the descriptors assigned by the participant and those assigned by the previous coders was 40.1% – almost identical to results of study one. For categories, again, only three of the 63 reports returned a percentage agreement of 100%. Twelve of the 63 reports returned a percentage agreement of 0% at category level. Percentage agreement at both descriptor and category level was well below the criterion of 70%.

Study Three

Studies one and two found both low inter-coder consensus at both descriptor and category level, with many or few incident reports and coders. A remaining unanswered question concerned whether coders could achieve consistency when recoding the same incident, i.e. intra-coder consistency. Arguably, this should be considerably easier to achieve than inter-coder consensus, and if intra-coder consistency could be achieved and understood, then this may be a basis for improving inter-coder consensus.

Table 3. Percentage agreement for descriptors and categories assigned by the same participants to 20 incidents (five each) over a period of time.

Participant	Time period between first and second coding attempts	Percentage agreement (descriptors)	Percentage agreement (categories)
1	20 months	35.3%	41.2%
2	16–18 months	26.7%	36.8%
3	12 months	43.8%	43.8%
4	4–11 months	41.6%	46.2%
Overall intra-coder percentage agreement		35.4%	44.6%

Method

This study tested intra-coder consistency with a single group of participants with varying degrees of training and experience. The time period between analyses ranged from four to twenty months.

A convenience sample of four members of the ATC section participated in this study. All participants had HFACS-ADF training, or experience, or both. Five incident reports previously analysed by each member, totalling twenty reports, were used in the study. Each participant was asked to read the incident reports (without the associated contributing factor descriptors) that he or she had investigated previously, between four and 20 months earlier. The participants were asked to assign the appropriate contributing factors from the HFACS-ADF taxonomies to the findings in the report. Participants were given a 24-hour period to complete the exercise, and asked to refrain from collaborating on their answers.

The descriptors and categories assigned by each participant were compared with the previously assigned descriptors and categories, and percentage agreement was calculated. For this study, the percentage agreement referred to the first and second coding attempts by each participant.

Results

The results are shown in Table 3. The results show that for each coder, and overall for all coders, intra-coder consistency was low – again well below the 70% threshold at the descriptor level. While agreement was higher at both descriptor and category level than was the case in Studies 1 and 2, participants were not able to assign the same codes to the same incidents over periods of between four and 20 months.

Discussion

The three field studies assessed the reliability of HFACS-ADF, an adaptation of HFACS, in a number of ways, with 'non-expert' users in the natural environment of use. The inter-coder consensus and intra-coder consistency achieved in the three studies at the descriptor and category level was low to very low, and well below the 70% threshold for reasonable reliability. Even at category level, where HFACS-ADF is very similar to HFACS, percentage agreement was only 40-45% over the three studies.

The results could be explained by a number of reasons associated with HFACS-ADF, its underlying model, and the conduct of the field studies. In terms of HFACS-ADF, at the overall taxonomy level, coders often had difficulty determining if a causal factor was at the level of 'Unsafe act and condition' or a 'Precondition for unsafe acts'. With respect to this, it is noteworthy that HFACS uses as its basis an early 'Mark 1' model of the Swiss cheese model, which may be outdated, or even unsuitable as a taxonomic structure.

At the category level, some categories caused confusion. Often, there is insufficient evidence to determine a category with certainty – even at this coarse level. This raises a question of whether coders can reliably code information processing concepts, particularly from incident reports, and whether they can reliably distinguish these from contextual concepts. This problem may apply to HFACS-ADF, HFACS and other techniques.

The underlying HFACS model of categories may contribute to the poor results. The systems are similar at taxonomy and category level, with 78% of categories in common, yet HFACS itself is, according to the authors, a reliable system (Shappell and Wiegmann, 2001). Many evaluation studies (e.g. those cited within Shappell and Wiegmann, 2001) are unpublished dissertations, and some published studies (e.g. Wiegmann and Shappell, 2001) appear to use HFACS' developers as participants. Indeed, most published reliability studies do not report the independence of the researchers and coders (from the developers of the techniques), the number of pairs of coders, and the 'cleaning' of data (e.g. discarding of codes where unequal numbers of codes are selected). Also, Cohen's Kappa is widely (mis-)used.

Other independent studies have found inadequate reliability at category level. O'Connor (2008) found inadequate inter-coder consensus at category level using HFACS-DOD – an adaptation for the US Department of Defense, "in stark contrast to those studies that have examined reliability of HFACS" (p. 604). Like O'Connor, we wonder how many other HFACS evaluation studies with negative outcomes have not been reported.

A limiting factor in the comparison of HFACS and HFACS-ADF is that the latter modifies the structure of the HFACS categories. However, analysis of the categories that are common to both systems suggests that reliability even among only these categories still fails to approach 70% for our studies. Other studies have also modified HFACS, but have not reported reliability (e.g. Reinach and Viale, 2006).

At the descriptor level, many descriptors are not mutually exclusive. Similar descriptors are also used within different taxonomies within HFACS-ADF, as are vague, overlapping, unverifiable descriptors such as 'Information processing' and 'Situation awareness'. Another possible reason for the results is the number of detailed descriptors within HFACS-ADF; a total of 155 descriptors in four taxonomies.

Such a large number of descriptors may contribute to reduced mutual exclusivity and consensus/consistency.

The field study design has several limitations that could have affected the results. For instance, participants had varying degrees of training and experience in human factors and in the use of HFACS-ADF. This realistically represents the context of use of HFACS-ADF and many other classification systems. Although training would probably be of some benefit, HFACS-ADF appeared to produce similar levels of low consensus and consistency regardless of HFACS-ADF expertise; it appears to be inherently unreliable. The number of coders was relatively small, but typical of coder reliability studies and of limited importance light of the descriptive statistical analysis (note that larger numbers of coders tends to return even lower consensus). More importantly, the participants could identify unequal numbers of codes. This again represents the realistic context of use. Since unequal numbers are treated as disagreements, this reduces inter-coder consensus and intra-coder consistency.

The nature of the field studies meant that several variables could not be controlled, in addition to the amount of training. For instance, in study one, participants may have taken different durations to complete the task. In study two, participants' second attempt may have been subject to practice effects. Also, the participants were all drawn from a convenience sample. While the study context may have had some effects on the findings, the very low consensus is unlikely to reach 70% at the descriptor level even in very controlled conditions. This may extend to other techniques with many descriptors. However, with further refinement, and a more rigid method to ensure an equal number of codes allocated to each finding, it may be possible to improve consensus at the category level. Comprehension testing of all terms used within taxonomies, a review of the taxonomic structure, removal of descriptors as classification terms, periodic refresher training, and regular reliability testing of HFACS-ADF could result in better consensus and consistency. Alternatively, a new approach may be needed that is more operationally- and contextually-oriented.

As a result of these studies, doubt exists as to whether HFACS-ADF can be used reliably, particularly in the absence of appropriate methods to ensure consistent application of the taxonomy. This issue is not unique to HFACS-ADF. Many other classification systems are used in organisations by multiple operational users with minimal training. The results found in these studies, are, we believe, likely to reflect those that would be found for other classification systems, were they tested in the same way. This has significant implications for the use of the safety data obtained from safety classification systems, which may be unreliable, and therefore invalid, potentially leading to safety interventions (or non-interventions) that are ineffective in addressing the real contributory factors.

Acknowledgement

This article is a shortened version of Olsen, N.S. & Shorrock, S.T. (in press). Evaluation of the HFACS-ADF safety classification system: inter-coder consensus and intra-coder consistency. *Accident Analysis and Prevention*, accepted for publication 3 September 2009.

References

Beaubien, J. and Baker, D. 2002, A review of selected aviation human factors taxonomies, accident/incident reporting systems and data reporting tools, *International Journal of Applied Aviation Studies*, 2 (2), 11–36

Cohen, J. 1960, A coefficient of agreement for nominal scales, *Educational and Psychological Measurement*, 20, 37–46

O'Connor, P. 2008, HFACS with an additional layer of granularity: validity and utility in accident analysis, *Aviation, Space, and Environmental Medicine*, 79 (6), 599–606

Reason, J. 1990, *Human Error*, (Cambridge University Press, New York).

Reinach, S. and Viale, A. 2006, Application of a human error framework to conduct train accident/incident investigations, *Accident Analysis and Prevention*, 38, 396–406

Ross, A., Wallace, B. and Davies, J. 2004, Technical note: measurement issues in taxonomic reliability, *Safety Science*, 42, 771–778

Shappell, S., Detwiler, C., Holcomb, K., Hackworth, C., Boquet, A. and Wiegmann, D. 2007, Human error and commercial aviation accidents: An analysis using the human factors analysis and classification system, *Human Factors*, 49 (2), 227–242

Shappell, S. and Wiegmann, D. 2001, Applying Reason: the human factors analysis and classification system (HFACS), *Human Factors and Aerospace Safety*, 1 (1), 59–86

Wallace, B. and Ross, A. 2006, *Beyond Human Error*, (CRC Press, Boca Raton, FL)

Wiegmann, A. and Shappell, S. 2001, A human error analysis of commercial aviation accidents using the human factors analysis and classification system (HFACS), *Aviation, Space and Environmental Medicine*, 72 (11), 1006–1016

TRANSITION FROM A HUMAN-CENTRIC 'SOFT' SYSTEMS APPROACH TO A BASELINE ENTERPRISE KNOWLEDGE MANAGEMENT ARCHITECTURE

Huseyin Dogan & Michael J. de C. Henshaw

Loughborough University, UK

This paper reports on an industrially sponsored research programme to develop an enterprise Knowledge Management (KM) system for Through Life Capability Management. It summarises the transition from a 'soft' systems requirements analysis approach to the development of a baseline architecture for an enterprise KM system within the defence and aerospace industry. The first part of this study captured, analysed and synthesised the key stakeholder requirements for such an architecture through a case study which consisted of a human-centric approach called Interactive Management. The second part developed a baseline use case diagram and domain model derived from this 'soft' system approach to guide the implementation of such a KM system. The usefulness of a methodical approach using both 'soft' and 'hard' systems techniques to contextualise a problem situation and create its systems architecture is demonstrated.

Introduction

This paper describes a methodical approach using 'soft' and 'hard' systems techniques to contextualise a problem situation and create its systems architecture. The problem situation concerns Knowledge Management for Through Life Capability Management (TLCM) and the results of a brainstorming exercise by a group made up of industrial and academic technical experts, which indicated substantial differences in their models of TLCM. This paper deals with explicit rather than tacit knowledge and considers the delivery of military capability. Through Life Management (TLM) is the philosophy that brings together the behaviours, systems, processes and tools to deliver and manage projects through the acquisition lifecycle (AOF, 2008) whereas TLCM is "*an approach to the acquisition and in-service management of military capability in which every aspect of new and existing military capability is planned and managed coherently across all Defence Lines of Development from cradle to grave*" (AOF, 2008). Jashapara (2004) defines Knowledge Management (KM) as "*the effective learning processes associated with exploration, exploitation and sharing of human knowledge (tacit and explicit) that use appropriate technology and cultural environment to enhance an organisation's intellectual capital and performance.*" In the emerging defence acquisition environment, commercial success will be significantly determined by the ability of companies to manage the value of knowledge within the various enterprises in which they participate. Knowledge of the systems, for which a company has through-life management responsibility, may be distributed throughout an enterprise that comprises several commercial organisations and the customer. Prosperity in such an environment requires clear understanding of the value of knowledge within the organisation and management of knowledge, in addition to effective systems architecture in the wider enterprise.

Soft Systems and Enterprise Architecture

In this section, we consider 'soft' and 'hard' approaches to analyse systems. An enterprise system, by its very nature, imposes its own logic on a company's strategy, organisation, and culture (Davenport, 1998). Kennedy *et al.* (2007) developed a model of 'soft' enterprise characteristics and determined how these characteristics may be included in an integrated model of an enterprise system. In this research a number of components of the enterprise were grouped using the four views: functional, information, organisational and resource. Employees in an enterprise need to have sufficient knowledge or competency (both tacit and formal) to carry out the tasks allocated to them (Kennedy *et al.*, 2007). The Role Matrix Technique (Callan *et al.*, 2005) is a method of mapping organisational structure to understanding where particular knowledge needs to reside in an organisation. It does, however, have the drawback that it tends to represent a snapshot in time, rather than mapping an organisation over time.

Checkland's *Soft Systems Methodology (SSM)* can also be used to analyse a problem situation for an organisation seeking a better way of working. This approach is appropriate to socio-technical systems, and is particularly considered when the means of improvement is a new information system and the analysis task is to understand the requirements for this system (Checkland, 1981; Checkland and Jim, 1990). The current research adopted a 'soft' systems initiative by using the Interactive Management approach and SSM rich pictures rather than a more engineered approach (such as Quality Function Deployment or Functional Modelling) to define the problem space. 'Soft' systems thinking is more appropriate in fuzzy ill-defined situations involving human beings and cultural considerations, whereas 'hard' systems thinking is appropriate in well-defined technical problems, and hence involve logical rules to engineer solutions. Our soft systems approach is not the formal SSM of Checkland, but draws some techniques from this.

Turning our attention now to hard systems approaches, the ones considered (use case diagrams and domain models) are examples of systems architecting techniques. Systems Architecture is defined by the IEEE as "*the organisation of the system components, their relations to each other, and to the environment, and the principles guiding its (i.e. the system's) design and evolution*" (IEEE Std1471-2000). Many systems engineering tools and techniques, specific to requirements analysis and design, include Systematic Textual Analysis (STA), Viewpoint Analysis and Quality Function Deployment (QFD) which can be used during the early phases of the research to guide the design of the systems architecture. Functional Modelling is another example, which has many types including IDEF0, SADT, Data

Flow Diagrams, Control Flow Diagrams, State Transition Diagrams, Flow Charts, UML and SysML (Burge, 2008). The 'problem statement' technique, developed by Newman and Lamming (1995), is an approach that can support Systematic Textual Analysis as it provides a clear understanding of the problem situation. An unclear understanding of the problem would lead to a project being in danger of grinding to a halt, or wandering off towards some random and inconsequential outcome.

The starting point for this analysis is a case study approach which is discussed in the next section.

Case Study: Requirements Analysis and Baseline Architecture

A 'soft' systems approach called Interactive Management (IM) has been adopted to capture the requirements and contextualise the problem. Three of the group methodologies were used in the Interactive Management workshop: Idea-Writing (IW), Nominal Group Technique (NGT) and Interpretive Structural Modelling (ISM) (Delbeq *et al.*, 1975; Broome and Keever, 1986; Warfield and Cardenas, 1995). Rich pictures were also constructed by the Subject Matter Experts (SMEs) to provide a diagrammatic representation of the systematic but non-judgmental understanding of the problem situation (Checkland and Jim, 1990; Lewis, 1992; Wilson 2001).

Method and Procedure

The IM session was conducted during two workshops with a total of four groups; Academics A, Academics B, Industrials A and Industrials B. The participants (n = 21) were chosen because they were either researching or managing within TLCM. There was no-one from the Ministry of Defence (MoD). The workshop consisted of the following four sessions:

- Session I: Idea Writing (IW) was used partly to produce the issues related to a given KM trigger question and partly as an enabling process to aid consensus decision-making (e.g. categorising ideas) amongst the SMEs during the workshop.
- Session II: Nominal Group Technique (NGT) was used to generate, clarify, edit and obtain a preliminary ranking of a set of KM objectives.
- Session III: Interpretive Structural Modelling (ISM) helped members to examine the inter-relationships between elements gained through the NGT process and provided a structure for tackling its complexity.
- Session IV: Rich pictures were constructed for the expression of the problem situation.

IM is designed to prevent groups from prematurely focusing on solutions before defining the problem situation and under-conceptualising design alternatives (Broome and Keever, 1986). This approach supports consensus decision making through idea generation, structuring and design. The advantages of this 'soft'

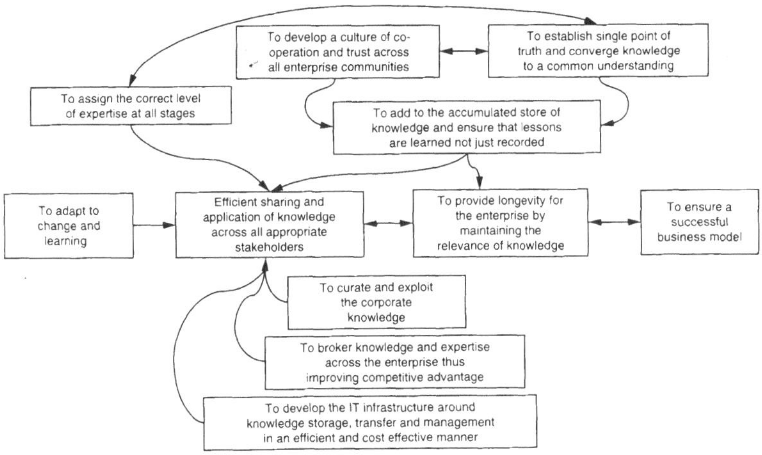

Figure 1. ISM results from Industrials B.

approach include the efficient use of participant's time; provision for iteration and documentation; and promotion of participation, hence effective communication.

Results

Through a set of trigger questions, the four groups generated models and requirements that they believed would provide effective KM for TLCM, i.e. a notional 'to be' model for KM. The total number of IW statements from the four groups (n = 21) was 268 and this led to 199 NGT objectives. All the academics participants were from the engineering discipline including human factors. The academics in this study used a more traditional Systems Engineering approach and categorised the IW statements under the 'people', 'process' and 'technology' headings, Industrials used the heading 'business' rather than 'process' and introduced a new heading called 'knowledge creation'. The NGT and ISM results indicate that

- the academics focused more on creating and adapting a KM vision and strategy within and outside the enterprise whereas the industrials didn't;
- both of the groups stated the importance of developing a culture of co-operation and trust across all enterprise communities that considers KM as a core discipline;
- the groups also touched on points such as enablement to become a learning organisation; adding to the accumulated store of knowledge; and ensuring that lessons are learned;
- all four groups emphasised the importance of exploiting, measuring and managing the corporate knowledge and information for the business benefit.

Figure 1 shows the interpretation of the objectives derived from Industrials B which can support a proposed enterprise KM system. The product of the ISM process in this example is an 'intent structure'; this shows how the objectives captured through NGT inter-relate with one another.

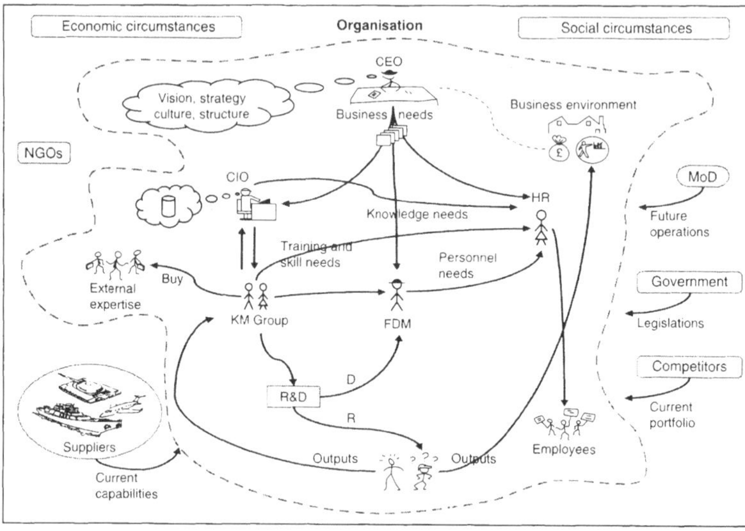

Figure 2. Baseline rich picture of through life enterprise KM.

Figure 2 shows a baseline soft systems rich picture for 'through life enterprise KM' derived through interactions with SMEs ($n = 21$). This rich picture provides a diagrammatic representation of the systematic but non-judgmental understanding of the problem situation. An architectural model containing a use case diagram and domain model derived from this rich picture is illustrated in later sections.

Figure 2 is an interpretation of the rich picture model generated by participants. The rich picture shows that SMEs focused on external actors such as competitors, government, MoD, NGOs and suppliers in addition to defining a boundary of the organisation. It also highlights that the organisational implications include the development of a vision, culture and structure within different business divisions. Social and economic circumstances were considered to be part of the external environment and knowledge flow part of the internal structure. An example of a knowledge flow is the transmission of business requirements from the Management Board to the Chief Information Officer (CIO), Functional Delivery Managers (FDM) and Human Resources (HR). This rich picture also illustrated that the CIO forwards these business requirements to the KM group which may start the Research and Development (R&D) process or even seek external expertise. The personal needs including the training and skill needs are perceived to be handled by HR that has access to the whole organisation.

Transitioning to a formal model

The transition is accomplished by building a baseline enterprise KM architecture. Consequently, a use case diagram and domain model were developed as a result of analysing the Interactive Management findings including the rich pictures. The external events and stakeholders shown in the rich picture (e.g. the government and competitors) were not considered to have direct involvement with the system. As part of the analysis, the researcher divided the actors in the rich pictures into *metal-level* and *direct* users of the system. This division is the researcher's interpretation and analysis of the rich picture to provide the structure for the use case model. Consequently, the meta-level users of this enterprise KM system are identified as:

- Chief Executive Officer (CEO)
- Chief Information Officer (CIO)
- Knowledge Management (KM) Group
- Human Resources (HR)
- Functional Development Manager (FMD)

The *direct* users involve the employees, researchers and external experts as they contain much of the organisation's knowledge. The use cases derived from the rich picture are

- *manage business needs:* considers the business environment as well as a vision to create a KM strategy, culture and structure;
- *manage knowledge needs*: comprise the key competencies and skill-sets required;
- *manage KM training or skill needs*: development of training material specific to KM;
- *manage KM personnel needs*: involves rapid on-boarding, secondments and subcontracting, consequently the recruitment of KM specialists;
- *buy external KM expertise*: acquisition of external KM expertise into the organisation;
- *administrate KM research and development (R&D):* this involves managing the KM research teams to fulfil the research requirements.

A sequence diagram was also derived from the rich picture to illustrate the knowledge flow between different stakeholders within the enterprise. This was simply a high level overview of the knowledge interactions that could support the basis of a more detailed UML sequence diagram development. There is also a need to validate these use cases and sequence diagrams through iterative interactions with SMEs. This can lead to refinements to the operational and functional KM concepts and to the use cases themselves in order to better align with the stakeholder requirements. The use case diagram is a potential model where the consensus rich picture was realised. The use case diagram is shown in Figure 3.

Domains are the distinct subject matters present in any system representing large reusable components and are depicted using a domain model which shows an organisation of UML packages and their dependencies (Dickerson, 2009). People's roles and responsibilities are considered from the rich picture when creating the domains. Functional decomposition is avoided when developing the domains and use cases. Research and development (R&D) is provided as a service in the following domain model, as the SMEs highlighted that the KM group delivers R&D through internal researchers and external expertise. HR are also provided as a service interface as the training and skill needs; personnel needs; and knowledge needs within the enterprise are forwarded to HR via CIO, KM group and FDMs. The HR and R&D service domains will have a user interface to enable the user interaction. Figure 4 shows a high level overview of the enterprise knowledge management domain model.

Figure 3. Baseline use cases for through life enterprise KM management process.

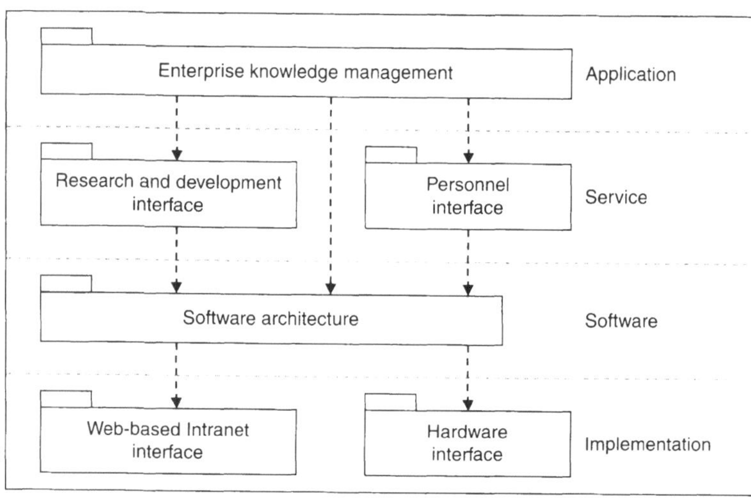

Figure 4. Enterprise KM domain model.

The actors that directly interact with the R&D interface in the domain model are CIO, FDA and the KM group as opposed to the actors that interact with the personnel interface; CEO and HR. This is derived from the information flow in addition to the roles and responsibilities illustrated in the rich picture. Furthermore, a web based KM system would be a useful support mechanism to this enterprise domain model.

Discussion

This paper tackled two aspects; (1) a methodical approach to enable transition from 'soft' to 'hard' systems and (2) the derived workshop results. We have discussed this transition as a means of identifying knowledge management requirements for TLCM. The use case diagram and domain model provide a baseline architecture for a future enterprise KM system. Brooks (1995) expresses the role of a Systems

Architect as "*the architect should be responsible for the conceptual integrity of all aspects of product perceivable by the use*". This baseline architecture supported by the soft systems approach can guide the focus on a more traditional Systems Engineering approach e.g. Quality Function Deployment (QFD). Cohen (1995) describes QFD as a "*method for structured product planning and deployment that enables a development team to specify clearly the customer's wants and needs...*" Consequently, QFD can be supported by such a human centric socio-technical 'soft' systems approach as illustrated in this paper. The benefits of this approach are complementary to the key business benefits of systems engineering as identified by Sillitto (2004). These include building the right system; ensuring stakeholder satisfaction; and avoiding overspend and over-runs in the expensive phase of the life-cycle, generally during integration, test and setting to work. The drawbacks of this 'soft' to 'hard' systems approach is complementary to Kestell (2004) findings which highlight that functional modelling on its own is not sufficient and other factors such as black-box specification; requirements management and traceability; and maintaining quality need to be considered when producing the functional model. Consequently, the baseline architecture illustrated in this paper needs to be applied to an ongoing project for validation and verification purposes. Kestell (2004) identified the key benefits of functional modelling as having clear, usable and verifiable specifications whereas Galliers and Swan (2000) argue that much more is needed than just standard hard methods when trying to generate a comprehensive set of representative requirements. Therefore, a 'soft' systems approach to requirements analysis was adopted as opposed to more traditional methods. Traditional requirement capture is much more systematic and formal in approach using such tools, techniques and methods as QFDs, formal modelling, document analysis and linguistics analysis.

Turning our attention to the derived IM workshop, the results illustrate that the academic participants focused more on knowledge flow within the enterprise through a role based approach e.g. having a Chief Information Officer, whereas the industrials emphasised how a defence company operates in the existing environment e.g. having an information repository to encapsulate and store corporate knowledge. As a result of analysing the data, especially the rich pictures, it can be concluded that for these samples the academics are more organisational oriented e.g. less constrained by time, whereas industrials are more product oriented hence considered more the through-life ('cradle to grave') product development. Additional research is necessary to further analyse and validate the models developed.

Conclusion

This paper illustrated how system thinking is applied in a systematic and repeatable manner through exploitation of current 'soft' and 'hard' systems tools and techniques to build a knowledge management architecture to support Through Life Capability Management. Transition from 'soft' to 'hard' systems is difficult and needs to be performed with a great deal of care to avoid leaving gaps in the coverage of considerations. The approach described in this paper is a possible technique for managing this difficulty.

References

Acquisition Operating Framework (AOF), 2008, UK Ministry of Defence (MoD), Available at: www.aof.mod.uk

Brooks, F. P. 1995, *The Mythical Man-Month: Essays on Software Engineering, 20th Anniversary Edition.* (Addison-Wesley, Reading)

Broome, B. and Keever, D. 1986, Facilitating Group Communication: The Interactive Management Approach, Paper presented at the *77th Annual Meeting of the Eastern Communication Association*

Burge, S. 2008, *Systems Approach, Analysis and Concepts*, ELP061 Course Learning Material, Engineering Doctorate programme, (Loughborough University, UK)

Callan, K., Siemieniuch, C. E. and Sinclair, M. A. 2005, A Case Study Example of the Role Matrix Technique, *INCOSE*

Checkland, P. 1981, *Systems Thinking, Systems Practice,* (John Wiley and Sons, London)

Checkland, P. and Jim, S. 1990, *Soft Systems Methodology in Action,* (John Wiley and Sons, Chichester)

Cohen, L. 1995, *Quality Function Deployment: How to Make QFD Work for You,* (Addison-Wesley, Reading)

Davenport, T. H. 1998, Putting the enterprise into the enterprise system, *Harvard Business Review*, Vol. 76, No. 4, pp. 121–31

Delbeq, A. L., Van De Ven, A. H. and Gustafson, D. H. 1975, *Group techniques for program planning: A guide to Nominal Group and DELPHI processes*, (Scott Foresman, Glenview)

Dickerson, C.E. 2009, *ELP072 Systems Architecture Course Learning Material*, Engineering Doctorate programme, (Loughborough University, UK)

Galliers, R. D. and Swan, J. A. 2000, There's more to Information Systems Development than Structured Approaches: Information Requirements Analysis as a Socially Mediated Process, *Requirements Engineering*, Vol. 5, pp. 74–82

IEEE Standards Association, Std1471-2000 Available at: www.standards.ieee.org/reading/ieee/std_public/description/se/1471-2000_desc.html

Jashapara, A. 2004, *Knowledge Management: An Integrated Approach*, (Prentice Hall, Harlow)

Kennedy, G. A. L., Siemieniuch, C. E. and Sinclair, M. A. 2007, Towards an Integrated Model of Enterprise Systems, *INCOSE International Symposium*

Kestell, A. G. 2004, Functional Modelling of Complex Systems, *In*: *Railway System Modelling – Not Just for Fun,* IEE Seminar

Lewis, P. J. 1992, Rich picture building, *European Journal of Information Systems*, Vol. 1, No. 5

Newman, W. M. and Lamming, M. G. 1995, *Interactive System Design*, (Addison-Wesley, Wokingham)

Sillitto, H. 2004, Systems engineering trends, vision and business benefits. *INCOSE UK Autumn Assembly 2004, 10th Anniversary Presentation*

Warfield, J. N. and Cardenas, A. R. 1995, *A Handbook of Interactive Management*, (Iowa State University Press, Ames)

Wilson, B. 2001, *Soft Systems Methodology: Conceptual Model Building and Its Contribution*, (John Wiley and Sons, Chichester)

THROUGH LIFE CAPABILITY MANAGEMENT: BENEFITS AND BEHAVIOURS

Esmond N. Urwin, K. Sofia Ahlberg Pilfold & Michael J. de C. Henshaw

Engineering Systems of Systems Group, Garendon Wing, Holywell Park, Loughborough University, UK, LE11 3TU

Many commercial and social endeavours require a multitude of socio-technical systems to work together effectively in what has come to be known as systems of systems. The Through Life Capability Management (TLCM) construct, currently being embraced by defence departments across the globe, is one such endeavour. TLCM demands changes in the organisation and culture of the defence supply chain in ways that fly in the face of traditional commercial wisdom. This paper reports on two workshops held with TLCM stakeholders in which they identified, and then prioritised, the benefits that they sought from TLCM. From an agreed set of priority benefits, the groups identified the behaviours needed to realise them; the results point to significant challenges in terms of culture and knowledge management. Ergonomists and systems engineers will need to support development of intervention strategies to effect these required changes.

Introduction

Recent years have seen a trend towards manufacturers and suppliers integrating products and services into capabilities to which customers can subscribe. This gives suppliers the advantage of better predictability of demand, while customers get the capability they need without the overheads associated with ownership. In defence, governments are integrating industry into overall management of military capability (especially in the UK), so that industry now takes responsibility for many roles that were previously the provenance only of Government owned entities (MoD, 2005). The aims are to make systems more cost effective, affordable through life, and responsive to the rapidly changing threat. It is believed that only by taking a whole system approach to managing capability, both in terms of large scale integration and over the long term that the benefits can be realised. Through Life Capability Management (TLCM) is an approach being developed to manage military capability (MoD, 2007; Stein, 2009). It considers not only the delivery of systems and equipment into service, but also the servicing, upgrade and support of those systems throughout their entire lifecycle. It is generally expected that TLCM will need new customer-supplier relationships; but how best to realise these and what are the steps to be taken?

This paper describes a workshop (held twice with different stakeholders) in which the benefits of TLCM to both customers and suppliers were identified and used to initiate a consideration of the behaviours needed to achieve them. These point towards the need for substantial changes in culture within the defence supply chain and advances in knowledge management. The results of this work provide a platform from which research will develop into how these changes may be achieved.

Systems of Systems and the TLCM Challenge

The interoperation of discrete systems within a wider environment has always been a feature of commercial, social, political, and technological endeavour. But as the interdependencies between systems increase, so does the risk of unexpected, or emergent, behaviours. Systems of systems (SoS) engineering (Delaurentis *et al.*, 2007) is a discipline that seeks to understand and predict the behaviours of interacting monolithic systems, and so design those systems for participation in SoS. Interoperability is fundamental to SoS and this poses significant technical and commercial challenges for the defence supply chain. TLCM is a SoS problem. Systems of Systems are distinguished from monolithic systems by the following properties, identified by Maier (1998):

- Operational independence of elements
- Evolutionary development
- Geographical distribution of elements
- Managerial independence of elements
- Emergent behaviour

The systems through which they are constituted generally have managerial and operational independence; the SoS undergoes evolutionary development (i.e. no clean sheet), the individual systems are spatially and/or temporarily distributed (i.e. it is information that travels between the individual systems); and, very importantly, they exhibit emergent behaviours that are not present in any of the individual systems, but only from the interoperation of the constituent systems. Emergent behaviours may be beneficial or detrimental and even – in some circumstances – catastrophic. We emphasise that the SoS include humans and organisations and it is very often the interoperation between these (or rather lack of it) that can lead to undesirable emergence. This is a problem of central importance to the ergonomics community.

TLCM is an evolving concept and in the UK organisational changes are occurring very rapidly as the understanding of TLCM develops. Daw & Sillitto (2009) suggest it is made up of three components: Capability (the ability to achieve desired military outcomes), Force Elements at Readiness (FE@R are the various components of military force that are ready for deployment at a particular time and may be combined to provide the required force), and the Defence Lines of Development (DLOD). The latter are Training, Equipment, Personnel, Information, Doctrine and Concepts, Organisation, Infrastructure, and Logistics. All of these aspects must be developed appropriately to realise capability. The formal definitions of all these terms for the UK MoD are provided in the Acquisition Operating Framework (MoD, 2009).

Some relevant characteristics of TLCM, for the results of the workshops reported herein, are that a systems approach is necessary for its realisation; systems engineering is a crucial skill for both customer and supplier (Tibbitt, 2009). Whole life costs (development, procurement, in-service operation, and disposal) and incremental delivery must be considered. Interoperability between systems is a critical feature. This poses significant challenges associated with the integration of legacy and new systems so that they work effectively, and safely, together. There is a strong implication of the need for open architectures and open systems implied in this characteristic (Tibbitt, 2009), which in turn implies that the supply chain must adopt more open working practices and be, itself, more interoperable. There are both commercial and cultural challenges to be overcome in this regard. Finally, it also demands systems to be designed for flexibility, so that they can be changed in an agile fashion to meet the demands of a rapidly changing threat. The need for a more agile supply chain implies the need for new cultural behaviours within it and changes in the range of, and relationships between, the stakeholders (Yue & Henshaw, 2009). New systems engineering approaches have also been recommended (Neaga *et al.*, 2009).

Case Study: Stakeholder Perceptions of Through Life Capability Management

A case study has been constructed by running two workshops with TLCM stakeholders. The workshop format enabled the collection of opinions and experiences concerning changing working practices and attitudes to TLCM from stakeholders immersed in those changes. Both workshops followed the same structure so as to enable consistent analysis of any information captured. The first of these involved members of the TLCM research & development community (industry and academic). Attendees were selected from a group of systems engineering practitioners and researchers from Loughborough University, BAE Systems and the Defence Science and Technology Laboratory (dstl),by way of interest and field of expertise. The second workshop was conducted as a part of an international conference on through life support and costing, attended by operational staff (industry and military) from the international Integrated Logistics Support (ILS) community. The workshops were both small groups (9 and 5, respectively) but provided useful data about stakeholder perceptions and form a foundation upon which further research will be based. The workshops included discussions about the following key points:

- Why TLCM needs new approaches to systems engineering.
- What is the impact of systems of systems engineering on the commercial environment in defence?
- The significance of information and knowledge management in a TLCM environment.

The stated objectives for the workshop were that attendees would:

- Appreciate a range of perspectives on the approaches and priorities for Systems of Systems Engineering.

- Be able to contribute positively to the transformation of the defence supply chain to deal with the challenges of TLCM in the future.

Method

The workshops used the mini-Delphi technique (Dalkey and Helmer, 1963; Cuhls; Rowe & Wright, 2001) to solicit and then prioritise stakeholder views on TLCM benefits. Such an approach is perfect for face-to-face discussion of future trends, decision making and forecasting, because it allows discussion, assessment, and reflection. In addition, the Single Transferable Vote (STV, www.electoral-reform.org.uks) method was chosen. It is a form of preferential voting for multiple criteria and has the advantage of providing a more representative ranking of candidates because it takes account of voters' ordered priorities instead of just their first choice. Preferential voting means that instead of casting a single vote for a single criterion, a voter provides a rank ordered list; every vote has equal value and the number of 'wasted' votes is extremely small. Attendees were encouraged to identify their own individual choice of TLCM benefits and behaviours through open questions. Discussions between the attendees enabled a consensus to be reached when identifying and selecting the top five TLCM benefits. TLCM behaviours were elicited by asking the attendees their opinions concerning future issues to be addressed. Data was collected by way of note taking and the STV software.

Procedure

The workshop was set out into three main sections, these being:

> Section 1: an initial overview of TLCM was presented. This provided the background to the problem domain (with a focus on the UK instantiation of TLCM) and touched on future issues and potential problems. This presentation was very interactive and served to establish an agreed characterisation of TLCM among the workshop members.
> Section 2: TLCM Benefits: working individually, workshop attendees wrote a list of their top three, perceived, TLCM benefits. The full list from all attendees was then collated and presented back to the group for discussion and reflection. This ensured that the intended meaning of the benefit descriptions was clear to all and some rationalisation of the full list took place to remove duplication (i.e. where two attendees had individually identified a benefit that was the same). The attendees chose and ranked their top five benefits from the full list and this generated the overall ranked list of benefits.
> Section 3: TLCM Behaviours: focusing on the top five benefits overall, the final session examined the behaviours that would be needed to realise those benefits. The behaviours and the discussion that took place during their identification was recorded using a flipchart to capture the main points from the group discussion and note taking to record the detailed information.

Table 1. TLCM benefits as identified in workshop 1 (academic and industry).

Academic Generated	Industry Generated
Through life costing for existence in the market	Rapidly changing organisation to compete in global market places
Ensuring interoperability through better management	Reduction of military manpower becomes okay
Geared towards customer & Organisation Satisfaction (relationship)	More flexibility to address more scenarios as required by politicians
Increase number of stakeholders - hence customers	Allows more opportunity to work more closely with private military organisations
Extend lifecycle of product	Availability at lower cost
Enhance competiveness of product	Continuous product development
Greater flexibility of use of resources (hence availability)	More opportunities for business (market opportunities)
Better requirement documents (involvement of users)	
Clearer drive to long-term strategic thinking in defence acquisition	
Less waste	
Risk reduction by working with customer	
A structured approach to management of capabilities	
Ensure consistent processes and means during development	

Results

Workshop 1 consisted of six academics and three industrial participants all working in R&D and with some level of understanding of TLCM. The rationalised list of benefits is presented in Table 1, with those from the academics and industry identified separately. Table 2 shows the top five TLCM benefits resulting from the STV ranking from workshop 1. Workshop 2 consisted of two industrial practitioners and three military personnel working in logistics with a very good understanding of TLCM. The results for workshop 2 are presented in Table 3 and Table 4.

Behaviours needed to realise and support TLCM benefits

The attendees were asked about their views on what behaviours should be exhibited by members of the defence supply chain (suppliers and customers) to realise and enable the aforementioned TLCM benefits. To some extent this discussion was structured around the behaviours that the customer community would wish

Table 2. Ranked list of benefits (top 5) from workshop 1.

Rank	Benefit	Origin
1	More opportunities for business (market opportunities)	Industry
2	Availability at lower cost	Industry
3	Continuous product development	Industry
4	Rapidly changing organisation to compete in global market places	Industry
5	Geared towards customer & organisation satisfaction (relationship)	Academic

Table 3. TLCM benefits as identified in workshop 2 (military and industry).

Military Generated	Industry Generated
Definition of goals	Managing capability development
Modelling for scenario effects	Cost savings
Enhanced operational effectiveness	Effective contracting
Enhanced theatre of operation (flexibility)	Reduced time to market of systems
Increased systems availability	Better risk management
Better life cycle management	Effective total cost of ownership
Manpower capacity increased	Better integration of logistics support
Better understanding of risk and uncertainty	
Better understanding of constraints and limitations of systems	

Table 4. Ranked list of benefits (top 5) from workshop 2.

Rank	Benefit	Origin
1	Managing capability development	Industry
2	Increased systems availability	Military
2(3)	Better integration of logistics support	Industry
2(4)	Effective contracting	Industry
5	Modelling for scenario effects	Military

to see from suppliers and vice versa. The five areas that were identified by workshop 2 (that included suppliers and customers) are reported here, these were (in no particular order):

- Retain knowledge better
- Openness of long term planning
- Availability
- Cooperation at all levels (commercial)
- Integration (lifecycle)

Discussion of Workshop results

Breaking the list of benefits into industrial and military generated (Table 3) reveals, unsurprisingly, the focus of military on the functionality and the service to provide the end effect, i.e. the sustainment and facilitation of military forces within theatre (be it peacekeeping, war-fighting, etc.). The benefits identified by the academic participants (Table 1) mostly have process and tool characteristics, which is unsurprising as the group was broadly drawn from the systems engineering and ergonomics communities. The top five benefits from both workshops reflect some operational aspects, but are mostly focused on benefits that require industry and Government to work effectively together. This indicates that TLCM offers benefits to industry as well as the customer, although we note that in these and similar workshops the industry representatives have tended to identify benefits to their customer and, ultimately, ranked those customer focused benefits most highly. Given the mutual benefit that is anticipated by the stakeholders, the final part of the workshop, in which the behaviours needed to realise the benefits were considered, was likely to focus on the enterprise aspects of the TLCM problem.

Retain knowledge better – it was recognised that TLCM required the whole supply chain to take a long term view, not just government, and that the length of some programmes (decades) meant that knowledge retention was a major challenge. From industry's point of view, investment in training and skills retention requires clear long term commitment from the government side. On the Government side, though, this reduces the ability to gain value through competition. Overall, the behaviours needed to support this challenge are collaborative planning of skills and knowledge needs, managed through enterprise-wide long term partnership contracting. Through such an approach, industry can have the confidence to invest significantly in training and skills retention.

Openness of long term planning – in general, the military plan future acquisition using scenarios that are not shared with industry. From this a set of requirements follows and industry is contracted against these. The group (workshop 2) felt that this led to dissatisfaction on both sides, because industry does not fully appreciate the way that the systems will be used. Overall, this was manifested as a strong request from industry members of the workshop to be given earlier involvement in the planning process in order to better prepare for capability development, by way of maintenance of skill sets and better understanding of requirements.

Availability – this feature was probably motivated, at least in part, by the current move towards availability contracting (e.g. Availability Transformation: Tornado Aircraft Contract - ATTAC), which could be regarded as a first step in TLCM. But importantly, this reflects a shift towards a service based environment, consistent with through-life concepts. Availability means that a particular service is available for use an agreed percentage of the time, where that agreement takes account of the type of use (e.g. for training, etc.). Participants noted the difficulty of contracting appropriately for availability and in particular the fact that so far availability has been platform specific, whereas the provision of an available service could be platform independent. The delivery of services is generally dependent on multiple organisations and relies, particularly, on the co-creation of value by customers and suppliers (Ng & Yip, 2009). This implies the need for a highly partnered environment.

Co-operation at all levels – although obvious, it was specifically identified by the participants in relation to commercial matters. Both workshops' participants were drawn from the technical community who felt that technical co-operation is frequently hindered by lengthy contracting processes and even restrictive commercial arrangements. The TLCM environment implies the need to manage long-term commercial arrangements and participants identified the need for those arrangements to be sufficiently flexible that they could develop without the need for fundamental renegotiation and the delays and interruptions that might entail.

Integration – the context of this need, as expressed by participants, was the integration of the various organisations involved in TLCM. Such an approach would need better cooperation and better understanding of one another's policies, procedures and modus operandi so as to harmonise planning and operations. In effect an holistic approach (common procedures perhaps) would be a distinct advantage and allow for a smoother approach to TLCM, and possibly remove obstacles to easier and effective co-working and system and service support.

Discussion

The two workshops described above were conducted with a small number of stakeholders located largely in the middle management and ranks of the defence supply chain. It was also restricted to the technical community; the stakeholders could be regarded as largely taken from the developers and implementers of TLCM (now or in the future), rather than those charged with its strategic direction. TLCM is a high priority for the UK MoD (Tibbett, 2009), the workshop revealed that similar strategies are being pursued internationally. The degree to which individual nations integrate the defence supply chain (i.e. to what extent industry assumes functions traditionally carried out by armed services personnel) varies, but the implications for all are that TLCM will rely on significant cultural and organisational changes.

Although we have identified the benefits as being industry-, or government-centric, in fact the prioritisation exercises and subsequent discussions identified the most important benefits as being those which were mutually beneficial. They all centre on greater levels of collaboration and, in particular, the need to share more information within the supply chain and for the earlier engagement of industry in the capability planning process. Areas of especial importance were the management of skills and knowledge and the risks associated with those. The risks within a commercial endeavour are different for different organisations: the industry participants were concerned with the risks associated with committing to retain, or create skills that might not be effectively utilised (and supported financially) in the longer term. The government participants were concerned with the implied reduction in competitiveness that long term contracting might bring about and possible lack of flexibility going forward in the systems that are long-term contracted. Both sides were concerned to have better knowledge of systems management for the future and better confidence in the costs associated with long term capability management. This suggests that the TLCM commercial environment must be one in which – as nearly as practical – risks are shared within the supply chain. Inevitably, this can only really apply to commercial risks. It was implied that greater sharing of information, especially earlier sharing of information, between customer and supplier (both ways) would be a vital ingredient in managing risk and achieving the overall objectives of TLCM. It was recognised by participants that this would require changes in contracting arrangements and changes in culture. It is certain that to achieve these changes, specific interventions will be required, although the nature of those interventions is not yet defined.

We have used the workshop format described above for a number of events involving industry and government stakeholders. The mini-Delphi technique and STV method for prioritisation is an effective approach for gaining agreement of the benefits in a particular situation and for helping the stakeholders to understand the role they must play (i.e. changes they must make) in order to realise those mutual benefits.

Conclusion

The work reported herein should be considered as 'work in progress' but, working with a small number of TLCM stakeholders, we have established agreement around some key mutual benefits that those stakeholders seek. TLCM is already a fact (not an ambition) in many nations' defence departments, but significant further development will be required to achieve the benefits that are sought. Future development will focus on a change in the commercial context and culture, towards a more collaborative environment in which customer and supplier share information more readily and suppliers are engaged earlier in the activity of capability planning. The ergonomics community will have a crucial role in supporting these changes.

References

Cuhls, K., Delphi Method, Fraunhofer Institute for Systems and Innovation Research, http://www.unido.org/fileadmin/import/16959_DelphiMethod.pdf

Dalkey, N. & Helmer, O. 1963, An experimental application of the Delphi method to the use of experts, Management Sci., 9(3), 458–467

Daw, A.J. & Sillitto, H. 2009, SE And TLCM Industry and MoD - Working Together, Defence Academy SE & TLCM Symposium, 18/19 May2009

De Laurentis, D., Dickerson, C., DiMario, M., Gartz, P., Jamshidi, M., Nahavandi, S., Sage, A., Sloane, E., & Walker, D. 2007, A case for an international consortium in systems of systems engineering, IEEE Systems Jour., pp 68–73

Maier, M.W. 1998, Architecting Principles for Systems of Systems, Systems Engineering 1(4), pp 267–284

MoD. 2005, UK Ministry of Defence, Defence Industrial Strategy, Defence White Paper Cmd 6697

MoD. 2007, UK Ministry of Defence, TLCM Workstrand, Capability Management Handbook (Interim Ed.)

MoD. 2009, Acquisition Operating Framework, www.aof.mod.uk/, (accessed August 2009)

Neaga, E.I., Henshaw, M.J.D., & Yue, Y. 2009, The Influence of the Concept of Capability-based Management on the Development of the Systems Engineering Discipline, Proc. 7th An. Conf. on Systems Engineering Research (CSER), Loughborough, UK

Ng, I.C.L., & Yip, N. 2009, B2B performance-based contracts in services: The attributes of Rowe, G., & Wright., G. 2001, Expert opinions in forecasting: The role of the Delphi technique, in Principles of Forecasting, J. S. Armstrong (ed.), (Kluwer), pp 125–144

Stein, P. 2009, System Engineering and Through Life Capability Management – Strategic Overview, Defence Academy SE & TLCM Symposium, 18/19 May 09

Tibbitt, I. 2009, Where does systems engineering fit and what skills are needed, Keynote Dstl Systems Skills Symp., http://www/dstl.gov.uk/systemsskills/presentations/2009/01.pdf

Yue, Y., & Henshaw M.J.D. 2009, An Holistic View of UK Military Capability Development, Defense & Security Analysis, 25(1), pp 53–67

DESIGN

HANGING TEN KILOGRAMS OF BRASS FROM MY NECK; A CASE STUDY OF POSTURE PROBLEMS IN BASS SAXOPHONISTS

Judith Hills

University of Glamorgan

This paper records an investigation into the issues arising from playing a bass saxophone in a jazz band. The majority of bands and orchestras use a variety of brass and wind instruments from the piccolo to the tuba which are usually played while seated. However the genre of jazz music has a tradition of playing while standing or even walking which is referred to as a 'marching gig'. Playing a bass saxophone involves supporting up to ten kilogram's of brass at an angle across the body, to accommodate the instruments length, suspended by a neck strap. Typically a gig or music session is comprised of two 45 minute sets with a break in the middle while a 'marching gig' would last at least an hour.

Many musicians using the bass saxophone experience pain in the neck, shoulder, left side of the rib cage as well as numbness in the left hand and arm as a result of playing. This paper discusses several issues which affect the experience and posture of bass saxophone players and reviews previous research into load distribution on the torso and into the biomechanical effects of musical performance. Other factors which can affect the experience of pain in musical performance are environment, arousal condition for performance and anthropometric variation. These are also discussed, alongside a review of medical research into muscular skeletal disorders caused by pressure from a neck strap.

In order to produce a rounded holistic picture of saxophonist posture, a single case study approach is used. This allowed the author to develop a design prototype which significantly improved the posture of the participant. The participant was a 58 year old male, who had been playing musical instruments all of his life; he had taken up the bass saxophone 10 years ago but experienced pain and discomfort. As a result of the initial assessment for this research, which included the collection of anthropometric data, a harness was developed. The use of a single participant meant that the problems common to many saxophonists could be discussed, whilst the solution (in the form of a prototype harness), could be tailored to the individual needs of the participant.

Figure 1. Suspension angle of the bass saxophone.

Introduction

The bass saxophone was the first saxophone created by Adolf Sax in 1844 and became very popular in the early and mid 20th century as a band instrument. It was made famous by players such as Harry Gold in jazz music but also was scored parts in musicals such as West Side Story by Leonard Bernstein and classical pieces such as Lincolnshire Posy by Percy Grainger. It is the second largest of the saxophone family, traditionally playing the bass line of a musical composition but it has also been used as a solo instrument and has a unique timbre. In the 1920's the bass saxophone was often used as it was easier to record than the double bass or tuba. Today it has been largely replaced by the smaller baritone saxophone since the baritones range was extended to concert low C.

The bass saxophone is a large and unwieldy instrument with each manufacturer having their own design details so that the weight and overall dimensions vary from model to model. However all bass saxophones are suspended at an angle across the musician's body from a hole in a bracket on the back face of the instrument. The bracket is positioned close to the centre of the instruments length with 48% of the length above the suspension hole. The suspension hole of the example used for this research gives a suspended angle of 133 degrees to the vertical as shown in Figure 1.

Existing straps for saxophones hang primarily from the neck, sometimes with additional support straps passing under the arms and across the shoulder blades. A leading saxophone player advises using a strap that provides instrument stability without much upwards and downwards movement, although unfortunately some contemporary straps have elastic sections, he also recommends a metal fixing hook as the plastic ones can shear and damage the instrument (McChrystal, 2007).

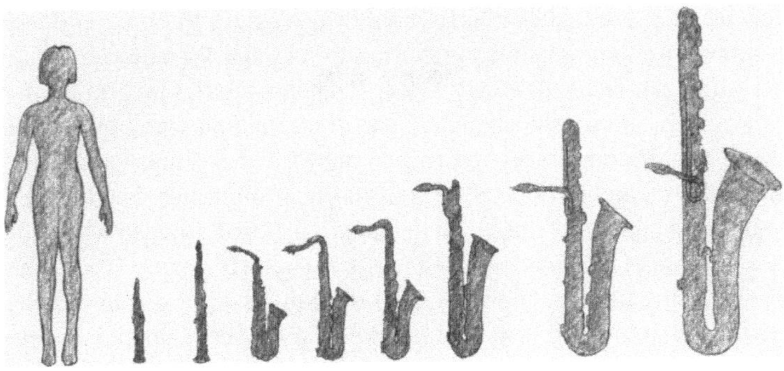

Figure 2. Relative scale of saxophone family, bass saxophone second from the right.

Supports for other large instruments include a chair strap or floor spike, such as for a bassoon and cello, but this is impractical for playing while standing or walking. The participant had tried playing the bass saxophone while seated; this resulted in discomfort in the mid/lower back region (4th–10th thoracic), possibly because of the additional strain in supporting the saxophone further away from the musician's torso.

Research by Manchester and Cayea (1998) rated instruments against music student injury rates over a 14 year period. The injury rates were calculated based on the number of injuries per 100 students giving a median injury rate of 8.3. The harpists had the highest rate at 17.1 in a 100 and horn players the least with 1.7 per 100. The saxophone came towards the lower end of the middle band injury rate (6.0–11.9) although the research is likely to have been undertaken with alto or tenor saxophones which are smaller. These are shown in Figure 2 as the 4th and 5th instruments from the right hand side. However, the instruments with the highest rates of injury were all large instruments (Manchester & Cayea, 1998) and there appears to be a significant positive correlation between the size of the instrument and the frequency of injury.

Extensive research has been undertaken during the last 15 years into the muscular skeletal disorders of orchestral musicians. Awareness amongst musicians has increased with the result that postural and well being classes are taught in many music conservatoires; the causes and treatment for muscular skeletal disorders are well documented. However the posture of jazz musicians can vary widely from orchestral musicians with the environment being typically less formal and a performance style that encourages movement and standing.

Previous Research

Investigation into a mainly static frontload includes research into a variety of load carrying areas incorporating hiking backpacks, mailbag users, apple harvesters and the loading of military equipment.

It has been established that standing balance is affected by size of load, load position and base support (Zultowski & Aruin, 2008) and that locating weight close to the body's centre of gravity gives lowest energy expenditure (Knapik, *et al.*,1996). A greater amount of weight can be supported by hip and leg muscles using a well designed waist belt (Knapik *et al.*, 1996). Transferring the load to the waist/hip area reduces pressure on the shoulders and trapezius muscle activity (Holewijn, 1990), with less discomfort reported when the load was supported mainly on the pelvis instead of shoulders (S. A. Reid, Stevenson, & Whiteside, 2004). Side muscle (latissimus dorsi and teres) fatigue in mailbag users, was reduced when additional support with a waist belt was included in the design. (Bloswick *et al.*, 1994). The addition of lateral stiffness, connected to the hip strap on a rucksack has been shown to transfer part of vertical load to the pelvic area from upper body/shoulders without increasing the internal shear force in lumbar region L3 and L4 (S. A. Reid *et al.*, 2004). Apple harvesting requires the workers to carry heavy loads in a frontal oblique position, traditionally with a single traverse shoulder strap and muscle strain is a common occupational injury amongst harvesters. A modified apple picking bucket, with one shoulder strap and a hip belt that took some of the load, was found to reduce discomfort in the shoulders, neck or back by 72% of the apple harvest workers who took part in the study (Earle-Richardson *et al.*, 2005).

Although a backpack is frequently the most practical solution for carrying loads, it has been shown that distribution of weight on the front and back of the body has some advantages (Lloyd & Cooke, 2000), presumably by bringing the load closer to the centre of gravity. Use of front and backpacks by children show improved posture, especially correction of the forward head angle and distance (Kim, *et al.*, 2008). Side muscle fatigue was reduced by using mailbag designs that divided the load between a front and back bag (Bloswick *et al.*, 1994).

When load is added to the back of the body, for example when using a backpack, the centre of mass of the body shifts backwards and the upper trunk and head compensates by leaning forward (Palumbo *et al.*, 2001). Logically the reverse may be true and, with a front load, the trunk is likely to compensate by leaning back increasing the lumbar lordosis. Little research has been undertaken into musicians who move; however Harman's research (1992) showed a high incidence of lumbar pain amongst players in marching bands.

Symptoms, posture and prototype

The bass saxophonist participant initially started playing his instrument with a neck strap. After a few months he started to experience numbness, tingling and pain in his hand, arm and armpit, whether playing or not, and especially at night when he could only sleep with additional padding to support the left arm and shoulder. He did not consult a medical practitioner but added extra straps to the neck sling around his chest to help support the weight of the saxophone. The straps needed to be quite tight to be effective and restricted lung expansion a little. The symptoms indicate a nerve or tendon restriction.

The supra spinatus tendon can be compressed by the head of the humerus when the arm is lifted to play; this can make the tendon swell and is known as shoulder impingement syndrome.

Typically the pain does not usually spread as far as the hand but the musician cannot lie on the affected shoulder at night and pain will continue, even when not playing. Any overhead activity such as removing clothes or brushing the hair will increase the pain immediately (Paull, 1997)

Thoracic Outlet Syndrome (TOS) can give pain, tingling, sensations of heaviness, fatigue, changes in weakness or sensitivity and can affect the arm, elbow, shoulder and hand. The three main nerves, radial, ulna and median, leave the spinal column at neck level and pass through a narrow aperture between muscle and bone in the brachial plexus area of the shoulder. Additional pressure from a neck strap, elevation of the arms and neck tension can compress and irritate the nerves and arteries resulting in one of three thoracic outlet conditions, depending on which nerve is affected. Susceptible anatomical body types that are tall and thin, have sloping shoulders, tight pectoral and neck muscles and a collapsed chest have been found to be more prone to the condition, as the thoracic outlet is more likely to be narrow (Horvath, 2002), (Watson, 2009). Thoracic outlet syndrome is difficult to accurately diagnose and according to Roos (1986) "depends on a high index of suspicion". However the Norris (1993) test, which requires raised arms bent above the head and exercising the hands for 3 minutes, produces pain for of all three types of thoracic outlet syndrome and is a reliable and easy indicator of the condition (Wynn Parry, 2005).

Posture

The participant bass saxophonist's posture was observed when playing and resting. Using the web-based version of Rapid Upper Limb Assessment resulted in a 3/3 score while the worksheet gave a score of 5. Both of these results indicate a need for further investigation and change. In addition this assessment did not include the participant's tilting of the pelvis. While playing, the participant's left shoulder was raised and his left foot was raised onto the ball of the foot, potentially curving the spine. The participant also tried to counterbalance the front load by leaning his upper trunk backwards, increasing the lumbar lordosis curve as shown in Figure 3.

The position for playing the instrument forces the suspension point 100mm left from the body centre line but the backward tilt of the upper body forces the participant to push his head forward to reach the mouthpiece, flattening the cervical lordosis. Viewed laterally the ear should be directly above the acromion bone, when the ear is in front of the shoulder it can cause stress and pain in soft tissue (Kember, J. 2005). Anatomically the participant is fairly tall, slim with sloping shoulders and a sunken chest. He undertook the Norris (1993) test and the pain started within about one minute so it is likely that he is susceptible to thoracic outlet syndrome.

Tolerance and excitement

Most performing musicians experience nervous tension, which can manifest itself in anxiety/stage fright, a volume of research has been undertaken on performance anxiety. The participant and fellow musicians used for this research reported that the 'excitement' and rush of adrenalin associated with performance actually helped them play better. The participant reported that performance arousal helped him play through the discomfort and neck pain he was experiencing. There is evidence of a

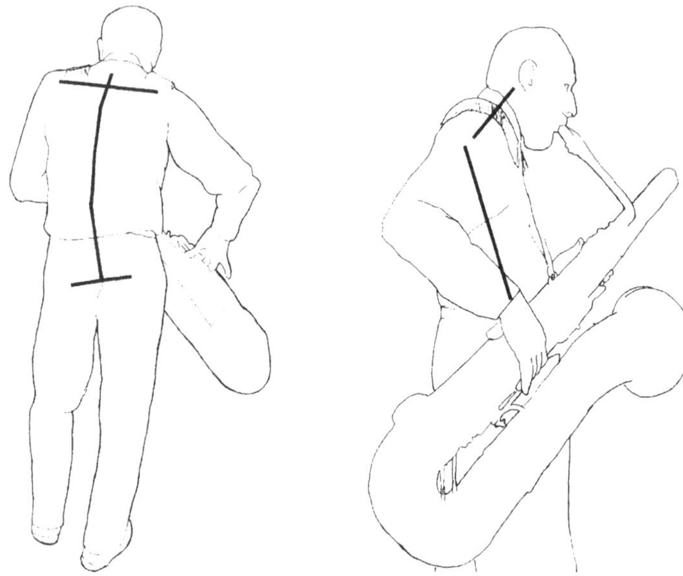

Figure 3. Recorded posture of the participant from back and side.

strong sense of team support and positive psychosocial aspects of playing in a band. Compared to the more formal orchestral setting the players seemed less constrained in their behaviour but some of that may be attributable to the consumption of beer while playing. An extension of this anecdotal piece of evidence may lead to an interesting avenue of further research into the effect of alcohol on jazz performance, particularly with regard to reduced inhibitions and reduced sensitivity to pain and fatigue.

Design objectives and prototype

The main design objectives were to:

- reduce pressure on the shoulders by transferring some of the weight to the hips
- hang the instrument so that the hips are level
- hang the instrument so that the shoulders are level
- reduce amount of lumbar lordosis curve
- get the head more upright and increase the cervical lordosis
- improve the mouthpiece position
- improve participant's awareness of posture.

A prototype harness was manufactured that had a waist/hip belt attached at the front and back to padded shoulder straps that were adjustable in length and crossed over at the back. The suspension hook was a strong, plated steel clip. Central and asymmetrical positions for the suspension were tested and it was found that the suspension point needs to be 100 millimetres to the right of the participant's centreline. The front panel above the waist/hip belt was reinforced with vertical stiffening at the front using an aluminium section inside canvas channels. The design semiotic associations for the harness were the cummerbund and braces, traditional male clothing accessories.

The participant reported that the harness was comfortable to wear; the suspension point required an asymmetrical position. He also estimated that the waist/hip belt carried just over half the weight of the instrument with definite reduction of shoulder pressure. The vertical stiffening at the front was necessary to transfer the weight downwards onto the waist/hip belt, without the stiffening the weight would have lifted the rear of the belt and increased the pressure on the shoulders. Observation of the participant using the harness with the saxophone showed aligned and level shoulders and hips, although some of this can be attributed to the participant's increased awareness of his posture. Laterally his stance was more upright and although his body was still in contact with the instrument he was not leaning backwards so much with his upper body.

Conclusion

The first prototype had some success; it reduced pressure on the shoulders by transferring instrument weight away from the shoulders. It allowed the pelvis to be level and perpendicular to the spine and the shoulders to be level with one another. The lumbar area of the spine had a more natural lordosis and the head was more upright. The extent of this which can be attributed to the participant's increased awareness of posture is hard to say. There is a need for additional research and development to assess the ability of the harnesses adjustment to accommodate a wider range of anthropometric sizes and also investigation into the ease of harness removal and size adjustment while being worn. The suspension hook offset adjustment range also needs to be established and will be affected by the anthropometrics of mouth position, torso width and arm length. Neoprene will be included for comfort and memory foam pads are specified for the shoulders on the second prototype. Consideration of appropriate aesthetics for male and female harnesses should also be taken into consideration.

Additional testing needs to be undertaken to assess the freedom of movement while walking for a marching gig. Future work will research the suitability and possible adaption of the harness for the baritone saxophone as several baritone instrumentalists have already expressed an interest in this research.

I would like to thank the Wonderbrass band for letting me observe them and especially Jim Barrett, the bass saxophonist, for his patience and help.

References

Bloswick, D. S., Gerber, A., Sebesta, D., Johnson, S., & Mecham, W. (1994). Effect of Mailbag Design on Musculoskeletal Fatigue and Metabolic Load *Human Factors, 36*(2), 210–218

Earle-Richardson, G., Jenkins, P., Fulmer, S., Mason, C., Burdick, P., & May, J. (2005). An ergonomic intervention to reduce back strain among apple harvest workers in New York State. *Applied Ergonomics, 36*(3), 327–334

Harman, S. E. (1992, Jul). *Medical Problems of Marching Musicians* Paper presented at the 10th Annual Symposium on the Medical Problems of Musicians and Dancers, Aspen, Co.

Holewijn, M. (1990). Physiological Strain due to Load Carrying. *European Journal of Applied Physiology and Occupational Physiology, 61*(3-4), 237–245

Horvath, J. (2002). *Playing (Less) Hurt: An Injury Prevention Guide of Musicians*: Janet Horvath

Kember, J. Specific Conditions, Neck and shoulders. In Winspur, I. and WynnParry, C. (ed.) *The Musicians Hand 2005*. (Taylor & Francis, London), 136–142

Kim, M. H., Yi, C. H., Kwon, O. Y., Cho, S. H., & Yoo, W. G. (2008). Changes in neck muscle electromyography and forward head posture of children when carrying schoolbags. *Ergonomics, 51*(6), 890–901

Knapik, J., Harman, E., & Reynolds, K. (1996). Load carriage using packs: A review of physiological, biomechanical and medical aspects. *Applied Ergonomics, 27*(3), 207–216

Lloyd, R., & Cooke, C. B. (2000). Kinetic changes associated with load carriage using two rucksack designs. *Ergonomics, 43*(9), 1331–1341

McChrystal, G. (2007) Sling Your Hook. *Clarinet and Saxophone, Winter2007,* 15

Manchester, R. A., & Cayea, D. (1998). Instrument-specific rates of upper-extremity injuries in music students (Medical problems). *Medical Problems of Performing Artists, 13*(1), 19–25

Norris (1993) *The Musicians Survival Manual.* MNP Music INC. St Louis

Palumbo, N., George, B., Johnson, A., & Cade, D. (2001). The effects of backpack load carrying on dynamic balance as measured by limits of stability. *Work, 16*, 123–129

Paull, B. (1997). *The Athletic Musician: A Guide to Playing without Pain*: Scarecrow Press

Reid, S., Bryant, J., Stevenson, J., & Doan, J. (2000). *Biomechanical Assessment of Rucksack Shoulder Strap Attachment Location: Effect on Load Distribution to the Torso*. Paper presented at the Soldier Mobility: Innovations in Load Carriage System Design and Evaluation

Reid, S. A., Stevenson, J. M., & Whiteside, R. A. (2004). Biomechanical assessment of lateral stiffness elements in the suspension system of a backpack. *Ergonomics, 47*(12), 1272–1281

Roos, D. B. (1986). Thoracic Outlet syndromes - Symptoms, diagnosis, Anatomy and Surgical Treatment. *Medical Problems of Performing Artists, 1*(3), 90–93

Watson, R. (2009). *The Biology of Musical Performance and Performance-Related Injury*: Scarecrow Press

Wynn Parry, C.B. Specific Conditions. In Winspur, I.and Wynn Parry, C. (ed.) *The Musicians Hand 2005*. (Taylor & Francis, London), 53–76

Zultowski, I., & Aruin, A. (2008). Carrying loads and postural sway in standing: The effect of load placement and magnitude. *Work, 30*, 359–368

PRACTICAL APPLICATION OF ERGONOMICS TO THE DEVELOPMENT OF POLICE BODY ARMOUR

Janette C. Edmonds[1], Cath Rogan[2] & Charlie Hartill[3]

[1]*Ergo Innovation (UK) Limited, www.ergoinnovation.co.uk*
[2]*Smart Garment People Limited, www.smartgarmentpeople.com*
[3]*Aegis Engineering Limited, www.aegis-eng.com*

It has become mandatory within several police forces across the UK for officers to wear body armour at all times when on duty. This carries a substantial burden for the wearer, particularly in terms of mobility, thermal strain, physical effort, comfort and musculoskeletal strain. The focus of this project was to improve the ergonomics and other aspects of the body armour design to reduce the burden for the wearer and to facilitate their performance. The ergonomics contribution was primarily in relation to the body armour garment within which the armour panels are carried. As part of the process, consideration was given to the carriage of officer safety equipment (primarily the baton, handcuffs, CS spray and radio). Various design options of the body armour garment and load carriage system were developed. The design options were user tested using a comprehensive test regime based on British standards and police test protocols. These included reach and range of movement, mobility, thermal comfort, physical workload, size/fit, and usability. The work was undertaken through the collaboration of a body armour manufacturer, a textile and garment consultancy and an ergonomics consultancy and was undertaken in response to a competitive tender for a nationwide project. The paper discusses the application of ergonomics to support and drive the design.

Introduction

Several police forces within the UK now stipulate that it is mandatory for front line officers to wear body armour at all times when on duty, or at least during specific core hours. This has been introduced as a result of both the perceived and actual threat posed by knife and gun crime and the recognition that officers require ballistic and stab protection. Few would suggest that body armour is not required. Indeed, during a study by Stubbs *et al.* (2006) all police officers interviewed stated that they would not wish to be without it during the performance of their duties. Despite the importance of providing body armour protection, there is recognition that there is a 'human' cost associated with wearing it. The human cost typically relates to reduced mobility, thermal strain, increased physical effort, discomfort and musculoskeletal strain. This has been reported in a number of studies, for example, Hooper (1999) and Stubbs *et al.*, (2006).

Under the Police (Health and Safety) Act (1997), there is a requirement to secure the health and safety of officers whilst on duty, and the Personal Protective Equipment (PPE) at Work Regulations (1992) require that clothing and equipment are appropriate for the job to be performed. In particular, the PPE Regulations clearly state:

- PPE must take account of ergonomic requirements
- It must fit the wearer correctly
- It must be compatible with other equipment
- An assessment of its suitability should be undertaken
- It should be properly and correctly used

Ergonomics as a discipline therefore has a key role to play to ensure that these requirements are met.

The aim of this study was to improve the ergonomics of a range of body armour products to reduce the burden for the wearer and to improve overall performance.

Methodology

The application of ergonomics to the design of the new range of body armour products involved a number of steps:

- User requirements analysis
- Design development
- User trials

User Requirements Analysis

The ergonomic requirements for the design were clearly stated within the tender documentation from the issuing authority. They were related to:

- Reducing the thermal burden suffered by wearers of the body armour product
- Improving weight distribution and ergonomic consideration of the positioning and carriage of equipment
- Appropriate size grading to protect the vital organs taking account of gender issues and officer fitness
- Reduction in mobility and agility restriction
- Usability of the product in relation to donning, doffing and adjustment.

Ergonomics studies previously undertaken for two UK police forces enabled a greater understanding of the context of police work and the requirements which the design would need to meet. The requirements stipulated by the issuing authority were therefore supplemented by work undertaken independently of this specific project. The first such study was undertaken for the Avon and Somerset Constabulary prior to the introduction of personal issue body armour, which was aimed at reducing discomfort related to load carriage for female officers. This study was reported by Edmonds and Lawson (2000). The second, more recent study, also undertaken by the author, and due for completion in 2009, was undertaken to resolve the issues related to load carriage for a Midlands based police force, taking account of the mandatory use of body armour for front line duties. Both studies involved the derivation of user requirements to support the design and/or selection of appropriate methods of carrying police equipment.

The requirements were based on:

- A literature review of relevant research
- Questionnaires to identify key issues and officer needs
- Interviews with front line officers
- Analyses of officers tasks

Design Development

Body armour typically has two components. The armour panels and the garment in which the removable panels are carried. The detailed technical design of the armour panels was undertaken by a separate, but linked, design team running in parallel with the garment design team. The technical design of the armour panels refers to the materials engineering aspects and test certification of the ballistic and knife protection level. The majority of the ergonomics contribution was in relation to the design of the garment, although the armour panel size, shape, coverage, and weight were naturally issues for both teams.

The garment design team included contributions from the body armour manufacturer in relation to technical expertise related to body armour and production design, a textile and garment consultant to advise on materials, size/fit and production, and an ergonomics consultant to advise on the biomechanical and physiological response, usability, user requirements and appropriate user testing methods.

Concept designs were derived through a series of design workshops focusing on the tender requirements presented by the issuing authority. At various stages during the development, 'mini' trials were undertaken to test and refine specific concept ideas using prototypes. Specifically, the 'mini' trials provided objective and reasoned justification for discounting or progressing design concepts, but also enabled the team to identify how the concepts should be refined. The issuing authority was also involved in specific design reviews to inform the refinement and selection of design options.

User Trials

A comprehensive test regime was developed on the basis of ergonomic test protocols provided by the issuing authority and British standards, specifically:

- BS EN 340:2003 – Protective Clothing. General Requirements
- BS EN 13921: 2007 – Personal Protective Equipment. Ergonomic Principles

- BS EN ISO 9886:2004 – Ergonomics. Evaluation of Thermal Strain by Physiological Measurements
- BS 7963: 2000 – Ergonomics of the Thermal Environment. Guide to the Assessment of Heat Strain in Workers Wearing Personal Protective Equipment

It would have been preferable to recruit operational police officers, ideally from different forces and operational roles to take part in the trials. This was not possible in the timescales available and was a limitation of this set of trials. However, further laboratory and field trials were subsequently undertaken by the issuing authority with police officers, which are not reported within this paper.

The trials were undertaken by staff from the Aegis Engineering factory in Warrington. The test pool included males and females, and a range of sizes including 'limited user' types, for example:

- A petite female with limited space available for the carriage of officer safety equipment – this was to include the potential affect on the 'usability'
- A large breasted curvaceous female who, it was considered, may be more affected by the coverage and shaping of the body armour, movement restriction and the ability to see officer safety equipment carried on the garment
- A large male officer who, it was considered, may find that there are difficulties with movement and coverage of the body armour in different postures.

The test regime included objective and subjective measurements to compare the wearability of the new body armour products with equivalent existing products currently in use. The test measures included:

- Reach and range of movement – measuring the effect of the body armour on the ability of the end user to perform specific and representative actions
  - Objective measures of cross body, forward and vertical reaches
  - Subjective measures related to accessing equipment and basic cross body, above head, behind body reaches
- Mobility – measuring the effect of body armour on basic postures and movements; including, for example subjective ratings related to walking, running, stepping, sit/standing and other basic motion tasks
- Thermal comfort and workload – measuring the effect of the body armour on the thermal comfort and physical workload of the end user, including:
  - Heart rate monitoring and recovery
  - Rate of Perceived Exertion (RPE) ratings
  - Measurement of the thermal aspects of the microclimate
  - Measurement of the thermal response in an environmental chamber
- Usability – measuring the effect of the body armour on being able to perform representative tasks, including subjective ratings relating to:
  - Donning, doffing and adjustment
  - Undertaking tasks related to the use of a vehicle
  - Key officer safety tactical tasks
- Size and fit – measuring the comfort and suitability of the fit of the body armour, based on directed observation and subjective rating.

In addition to testing the core garment design (with body armour panels), there were a series of tests which were used to compare the wearability of the body armour in conjunction with wearing the primary officer safety equipment (handcuffs, baton, CS incapacitant spray and radio).

The user trials did not include any testing related to the performance of the body armour in terms of ballistic or stab protection as these tests were undertaken separately. Neither did the tests include performance measurement of the materials. It focused solely on the ergonomic wearer aspects of the design.

Outcome of the Design Process

The ergonomic requirements from the issuing authority enabled a specific focus on key issues for the wearer. The previous studies, which included the involvement of nearly 1000 police officers, revealed the wide variety of requirements for a project of this nature. Besides great variability in body size and shape, there are also differences in the way the equipment is worn, the way it is used for different tasks, job roles, operational circumstances, and conditions, even before personal preferences are considered. This information was used to inform the design and the test regime.

The resultant range of body armour products included covert armour, overt armour and load carriage options. It also included different colourway options, including

Table 1. Male Version of the Tor-Flex™ Design.

- Lighter weight, softer and more flexible construction of armour panels
- Shaping of armour panels to avoid movement restriction
- Flexible shoulders and waist to support upper body movement, to maintain fit and to reduce the need for separate adjustments
- Fewer, thinner, breathable and lighter fabrics with treatments for water proofing and heat reflectance to improve heat management
- Fabric selection and treatments for ease of maintenance
- Development of a fitting device for greater sizing accuracy
- Simple 3 step donning method
- Fixings and fastenings selected for ease of use and simplicity, including docking ports for lightweight primary equipment (specifically the radio in the upper chest region).

Table 2. Female Version of the Tor-Flex™ Design.

- The same construction as the male version
- Flexible armour panel structure enables moulding to the female shape
- Close fit tailored garment design with 3 dimensional shaping complements the armour panel design to accommodate the full range of female bust sizes and shapes
- Additional side coverage for women with larger bust sizes
- Special women's fit patterns
- Flexible panels provide flattering and 'slimming' shape profiling
- Shaping at the hips on longer length covers avoid tightness and restriction
- Binding aids snug underarm fit
- Pockets with vertical zips positioned, shaped and sized for comfort

Table 3. Design of the Hip Belt.

- Sizing to suit a range of users.
- Shaping to complement body movement.
- Materials chosen on the basis of aiding flexibility, strength, heat management, maintenance and relevance for the operational environment.
- Improved compatibility with body armour.
- Safety features to prevent accidental unfastening or deliberate removal by an assailant.
- A secure docking system to attach equipment holders to the hip belt and allow flexible positioning of equipment.

high visibility and a variety of style modifications. Information on the full range is available from the manufacturer but, for brevity, only the key highlights of the Tor-Flex™ design are presented in this paper. The 'male' version of the Tor-Flex™ design is described in Table 1 and the 'female' version is described in Table 2.

The design intention was to limit the carriage of equipment on the torso, particularly at the front of the body. A hip belt was designed (described in Table 3) to complement the armour design and to satisfy the requirement for load carriage. The hip belt was designed for the carriage of the heavier items of equipment, in particular the handcuffs and baton. The key ergonomic reasons for this design option were to:

- Maintain a lower centre of gravity and reduce the loading at the shoulders (high loading of the body armour (and therefore the shoulders) increases vertebral loading of the spine)
- Reduce the energy required to stabilise the load (due to load placement)
- Improve the movement flexibility of the shoulder girdle (reducing the load on the shoulders)
- Position the greater loads in the hip region (better suited to load carriage)

The data from the user trial were analysed for each of the tests, both with and without wearing officer safety equipment. Subjective ratings were scored and averaged to enable a comparison between the existing design and the new design. This was also undertaken for some of the objective measurements, albeit that for some of the objective test measurements, the new and existing designs were compared with the control condition of 'no armour'.

The results demonstrated that there had been an improvement of the new design over the performance of existing body armour. In particular, the results relating specifically to the body armour without officer safety equipment showed:

- An improvement in basic stretches, reaches and mobility tasks
- An improvement for the objective reach tests with less restriction in the shoulder/chest areas. Subjectively, those tasks were also easier to perform and more comfortable
- The thermal comfort and physical workload results suggested some improvement but they were insufficient to draw conclusions

- There the performance related to donning, doffing and adjustment revealed good results for both conditions
- The overall fit was judged to be better with particular improvements for females
- The vehicle evaluation showed little or no impact of the body armour on driving in the new armour, but that there were difficulties and restrictions within the existing armour.

The results for users wearing the body armour with officer safety equipment showed:

- An improvement for the basic stretches and reaches and the mobility tasks over the existing armour / duty belt system
- An objective and subjective improvement for new armour for the reach tests
- Access to equipment was better, albeit that access for the existing system was also good
- An improvement in the performance of officer safety tactics
- An improvement over the existing armour / duty belt system for the vehicle evaluation, although there were some issues remaining related to the fit of the seat belt.

Conclusion

Front line police duties are undertaken in the context of the potential for injury from weapons, both ballistic weapons and knives (amongst other threats) which has lead to the mandate for body armour to be worn at all times. The focus of this project was to reduce the 'human' costs associated with wearing body armour whilst maintaining a high degree of protection for the user.

The overall conclusion of the study was that in addition to an improvement in the coverage of the armour (and therefore protection of the vital organs) there were also particular improvements in reducing the 'human' costs associated with its use.

The key improvements were:

- Reach and range of movement
- Mobility
- Size/fit, particularly for females
- Use in a vehicle
- Access to and use of officer safety equipment

It was also concluded that the performance of both armours for donning, doffing, adjusting was good, so there was little to distinguish between them. The objective results of the thermal tests revealed no difference, although subjectively the new armour was rated as more comfortable.

Many of the improvements have been made possible through major technological advancements in armour panel design and textiles. However, the realisation of the benefits has been made possible by having a clear focus on the end user throughout the design process.

The incorporation of an ergonomics approach has enabled the requirements of the issuing authority and the PPE Regulations to be met. The design has taken account of the ergonomics requirements, and improved the fit and compatibility with other equipment. The new design was comprehensively tested and its usability will enable the body armour to be worn correctly. This study has clearly demonstrated how ergonomics can be applied to improve the design of personal protective equipment.

References

BS 7963: 2000 - *Ergonomics of the Thermal Environment. Guide to the Assessment of Heat Strain in Workers Wearing Personal Protective Equipment*

BS EN 340:2003 - *Protective Clothing. General Requirements*

BS EN 13921: 2007 – *Personal Protective Equipment. Ergonomic Principles*

BS EN ISO 9886:2004 - *Ergonomics. Evaluation of Thermal Strain by Physiological Measurements*

Edmonds, J. and Lawson, G. *The Design of Methods for Carrying Police Equipment.* Contemporary Ergonomics, 2001, Eds. Hanson, M.A., Lovesey, E.J. and Robertson, S.A. (Taylor & Francis, London)

Health and Safety Executive. 1992, *Personal Protective Equipment at Work Regulations.* (HMSO)

Home Office. 1997, *Police (Health and Safety) Act.* (HMSO)

Hooper, R.H. *Why Do Police Officers Leave Their Body Armour in the Cupboard?* Contemporary Ergonomics, 1999, Eds. Hanson, M.A., Lovesey, E.J. and Robertson, S.A. (Taylor & Francis, London)

Stubbs, D., David, G., Woods, V. and Beards, S. *Problems Associated with Police Equipment Carriage with Body Armour, Including Driving.* Contemporary Ergonomics, 2008, Ed. Bust P. (Taylor and Francis, London)

HUMAN FACTORS INTEGRATION

BARRIERS TO SUCCESSFUL HUMAN FACTORS INTEGRATION WITHIN THE MILITARY DOMAIN

Paul Sirett & Jenny Sutcliffe

Systems Consultants Services Ltd (SCS)

As military systems become more complex the human element needs to be considered as part of the total system in order that the capability can be realised to its full potential. Human Factors Integration (HFI) attempts to smooth the integration of the system into service by taking account of all human related aspects and provide mitigation in key areas. This paper examines the issues surrounding the integration of military technologies from concept and assessment phases through to development and manufacture phases. These observations are taken from three important military programmes, focused on networked capability. The HFI process is examined in terms of its utility and effectiveness to integration. This paper sets out the problematic barriers to HFI and highlights the practical issues faced by HF practitioners borne from subjective observations by the authors. The issues have practical validity and are presented here as areas for HF practioners to keep an eye on throughout the design and development cycle. It is concluded that HF has added value throughout the design and development cycle, however, some HF related issues remain overlooked and are not attended to or implemented by the time the system is placed into service and deployed on military operations. Therefore the capability of the system is often not fully realised at the time of implementation.

Introduction

Ergonomics, now more commonly known as Human Factors (HF) is a relatively new discipline that started to be considered in the defence industry around World War II by people such as Hywel Murrel, through the Naval Motion Study Unit (NMSU) in 1947. Murrel emphasised the scientific basis of ergonomics and helped establish its use in system design and development (Stammers, 2007). More recently, in an article in the UK MoD Defence Equipment and Support journal, 'Desider', Air Chief Marshal Sir Jock Stirrup, Chief of the Defence Staff, stated that "Evidence shows that more than half our accidents and incidents are down to human factors; in other words it is our people who are causing most damage to our fighting capability. We must do something to understand these factors and drive down the number of accidents and incidents." (Stirrup, 2009). Despite this top level acknowledgement of the growing importance of HF, historically, within the UK defence sector, the integration of HF into design has been poor. When HF is taken up correctly it has a significant impact on system design, as Stone (2009) demonstrated it improved the level of fidelity for an IED search and disposal simulator.

The defence industry is not the only industry where HF is being taken more seriously but where integration has been poor. In the rail industry the application of HFI has been seen to fail even when there have been detailed models to support the integration of HF into design (Grimes, *et al.*, 2009). Grimes *et al.* highlighted that, at the programme level, the HF advice was still often overlooked or not fully realised leaving problematic issues for the end user or operator to overcome. The conclusion to the paper was that HF practitioners need to become smarter at integrating the HF advice into the engineering cycle.

This paper discusses the availability of HFI guidance information and the integration of HF into three MoD procurement programmes throughout the CADMID[1] cycle, and provides a subjective evaluation of possible reasons for its poor uptake based on the experiences of the authors.

MoD Guidance

To-date, much literature has been produced and organisations have been set up to aid the integration of HF into defence procurement programmes. The overarching Acquisition Operating Framework (AOF), which is "an authoritative source of policy and good practice for all members of the UK Ministry of Defence (MOD) and our Industry partners concerned with acquisition" contains a dedicated HFI section that provides "an authoritative source of policy and best practice on Human Factors Integration aspects ... concerned with Defence Acquisition" (see www.aof.mod.uk).

The AOF HFI section is designed to assist the HFI practitioner and the MoD customer with guidance information on the processes and procedures that should be carried out to implement HF activities. Similarly, a number of HF standards exist that are not only used to assist with conducting the HF activities, but also with integrating HF into programmes. Examples include, but are not limited to:

- Defence Standard (Def Stan) 00-250 "Human Factors for Designers of Systems", specifically Part 0 "Human Factors Integration";
- NATO Standard Agreement (STANAG) 3994AI "Application of Human Engineering to Advanced Air Crew Systems;
- MIL-HDBK 46855A "Department of Defence (DOD) Handbook".

It is important to note, however, that the literature and guidance information requires significant tailoring for each programme to which it is applied. Further, as with

[1] The CADMID Cycle is the process currently adopted for acquiring equipment, systems or services for the UK armed forces. CADMID stands for Concept, Assessment, Demonstration, Manufacture, In-service, and Disposal; each being a different phase in the cycle. For further details see: http://www.aof.mod.uk/aofcontent/tactical/ppm/content/lifecycles/cadmid.htm

many guidance documents, often the content of one document does not agree with another and the approach to be used is often vague.

The UK MoD currently has four UK Defence Technology Centres (DTC) that are a collaboration between the MoD and the broader UK science and technology community. These DTCs focus on emerging technologies that aim to create a high payoff when integrating technologies. They do this by conducting scoping studies, experiments and provide general support to system integration. HF plays a key part in these consortia, specifically the HFI DTC, which is predominately focussed on supporting the integration of HF into programmes. The HFI DTC aims to reduce whole-life costs through the more efficient use of equipment and personnel, and, through a number of initiatives, provides high level guidance to HF practitioners on how best to overcome barriers to change. Other MoD initiatives are research consortia, such as the Haldane-Spearman Consortium (H-SC), which aim to capture and exploit knowledge from research; focusing on cognition, command teams, technologies, organisations, personnel, training and medical issues. This research provides important guidance on the HF issues relating to future technologies and the associated people issues. The latest emerging initiative, planned to be created towards the end of 2009, is the HF Centre for Defence Technology (HF CDT), which aims to merge the HFI DTC, HS-C, and similar initiatives into a single centre. These initiatives and supporting outputs and guidance are valuable in order to ensure that appropriate change is enabled in the design process. Through these organisations and initiatives there is now a wealth of information available to HF practitioners and customers that help improve HF integration into technology development throughout the whole CADMID cycle.

HFI is no longer an optional activity in UK defence procurement programmes; as acceptance of HF at each of the system readiness levels (SRLs) is required to support qualification and verification for customer sign off. This demonstrates that HF is now taken seriously by the MoD customer and requires as much thought and management as the other areas of a programme, such as system engineering, Integrated Logistics Support (ILS), Safety, Security, and Training.

As part of this paper, a review of three defence programmes that have taken place in the last two years was carried out. The aim of this review was to examine the HFI activities undertaken and identify where barriers existed in the successful implementation of HFI. The review carried out is based on subjective observations by the authors of the HFI management and technical activities conducted on each of the programmes.

Review of Three Procurement Programmes

A full description of the three programmes reviewed is not possible due to the security classification placed upon the programmes. However, they relate to different aspects of air defence and reconnaissance. Programme1 (P1) had a three year D&M phase, whereas Programmes 2 and 3 (P2, P3) were Urgent Operational

Table 1. Key HFI Management and Technical Activities.

	P1	P2	P3
HFI Management Activities			
Provided with IPT HF Contact	Yes	Yes	Yes (partially)
HFI Requirements (and acceptance)	Yes	Limited	Yes
Input into ITEAP	Limited	Limited	Limited
Input into TLMP	No	No	No
HFI Strategy	No	No	No
HFI Plan	Yes	Yes	No
HFI Working Groups	Yes	Yes	No
HF Risks/Issues Register	Yes	Yes	Yes
Use of relevant Standards	Yes	Yes	Yes
HFI Technical Activities			
Utilised SME Involvement (i.e. User Centred Design)	Yes	Yes	Yes
Human Computer Interface (HCI) Specification	Yes	No	No
HCI Style Guide	Yes	Yes	Yes
Early Human Factors Analysis	No	No	No
Functional Analysis	No	No	No
Task/Mission Analyses	Yes	Yes	Yes
Target Audience Description and job design	No	No	No
Human Reliability/Safety Analysis	Yes	Yes	Yes
Training Needs Analysis	Yes	Yes	Yes
HCI/Usability Assessments	Yes	Yes	Yes
Ergonomic/Workspace Design	Yes	Yes	Yes
Workload Analysis	Yes	Yes	Yes

Requirements (UORs). An UOR is a result of a request for additional support from theatre, and therefore have a shorter development process than normal procurement programmes; typically a few months to a year, and do not necessarily undergo all activities that are usually carried out in normal procurement programmes.

Table 1 lists the key HFI Management Activities and Technical Activities that are required to be carried out as part of the HFI process for a MoD procurement programme. For each of the MoD procurement programmes reviewed the key HF activities that were actually undertaken were identified.

As Table 1 shows, in all three programmes reviewed the majority of the key Management and Technical activities were carried out. For all the HF activities undertaken and deliverables produced the HF practitioners delivered to both the time and quality expected by the customer. It is therefore not the content of the output that is questioned throughout this paper, rather, the process in which change is considered and implemented.

Barriers to HF

It is of note that P3, which was a UOR, had the same initial limited HFI requirements as the other programmes, but had no working groups, due to the timeframe, and had no initial HFI Plan (HFIP). However, the impact HF had to P3 was greater and resulted in changes occurring more frequently than the other programmes. Interestingly, this was not the case for P2, which was also a UOR. There are a number of reasons that change may have occurred at a higher rate on P3 and not the other programmes. These possible reasons, or barriers to change, are discussed below.

Poor Integration into Project Team

One significant issue identified across all three programmes reviewed was that the HF practitioner and associated outputs were poorly integrated into the project team and were not on the critical decision path. In Programmes 1 and 2 this created subtle problems with the integration of key HF outputs resulting in change that was either slow or did not happen. This lack of integration often meant that the HF advice provided was based on incomplete information received, was out-of-date, was inaccurate, or did not feed into the discussions and decision making process of the engineering design cycle. Consequently, much of the HF advice was not actioned and/or was superseded by other events.

Unsuitable Integrated Project Team HF Contact

For all programmes reviewed the Integrated Project Team (IPT) had a nominated HF point of contact (POC) at some point during the programme. However, the POC was not always provided throughout the duration of each phase of the programme and in P1 and P2 the POC changed throughout the life of the programme. Further, in each of the three programmes, the POC had minimal or no training in HF and had other commitments to fulfil. Consequently, only some HF issues were followed up whilst others remained unaddressed. Interestingly in P3, where there was no designated IPT HF POC, the customer programme manager had taken a keen interest in all HF activities and supported the prioritisation of change issues. In this programme, it was evident that the key issues were actioned more successfully than the other programmes observed. It would seem that even without an IPT HF POC, as long as there is a person with sufficient authority to champion the HF issues then appropriate change can still happen. The HF champion needs sufficient authority to enforce consideration of the HF issues and needs to take an active role in managing, as a minimum, the critical HF issues.

Limited Understanding of HFI Plan

Theoretically a HFIP is required and is relatively straightforward for the HF practitioner to write. It should contain a proposed plan of how the HFI activities will be conducted, in relation to the rest of the programme, and discuss any risks, limitations, assumptions, and Government Furnished Assets (GFA) required. The aim of such a document should be to help the IPT and design team understand what the HF activities to be conducted are, so as to accommodate them into the design cycle. Further, outlining plans for an activity helps ensure all aspects are considered and all related disciplines are aware of what will be involved.

HFIPs were delivered on P1 and P2 but there was no plan for P3, given the extremely short timeframe that was required to implement the system into service. It is of note that the HFIP on P1 and P2 were formally delivered but neither was fully understood or read by the engineering team or the customer. For example on P1, the project prime contractor requested that a synopsis of the plan be written for ease of use and understanding. Similarly on P3, where there was no formal HFI plan, an outline of the key tasks and associated dates were identified with only the Chief Engineer being informed of the impact of the deliverables. It was observed on P3, that following the critical task analyses (CTA) and HCI and usability testing, there was immediate change to the software design in the subsequent weeks. Therefore, it appeared that although an HFIP is useful in providing the project team with an outline of the HFI activities that are to be conducted and the proposed timescales, if it is too detailed and not understood it is likely not to be attended to by the project team.

The points raised here indicate that a HFIP is not, in itself, the critical activity in ensuring change happens. Rather, change will occur when HF issues are grasped and understood by the Chief Engineer (or similar role) and customer who has the authority to implement the necessary change. Of course, a HFIP is still necessary and required, however, it needs to be easily understood in terms of the design process, and it needs to be managed properly and updated appropriately with tasks and outputs within the plan placed in the engineering critical path. There needs to be at least one 'champion' in the design and integration teams involved in understanding and ensuring the HFIP, and HF issues and activities, are sufficiently considered. This champion is best placed in the Systems Engineering and Integration Team (SEIT), however, depending on the programme and lead company, other champions may be identified.

Poor/Incomplete HF Requirements Generation

On all programmes reviewed, the most significant activity that appeared to have had limited HF involvement was the initial generation of the HF System Requirements (HFSRs). In addition to the physical lack of HFSRs, those requirements that existed were considered high level and vague, for example, *'the system shall have a useable interface'*. It is therefore up to the design team to ensure that a Style Guides and HCI Specifications are implemented and followed. On P1 and P2 there was a lack of compliance with the Style Guide and HCI Specification by the software engineers as they were not contracted to follow such documents. Reasons given for non compliance varied between programmes, but the creation of HFSRs is of no value if the software team, or other teams, are not obliged to comply with them. The lack of HCI requirements, specifically, on all programmes reviewed resulted in lengthy detailed HCI issues lists.

HF requirements are a significant activity as system acceptance, during qualification and verification, is measured against whether the system meets the requirements. Poor or incomplete HFSRs mean that it is harder to test the system against crucial functions and accept them as being suitably designed, or that some aspects of the system are not tested. This ultimately leads to a lack of acceptance by the customer and a likely costly re-design, or worse still, future problems in-theatre. It is important that the development of the HFSRs and further iterations are carried out with the HF Engineer and customer at the appropriate phase of the CADMID cycle. This will ensure the correct requirements are generated and the subsequent HF activities are appropriate. Incorrect, incomplete or high level HFSRs will add very little or no value to the programme and will consequently mean HF is not given the time or acknowledgment it requires. As such, once baselined in the System Requirements Document (SRD) a review of the HFSRs should be carried out in subsequent phases of a programme to ensure continued relevance of existing requirements and allow the addition of new HFSRs, should they be required. When HCI Specifications and Style Guides are developed an inclusive approach with the software engineers is recommended and the customer has to decide early on as to the level of compliance required to these documents.

Lack of Informal Feedback

For Programmes 1 and 2, the impact of the lack of use of the Style Guide and HCI Specification was that when HCI assessments were conducted there were a large number of specific HCI issues placed onto the issues register. Given the large number of HCI issues, there was a reluctance to initiate change since the effort, and associated cost, required for the redesign and software programming would have been high. This meant that issues were either overlooked or the software team implemented those changes that were easily resolved, but which were not necessarily high priority. For Programme 3, the issues were handled differently. Reports were initially delivered informally, were short and easy to read, and were given to both the appropriate project team members and the IPT within the same week. Actions for change were taken quickly in the subsequent week(s). The formal HCI and usability reports were delivered at a later date and placed in a networked shared area to keep a formal record and for audit purposes. Although formal reporting was carried out for the other programmes, less notice was taken of the recommendations as there had been no informal delivery previously.

The essence of change appears to be with an informal delivery in the first instance with a more formal delivery of outputs that follows. Keeping the reporting informal helps ensure a timely delivery of the issues and allows time for changes to be actioned; keeping pace with the engineering build and critical path and ensuring both time and cost issues are kept to a minimum. The delivery of informal outputs needs to be followed up by question and answer sessions either by telephone, email, or face-to-face meetings. However, care must be taken, if issues are fed back too informally then the reverse can happen and actionees may fail to appreciate that the information they had received was important and required actioning, as appears to have happened on some occasions.

Poor Timeline for Critical Decisions

Decision points on complex longer term programmes, such as for P1, appear to be more subtle and appeared to be 'dragged out' over a protracted period of time.

For example, HCI issues identified from assessments and the issues discussed in the working groups tended not to be resolved between one HF working group and the next; even though significant effort was made by the HF team to monitor the actions. On P3, however, time was tight and decisions had to be made very quickly. In a dynamic system development environment that requires aspects of a system to be changed quickly, then clearly the most immediate route of communication should be followed, which is often not via quality departments or formal reports placed onto central working areas. However, it is fully understood that formal deliverables are required for quality and audit purposes. It is *not* recommended that these be removed or de-valued, simply they should used for the purposes with which they are intended and not to support activities and decisions in such a dynamic and changing environment.

Conclusion

The review of the thee programmes indicated that although most of the key aspects of HFI were considered and undertaken, the implementation of the HF recommendations made was, in the main, limited; appearing to be due to a number of organisational and working style barriers. Therefore, even with sound advice and knowledge of the HF process, techniques, and standards there are still practical barriers to overcome to ensure change happens when required.

Programme management complexity and the way in which deliverables are implemented appeared to be key aspects in influencing whether HF advice is implemented. In all programmes observed the programme manager had, understandably, a number of competing priorities that needed to be attended to within the design and development process and thus HF is just another area for consideration. However, that does not mean it does not deserve its place in the queue; the impact of poor HFI can be considerable and consideration of its issues must be undertaken by the relevant personnel in the project team; typically the programme manager.

HF appeared to work well and resulted in beneficial change when:

- The HF team outputs were well integrated into the critical path and acted on immediately;
- Key POCs or 'champions' were apparent; one being from within the project team (e.g. the Chief Engineer) and the other from the IPT (e.g. Customer).
- Deliverables were initially informal short, easy to read documents and were supported by larger more formal documents produced and delivered at a later date;
- Reports were followed up with timely workshops, phone calls and emails, where relevant personnel had input and were able to discuss and make decisions on the recommendations made;
- Information was 'pushed' rather than 'pulled' (i.e. sent to the relevant parties rather than being put on a shared working environment for them to go and find).

With regards to guidance information available, although a wealth of guidance information exists, and many HF consortia are aimed at improving HFI, there remains in inherent aversion to its proper uptake. It would seem, from a practical stand point, that it is not always possible to incorporate all the guidance that has been generated into a programme of work; given the cost associated with the effort and the competing priorities of other disciplines. Although the guidance developed is extremely thorough and useful, it usually requires tailoring to each programme to which it is applied. Further, this policy and best practice guidance is as it says – best practice and guidance – and as such it is very much reliant on the individuals involved in the programme to embrace and apply all that it describes. This requires the HFI practitioner to be sufficiently integrated into the project team in order to successfully carry out the activities. In addition, since the information is guidance, factors such as timescales, costs, and stage of programme also affect the amount of resources that are allocated to implement HF advice. Without sufficient consideration of these factors, HFI can end up being a simple 'tick in the box' exercise.

HFI has come a long way within the defence sector and is now recognised as a necessary activity in procurement programmes. With so many factors affecting the success or failure of HFI up-take, in order to maximise the value it brings to a programme, HFI as a discipline needs to become more flexible in its approach and smarter at integrating outputs into the design cycle.

The MoD Architecture Framework (MoDAF) is beginning to take this forward with the introduction of Human Views (HVs), which express (high-level) HFI concerns in a Systems Engineering language. It is important that HFI practitioners are made aware of such initiatives and are able to learn how to use them successfully. This relates to adequate HFI training, which is a whole different area for consideration outside the scope of this paper. For the defence industry, the gains that can be made in terms of efficiency, effectiveness and safety make it imperative that HFI is improved and made to have a greater impact on procurement programmes. As cost is always a key driver, it is up to the HF practitioner to ensure that HF is integrated in a manner that allows for a cost effective approach.

References

Stammers, R. B. (2007). Hywel Murrell at the Naval Motion Study Unit and the Development of Ergonomics. *Contemporary Ergonomics, 2007*, pp 93–98

Stirrup, J. (2009). The Cause is Just – Culture Must Be Too. In *Desider: The Magazine for Defence Equipment and Support*, Issue 9, Jan 2009

Grimes, E., Wright, K., and Hillier, G. (2009). Integration In Design Projects – Delivering The Return On Investment: *3rd International Rail Human Factors Conference*

http://www.aof.mod.uk/. Cited August 2009

http://www.aof.mod.uk/aofcontent/tactical/hfi/index.htm. Cited August 2009

Stone, B. (2009). How Human Factors helps to drive down the casualties of war. *The Ergonomist*, No 470, pp 4–5

ERGONOMICS OF UBIQUITOUS COMPUTING: HOW SENSOR-BASED INTERACTION INFLUENCES USER PERFORMANCE IN SIMPLE TASKS

Chris Baber & Anthony Schwirtz

Electronic, Electrical and Computer Engineering, The University of Birmingham, Birmingham, B15 2TT, UK

The notion of attaching sensors to everyday objects and using this to support Human-Computer Interaction is central to the field of pervasive computing. Typically, the interaction between human and computer is mediated by the object in such a way as to allow the person to maintain focus on the physical rather than digital world. In this experiment, we were interested in considering how such an interaction could shift the locus of control between person and 'system'. A simple painting task was used, in which all physical objects were fitted with radio frequency identification (RFiD) tags. When the user picked up one of these objects, this was responded to by a computer which offered prompts and advice. The advice offered was based on a context-aware model of the task and interpretation of the user's current activity. The objective was to copy a painting as accurately as possible. Participants completed the task with or without the sensor-based interaction. A second study, comparing performance with and without the sensor-based system, involved participants working on parts of a painting (in a 'production line' as might be found in an industrial setting and in which one might expect automated support). In both studies, there were significant main effects of condition on time and quality (measured by the accuracy of colour mixing as assessed through image analysis), with the participants using the sensor-based support interaction performing much better than the unsupported or manual access conditions. However, subjective response to workload (measured using the NASA-TLX) revealed, while the sensor-based interaction led to measurably superior performance, the subjective impression was that it increased mental demand. One explanation of this is that the 'system' shifted locus of control away from the user, who then found it difficult to interpret the quality of their activity.

Introduction

A key aspect of ubiquitous computing concerns sensor-based interaction with computers (Rogers and Muller, 2006). Obviously, one can use sensors as a means of tracking the movement of an object and representing this movement on the computer screen; thus, the 'tagged' object becomes simply another form of indirect pointing device, and would require the user to manipulate the object in order to control whatever is appearing on the screen. This can be seen in a variety of video games, such as the Nintendo Wii™. An alternative approach is for the person to use an object according to its everyday actions, say painting with a paintbrush, but for the object to be tagged so that it has a 'digital representation'. This could, for example, provide a means of logging the activity of an engineer in safety-critical maintenance tasks (Nicolai *et al.*, 2004; Schwirtz and Baber, 2005). In these examples, the use of the tagged object is exactly the same as its non-tagged counterpart, but the outcome of the action occurs in both the physical and digital domain. These examples imply that the action in the digital domain could arise without the person's awareness, e.g., the logging could be covert. Of course, one could provide some form of feedback when measurements were being taken (perhaps in the form of lights on the object or some display in the environment), but there would remain some *disassociation* between the person's action in the physical world and the resulting outcome in the digital world. A potential problem with sensor-based interaction is the apparent shifting of the control from the discretion of the user to the activity of the sensor (Bellotti *et al.*, 2002). Having the system change state or alter settings can be distracting or frustrating (particularly if the reason for such change is not immediately apparent to the user). One solution to this problem could be to provide sufficient information to the user concerning system state. An alternative would be to have the system offering advice to the user, rather than necessarily performing an action; in this way, the user could retain control, but advice could be tailored to the task-at-hand. An obvious question, therefore, is whether providing someone with feedback on the basis of their interaction with physical objects affects their performance?

In this paper, we report studies in which participants performed a familiar task using tagged objects which could result in task-relevant information being presented to them. The studies compare participants who received task-relevant information as a direct result of their interaction with tagged objects, with control conditions in which participants either had no information or had to manually request information. One might expect to see the task of consulting information to be an interruption to the task-at-hand, but one might expect the level of disruption to be lower for the people using the tagged objects because the information is presented in parallel with their activity, whereas those using the requested information have to pause their task-at-hand. On the other hand, one might expect participants in the tagged condition to feel that they had less control over the provision of information. The definition of what information to present was made on the basis of two forms of recognition: use of accelerometers to recognize the activity being performed and use of Radio-Frequency Identification tags to recognize the objects being used.

Recognising objects and activity

Simply by defining thresholds for activation in one of the 3 axes of an accelerometer, it is possible to define postures such as Sitting, Standing, Walking, and for these to be used to determine the level of information a person is ready to be shown, Bristow *et al.* (2004). To develop more reliable recognition of activity, one can use Hidden Markov Models (Westeyn *et al.*, 2003). On the basis of such modelling, it is possible to define a set of everyday actions. For example, Amft *et al.* (2005) used four arm-mounted accelerometers to distinguish between, eating with a spoon, hands, cutlery, or drinking with a glass. In the current work, the accelerometers, positioned on the users' wrists allow categorisation of a small set of actions, such as lifting, stirring / mixing, painting. While some of these tasks could be ambiguous, e.g., stirring and painting, the use of Radio Frequency Identification tags on objects helped to resolve this ambiguity. Thus, knowing that the person was holding a /paintbrush/, the system would identify the action 'paint', and if they had held /paint-bottle/ and /paintbrush/, the system would identify the action 'mix'.

Radio Frequency Identification (RFiD) tags are similar to bar-codes (in that they are read using a special reader and contain codes identifying the product), but do not rely on line-of-sight for reading. Rather, the reader emits a small electromagnetic field which can activate tags when they are in range; once activated, the data on the tag can be read by the reader. This means that it is possible, for instance, to embed a tag in the handle of a tool and for the tag to be read when the tool is picked up (by an antenna fitted into the cuff of the user's jacket). Thus, a tagged environment can help to extract patterns in object interaction as the user performs everyday activity. Phillipose *et al.* (2004) apply Hidden Markov Models to sequences of object interaction for task recognition.

In the current studies, identification of activity relies on the history of object interactions. Starting at the bottom of figure 1: the User performs an action using an object. Having identified the object (through RFID) and the action (using accelerometer), the 'object-in-use' is added to the history-trace in a SQL server. The 'object-in-use' defines the particular manner in which the object and action relate to each other, e.g., a /paintbrush/ being used to <paint>. Objects 'evaporate' from the history-trace if they have not been used for a period of time (typically 2 or 3 minutes). If the object is defined in terms of its weighting. Tools, such as paintbrushes, have a high weight for 'action' (because one can do several actions with these) but a low weight for 'goal' (because the overall goal will be defined by the colours being used), whereas consumables, such as paint-bottles, have a low weight for 'action' (because one can only pick up or pour using these objects) but a high weight for 'goal' (i.e., mixing specific colours of paint). Finally, a database of Cases is defined. Each Case combines objects and actions to define goals, in a hierarchical structure. This helps to determine the most likely current activity; if there are several activities which are equally likely, then the system will wait for more information before making a decision (but will indicate which activities it is considering). From this definition of current activity, a second database is used to provide information to the user on the task-at-hand. In this version, the task information database is populated with guidance on paint colour mixing (see Figure 2).

The User Interface, in this application, uses a visual display screen, showing the proportions in which colours can be mixed from the paints being used. Figure 2

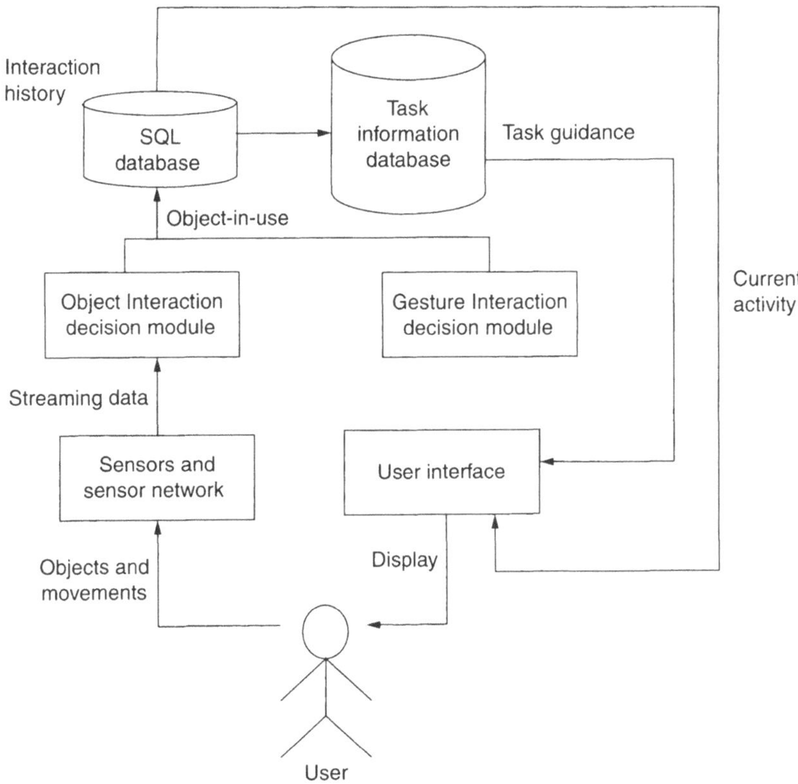

Figure 1. System Schematic.

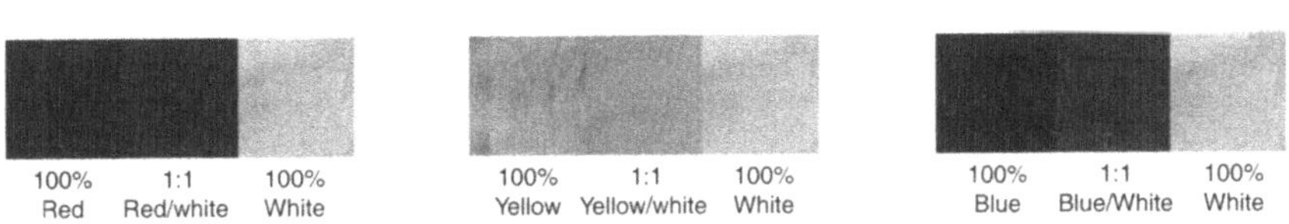

Figure 2. Mixing patterns.

shows examples of these displays, which are presented to the user when the Red and White paint bottle is picked up (or Yellow and White, or Blue and White). The User Interface also uses the 'current activity' to provide feedback on the user action, in the form of simple icons, e.g., it shows a brush in a pot when the user is washing a brush, or a brush pointing downwards when the user is painting.

Experiment design

Two experiments were designed to explore the impact of real-time feedback on users' activity. At first, the mixing of chemicals was suggested after talks with a

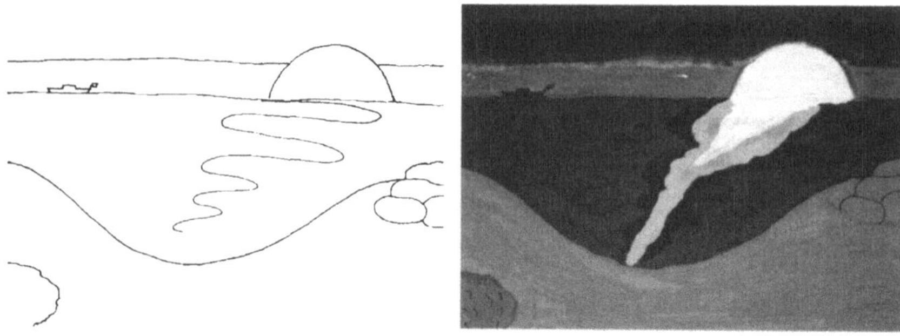

Figure 3. Sketch to be completed and painting to be copied.

colleague in the pharmaceutical industry. This however would require participants to be pharmacists with appropriate training, and access to chemicals. This would also require special tests to be conducted to ascertain if the correct chemical compounds had been created. The idea of mixing chemicals led to the experiment design: mixing primary colour paints to achieve a colour match while copying a painting. Help can be provided to instruct users how to mix the paints to achieve a shade of colour in order to complete simple pictures, as shown in the right on figure 3. This required the system to identify: what paints the user is using, and what the user is doing, i.e., painting, mixing, cleaning the brush, and selecting the paints.

Dependent variables

In terms of the user performance, the time to complete the painting, the relative time spent performing each of the actions, and an objective measure of the accuracy of the completed painting were used. In terms of 'time', the output from the recognition system was used to identify the sequence and duration of tasks. In a pilot study, this output was compared with manual activity-sampling from of videos of performance, and a correlation >95% was found. In terms of 'accuracy', sections of the individual paintings were scanned and analysed using PaintShop Pro to determine what proportion of red, blue and green was present. Comparing these values against the baseline defined by the original painting, provided a score of matching between the two paintings. Subjective opinion on task performance was collected using the NASA-TLX workload assessment form. Tests for normality, using Skewness and Kurtosis, were conducted and ANOVA is conducted for parametric data and main effects were explored using Tukey tests.

Experiment one

In the first experiment, participants were asked to copy the complete painting. Participants were assigned to conditions of 'No Support' (in which they completed the task with the objects but had no computer support), and 'Support' (in which they completed the task with computer support, automatically in response to recognition of objects and actions). This provides a comparison between providing automatic help and performing the task completely unaided. If there is not benefit in terms of providing help for the task, then there is little point in continuing development.

Participants

Experiment one required 10 participants to produce a complete painting each. This provided 100 individual paint mixes for assessment. 5 participants performed in the 'Support' condition and used the help provided by the recognition system, and 5 participants performed in the 'No Support' condition and received no help.

Results

Time

ANOVA of the results showed a main effect of experimental condition [$F(1,42) = 5.812$, $p < 0.05$], and a main effect of activity [$F(6,42) = 16.427$, $p < 0.0001$]. Painting time was reduced from 984 seconds in the 'No Support' condition to 507 seconds in the 'Support' condition; a reduction of 51%. Post-hoc tests revealed that 'painting' took significantly longer than other activities. ANOVAs revealed main effects for 'painting' [$F(1,6) = 8.391$, $p < 0.05$], 'holding paints' [$F(1,6) = 6.614$, $p < 0.05$] and 'consulting display' [$F(1,6) = 18.451$, $p < 0.01$]. Holding paint bottles was increased in the 'Support' condition because participants interacted with the system by holding the paint bottles to request information.

Accuracy

In terms of 'accuracy' of painting, there is a better match between the 'target' colour and the painted colour for the 'Support' condition, when compared with the 'No Support' condition. This is particularly evident for the set of colours {red sky, yellow sun, grey stones, green bush, brown sand}. Overall, this led to significantly better colour matching in the 'Support' condition, i.e., the 'Support' condition reduced average error of colour mixes 31%, when compared with the 'No Support' condition. ANOVA revealed a main effect of condition with [$F(1,118) = 4.890$, $p < 0.05$].

Subjective workload

There was no effect of condition on workload [$F(1,60) = 0.264$, $p > 0.05$], though a main effect did exist between NASA Categories [$F(5,60) = 6.705$, $p < 0.0001$]. Post-hoc analysis, using Tukey tests, showed mental effort to be significantly higher for the 'Support' condition ($p < 0.05$), implying that the participants thought that using the real time help system based on object interaction took more mental effort than completing the task in the 'No Support' condition.

Conclusions

Experiment One showed clear differences between using the object interaction system to access help and not using it. The colour match was much closer in the 'Support' condition and the time taken to complete the painting was reduced. This shows that automatic provision of help improves performance. However, participants rated the 'Support' condition as being more mentally demanding than painting without a help system.

Experiment two

Experiment two uses a part-task approach. It is assumed that in many industrial applications, automatic support could be provided to unskilled operators (either due to limited training or to unfamiliarity of the specific task at hand). In this study, participants were asked to complete parts of the painting under either automatic support or manual access to the same information.

Participants

18 undergraduate students from different backgrounds were required to produce 4 paintings working with a subset of colours (each participant worked on 2 paintings before passing them on to the next participant). This provides 44 individual paint mixes for evaluation. Each participant is given areas of the painting to mix the colours and paint. In this way the paintings are produced in a production line manner. 9 participants performed in the 'Sensor' condition and used the help provided by the recognition system while 9 participants performed in the 'Manual' condition and manually navigated the help. Both conditions provided the same information in the same format.

Results

Time

ANOVA showed a main effect of condition [$F(1,12) = 4.868$, $p < 0.05$]. It took 5573 seconds to complete two paintings in the 'Sensor' condition and 7123 seconds in the 'Manual' condition. This gives us a decrease in time of 21.76% for the 'Sensor' condition.

Accuracy

ANOVA showed no main effect of condition [$F(1,38) = 0.750$, $p > 0.05$]. However, ANOVA of the data partitioned by colour revealed a main effect of colour

[$F(9,20) = 2.880$, $p < 0.05$]. The analysis of colour reproduction found that certain colours tend to be more difficult to mix than others. 'Hard-to-mix-colours' were defined as having an error value in the upper quartile, and 'easy to mix colours' as having an error value in the lower quartile of the data. Thus, 'easy to mix colours' are {Dark Blue water, Grey Water, Yellow Water, Yellow sun, Green Bush, and Red Sky} and 'hard to mix colours' are {Grey Stones, Light Blue Water}. The 'Sensor' condition lead to a significant reduction in mean error for 'hard-to-mix-colours' [$F(1,2) = 100.119$, $p < 0.01$].

Subjective workload

There was no main effect between the two experimental condition using an ANOVA [$F(1,96) = 1.043$, $p > 0.05$], though a main effect of NASA Categories [$F(5,96) = 6.112$, $p < 0.0001$]. Post-hoc tests, using Tukey tests, revealed higher rating of that mental effort for the 'Sensor' condition.

Conclusions

The participants had access to the same information in the same presentation format in both conditions. In the 'Sensor' condition, access to the information required the system to recognize user activity and provide feedback in the form of the mixing charts. The results show that the 'Sensor' condition led users to complete the task in less time than the 'Manual' condition, and to produce better colour mixes for the 'hard-to-mix-colours'. As with the first study, mental workload was rated significantly higher for the 'Sensor' condition than for the 'Manual' condition.

Overall conclusions

The experiments have explored how a sensor-based interaction affects the way in which participants complete a task. The metrics that were measured during the experiments were: Time, Colour mix accuracy, and Work load.

Time: Experiment One showed a main effect of time, with participants in the 'Support' condition showing significant reduction in time spent painting, while increasing the time spent holding paint bottles and consulting the help screens. The 'Support' condition also showed a general trend to reduce the time spent mixing and pouring paints, presumably because help was provided and pro-actively pushed to the user. When both conditions had help available in the same presentation format, Experiment Two , the 'Sensor' condition was faster.

Colour Mixing: There was a main effect of colour mixing for Experiment One, such that the 'Support' condition reduced colour match error by 31%. In experiment two, the same help is provided in the same presentation format and so this effect is lost. However, there was a significant main effect showing that the 'Sensor' condition significantly reduced the error for 'hard-to-mix-colours'.

Workload results: Overall, the NASA TLX workload showed no main effect for either experiment, which shows that the workload was similar for all conditions. However, the highest scoring category of the NASA TLX for both experiments was 'mental effort', showing that the painting task was viewed as mentally demanding, and that this perceived 'mental effort' increased with computer support.

Discussion

A system using activity and object recognition was evaluated through a simple task. Two experiments were presented, in which participants were provided with help when the system recognized their activity and the objects being used ('Support' condition). This condition was compared with either 'No Support' and 'Manual' access to the help. The experiments suggest that participants enter into a negotiation with the system for both experiments during colour mixing. Participants realised that the more they interacted with the paints they believed were required to mix a colour, the more likely the colour mixing chart they required would be shown. If a chart was shown they did not want, or they had guessed the colours incorrectly they would use the chart to mix another colour on the painting. This interaction arose because the help was being pushed to them. Participants in the 'Manual' condition of experiment two had to actively search for the appropriate help pages and seemed reluctant to change their attention from the primary task of painting, and would not deviate from the colour they wished to mix next. This reluctance meant that they took longer to complete the painting. The user develops an understanding and can see a link between help provided and their interactions with the objects. The results indicate superior performance, in terms of time and colour match, with 'Support'. Participants, however, did rate mental demand as higher in the 'Support'. One explanation of this was that the system does not always get it right, users do feel a level of frustration as they cannot manually override the help provided and go in search of a closer matching mixing chart.

Overall these two studies indicate a value in using sensor-based interaction to embed information provision into the task-at-hand. As the person performs a particular activity, so appropriate information is made available to assist them. The differences in rating of mental demand suggest that the potential shift in locus of control (to the 'Support' being automatically generated) requires the user to determine the relationship between the information and the task.

References

Amft, O., Junker, H., Troster, G., 2005. Detection of eating and drinking arm gestures using inertial body-worn sensors, *Ninth IEEE International Symposium on Wearable Computers*, Los Alamitos, CA: IEEE Computer Society, 160–163

Bellotti, V., Black, M., Edwards, W.K., Grinter, R.E., Henderson, A. and Lopes, C., 2002. Making sense of sensing systems: the five questions for designers and researchers, *Proceedings of SIGCHI Conference on Human Factors in*

Computing Systems: changing our world, changing ourselves, New York: ACM, 415–422

Bristow, H. W., Baber, C., Cross, J., Knight, J.F., Woolley, S.I., 2004. "Defining and Evaluating context for wearable computing", *International Journal of Human – Computer Studies, 60,* 798–819

Nicolai, T., Sindt., T., Holger, K., Witt., H., 2004. Case study of Wearable Computing for Aircraft Maintenance, *2nd International Forum on Applied Wearable Computing*, Zurich, Switzerland

Philipose, M., Fishkin, K. P., Perkowitz, M., Patterson, D. J., Fox, D., Kautz, H., Hahnel, D., 2004. Inferring Activities from Interactions with Objects, *IEEE Pervasive Computing, 3,* 50–57

Rogers, Y. and Muller, H., 2006. A framework for designing sensor-based interactions to promote exploration and reflection in play, *International Journal of Human Computer Studies, 64,* 1–14

Schwirtz, A. and Baber, C., 2005. Smart tools for smart maintenance, *The 2005 IEE and MoD HFI-DTC Symposium People and Systems – who are we designing for?* London: Institute of Electrical Engineers

Westeyn, T., Brashear, H., Atrash, A., Starner, T., 2003. Georgia Tech Gesture Toolkit: Supporting Experiments in Gesture Recognition, *ICMI'03*, New York: ACM Press, 85-92

A STUDY OF THE FACTORS LEADING TO SUSTAINABLE GROUP DECISIONS DEVELOPED IN A CREATIVE DECISION SUPPORT SYSTEM

Linda Johnstone Sørensen[1] & Garrick Jones[2]

[1] *University of Southampton, Transportation Research Group, School of Civil Engineering, Southampton, SO17 1BJ*
[2] *London School of Economics, Institute of Social Psychology, London, WC2 2AE*

Problems facing modern society are unstructured and complex rendering traditional decision support with the formation of single solutions inadequate. Decision support systems (DSS) have recently sought to address this issue by providing creative decision support to facilitate the generation of several alternative solutions to a problem, with particular emphasis on groups. Although it is recognised that decision support can improve the quality of decision making, as of yet there has been no exploration of the sustainability of these decisions. Is group decision support akin to management fads that over time fade through a lack of durable results? This paper presents a study which sought to take the first steps to inform this question by investigating retrospectively the sustainability of the decisions made by a group supported through a Group Decision Support System (GDSS). Through an explorative, qualitative research process we sought to understand whether group decision support provides fundamentally short-term gain, or if long-term effects are possible. In-depth interviews were conducted firstly with experts in decision-making theory and support; and, secondly, with a group that had received decision support in a creative group decision support environment. We found that to ensure decisions made in group decision support systems the decisions must be reinforced also after the initial decision has been made to ensure sustainability. This has implications for all decision support aimed at supporting complex decisions where the decision is not an isolated occurrence.

Introduction

Most support for decision making has treated the problem as structured and the decision maker as a single individual. Yet, majority of the challenges facing society and organisations are unstructured and complex involving a diverse range of stakeholders (Meredith, 2006). For this class of problems there might not be a pre-existing solution or historical data which can help to predict the future (Meredith, 2006). Such problems require creative decision making.

The emphasis in the DSS literature has been focused on how informational technology can improve the efficiency of making a decision (Shim *et al.*, 2002). Carlsson and Walden (2000) argues that standard DSS methodology resembles operations research and hence it can be criticised for being to analytical and rational in its approach to decision problems and not fully taking into account the context within which the problem resides. DSS' were originally defined with process statements that lead to an answer, i.e. "what if" questions that allowed the decision-maker to test alternatives and evaluate parameters (Hatcher, 1995). By the beginning of the 1980s there was wide recognition within organisational theory that decisions were better if made in groups rather than by isolated individuals. Group decision support systems (GDSS) evolved in response to this recognition (Shim *et al.*, 2002); and combined DSS theory with psychology (Hatcher, 1995).

GDSS' support the generation of fresh insight to problems and aim to create a sense of common purpose (Phillips, 1988). Although the technical tools the group enjoys in GDSS' supports the group in making explicit that which is often taken for granted, such as experience and intuition, these may also limit the groups thinking about the problem in that software tools can be over-prescriptive (Jones & Humphreys, 2006; Phillips, 1988).

The use of a physical environment aimed at supporting creative exploration of decision alternatives have been in use at the London School of Economics, and elsewhere, since 2004 (Harrison, 2009). Research has established that collaborative authoring of solutions enriches the breath and depth of possible decisions (Humphreys & Jones, 2006). This study aimed to answer the question of whether or not creative GDSS' are merely a management fad or if the decisions made in these environments are sustainable. Collins (2001) concise dictionary define 'to sustain' as 'to maintain or prolong' (p.1518). Sustainable decisions are for the purpose of this study defined as the continued pursuit of the decisions made in the GDSS event. A sample of 11 individuals were selected and interviewed in-depth. The GDSS the participants had experienced was an Accelerated Solutions Environment. Using a grounded theory approach interview transcripts were coded into categories and compared to decision making theory and reported actions to judge level of decision durability. In the following the main points of the theoretical context is elaborated.

The 'making' of a decision

The decision making process begins with an identification of a gap between the status quo and a desired future state (Humphreys & Jones, 2006). Nappelbaum (1997) disagrees with the notion put forward by Herbert Simon that decision making occur in a linear fashion. Nappelbaum (1997) argues in contrast that decision making in fact takes the form of a *circular logic of choice* (see Figure 1).

An individual or group of decision makers move from describing the problem to discussing the means by which it can be solved; to assessing the value of the options generated; through to an evaluation of these options in light of the described problem, thus starting the process once more (Nappelbaum, 1997). The circular

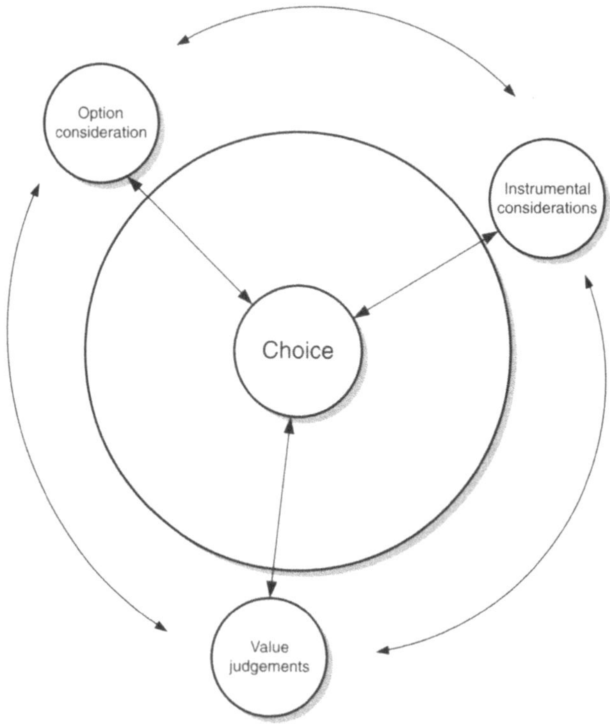

Figure 1. Circular Logic of Choice (adapted from Jones and Humphreys, 2004).

logic continues by narrowing down the options, by going through a *decision spine*, a decision process with five levels of increasing constraints (Humphreys & Jones, 2006). Please see Figure 2 for an illustration.

At the outset of the process all imaginable courses of action are possible. Decision making, and in particular DSS, is normally focussed on developing a *single representation* of the problem, which is done by constraining the representation of the problem through the five levels until only one course of action is left (Humphreys & Jones, 2006). At level five discourses is used to explore imaginatively about possible solutions, while at level four discourse express what the problem is (Jones & Humphreys, 2004). These lead to claims concerning what should be included in the context of the problem. At any given step in the decision making process the part of the context which is relevant must be distinguished from that which is irrelevant (Brezillion and Pomerol, 1999). The latter becomes external knowledge while the former is contextual knowledge. As the decision making process spirals down the spine through the circular logic of choice a part of the contextual knowledge is structured, or proceduralised (Brezillion & Pomerol, 1999). Proceduralised knowledge is built through interaction by extracting information from each individual

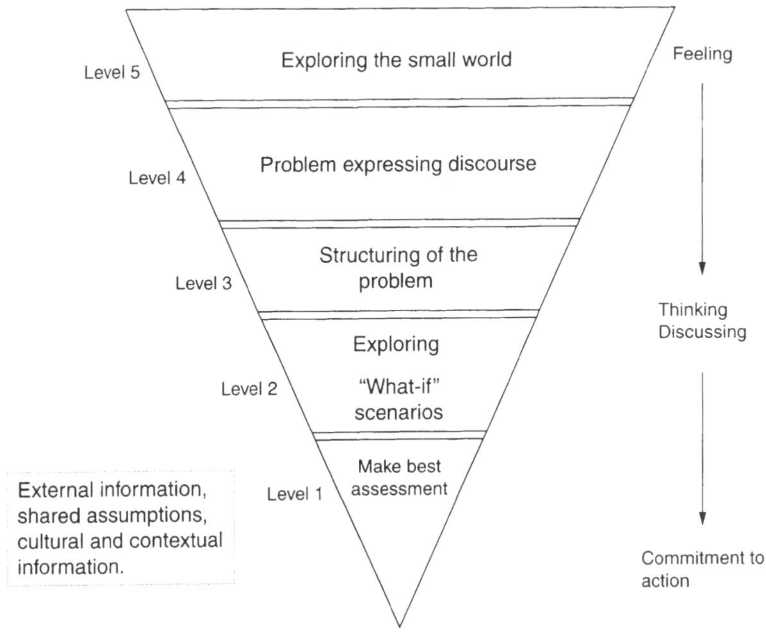

Figure 2. The Decision Spine (adapted from Humphreys and Jones, 2006).

and is jointly structured by them (Brezillion & Zarate, 2004). This interaction context leads to shared knowledge (Brezillion & Zarate, 2004). In a group decision the proceduralised context contains all the pieces of knowledge that has been discussed and accepted (Brezillion & Zarate, 2004). In other words the group comes to an understanding and consensus of the boundaries for the problem; of 'what's in and what's out' (Humphreys & Jones, 2006). Traditional forms of DSS are prescriptive and seek to determine what a decider *should do* and drives the decision makers towards making a choice, to ensure that a decision is made (Brown, 2005). In contrast, creative decision support, such as the ASE, has been developed to prevent a group from narrowing down their options too early and aim to support the group in exploring a multitude of options and arrive at multiple solutions.

Accelerated Solutions Environment

The Accelerated Solutions Environment (ASE), a form of GDSS, enables decision support through its physical space (Jones & Lyden-Cowan, 2002). The ASE was developed in support of Eden's (1990) emphasis on the importance of ensuring that a group can work without being hindered by their physical environment (Philips, 1990). The ASE is a purpose built environment, supporting information systems, design led processes and multimedia platforms to enrich group decision making (Philips, 1990; Jones & Humphreys, 2004). The emphasis on supporting exchange of information within the group is essential to achieve high quality solutions. Significantly, the information exchanged by the group is instantly made available to them by facilitators. A team of facilitators work with the environment and the group to support the decision making process.

Method

Although a significant number of studies has established how group decisions are made and how GDSS' can be utilised to aid group decision making (Jones & Lyden-Cowan, 2002; Philips, 1998), as of yet the authors have not identified any which investigate whether these decisions stand the test of the 'real world' outside the ASE environment and whether the decisions subscribed to are carried out. Due to the lack of literature in this area an explorative qualitative approach was taken. The aim of the research was to explore the participants' experience of the ASE and their subsequent decision related activity. A grounded theory approach was deemed particularly useful to achieve this aim (Strauss & Corbin, 1998). Two samples were drawn for in-depth interviews. The first sample contained 5 subject matter experts who between them have more than one hundred years of experience of both academic and practitioner application of decision making theory and decision support systems. The second sample of 6 was drawn from a group who had taken part in an ASE decision support event in 2003. These were all key decision makers and problem owners in their area as private and public leaders in large corporations. The aim of the decision making event was to establish rejuvenation and redevelopment goals for a particular region of the United Kingdom in which large resources would be distributed across the different stakeholders. The regeneration was to be of a social and economical character. Interviews were transcribed and coded using a thematic analysis. The codes were compared for similarities and grouped in categories. A randomly selected set of codes and transcript material was tested for inter-rater reliability by an objective researcher. The findings were interpreted using the decision making literature and subject matter expert interviews. As the participants were largely senior and held public or private positions of authority their subsequent actions with regards to the decisions made in the ASE event were considered in media and public records. These, however, are not reported in detail here to honour confidentiality agreements signed with participant interviewees.

Limitations

The sample of 11 is an obvious weakness of this research, as any small sample would be; however, as the individuals who took part in this GDSS process all held positions of authority in their areas they were extremely difficult to reach and it is difficult to see how a larger size could reasonably be drawn. A further limitation is the retrospective consideration of the event with obvious implications for the participants memory of what actually transpired. Participant's accounts may also be influenced by achievements made after the GDSS which may not be due to the decision support. The retrospective nature of the study was considered necessary, however, to allow adequate time for the group to act on their decisions. The limited sample size was sought compensated by considering the decision material created by the group in the ASE and by investigating the activities of the group as they were represented by media and other public records. In the following we discuss the main findings of this qualitative study.

Discussion

The main findings were the following categories of codes; the *environment* in which the decision support is delivered, the *participants* interaction, and the *group processes*. These categories were found to impact on the quality of the *decision support* and the decision making content. These in turn influence the *outcomes*, or decisions arrived at, and finally the *sustainability* of these. The interview material was considered together with media and public records to establish whether or not the decisions made in the ASE had been sustainable. The main findings formed a three layer pattern, as presented in table 1 below. Table 1 contains sample statements of interview transcripts which were coded and arranged into the categories presented here. In the following these findings and their implication for decision making theory is discussed.

We found the significance of the environment as part of the ASE to be related to how it supports the group in their discussions by adapting rapidly to their needs. This was highlighted by both subject matter experts and the participants. For instance if group discussions were becoming more focused around a few individuals walls would be moved with tables and chairs to accommodate their work. Similarly, if the group discussion was stalling the physical environment of the room would be altered by the facilitators to guide these individuals back to the main discussion. A further environmental control is temperature and light, both of which were actively adjusted to aid concentration and energise the group when needed. As such the environment played an active part in realigning the group activity. The flexibility of the environment removes physical barriers and allows the facilitation team to direct the attention of the group effectively to different aspects of the problem they are seeking to solve. Please see Figure 3 below for an illustration of the physical ASE environment.

Equally, the participant's willingness to contribute was essential to the success of the decision support. The subject matter expert's emphasised the importance of enabling the group, and the individuals within it, to author their solutions through a bottom-up process of solution exploration. Not surprisingly this ensures commitment both on an individual level and at the group level to the solutions arrived at, regardless of whether the solutions agreed upon are the ones each participant were seeking. The GDSS process, whereby the group is asked to develop and present material ranging from traditional PowerPoint to physical models breaks down barriers between them and open up for a different way of communicating for the group. It creates a shared language. All material created by the group, either as individuals, in smaller discussion groups or for the group as a whole was recorded in video, images, text and sound. In the ASE artists graphically represent problem explorations by the group and video record discussions. This allows important elements of decisions to

Table 1. Summary of main findings.

Level	Category	Sample statement
1	Environment	*– [it was] all environmentally controlled... if you were starting to look sleepy food appeared, and the light level went up and cold air blew, and if you were obviously getting over agitated or bored somebody came around and moved you on to something else, you know, re-aligned your activity. And at the point at which the bodies were starting to splay out a bit, they came and moved the walls.*
	Participant	*you see your input is taken into the process... even if the solution which is reached upon isn't the solution you would have wanted, your insight into how it was reached create a sense of loyalty to the solution and to the decision of implementing it.*
	Group process	*the process quickly bangs you together in all kinds of different formations* *– The mobilization of individuals, the involvement of individuals, the fact that they are involved with creating a solution.*
2	Facilitation	*[They] keep challenging blocks and the barriers that you keep trying to put in [to the solution]. When you're talking about thinking inside boxes or thinking beyond boxes, you're still in a box. You're just moving into another box. Whereas [what we should do is] take the walls away* *– Bringing in different SME's from outside the group to influence their discussions and challenge their views.* *The facilitation team capture the groups discussion in multimedia, as written material, and through graphical representations*
	Rich information	*– it was [all] minuted and noted, there were things going on to whiteboards, there were things going straight on to computers [for presenting back to the group later or for the records], there was a man who was like making cartoons, and he was pictorially notating the conversations, which was just brilliant*
3	Outcomes	*– You create a large degree of ownership. And the ownership which is created there is an ownership which is taken further when the decision is going to be implemented*
	Sustainability	*– working bottom-up to find the solution*

be played back to the group to reaffirm these and to ensure all conflict is resolved at all stages of the process. The material is also made available to the group through web pages and CD's. The subject matter expert explained that the aim of this is to ensure that all problem exploration content is captured and drawn upon by all members of the decision making group before it is proceduralised as a solution in the final stages of the decision support process.

The participants all highlighted how the decision support served to challenge the groups thinking. This is exemplified by interviewee5' who state: *they give you*

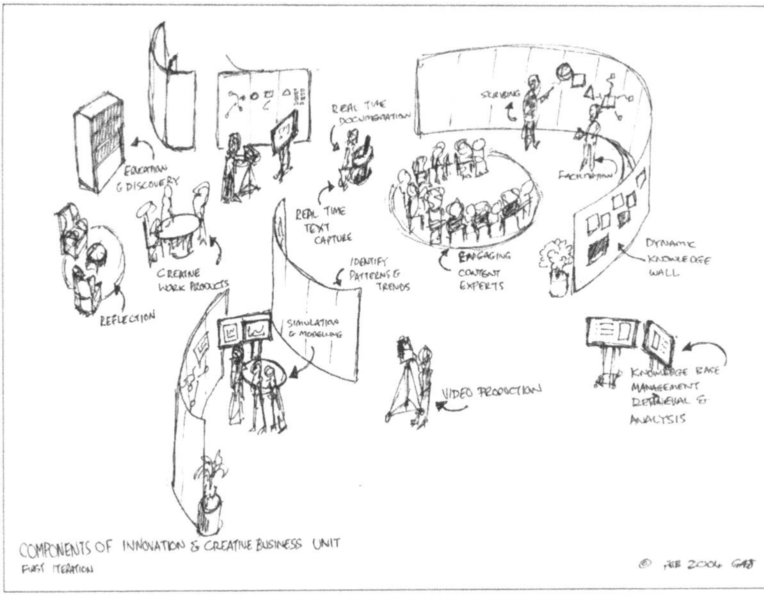

Figure 3. ASE environment, Ludic Group, 2007 © Used with permission.

the opportunity [to create alternative solutions] and then keep challenging blocks and barriers that you keep trying to put in. There is a tendency at the individual and group level to spiral down the decision spine too quickly, which leads to a premature conclusion of the decision making process and prevent a consideration of alternative solutions (Humphreys and Jones, 2004). Hence, an important part of the decision support is aimed at preventing the group from making decisions prematurely. Only after significant exploration of all possible alternatives does the decision support shift focus to ensure that decisions are made. The subject matter experts highlighted that the quality of the decisions is influenced by the degree to which the group's thinking has been challenged. While traditional DSS provide support for the decision-maker primarily at level three and two in the decision-spine creative GDSS additionally provide more support at the fourth level of the decision-spine. This allows the group to obtain as much information about the problem as possible to provide a sound representation of it, as well as to explore "what if" scenarios (Shim *et al.*, 2002; Phillips, 1998).

Jones and Lyden-Cowan (2002) asked whether it is possible to link personal objective with those of the system, in a way that enables the individual to be engaged, to track process and behave in a manner that enables the system to achieve its objectives. Indeed, why does the group take it upon itself to take forward that which is decided? Once decisions have been made are they actually acted upon? Little, if anything, is mentioned in the decision-making literature with respect to long-term effect of group decision making support. We set out to investigate whether decisions made in GDSS environments, such as the ASE, provide sustainable decisions through exploring the experiences of participants and subject matter expert's opinions. Although initially expecting to find similarities with management fads we found that the experiences of the participants of the GDSS event and their later actions, individually and as a group, sustained the decisions they had made in the ASE. This view was strengthened by considering media articles and public records in the areas in which the participants had authority. Many of the solutions they had set out were reflected in concrete actions taken to meet the aims of the decisions arrived at. Based on the definition offered above, and the qualitative data presented above, we therefore conclude; perhaps surprisingly, that the decisions made by the group were implemented and reinforced over the course of the three years that had passed.

Summary

This research investigated the level of implementation of problem solutions arrived at through group decision support systems, such as that provided in the ASE. The study identified that durable decisions were enabled by the physical and media rich environment in which the support was provided, the participants and the group processes evolving among them. The decision implementation related activities, media and public record material as well as the participants own statements lead us to conclude that the decisions they made were sustainable. This, though, may not be the only interpretation as the later success of the participants in their own field may have led them to place greater weight on the decision support event than it would otherwise have been given. We would therefore caution that further research should subject the ASE and other types of GDSS' to further investigation to establish whether or not these are mere management tools or give measurable benefit for decision implementation. Further work should also seek to address the limitations of this study. Although firm conclusions as to the sustainability of decisions created in the ASE are premature these are interesting findings which we believe have implications for how group decisions are addressed and more broadly for complex unstructured problem solving and organisational theory.

References

Brézillion, P. and Pomerol, J.-Ch. 1999. *Contextual knowledge sharing and cooperation in intelligent assistant systems*, Le Travail Humain 62 (3) p.223–246

Brezillion, P. and Zarate, P. 2004. *Group Decision Making: A Context-Oriented View*. In Meredith, R., G. Shanks, D. Arnott, S. Carlsson & I. Prato (Eds) Decision support in an uncertain and complex world. Proceedings of the 2004 WG8.3 International Conference on Decision Support Systems DSS2004, 1–3 Juliet, pp.123–133

Brown, R., 2005. *Rational Choice and Judgement: Decision Analysis for the Decider.* (New Jersey: John Wiley & Sons Inc)

Carlsson, C. and Walden, P. 2000. *Intelligent Support Systems – The Next Few DSS Steps*, Human Systems Management, Vol. 19, 2000, p.135–147

Collins, 2001. *Concise Dictionary*. 21st Century Edition, (Collins: Glasgow, p.1518)

Eden, C. 1990. *Managing the Environment as a Means to Managing Complexity*. In Eden, C. and Radford J. (Eds), Tackling Strategic Problems. The Role of Group Decision Support. (London: Sage Publications) p.154–161

Hatcher, M. 1995. *A tool kit for multimedia supported group/organizational decision systems (MSGDS)*. Decision Support Systems, 15(3), p.211–217

Harrison, A. 2009. *London School of Economics: Box*. EDUCASE http://www.educause.edu/learningspacesch23. Webpage accessed 21.09.09

Humphreys, P and Jones, G. 2006. *Contexts and Process Design in Group Decision Authoring and Communication Support*. In F. Adam, P. Brézillion, S. Carlsson and P. Humphreys, Creativity and Innovation in Decision Making and Decision Support. Vol. 2 (London: Decision Support Press) p.523–545

Jones, G. and Humphreys, P. 2004, *The evolution of Group Decision Support Systems to enable Collaborative Authored Outcomes*. Research Report LML-LUDIC-RR-001, (The London Multimedia Laboratory for Audiovisual Composition and Communication)

Jones, G and Lyden-Cowan, C. 2002. *The bridge: Activating potent decision-making in organizations*. In: F. Adam, P. Brezillon, P. Humphreys and J-C Pomerol (Eds) Decision Making and Decision Support in the Internet Age. (Cork, Ireland: Oaktree Press)

Meredith, R. 2006. *Creative Freedom and Decision Support Systems*. In Adam, F. P. Brézillion, S. Carlsson and P. Humphreys, Creativity and Innovation in Decision Making and Decision Support. Vol. 2. (London: Decision Support Press) p.30–46

Nappelbaum, E.L. 1997. *Systems logic for problem formulation and choice*. In P. Humphreys, S. Ayestaran, A. McCosh, and B. Mayon-White Decision Support in Organizational Transformation. (London: Chapman & Hall)

Phillips, L. D., 1990. *Decision Analysis for Group Decision Support*. In Eden, C. and J. Radford (Eds), Tackling Strategic Problems. The Role of Group Decision Support.(London: Sage Publications) p.142–150

Phillips, L. D. 1988. *Conferencing to Consensus*. In J. Hawgood and Humphreys P. Effective Decision Support Systems. (Gower: Farnham) p.1–12

Shim, J P, Warkentin M., Courtney J.F., Power, D.J., Sharda, R., Carlsson, C. 2002. *Past, present, and future of decision support technology*, Decision Support Systems 33(2), p.111–126

Strauss A. and Corbin J. 1998. *Basics of Qualitative Research. Techniques and Procedures for Developing Grounded Theory*. 2nd Edition (California: Sage Publications)

DEVELOPMENT OF A TOOL TO PREDICT ORGANISED TEAM PERFORMANCE IN PROCESS EXECUTION

M.A. Sinclair[1], C.E. Siemieniuch[1], Michael J. de C. Henshaw[1], R.A. Haslam[1] & L. Evans[2]

[1]*Loughborough University,* [2]*BAE Systems*

The paper describes the development of a tool to predict the success of a team executing a process. It is expected to be used by systems engineers in initial stages of systems design, when concepts are still fluid, including the team(s) who are expected to be operators within the system. Currently, the tool is undergoing verification and validation; to date, the tool predicts well and shows promise. An unexpected finding is that the tool creates a good argument for Human Factors Integration in projects.

Introduction

Slowly, we have developed a tool that is able to predict the performance of teams that are executing a process. Verifications and validations of the tool indicate that it performs satisfactorily, and this paper outlines both the development and the evaluations of the tool to date, though it should be noted that further verifications and validations are planned. These may adjust some of the statements below; in view of this, the paper is conservative in tone.

The original purpose of the tool was to provide designers of military systems with a simple tool to use in the conceptual stages of design (when variables are still variables and not parameters) to help in risk reduction exercises when considering the staffing of processes. Some sample questions for which the tool could help in providing answers are:

- What is the likelihood that this team will be successful in executing the given process?
- By how much can the team size be reduced, before the likelihood of success becomes unacceptable?
- By how much can the attributes of the individuals in the team be reduced, before the likelihood of success becomes unacceptable?

It will be noted that these questions are phrased in terms of success. This is a significant point; 'success' is defined here as executing the process correctly and attaining all of the goals of the process; no reworking, no extra resources, no extra time. 'Likelihood', as usual, is expressed as a probability.

There is still debate about a name for the tool. Hence, it will be called 'the tool' in this paper.

Development of the tool

From the outset, three constraints guided the development of the tool:

- A user-defined constraint: If the tool needs more that two pages of A4 to explain it, it will not be used. This was a unanimous view among a target group of engineers who were interviewed at the beginning of the project.
- A business process constraint: Systems designers are already familiar with, and may be using, techniques such as HEART and CREAM for assessing the reliability of individuals. The tool should incorporate these techniques, to enhance ease of acceptance into design processes.
- A design constraint: At the conceptual stages of design, little will be known about the individuals in the team that executes the process, apart from generic attributes. Equally, the process will be undefined – perhaps just a single flow diagram sketched on a sheet of A4. Hence, the tool must make a minimum demand for input data.

The first constraint has been met, without recourse to minuscule script. The second constraint has been met by creating a three-stage tool; Stage 1 collects data about the attributes of the individuals in the team and the intercommunications necessary for execution of the process. Stage 2 collects an analysis of the process environment, using either HEART or CREAM, or the organisation's in-house technique. Stage 3 convolves the outputs of Stages 1 and 2, and produces a likelihood of success, depending on the binding of the team to the process.

Development of the tool was based on an iterative, user-oriented design process, similar to Extreme Programming. Paper-based versions were developed initially, with subsequent development in spreadsheets. At each step forwards in development, a small-scale study was conducted, usually in real-life situations (e.g. station staff in London Underground, yacht crew, helicopter crew, student project groups). In all, 8 identifiable versions of the tool were developed. The description below is for the latest version.

Starting conditions for the use of the tool assume that the user has a process description available, though this does not have to be in detail. There will be some team assigned to execute the process; its size, and the expected binding of individuals to the process activities that have been identified so far should be known. Finally, there needs to be some level of knowledge about the environment in which the process will be executed.

Structure of the tool

Input data on individuals cover five variables. These variables map to Thurstone's 'Five Factor Model (Thurstone 1934) and to Salas *et al*'s. 'Big Five' model (Salas, Sims *et al.*, 2005).The first is an identifier for each person, to aid the user to understand the intermediate steps. The second is a trustworthiness variable; the dependability of the person to deliver results. The third measures team skills; how constructive the person is in aiding the team to its goals. The fourth assesses the knowledge and skills that the person brings to the process. Finally, the fifth variable assesses the authority of the person within the team. For the first 4 variables, 4-point rating scales are provided. For the last, there is a 5-point scale. Albeit simple, these scales provide adequate sensitivity, as discussed later.

In addition, the proposed communication structure (who talks to whom) is captured, as a matrix of one-way links.

From the trustworthiness, team skills, and knowledge ratings, the tool produces a Performance Shaping Factor (PSF). High-performance people will have a PSF less than 1.0; for it poor performers will be greater than 1.0.

Next, the user analyses the process environment for each individual; this allows for distributed teams; if the team is co-located, only one analysis needs to be performed. This analysis may be performed using either HEART (Williams 1986) or CREAM (Hollnagel 1998). The probabilities of failure that emerge from this analysis are now multiplied by the PSFs from Stage 1 to produce a probability of error for each individual in the team in his/her environment.

Stage 3 happens without necessitating user input. It now combines the values obtained above, the authority ratings, and the communications matrix to arrive firstly at what is called 'interactive probabilities of error' for each individual; in other words, acknowledging peer effect. The initial probabilities of error from the paragraph above are now adjusted by these extra variables. The assumption here is that a person's performance will be influenced by the performances, knowledge, and teamworking capabilities of those who communicate with that person. This might be feedforward, feedback, or observable quality of work.

An algebraic equation performs this step, with the property that if a person communicates with very good people, then that person's probability of error will decrease, and vice versa. Hence, we now have an 'interactive probability of error' for each individual.

Secondly, the team's binding to the process is addressed. It was discovered that three classes of binding cover a range of real teams. The three classes are described below.

- The 'aircrew' team. Consider a helicopter, flying from A to B over hostile territory. The pilot executes the process of flying; the rest of the crew act only as advisors (e.g. 'A SAM has been fired'), but do not play a part in flying the helicopter.
- The 'Boatcrew' team. Consider a rowing eight (nine, with the coxswain). From start to finish, each person has a specific task, and cannot perform anyone else's task. Consequently, the absence of any crew member ensures failure.
- The 'Omnicompetent' team. Here, anyone can perform anyone else's task, and may do so in executing the process.

Likelihoods for each of these teams are presented for the choice of the user.

An example of the tool in use

A verification study concerned the preparation of 26 Tornado aircraft for active service in the Middle East. A team of 5 RAF personnel prepared the radiation subsystems; radar, IFF, ECM, radio, etc. Because of the compressed timescale to deliver these aircraft, each member of the team was allocated a 'go-fer', drawn from maintenance personnel qualified for other aircraft who fetched, carried, held, prepared paperwork, etc. for his/her team member.

The team worked as an omnicompetent team. All were rated highly on the tool's variables, with equal authority, save for one who exhibited symptoms of 'burn-out', and whose ratings were downgraded accordingly.

The process took place in a hangar, round the clock with the team and go-fers catnapping until all 26 aircraft were delivered, on time. This environment had the effect of increasing the basic error rate for each individual in the whole team, to p(error) = 0.02 for team members and 0.4 for gofers.

Combining all the information above, the tool predicted a p(success) for the whole team of 0.1; in other words, 2.6 aircraft should have undergone the process with no rework, no extra resources, etc. In reality, the success rate was 3 aircraft (the rest required some degree of rework).

Current state of the tool

Due to the adoption of a form of Rapid Prototyping as a development methodology, the tool is now in spreadsheet form, in its eighth version. Version 6 was in a testable state; verifications and validations have taken place with versions 6, 7, and 8, and these are described below.

Functionally, the tool is being made more presentable and understandable to the intended user. This is necessary for testing of the user manual, currently being written. In accordance with the first constraint above, a two-page version and a more detailed version are being produced.

At the moment, the tool is constrained to deal with teams of ten or less members. This is a developmental constraint; it is expected that when the tool is deemed to be fully tested, the constraint will be removed. With cognizance of this constraint, the tool appears to be able to deal with a number of different scenarios; as shown later in the verifications, the teams considered range from military teams to system development teams, to the Board of Trustees for a pension fund. It can accept distributed teams, and teams where some members do mainly physical tasks whereas others do mainly cognitive tasks.

Verification and validation of the tool

In other words, 'Does it work?'. The evidence so far is that it does. A second question is, 'Is it sensitive enough?' Again, the indications are that it is. This sensitivity issue is discussed first below Normally, one would then describe the verification (does it do what we say it does? Does it behave as we might expect?) of the tool first, then its validation (does it predict in the real world?). Because of an unexpected finding during the validation phase, it seems better to discuss the validation first. then the verification.

Sensitivity of the tool

For a team of two, using CREAM to categorise the environment, 1360 values may be obtained between 0.0 and 1.0. For a team of 10, using HEART the range of values is approximately 57 billion. There appears to be sufficient sensitivity for most purposes.

Validation of the tool

To date, 14 validation exercises have been carried out, summarised in Table 1. All are historical cases, with the results known; i.e. validation-by-criterion. However,

Table 1. Aggregated data for validations. Minimal information is included for commercial and individual confidentiality.

#	Team process	p(success)	Comment on result
1	Insert new FCS in tank	0.7552	"That looks good."
2	Dev't of control system for UAV	0.5906	"OK, if a little generous"
3	Dev't of systems Health Mgmt System	0.8713	"I'm happy with that result – perhaps a little bit high"
4	Create engineering Technical Demonstrator	0.9195	"Result is OK; perhaps a bit high"
5	Dev't of comms system for navy ship	0.9464	"Rings reasonably true"
6	Bid prep'n for US DoD ITT	0.4541	"Result looks OK; wouldn't want to argue with it; perhaps a little bit high."
7	Execution of a Des & Build project for M.Eng degree	0.4875	"Result is a bit low"
8	Deliver HFI to manufacturing Technical Demonstrator	0.9234	"OK; but doesn't account for a weak team member. Got it together because of the efforts of the rest of the team"
9	Dev't of guide for UK Gov't dept.	0.9007	"That's OK"
10	Software development for UK NHS	0.5984	"Estimate is a bit low; would have expected about 0.75"
11	Development of a UAV ground station	0.4831	"A bit low – would have expected around 0.7"
12	Management team in University	0.8287	"Result is OK, but this isn't a normal team"
13	Pension Bd of Trustees managing fund	0.9968	"That's about right"
14	Preparing aircraft for Gulf War	0.1032	"About right – only 3 of 26 went through without rework"

Table 2. Kolmogorov-Smirnoff test for departures from a ideal distribution. Getting 3 predictions wrong represents a proportion of 0.214. According to table E of Siegel (Siegel and Castellan 1995), this indicates that the Null hypothesis (no departures from Ideal) is not rejected (p > 0.05).

Classes	Ideal cumulative	Actual cumulative	Proportional difference
Miss	0	3	0.214
Correct predictions	14	14	0

Table 3. Binomial test for bias in predictions. Only the first two columns were used for this test. According to Table D of Siegel (Siegel and Castellan 1995), this indicates that the Null hypothesis (no bias in predictions) is not rejected (p > 0.05).

Prediction below user's opinion	Prediction above user's opinion	Predictions deemed exact
3	4	7

in all cases but the last the criterion was subjective, since the processes were not repeated.

A Kolmogorov-Smirnoff test (Siegel and Castellan 1995) was carried out to discover any significant departure from accuracy (note that a Chi-squared test is not appropriate due to Cochran's criterion (Cochran 1954)). This is shown in table 2 below.

A further, Binomial test (Siegel and Castellan 1995) was executed on those estimates deemed above and below, to test for a bias in the predictions. Table 3 shows the results of this.

We may conclude from these tests and tables that the tool can produce reasonable predictions.

Verification of the tool

To date, some 350 verification tests have been carried out on these teams, changing at least one variable for each test. Further verification tests are under way. To date, two particular teams have been investigated in some detail. Both are teams of four; the difference between them is in the communication patterns. In a 'Linear' team, the members are arranged linearly, with two-way communication between adjacent members (similar to a production line). In a 'Cocktail' team the arrangement is that of a star, with a central person communicating with all the others. The others have partial communications between them; the net effect is that each team member has a different number of communication links. A further restriction on these tests is that all the tests have been performed using CREAM to characterise the environment.

From all of these tests, it seems that some general conclusions can be drawn.

1. Given that the tool has been constructed as a simple technique, with no feedback loops and no 'if-then' rules, it is gratifying that the behaviour is as expected; trends are consistent, and no test produced a prediction outside the range 0.0–1.0.
2. It is striking that in all sets of tests, a well-designed working environment is the biggest contributor to p(success). The other variables in the tool become important as the working environment degenerates, but they cannot make up for it. In a good working environment, team variables are relatively unimportant; even when the team is of poor quality, p(success) > 0.9. For a military environment, this is equivalent to saying that if the Defence Lines of Development have delivered in full, taking full account of human factors integration, then successful performance of the process is assured (note that this is not saying, 'Victory is assured').
3. In poor working environments, the most important variable is the quality of the team members. As long as there are several high quality members in central roles, able to communicate with the other members, then a level of performance (p(success) ≈ 0.25) can be achieved. However, even with a high quality team, it is not possible to lift p(success) above 0.35.
4. A good team is always better than 4 individuals.

Of course, the truth content of these statements depends on the validation studies above, and more of these must be carried out. But, insofar as there is truth in these statements, findings 2 and 3 together have a significant corollary; they provide a strong argument for the importance of Human Factors/ Ergonomics in engineering projects. One could state that, 'To the extent that HFI is ignored in systems design, greater rates of failure for those human processes are guaranteed'. Furthermore, this tool could provide a probability estimate for failure.

The exploitation of this is evident; currently, attempts to justify the need for resources to be devoted to Human Factors Integration are largely made by appeals to instances from the past; Hendrick's article is a good example ..(Hendrick 1997), as is Oxenburgh, 1991. However, the argument above does not rely on instances; it is an *ab initio* argument, in theory applicable to any team scenario. Its validity comes from the validation tests, discussed earlier. It is likely that those responsible for bids for work will be able to turn the statements in the bullet points immediately above into risk and cost estimates that will provide strong arguments for improved funding of HFI.

Conclusions

In recognition of the fact that further testing of the tool is required (particularly validation tests), conservative conclusions are in order.

- A tool with some predictive power has been produced. However, it is not yet clear how powerful the tool is, nor how extensive is its range of application.

- It is believed to be the only tool able to provide estimates of team performance available for designers and human factors experts to use in the early stages of design.
- Planned developments to the user interface should make the tool usable by non-experts in human factors. This is believed to be an asset; if the engineer's own use of the tool shows that human factors issues must be addressed, this is likely to be a convincing argument.
- That the tool is in EXCEL spreadsheet form is good, but insufficient; it is necessary that a web-based version is available, with a management process associated with it. Some plans exist for this.

Acknowledgements

This tool has been developed within the ambience of an Engineering Doctorate. Thanks are therefore due to the Engineering and Physical Sciences Research Board and the sponsor, BAE Systems, without whose funding this would not have happened. Many, many other have provided insight and knowledge to this exercise; those most directly involved in this are Dr Siemieniuch, Prof Henshaw, and Prof. Haslam, all of Loughborough University, and Mr. Evans of BAE Systems. Their contributions are saluted.

References

Cochran, W. G. (1954). "Some methods for strengthening the common Chi-squared tests." *Biometrics* 10: 417–451

Hendrick, H. (1997). "Good ergonomics is good economics." *Ergonomics in Design*(April): special insert

Hollnagel, E. (1998). *Cognitive Reliability and Error Analysis Method (CREAM)*. Den Haag, Elsevier Science

Salas, E., D.E. Sims, *et al.* (2005). "Is there a "Big Five" in teamwork?" *Small Group Research* 36(5): 555–599

Siegel, S. and N.J. Castellan (1995). *Nonparametric statistics for the behavioural sciences*. London, McGraw-Hill

Thurstone, L.L. (1934). "The Vectors of Mind." *Psychological Review* 41: 1–32

Williams, J.C. (1986). *HEART – a proposed method for assessing and reducing human error*. 9th Advances in Reliability Technology Symposium, University of Bradford, UK

INCLUSIVE DESIGN

A FRAMEWORK FOR ESTIMATING DIFFICULTY AND EXCLUSION FROM TASKS DURING INCLUSIVE DESIGN

David Nery, Pat Langdon & John Clarkson

Engineering Design Centre, Department of Engineering, University of Cambridge, Cambridge, CB2 1PZ, UK

Measuring capability variations within the population is an important process that can be used to support decision making regarding the inclusivity of design for all users, thus allowing the level of exclusion to be defined very early and throughout the design process. Our hands often represent a central feature of the human-task interaction, and therefore, variations in the capabilities of the hands has the potential to exclude people from all or part of the tasks they perform. Data is presented from the performance of 15 people in one of three age groups (18–40, 41–64 and 65+). Using a classification system for defining hand actions the prevalence of different grips in response to a range of physical task demands was mapped in a way that allowed capability to be measured against other variables such as task quality. This was found to enhance the granularity with which exclusion could be both measured and predicted.

Introduction

Inclusive design research and practice seeks to make work and everyday tasks more accessible to all. The analysis of human characteristics for the purpose of user centered design is a central theme within "inclusive design" research and practice. Using this inclusive design approach requires a detailed understanding of the functional capability of the users relative to the particular task demands. Within the context of inclusive design, exclusion occurs when the task demands exceed a person's capability to effectively respond to them. Variations in capability have been identified across the population which has included age related reductions in a range of capabilities, including muscular strength (Voorbij & Steenbekkers, 2001) and injury responses to particular physical task demands and poor work postures (Tuomi *et al.*, 1991). It would appear from this research that the older workers did have reduced capability relative to the younger workers in terms of their muscular strength and injury experiences. In contrast, there are equally significant injuries and conditions which impact on capability which are not age dependent. Examples of this have included arthritis, strokes and cerebral palsy (Sears & Young, 2003). These seemingly contrasting results tend to support the view that capability changes across the population are not driven by a one dimensional factor of ageing. Rather, that there are a range of factors which may impact on the capability ranges of the population. The notion of inclusive design within the context of the current work is to examine hand functional capability across a wide range of the population to appreciate the variations in capability for the whole population with a view making tasks more accessible to all. There is an increase in the proportion of older people in the workplace and the community in general and this trend is predicted to continue for many nations with the global increase in the proportion of people over 65 years set to triple by 2051 (United Nations Population Division, 2001). This global trend of an increase in the proportion of older people requires an analysis of the functional capability of this changing population to provide an objective assessment of the capabilities of the whole population. The reduction in capability for many people within the population contrasts sharply with the broad proliferation of many products and task demands that are beyond the capabilities of these people. This mismatch between user capability and task demands can result in people being excluded from part or all of their tasks that have demands that are in excess of their capabilities despite these people providing enormous benefit to industry in terms of their skills and knowledge. This issue has the potential to have a high personal and financial cost to the individual if they are marginalized and cannot safely and effectively complete their work and everyday tasks. This will also have a profound impact on the business in terms of retaining its "knowledge capital" which is vital for sustaining business operations and retaining the growth potential of an organization.

In addition to the multi-faceted nature of the capability-task demand relationship, there are significant capability variations both between people and within people over time. That is, the nature of the capability-task demand relationship is multi-faceted and the rate and rate and type of changes being highly variable between people over time. The current research has focused on the capability of the hands as a mechanism for focusing the discussion a specific set of functional capabilities. The hands are the part of the upper limbs that provide the opportunity to have tactile contact between us and the tasks we are performing. The fine bony architecture of the hands provides a high degree of dexterity whilst also being capable of forceful gripping actions. In many cases the hands represent a central feature of the human-task interaction. Being able to measure the hands functional capability at the individual and population levels is a significant issue for predicting exclusion from work and everyday tasks. The extent of an individual's hand functional capability has the potential as a measure to determine if they can perform a particular level of task demands. The hands are remarkably malleable which creates challenges if one were to measure each degree of freedom of movement for varied hand movements (assuming this were technically possible) during a task where there could potentially be many changes in the capability of the hands. That is, the very things which provide the hands with their highly flexible and variable capabilities also create challenges for the measurement of those aspects of the hands performance. Some of the early work undertaken in the area of classifying hand movements was undertaken by Napier (1956) who qualitatively differentiated types of grips according to the functional capabilities of each grip type. More recent work has identified other attributes of different classes of hand actions such as maximal grip forces in different hand grips (e.g. Kroemer, 1994). In contrast some research has used epidemiological data to determine hand / arm capability limits (e.g. Silverstein, *et al.*, 1987) and this data has in turn been used to estimate task demands limits beyond which people could become excluded through injury. The difficulty with using hand capability injury limits as a basis for defining exclusion is that the injury limits in many ways represent the extreme level of exclusion. In reality people could be excluded from a task based on their limiting capabilities (e.g. having arthritis and not being able to push the buttons on a mobile telephone), but, this level of activity may not be sufficiently demanding for them to become injured. Therefore, a more granulated metric of exclusion is required which can be used to identify exclusion for sub maximal task demands that may not necessarily cause injury but are sufficiently demanding to cause exclusion from the effective performance of a task.

Given different grips have a range of performance-based attributes (e.g. strength and dexterity) the selection of a particular type of grip has the potential to modulate the hands' performance. Figure 1 illustrates a two dimensional framework for

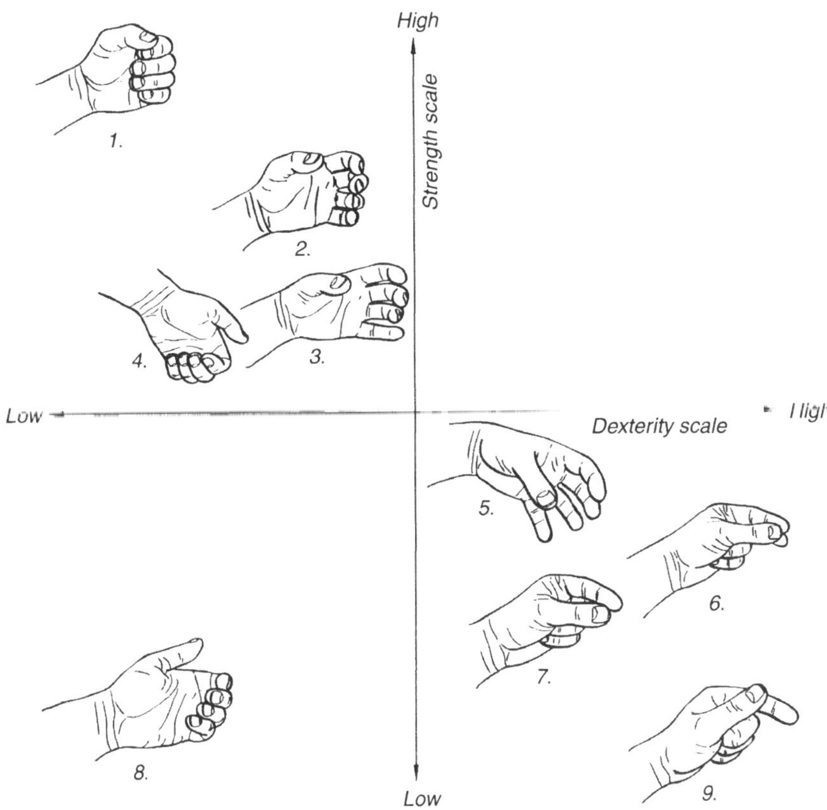

1. Power grip closed *2. Power grip open* *3. Power grip open - finger tips*
4. Non-gripping - hook action *5. Pincer grip - three plus ginger pads* *6. Pincer grip - two finger tips*
7. Pincer grip - two finger pads *8. Non-gripping - cradle action* *9. Non - gripping - finger push*

Figure 1. Hand actions differentiated by their strength and dexterity attributes.

differentiating hand actions based their dexterity and strength attributes. The analysis of this relationship will determine if there are patterns associated with the selection of grips by people and to establish if this selection of particular hand actions is systematically related to the age and capability of people and the tasks they are performing. The patterns of selected hand actions will ultimately modulate performance in ways that can allow us to estimate if people can perform a task to a particular level. The analysis of "performance" will be examined at a more granulated level than just whether the task can be completed or not. The capability-task demand relationship will be supplemented with measures of "Task Quality" (TQ) to provide some clarity about the actual performance level associated with a certain type of hand action in response to a particular level and type of task demand. The measure of TQ is intended to provide more information about the task demand-capability relationship than just whether a person can perform a task or not. That is, as a person approaches the capability boundary for a particular hand action there maybe some decrement in performance to the extent that the task is still completed but the quality of the task performance deteriorates. For example, people can still pour fluid from one container to another but as the task becomes more demanding (e.g. the load being lifted becomes heavier) the quality of that task may deteriorate (e.g. they may spill some water during the pouring activity). This approach will provide more insight into the actual exclusion level which may occur during the performance of a task since the measures of TQ will reflect the quality of performance for particular hand actions irrespective of whether the task is completed or not. The current research is concerned with developing different categories of hand actions that are differentiated according to the capabilities of the different categories. The frequency with which these hand actions are used by people of different ages and capabilities in response to a range of task demands will form the central feature of this work. The task demands vary according the to the strength and dexterity elements of the conditions which is determined by the weight and dimensions of the load and the precision requirements when manipulating the load being lifted. The primary research question being addressed is, whether the identified classes of hand actions account for the actual hand actions used during the various task conditions in this study and how will the interaction of these task demands with individual's capabilities modulate the level of TQ that can be achieved?

Method

There were 15 participants involved in this study which had 5 people in each of the following age groups: 18-40 years, 41-64 years and 65+ years. The median age of people in the three groups was 29, 58 and 72 years respectively. Participants were randomly selected for gender distribution which was 3/2, 2/3 and 2/3 respectively. Each of the participants completed 7 task activities with three trials of each task being completed and the task sequence being randomized for each subject. The individual trial conditions are summarized in Table 1.

Individual measures of hand strength and dexterity were measured using a Jamar Lafayette grip dynamometer and Jamar pincer grip dynamometer (Lafayette

Table 1. The seven hand function testing conditions.

Condition 1	Condition 2	Condition 3	Condition 4	Condition 5	Condition 6	Condition 7
Small Cup (Handle)	Large Cup (Handle)	Large Cup (No Handle)	Container (Light)	Container (Heavy)	Box (Light)	Box (Heavy)

Table 2. The different types of hand capability testing conditions.

Hand Strength	Hand Dexterity
Pincer Grip Strength. (Index finger & thumb)	Left hand dexterity measure
Power Grip Strength (Full power grip grasping)	Right hand dexterity measure
	Left + right hands dexterity measure

Instrumentation Co.). The dexterity measures were derived from a Purdue Peg Board Dexterity Test. The specific testing conditions for the strength and dexterity measures are summarized in Table 2. The order for the performance of the strength and dexterity measures was randomized as was the individual tests within each measure. Measures ofTQ for the container pouring tasks were measured according to the spillage of water during the pouring tasks and the TQ for the container lifting tasks was measured according to proportion of doweling that fell when they were balanced on the object that was being lifted. Participants completed three trials for each pouring condition and 12 lifting elements for each of the plate and box conditions.

The experimental arrangement is illustrated in Figure 2. All tasks were completed in the seated position as illustrated in Figure 2. The box and plate conditions involved lifting the item between four quadrants drawn on the desk (i.e. 12 lifts for each item lifted) with the requirement being to lift (not slide) the item between each quadrant. The experimental arrangement is illustrated in Figure 2. Nine dowelings were placed on the box conditions and 5 on the plate conditions with the task requirement to lift the object between quadrants without the dowelings falling. If a doweling(s) fell during a lift they were repositioned before the start of the next lift. The pouring conditions involved holding the selected item in the dominant hand and pouring the water contents into a standard cup size that was used in all conditions and could not be touched during the pouring task. There were no constraints on how any of the lifted objects could be gripped or handled (e.g. grip type or number of hands being used for gripping during the task).The frequency and magnitude of any spilling was recorded on a table covering that was placed under the containers.

Results

Figure 3 illustrates the prevalence of grips that were used by the 15 participants in each of the experimental conditions of this study. The results indicate that the cup (with a handle) conditions resulted in an increased prevalence of pincer grip actions. In contrast, the containers with no handles, that is, the water containers and large cup (no handle) had an increased proportion of power gripping hand actions.

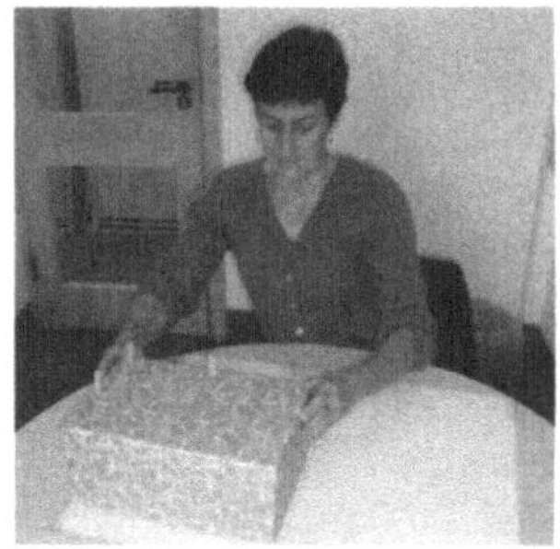

Figure 2. The experimental arrangement.

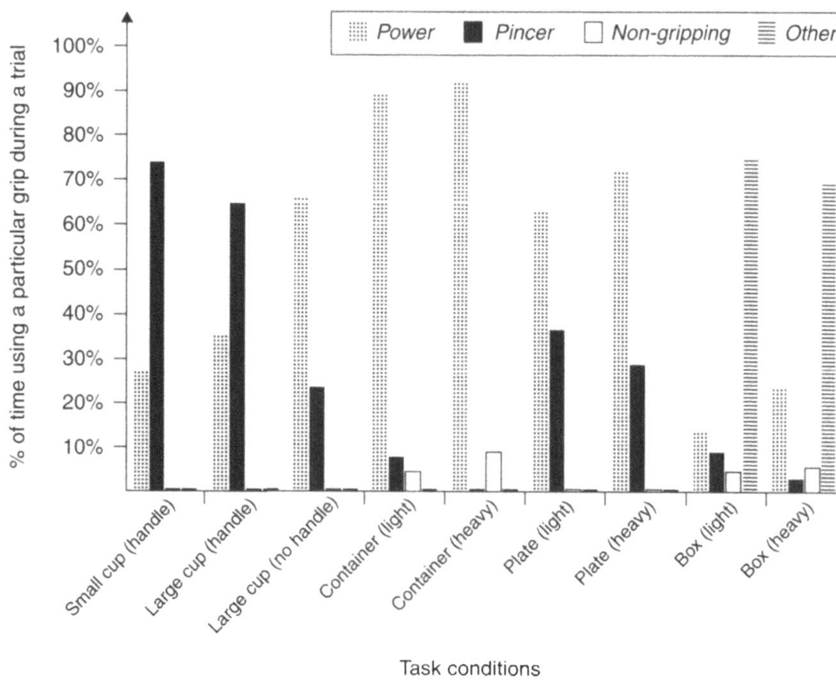

Figure 3. Proportion of time spent using a particular class of hand action for each of the task conditions.

Furthermore, the prevalence of power grips increased for the water containers with no handles as the weight and dimensions of the load being lifted increased. The hand actions in the 'lifting the plates' condition in Figure 3 clearly illustrates the high prevalence of the power gripping action used for this type of task condition, with that action being more frequent with the heavier plate condition. The box handling condition resulted in a predominant hand action that was not classified by one of the existing 9 actions being used in this study and was therefore classified as the "other" type of hand action. In all cases the hand action that was classified as 'other" reflected a flat palm push up against two sides of the box in a "vice' like grip.

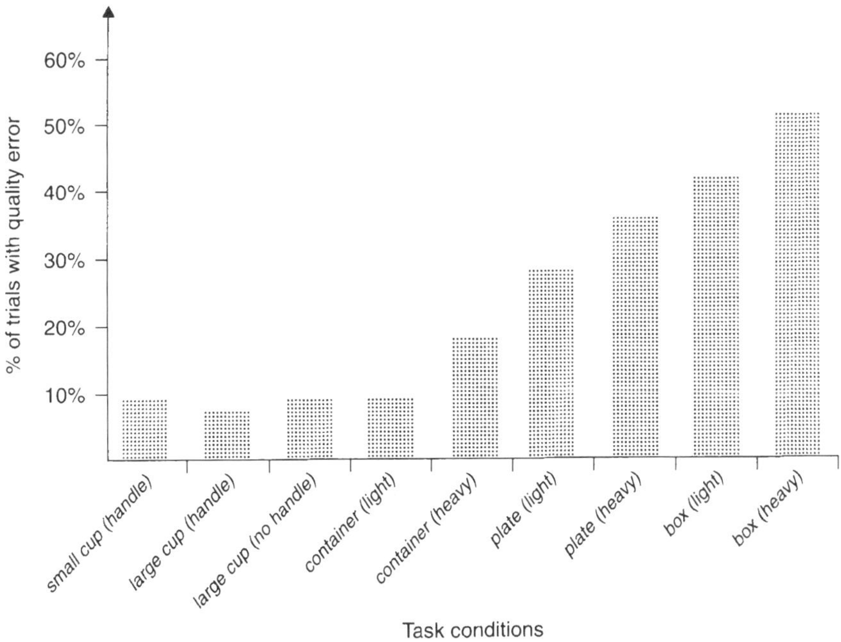

Figure 4. Proportion of trials with a TQ error as a function of the 9 tasks.

Figure 4 illustrates the 'Task Quality' (TQ) results as a function of the 9 experimental conditions. The TQ measures represent the percentage of trials where a TQ error occurred (spilling for the pouring tasks and dowels falling for the lifting tasks). The data in figure 4 illustrates that the level of TQ decreases with the increased weight in the lifting tasks. In contrast, there is no meaningful difference in the level of TQ between the pouring conditions with the exception of the heavy pouring container and larger water containers were the only pouring conditions where some people used two hands when lifting the container.

Discussion

Figure 3 clearly illustrates that the nature of the task demands associated with the different task conditions produced a varied, but, in many ways a consistent set of responses from people. That is, the lighter and smaller cup conditions where a handle was available resulted in people using a "pincer" grip to perform this task. This reflected the low force characteristics of the "pincer" grip (as contextualised in the strength-dexterity framework illustrated in figure 1) being adequate to perform this task. However, as the weight and dimensional characteristics of the object increased, and hence strength capability demands on the person increased with the larger and heavier containers (which also have no handles), there was a shift towards the "power" gripping action. This type of grip has a more compatible "fit" with the cylindrical shape of the container being easier to hold with the closing grasp of the power grip. Furthermore, the power grip has the mechanical properties

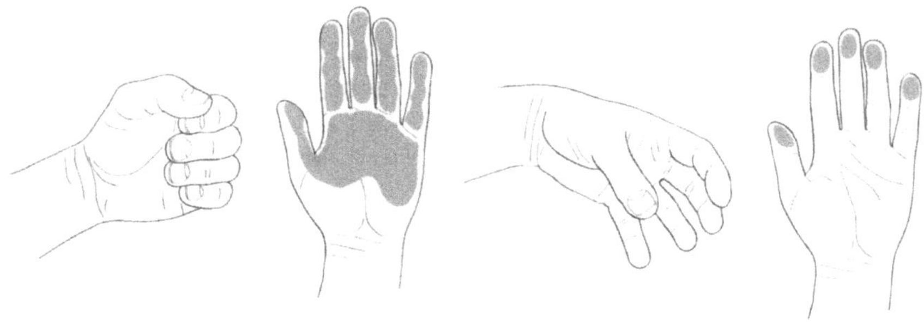

Figure 5. Surface area contact using a power grip (left) and a pincer (right).

to produce more grip strength than a comparable pincer gripping action so the increase in the dimensions and weight of the object would appear to the driver for changing from a pincer to a power grip. This shift in gripping action has a strong effect with the sample and the reduction in pincer gripping action increase in the prevalence in power gripping actions is a clearly visible effect for the pouring conditions illustrated in figure 3. The implications of this result are that a relatively small increase in the size and weight of the object has had a powerful effect on determining the prevalence of gripping action being used. Figure 3 illustrates that in the box handling conditions the predominant hand action was not one of the 9 classes of hand action selected at the start of this experiment. The nature of the box shape meant that utilizing a traditional power gripping action (refer to figure 1) was not possible given there are no handles on the box and one of the key task requirements was to keep the box horizontal to the table so the dowels would not fall off. Given these constraints, the "other" hand action category was made up exclusively of a technique where the palm of both hands pushing the side of the box in a "vice" like grip. Whilst this gripping action was different from the 9 classes of hand action that were identified at the outset of this experiment, there are some functional similarities with this grip compared to the traditional "power" grip action illustrated in Figure 5. That is, the flat pushing hand action described above provides a maximal surface area contact between the hand and the object thus providing a high level of stability and potential for force transference between the hand and the object. Therefore, whilst the power grip and "flat handed" grip are different in terms of the biomechanics of the hand action, they have a similarity in terms of maximising the surface area contact between the hand and the object which provides an opportunity for force transference and stability.

The heavier load in the container, plate and box conditions had a lower TQ compared to the respective comparison with the lighter version of the same object. Whilst there was no significant variation in TQ for the cups and light container conditions, people were more likely to use two hand to lift the container and this effect was highlighted for the heavy container condition. This illustrates that even though TQ was similar people had to utilize different strategies (i.e. use two hands) in the more demanding (heavier) tasks to achieve the same level of TQ. This indicates the increased level of capability function that is required (i.e. using two hands), when the capability limits are reached for a person if they are going to achieve the same level of TQ. Future developments will include validating the current experimental approach using subjective measures of difficulty and exertion together with expanding the range of classes of hand actions that are measured to provide a more granulated assessment of the classes of hand action that are used in response to a range of task demands.

Conclusions

The framework of 9 hand actions accounted for the variance in hand actions used by people in all conditions except the box handling conditions which indicates that the current framework should be expanded to include an increased range of non gripping actions such as the flat palm pushing action. The prevalence of power and pincer gripping actions was driven by the weight and dimensional task demands of the lifted objects with the heavier levels of these conditions producing an increase in the prevalence of the power gripping actions. The TQ measure is sensitive to changes in performance even when the task has been completed. This measure will assist with predicting exclusion in boundary conditions where people are operating at the boundary of their capability for a particular hand action.

References

Kirkland, R.I. (1994) Why will we live longer and what will it mean, Fortune, pp. 66–78

Kroemer, K.H.E. (1994) Coupling the hand with the handle: An improved notation of touch, grip and grasp. *Human Factors*, 28, 337–339

Mattila, M. & Vilkki, M (2003) OWAS Methods. In Karwowski, W. & Marras, W. (2003) Occupational Ergonomics: Principles of work design. CRC Press: London

Napier, J.R. (1956) The Prehensile movements of the human hand. *The Journal of Bone and Joint Surgery*, 38, No. 4, 902–913

Sears, A. & Young, M. (2003) Physical Disability and Computing Technologies: An Analysis of Impairments. In Jacko, J.A., Sears A. (Eds.) The Human-Computer Interaction Handbook: Fundamentals, Evolving Technologies and Emerging Applications. LEA: New Jersey, pp 482–503

Silverstein, B.A., Fine, L.J. & Armstrong, T.J. (1987) Occupational factors and carpal tunnel syndrome. *American Journal of Industrial Medicine*, 11, 343–358

Tuomi, K., Eskelinen, L., Toikkanen, J., Jarvinen, E., Ilmarinen, J. & Klockars, M. (1991) Work load and individual factors affecting work ability among aging municipal employees. *Scandinavian Journal of Work Environment & Health*, 17, 128–143

United Nations Population Division, 2001. World Population Prospectus: The 2000 Revision United Nations: New York

Voorbij, A.I.M. & Steenbekkers, L.P.A. (2001) The composition of a graph on the decline of total body strength with age based on pulling, twisting, and gripping force. *Applied Ergonomics*, 32, 287–292

THE IMPORTANCE OF CONTEXT IN INCLUSIVE DESIGN

E. Elton & C. Nicolle

Ergonomics and Safety Research Institute (ESRI), Loughborough University, Loughborough, UK

Capability data used in current inclusive design tools fail to take into account context of use. Two experiments were conducted with older users to determine what effect the physical environment has on two essential product interaction capabilities (vision and dexterity). For vision ($n_1 = 38$) everyday lighting levels were investigated, and for dexterity ($n_2 = 14$) warm (19°C–24°C) and cold temperatures (5°C) were investigated. Results from the vision study showed that when the lighting level decreased from daylight to street lighting, there was a decrease of up to 44% in the number of participants able to correctly read particular rows of letters. Findings from the dexterity study indicated that fine finger dexterity is significantly reduced ($p < 0.05$) when exposed to average winter temperatures (5°C). Failure to consider the capabilities of users in these everyday contexts of use could result in products excluding or causing difficulties to those intended to be included.

Introduction

The ageing population has become such an issue that the UK Government along with Japan and the US are reviewing their strategies for meeting the challenges of our ageing society. In the UK alone there are more people aged 65 and over than there are children under the age of 16 years (HM Government, 2008). In relation to design this presents a problem, some may say a challenge.

As the human body ages, especially beyond the age of 65 years, there is a significant reduction in functional capability (motor, sensory and cognitive capabilities). Capability is one of the fundamental attributes a person needs to access and use everyday products. Failure to take account of this reduced functional capability in the design process results in users becoming excluded from product use. Inclusive design is the philosophy that aims to consider this reduced functional capability during the design process, with the aim of making products functionally accessible and usable to as many people as reasonably possible. However, designers have to understand users' capabilities and needs if inclusive design is to take place (Keates and Clarkson, 2004).

Two inclusive design tools have been developed in order to help designers achieve this. They are HADRIAN (Human Anthropometric Data Requirements

Investigation and Analysis) (Marshall *et al.*, 2002), and the Inclusive Design Toolkit (Clarkson *et al.*, 2007).

For HADRIAN the capability of 102 older and disabled adults was measured in a laboratory environment (Gyi *et al.*, 2004), whilst the exclusion calculator in the Inclusive Design Toolkit utilises data gathered from the 1996/97 disability follow up survey (Grundy *et al*, 1999). Data used in the Inclusive Design Toolkit was initially gathered to help plan welfare support, thus not intended to cover all aspects of product interaction such as context of use (Waller *et al*, 2008). Both tools are a big step forward in aiding the uptake of inclusive design in industry. However, one serious weakness with the data is its failure to consider capability in real-life everyday contexts of use. Designing a product to minimize exclusion requires knowledge of the demands made by the product on its users, knowledge of the range of capabilities of the target users and their prevalence, and knowledge of context of use (Clarkson, 2008).

Recent evidence suggests context of use can have a multi-faceted impact on product interaction (e.g. increasing or decreasing user capability, and/or increasing product demand) particularly with older adults who have significantly reduced capability due to their age (Elton *et al*, 2008). Thus, when a mismatch between context and a product occurs, it is unlikely that the benefits of a product will be realised (Maguire, 2001). In real world terms this could mean inclusively designed products in fact become unusable in everyday contexts.

The overall aim of this research is to understand the extent to which everyday contexts of use can exclude people from using products in their daily lives, especially those who are older or disabled and may already be working to the limits of their abilities.

Context of use and capability

A considerable amount of literature has been published on the subject of context of use. It is not the intention of this research to develop these definitions further. However, it is important to understand what is meant by it, before it is discussed.

Context is defined in the Oxford English dictionary (2000), as "*the circumstances that form the setting for an event.*" The British Standard BS EN ISO 9241-11 (1998) defines it in more detail as: users, tasks, equipment and the physical and social environments in which a product is used. Other disciplines such as computing and mobile communication consider context, expressing the same concept but with their own vocabulary or emphasis. Detailed in Table 1 are the main factors relevant to each of the identified context characteristics.

As can be seen from Table 1, context is vast and covers a wide selection of circumstantial factors. Identifying which of these characteristics are experienced on a frequent basis during product interaction and which are likely to have the greatest impact on capability is discussed in the following section of the paper.

Table 1. Context characteristics and their associated factors (adapted from Thomas and Bevan (1995).

Context characteristic	Associated factors
User (People intended to use the product)	Knowledge, skill, experience, education, training, personal attributes, physical, motor and sensory capabilities.
Task (Activities undertaken to achieve the goal)	Frequency, duration, flexibility, sub-tasks, physical and mental demands, goal/output, procedure, and criticality.
Equipment (Product/s used to achieve the desired goal)	What's used, where located, duration used, choice, anything not used.
Physical Environment (The external surroundings/ conditions)	Visual conditions, atmospheric conditions, temperature, auditory conditions, vibration and the built environment.
Social Environment (The culture and people that surround the user)	Location, co-location of others, group working, assistance, number of people nearby, customs of users and surrounding people.

Product interaction capability

When interacting with a product, demand will typically be made on up to seven user capabilities (Waller *et al.*, 2008). These capability categories have been identified by Clarkson *et al.* (2007) as:

- Vision
- Hearing
- Thinking
- Communication
- Locomotion
- Reach and Stretch
- Dexterity

The demand placed on each capability is dependent upon the characteristics of the product being used and the task being accomplished. For example, making a mobile phone call requires a user to get their phone from their pocket/bag (reach), grasp the phone with their hand (dexterity), visually identify the buttons they need to press (vision), use their finger(s) to press the buttons (dexterity), and lift the phone to their ear (reach). During this process the user needs to recall a sequence of actions and numbers (thinking), and once this has been achieved the user then has to speak (communicate) and listen to the other person (hear).

Although demands are made on up to seven user capabilities, the vast majority of products make demands on the visual and dexterous (arm, hand and finger) capabilities of the user. For example, entering information into an electronic device (such as a computer, mobile phone, MP3, ATM), operating products around the home (kettle, toaster, vacuum cleaner, dishwasher), and using products outside the home (such as ATMs, gardening equipment, leisure equipment, devices for work) all require the perception of visual information and manipulation by hand for them to be successfully used.

Thus, identification of which everyday contextual characteristics frequently impact on these two capabilities will help to ensure products are usable in real world contexts of use.

Identification of relevant contextual characteristics

The Inclusive Design Toolkit (Clarkson *et al.*, 2007) details the contextual characteristics that impact upon the seven user capabilities relevant to product interaction. For visual capability, ambient illumination and glare are identified as having an impact (Persad and Langdon, 2007). For dexterity, cold temperatures, sweaty/wet/lubricated hands, wearing gloves and vibration are identified as the contextual characteristics that can impact this capability (Persad and Waller, 2007).

Whilst glare is a significant factor that affects vision, it is mainly an issue on fixed surfaces/displays that cannot be moved. However, the ambient illumination is constantly changing from morning until night on a daily basis. On a bright sunny day lighting levels can reach 100,000 lux, and drop as low as 0.001 lux at night. The greatest variation in ambient illumination occurs outdoors; however ambient illumination can only be controlled to a degree when indoors, i.e. through lamps or mains lights that illuminate the room to a specific level. Detail on everyday products has to be perceived in both of these environments where the ambient illumination constantly varies. The issue is that the human eye automatically adjusts itself to changes in ambient illumination. At low light levels the eye loses its ability to perceive detail and increases its ability to detect light. However, when it gets bright the reverse happens, and the eyes can detect detail around 10 times as fine in daylight as they can at night under starlight (Hopkinson and Collins, 1970).

A greater number of contextual factors are identified as having an impact on dexterity. For example, vibration can impair a person's ability to make fine movements with their hands and fingers. Sweaty/lubricated/wet hands can require a person to exert a greater level of strength whilst gripping. Gloves can make it harder to operate controls on products, and the cold can reduce the flexibility and sensitivity of the hands and fingers (Persad and Waller, 2007). Although all of the aforementioned contextual factors could affect product interaction, it is only the ones experienced regularly that are of concern. Cold temperatures are experienced annually for up to 3-4 months at a time. Again, with products being used both inside and outside the home, it becomes a contextual factor relevant to a large number of product interactions. Furthermore, dexterity (both manual and fine finger) is significantly reduced due to physiological effects of the cold on the human body (Heus *et al.*, 1995).

Whilst a selection of everyday products are used within the home environment where such variations in the physical environment can have a lesser effect, there are still activities of daily living, leisure and work that require the use of everyday products outside or in less than optimum environments. Furthermore, developments in technology are enabling the use of more products whilst mobile and out of the home.

In summary, it would appear that the factors associated with the physical environment, in particular ambient illumination and temperature, could affect the two main product interaction capabilities (vision and dexterity) on a frequent basis. What we know is that both of these contextual factors affect the physiology of the eyes and the hands, therefore capability can vary dependent upon the environmental conditions.

However, to what extent does capability vary under everyday contexts? Is variation in ambient illumination and/or temperature enough to push a user from being included to excluded?

Experimental investigations

Two experiments were conducted with older adults (65+ years) to determine what effect variations in everyday lighting and temperature conditions have on vision and dexterity respectively. A minimum age criterion for the sample in both experiments was set at 65 years old. Hand function becomes significantly reduced at this age and the vast majority of people with sight problems are this age and older (Shiffman, 1992; www.rnib.org.uk). A dataset that details this reduction and variation in capability will allow for the design of mainstream products that are accessible to, and usable by, as many people as reasonably possible, without the need for special adaptation or specialized design (Clarkson *et al.*, 2007).

The first experiment ($n_1 = 38$) measured the effect everyday lighting levels had on visual acuity (ability to perceive detail) of older adults (mean age = 74 years, SD = 6.1). Visual acuity was measured under four light levels at the distance of one meter. The conditions represented typical everyday environments where product interaction frequently takes place:

1. 7.5 lux (e.g., street lighting)
2. 150 lux (e.g., in-house lighting)
3. 6000 lux (e.g., optimum lighting/cloudy day)
4. 40,000 lux (e.g., daylight)

Under each of these lighting conditions visual acuity was measured at four different contrast levels i.e. 90%, 50%, 25% and 10%. For full details of methodology and results please see Elton and Nicolle (2009).

The second experiment ($n_2 = 14$) measured the effect everyday cold temperatures had on the dexterous abilities of older adults (mean age = 69.57, SD = 3.756). Fine finger dexterity (ability to manipulate objects with the fingertip part of the hand), power grip (maximum grip strength a person can exert with their hand) and pinch grip (maximal force that can be exerted between the index finger and thumb pulps) were measured in the warm (19°C–24°C, an environment that keeps the body at an optimum point) and in the cold (a climatic chamber regulated at 5°C). The temperature for the cold environment was based on the average UK winter temperature and the temperature threshold used by the Met Office to issue a cold weather warning, which is 5°C. For full details on methodology, rationale and results please see Elton *et al.* (2010).

Results

In experiment one (vision/ambient illumination) results showed that the mean readable letter size, across all contrast levels, generally decreases as the level of ambient illumination increases. Mean acuity scores, measured at a distance of one meter, are detailed in Table 2.

On average, participants could read letters 1.4 mm smaller in daylight compared to street lighting. Lighting appeared to have the biggest effect on letters printed at 25% contrast, as on average participants could read letters 1.9 mm smaller in daylight compared to street lighting. When the lighting level changed from daylight to street lighting, there was a decrease of up to 44% in the number of participants able to correctly read particular rows of letters.

For experiment two (dexterity/cold), a reduction in mean fine finger dexterity was observed across a number of tests when in the cold environment. Results from the experiment are detailed in Table 3.

Table 2. Smallest letter size 50% of participants could read in all conditions.

	90% Contrast	50% Contrast	25% Contrast	10% Contrast
Street lighting	2.9 mm	2.9 mm	3.7 mm	3.7 mm
In House	2.3 mm	2.9 mm	2.9 mm	3.7 mm
Optimum	1.8 mm	1.8 mm	2.3 mm	2.3 mm
Daylight	1.8 mm	1.8 mm	1.8 mm	2.3 mm

Table 3. Average dexterous performance in warm and cold environments.

Dexterity Test	Warm (19°C–24°C) Mean score and SD	Cold (5°C) Mean score and SD
Perdue Pegboard (no. of pins inserted)	Mean = 35.50 SD = 1.46	Mean = 33.14 SD = 1.02
Moberg Pickup Test (sec)	Mean = 13.79 SD = 2.21	Mean = 15.74 SD = 4.84
Mobile Phone (sec)	Mean = 13.35 SD = 4.28	Mean = 14.20 SD = 3.60

Mean performance on the Perdue Pegboard (picking up and placing as many pins as possible into holes on a pegboard in 30 seconds) was reduced by 7%, entering an 11 digit mobile phone number took on average 6% longer, and performance on the Moberg Pick-up test (picking up and placing 12 everyday items into a container as quickly as possible) took on average 14% longer in the cold. Results indicated that reductions in performance on the Perdue Pegboard and Moberg Pick-up test were significant ($p < 0.05$) when participants were exposed to the cold (5°C).

Discussion

In reviewing and analysing the literature, it was possible to identify the key capabilities involved in the majority of product interactions (vision and dexterity) and the aspects of context of use that affect them. Factors associated with the physical environment, i.e. ambient lighting and temperature, were identified as aspects of context frequently experienced when interacting with products.

Whilst inclusive design capability data already exists for vision and dexterity, there is currently no data that takes into account the effects these aspects of context of use have on these capabilities.

Results from the two experiments have shown that contexts of use (i.e. ambient illumination and temperature) can have a significant effect on capability. For vision, the size of detail capable of being perceived/recognised can vary up to 1.9 mm in a typical day. Thus, detail on products that can be successfully perceived during the daytime may be illegible at night under street lighting, resulting in the user becoming excluded. In certain scenarios this could be up to 44% of the target population, as results from experiment one showed that the number of participants able to correctly read particular rows of letters decreased by this amount when the lighting conditions changed.

The results from experiment two suggest that fine finger dexterity is affected by everyday cold temperatures. On average, performance on such tasks was reduced between 6%–14%. This meant tasks either take substantially longer (e.g. entering a mobile phone number) or the same work rate is not possible (e.g. placing the same number of pins into a board) when in the cold at 5°C. In relation to product interaction it is likely that such tasks as using a mobile phone, pressing a sequence of buttons on a screen or keypad, using a stylus to interact with a touch screen, and picking up and placing small objects are likely to be affected.

Conclusion and future work

The aim of this research was to understand the extent to which everyday contexts of use can exclude people from using products in their daily lives, especially those who are older or disabled. This research has shown that everyday contextual factors such as ambient illumination and temperature levels can significantly reduce capability. Evidence presented in this paper suggests that failure to take account of the effects everyday contexts have on capability could result in the target population becoming unintentionally excluded, especially those already working to the limits of their ability. Thus, if design decisions are based on capability data only measured in a neutral environment, it could result in everyday products becoming unusable in everyday contexts of use. Further experimental work will be conducted to increase the reliability and validity of the dexterity cold temperature data. Following this, the data will be translated into a usable form for product designers.

The authors wish to acknowledge and thank the Engineering and Physical Sciences Research Council for funding this work as well as our collaborators on the i~design3 project, including the Engineering Design Centre and the Department of Psychiatry at the University of Cambridge and the Royal College of Art Helen Hamlyn Centre.

References

BS EN ISO 9241-11, 1998, Ergonomic requirements for office work with visual display terminals (VDT's). Part 11 – guidelines for specifying and measuring usability. (British Standards Institute, London)

Clarkson, P.J., Coleman, R., Hosking, I. and Waller, S. 2007, *Inclusive Design. Toolkit,* (Engineering Design Centre, University of Cambridge)

Clarkson, P.J. 2008. Human capability and product design. In H.J. Schifferstein, and P. Hekkert (eds.) *Product experience,* (Elsevier)

Elton, E., Nicolle, C. and Mitchell, V. 2008, Identifying contextual factors in inclusive design. In: *Designing inclusive futures: Poster proceedings of CWUAAT 2008,* (Springer-Verlag, London)

Elton, E. and Nicolle, C. 2009, Now you see it, now you don't. In: *Proceedings of Include 2009,* (RCA, Helen Hamlyn Centre, London, UK)

Elton, E., Dumolo, D. and Nicolle, C. 2010, Have I just pressed something? The effects of everyday cold temperatures on dexterity. Under review for CWUAAT 2010

HM Government. 2008, Preparing for our ageing society: A discussion paper. (Department for Work and Pensions)

Heus. R., Daanen, H.A.M. and Havenith, G. 1995. Physiological criteria for functioning of hands in the cold: A review, *Applied Ergonomics,* **26**: 1, 5–13

Hopkinson, R. and Collins, J. 1970, *The Ergonomics of Lighting.* (Macdonald & Co, London)

Grundy, E., Ahlburg, D., Ali, M., Breeze, E. and Sloggett, A. 1999, Research Report 94: Disability in Great Britain. (Corporate Document Services, London, UK)

Gyi, D.E., Sims, R.E., Porter, J.M., Marshall, R. and Case, K. 2004, Representing older and disabled people in virtual user trials: data collection methods. *Applied Ergonomics*, 35: 443–451

Keates, S. and Clarkson, P.J. 2004, *Countering design exclusion: An Introduction to inclusive design.* (Springer-Verlag, London, UK)

Marshall, R., Case, K., Gyi, D.E., Oliver, R. and Porter, J.M. 2002, HADRIAN: An Integrated Design Ergonomics Analysis Tool. In: Proceedings of the XVI Annual International Occupational Ergonomics and Safety Conference, Toronto, Canada

Maguire, M. 2001, Context of use within usability activities, *International Journal of Human Computer Studies,* 55: 4: 453–484

Mansfield, N.J. 2005, *Human Response to Vibration.* (CRC Press, London)

Oxford English dictionary – Second Edition. 2000. (Oxford University Press, UK)

Persad, U. and Langdon, P. 2007, Vision. In: P.J. Clarkson, R. Coleman, I. Hosking, and S. Waller (eds.) *Inclusive Design Toolkit* (Engineering Design Centre, Cambridge, UK). 428–451

Persad, U. and Waller, S. 2007, Dexterity. In: P.J. Clarkson, R. Coleman, I. Hosking, and S. Waller (eds.) *Inclusive Design Toolkit* (Engineering Design Centre, Cambridge, UK). 4138–4159

RNIB (Royal National Institute of Blind People): Facts and figures about sight loss in the UK. Available at: http://www.rnib.org.uk/xpedio/groups/public/documents/publicwebsite/public_researchstats.hcsp. Accessed: 03 Feb 2009

Shiffman, L. 1992, Effects of ageing on adult hand function, In: The *American Journal of Occupational Therapy*. 46: 9, 785–792

Thomas, C. and Bevan, N. 1995, *Usability context analysis: A practical guide, Version 4.* (National Physical Laboratory, Teddington, UK)

Waller, S., Langdon, P. and Clarkson, P.J. 2008. Converting disability data into a format suitable for estimating design exclusion. In P. Langdon, P.J. Clarkson, and P. Robinson, (eds.) *Designing inclusive futures: Proceedings of CWUAAT 2008,* (Springer-Verlag, London). 3–13

MEDICAL

REVIEW OF GUIDELINES ON ENTRAPMENT AND FALLS FROM THE USE OF BEDRAILS

Amy Jones[1] & Simon C. Monnington[2]

[1]*Health and Safety Laboratory, Harpur Hill, Buxton, Derbyshire, SK17 9JN, UK*
[2]*Corporate Specialist Division, Health and Safety Executive, Cardiff*

Bedrails are designed to prevent users inadvertently falling from their bed but accident history shows that they sometimes can fail to do this. They have also been the cause of user entrapment, which can have fatal consequences. A review of three issues raised by HSE inspectors has been carried out: 1) The gap between mattress side and bedrail; 2) The minimum height of a bedrail when used with a range of mattress types in order to provide protection to bed occupants; 3) The risks associated with the entrapment zone between the bottom of a bedrail and mattress support platform when easily compressible (e.g. air flow) mattresses are used. Additionally, areas of possible further research have been identified that could potentially aide those carrying out bed rails risk assessments.

Introduction

Bedrails are designed to prevent users inadvertently falling from their bed but accident history shows that they sometimes can fail to do this. They have also been the cause of user entrapment (where a patient's neck, chest or limbs become trapped within the bedrail or between the bedrail, bed or mattress), which can have fatal consequences. An individual case study cited by the National Patient Safety Agency (NPSA, 2007) suggests that a patient can move from the centre of the bed into a trapped position in less than two minutes and asphyxiate within 14 minutes. Bedrails are not suitable for all patients/service users who are at risk of inadvertent falls from bed. Assessment of the individual by a competent member of staff is required to determine the risks and appropriate control measures before selecting bedrails. If bedrails have been chosen as a control measure to prevent falls from bed (rather than other control measures, such as extra low beds) then duty holders need to manage their installation and use effectively.

Information on the safe use of bed rails for users, carers and staff with responsibility for the provision, prescription, use, maintenance and fitting of bed rails is given by the Medicines and Healthcare products Regulatory Agency (MHRA, 2006). Additionally, Health and Safety Regulators make use of advice in the HSE Sector Information Minute (SIM) 07/2007/06 Bedrail Risk Management (HSE, 2007). The HSE SIM draws on information in the British Standards (BS EN 60601-2-38 1997 and BS EN 1970:2000).

The HSE carried out a series of inspections to look at the management of some specific health and safety risks at nursing homes across Wales during March and April 2008. The results that were released by HSE (2008) documented that many nursing homes had poorly fitted bed rails. Numerous nursing homes had ill fitting mattresses that either raised the height of the bed occupant such that there was a risk of them rolling over the top of the bedrails or were too narrow giving rise to a risk of entrapment between the side of the mattress and the bedrail. Fifty four percent of the nursing homes visited received Improvement Notices (IN) on bed rail risk management.

As a result of these inspections, HSE Inspectors identified three issues on which they would like further information:

1. Guidelines for the height of a bedrail above a mattress top particularly where additional or thicker mattresses are used.
2. The horizontal gap between the edge of the mattress and the side of the bed rails;
3. The appropriate dimension between the mattress support platform and the bottom of the bedrail.

As a result, a review was undertaken of literature on key risk factors for entrapment and falls relating to the above issues, with the aim of identifying existing guidelines and information, and where further information or research could be used to amend the HSE Bedrail Risk Management SIM.

Method

A review of UK and US literature covered the following sources; Medline, Embase, CINAHL, Emcare, British Nursing Index, Applied Social Sciences Index and Abstracts, Ergonomics Abstracts, British Standards and a general Internet search to include the Food and Drug Administration (FDA), National Patient Safety Agency (NPSA) and MHRA. Other methods used were observations and measurements of bedrail configurations at nursing/care homes, information provided by HSE Inspectors and details of bedrail related fatalities from HSE investigations.

Findings

Why is it an important issue? A look at the statistics

Anecdotal evidence indicates that bed rails may contribute to serious injury and death resulting from falls over, under, between and around bed rails (Braun and Capezuti, 2000 cited in Hoffman *et al.*, 2003). This anecdotal evidence is supported by some statistics.

Bedrail related incidents

In the UK, the HSE (2007) states that for the period April 2001 to March 2005 Reporting of Injuries, Diseases and Dangerous Occurrences Regulations

(RIDDOR) statistics identified at least 10 fatal accidents and a number of major injury incidents in which the use of bed rails was implicated. The NPSA (2007) affirms that the MHRA have had reports of 21 bedrail related deaths over the seven-year period of 2000–2006.

In the USA between January 1st 1985 and January 1st 2008, the FDA received reports of 772 incidents of patients caught, trapped, entangled, or strangled in beds with rails. The reports included 460 deaths, 136 non-fatal injuries, and 176 cases where staff needed to intervene to prevent injuries. This represents an average of around 23 deaths a year occurring through bedrail entrapment or bedrail failure in US nursing home settings. In the UK, an average of around 3 deaths a year have occurred in care homes and patients' own homes. Deaths in UK hospitals account for 3 cases in 7 years reported by the MHRA (NPSA, 2007).

Falls

The NPSA (2007) reports that between September 2005 and August 2006, there were around 44,000 reports to the National Reporting Learning Service (NRLS) of patients who appeared to have fallen from beds in acute and community hospitals, mental health, and learning disability units. Many studies do not include whether a bed rail was present or not, but the study based on reports to the NRLS in 2006 from hospitals in England and Wales found that bedrails were recorded as raised in 8% of reports of falls from beds. A survey of a random sample of hospitals reporting to the NRLS in the same year indicated that 35% of patients had bedrails in use with 26% having bedrails raised on both sides of the bed.

Under-reporting

The extent to which these statistics accurately represent the occurrence of entrapment and falls is difficult to determine and also, near misses of any type are likely to be under-reported. As well as being under-reported, Todd et al (1997) also comment that causality cannot sometimes be inferred from individual reports due to a lack of detailed description of the adverse event, for example, the condition of the bed, rail, and mattress, exact location of entrapment and patient behaviour etc.

Issue 1: Falls risk associated with the height of a bedrail above a mattress

The presence of bedrails can be a contributory risk factor for a bed occupant falling from bed. MHRA (2001) found that the most frequent cause of injury arises from the bed occupant attempting to climb over the bedrail. The presence of a bedrail may actually increase the distance of a fall from the bed and therefore the severity of injury. This work is based on the assumption that bedrails have been chosen as the appropriate control measure to prevent falls from bed and therefore the bed occupants are not assessed to be at a high risk of attempting to circumvent the bedrails. Consequently, the key to the effectiveness of a bed rail in preventing an inadvertent fall from a bed will be its height above the level of the compressed mattress. The British Standard provides a dimensional guideline of 220 mm that is measured vertically from the edge of an uncompressed mattress to the top of the bedrail. This guideline, however, does not state specifically that airflow mattresses are considered in this criterion or how to account for them. Placing one mattress on top of another (or the use of mattress overlays or airflow mattresses) may reduce the effectiveness of the bed rails because the overall height of the rail is lowered in comparison to the top of the mattress.

This issue is highlighted in the literature. The HSE (2007) states that the use of a mattress overlay reduces the effective height of a bed rail. The MDA/2004/014 Alert "Replacement mattresses for hospital beds used in hospitals and the community" (MHRA, 2004) states that a mattress that is significantly thicker than that intended by the bed manufacturer can put the patient in a position, which adversely reduces the effective height of the bedrails. This could increase the risk of a patient involuntarily falling over the top of the bedrail.

However, no information could be found in the literature that gives explicit guidelines on what the height of bed rails should be in relation to airflow/overlay mattresses as the sources simply refer to "mattress" as highlighted in Table 1.

Table 1. Comparison of guidelines for the height of a bedrail above a mattress.

Advisory Body	Guidelines
HSE SIM 07/2007/06 (advice for H&S Regulators)	≥220 mm (standard mattress)
British Standards (BS EN 60601-2-38 1997 and BS EN 1970:2000)	≥220 mm (standard mattress)
MHRA Device Bulletin; Safe Use of Bedrails (2006)	Are the bed rails high enough to take into account any increase in mattress thickness or additional overlay? Extra height bedrails may be required
US FDA (2006) Hospital bed System Dimensional and Assessment Guidance to Reduce Entrapment	Does not provide a dimension as the rail height is not related to entrapment
US Consumer Products Safety Commission C.F.R part 1213 and 1513 Consumer Product Safety Standard for Bunk Beds	The top of the bedrails must be no less than 127 mm above the top of the mattress *This standard is for bunk beds intended for use/used by children (not adults) and therefore may explain the lower dimensional requirement.*
Standards Australia AS/NZS 4220:1994 Bunk Beds	The height of the guardrail should always be at least 160 mm above the top of the mattress. *This requirement is to reduce the likelihood of deaths/ injuries to children and therefore may explain the lower dimensional requirement*

There is scope to consider the term mattress to include any combination of system/assembly of mattresses constituting the patients sleeping surface. Further questions have arisen as a manufacturer has stated that the compressibility of airflow mattresses means that the body will sink lower into the mattress and it therefore does not need to meet the standards. How this relates to overlay mattresses is unknown.

The British Standard 220 mm dimension is based on the mid point (centre of gravity) of an adult male torso (95th percentile) lying on the shoulder. Pheasant and Haslegrave (2006) state shoulder breadth (bi-acromial) for a 95th percentile male aged 19–65, as 430 mm (midpoint of 215 mm). PeopleSize (1993–2008) states shoulder breadth (acromion) for a 95th percentile male aged 18–64, is 442 mm (mid point of 221 mm). The British Standard 220 mm dimension lies in-between these two data sources.

Perhaps more suited to this issue is anthropometric data for the age group 65–74 years old (PeopleSize 1993–2008) or "elderly people" as defined by Pheasant and Haslegrave (2006). This ranges from a mid point of 152.5 mm for a 5th percentile female to 221 mm for a 95th percentile male. Consequently, an assessment of each patient for bedrail height suitability would seem difficult aside from measuring each individual's biacromial dimension to find the consequent bedrail height needed.

Although the majority of the population would be protected by a bedrail height of 220 mm as stated by the British Standard, these figures are based on static anthropometry and dimensions are likely to be different when an individual is lying down with consequent tissue compression and compression of the surface they are laid upon. In effect, the 220 mm dimension may be overprotecting as the majority of biacromial dimensions are below this figure. The British Standard also assumes that this dimension relates to the centre of mass and that to be effective, a bed rail must be at least as high as the centre of gravity of the bed occupant.

Possible further research for issue 1

The information in the literature on the height of a bedrail above a mattress is that additional height bed rails should be used with overlay mattresses to help prevent falls. Given the identified shortcomings in the British Standards on which the HSE SIM is based, the authors propose that further information in this area could potentially be used to improve the HSE SIM. This could be provided through empirical research attempting to simulate falls from beds to ascertain the minimum height of bedrail required to protect individuals. Airflow mattress companies claim that owing to the way the airflow mattress works, the bed rail does not need to meet the British Standard in terms of height above mattress as they suggest that the body will sink lower into the air flow mattress. This could be investigated by testing a range of mattresses, people sizes and bed rail heights in order to ascertain a height of bedrail that will provide sufficient protection to the individuals in bed where bedrails have been assessed as an appropriate means of falls prevention. It may be possible to obtain sufficient protection with a lower height bedrail and then the use of bedrail extensions may not be needed.

Issues 2 & 3: Entrapment within the bed/rail system

Table 2 illustrates the two areas of entrapment that relate to issues 2 and 3.

Issue 2: Entrapment risk associated with the horizontal gap between the edge of the mattress and the bedrail

Currently, there is no dimension specified for the horizontal distance between the edge of the mattress and the bed rail in the British Standards. Consequently there are limitations in the application of the British Standards when assessing bedrail set ups for entrapment. MHRA guidance refers to large gaps between the mattress and bed rail as potential areas of entrapment and that measures should be taken to ensure that the space does not allow entrapment of the occupant's head or body but does not provide a dimensional limit.

However, the US FDA highlights seven potential entrapment zones (based on research and investigations) within the bed/rail system that have been identified and outlined by the Hospital Bed Safety Workgroup (HBSW) and FDA. One of these seven entrapment zones is entrapment between the rail and mattress with a recommendation of a gap of less than 120 mm (see Table 2). A dimension of 120 mm

Table 2. Recommendations in the literature for the two entrapment zones.

Entrapment Zone →	**Entrapment between the rail and mattress (Issue 2)**	**Entrapment under the rail at end of rail (Issue 3)**
Literature recommendations ↓		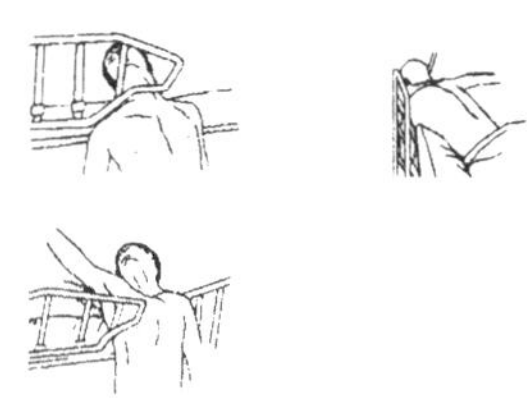
HSE SIM07/2007/06	No dimensional limit given	Refers to the British Standard
BS EN 1970:2000	No dimensional limit given	If distance between the end & rail is >250 mm then the smallest opening between rail & mattress support platform can be <60 mm. If <60 mm, then <120 mm
MHRA Device Bulletin; Safe Use of Bedrails (2006)	.... May be due to a poorly fitting mattress/ bedrail system	
Boocock (2006)	<60 mm *Modes of entrapment would suggest a dimension to prevent neck entrapment*	<60 mm
US FDA (2006).... Guidance to Reduce Entrapment	<120 mm *The space should be small enough to prevent head entrapment*	<60* mm AND > 60° ** angle

encompasses the 5th percentile female head breadth in all data sources that were used to develop this recommendation by the FDA.

Parker and Miles (1997) conducted a detailed review of all US reported adult deaths and injuries involving bed rails between 1993–1996. The study revealed that during the 3-year period, 70% of 74 deaths stemmed from entrapment between the mattress and rail resulting in pressure of the face on the mattress.

The replacement of original mattresses over time with mattresses of different dimensions is highlighted as likely to cause an entrapment risk - an important consideration for this zone of entrapment. Replacement mattresses may be narrower than the original mattress or smaller specialty mattresses used on top of existing mattresses may result in an excessive gap being created between the mattress and the bed rail. Some bedrails may be too flexible or have too much play in their mechanism and may deform under force thus making a gap between the mattress edge and bedrail also.

Possible further research for issue 2

Recommendations in the literature to reduce the risk of entrapment indicate that staff need to be able to set up the bed/rail/mattress system correctly to minimize the risk of entrapment and to know whether standards have been met. As already mentioned, the British Standards, on which the HSE SIM is based, do not reference this zone of entrapment. There are accessories for use in conjunction with bedrails such as bedrail bumpers, gap fillers and fastening the mattress to the support platform to reduce the risk of entrapment in this zone.

It is the authors' opinion that this zone (which is not referenced in the standards) is a critical area of entrapment as evidenced by fatal entrapments occurring in this gap and consequently it needs to be addressed by duty holders. This is in agreement with Boocock (2006) who, based upon evidence within the literature (Parker and Miles, 1997), suggested a distance of less than 60 mm. Anthropometric data and possible modes of entrapment would suggest that a distance of $\leq$60 mm is an appropriate dimension by which to prevent neck entrapment between bedrail and mattress rather than the 120 mm to prevent head entrapment suggested by the FDA.

Additionally, the International Bed Safety Group says, "the horizontal distance (gap) between the mattress and the bed rail must be less than 60 mm when the mattress is pushed to the opposite side of the mattress deck" (Veterans Health Administration Warning System, 2001). If lateral movement is probable, then a dimension of $\leq$60 mm becomes even more important.

In conclusion, it is suggested that there is a need for further information on this issue. This is providing it could include research on dynamic testing, i.e. putting the bedrail in its most onerous position in order to establish an appropriate guideline for the maximum horizontal gap between edge of mattress and bed rail.

Issue 3: Entrapment risk associated with the gap between the mattress support platform and the bottom of the bed rail

The dimension suggested in the British Standards for the entrapment zone between the mattress support platform and the bottom of the bed rail is described as being acceptable when less than 120 mm in some instances or less than 60 mm in others. However, the information presented may be difficult to understand as explanation of the issue is only via a one-line description in the British Standards table with a corresponding diagram.

There are basically two elements to this zone, differentiated by the means of access to the space:

1. Via the gap around the end of the bedrail (or between segmented bed rails) to the space beneath the bottom of the bedrail, if indeed it is accessible by this route. It is not with many bedrail types.
2. Directly between the edge of the mattress and the bedrail (i.e. via issue 2)

The applicability of these criteria depends on the type of bedrail in use and would appear to be aimed at winged bedrails for the first point and all for point 2. Understanding of this issue is that for the former condition above, if there is a gap of greater than 250 mm between the head/footboard and the end extremities of the rail, then the recommended measurement between the bottom surface of the rail and the mattress support platform would consequently need to be less than 60 mm to prevent neck entrapment. The allowance for a larger gap of 120 mm between the mattress support platform and the bottom of the rail is for the situation when the bed rail is positioned less than or equal to 60 mm away from the head/foot end, i.e. access for the neck is prevented.

Although the British Standard and FDA dimensions are similar, the FDA refers to the measurement being made from the compressed mattress to the bottom of rail, rather than the mattress support platform as the British Standard states.

It has been commented that the use of airflow mattresses or additional mattresses has resulted in thinner (often 2 inch) base mattresses being used with overlay mattresses on top to ensure that the distance from the top of the mattress combination to the top of the bedrail meets the British Standard guideline dimension of 220 mm. However, airflow mattresses are usually more compressible at the edges and consequently could leave large gaps under the bedrail. Therefore, whilst requesting that the height issue is addressed, there is a potential for the bedrails to have too large a gap underneath where easily compressible or thin mattresses are used to comply with the height dimension, which can increase the risk of entrapment in this zone. Consequently, bedrails would then need to have a lower bottom rail to narrow the gap to 60 mm or less, similar to the British Standard requirement where bedrails are fitted more than 250 mm from the headboard or footboard.

Possible further research for issue 3

US FDA incident statistics show that the most frequent zone of entrapment is between the mattress and the bottom of the rail. Additionally, in the authors' opinion, its significance is understated in the British Standards on which the HSE SIM is based. Therefore, this matter would seem to need further research to investigate how this entrapment zone correlates to mattress compression. This may require future work of laboratory testing for a range of mattresses and mattress combinations.

Conclusion

Three main areas have been identified from this literature review where further research could potentially be undertaken in response to the issues raised by HSE inspectors. In summary this is to address: the potential zone of entrapment between mattress side and bedrail with a dimensional limit; the minimum height of a bedrail when used with a range of mattress; and the entrapment zone between the bottom of a bedrail and mattress support platform when easily compressible (e.g. air flow) mattresses are used. There is a wide range of compatibility issues to consider when assessing a bedrail configuration (for example the bed, mattress, rail, individual characteristics of the user) and consequently the outcome of further research may potentially be used to assist duty holders when taking account of this in their risk assessments.

Additionally, we note that HSE and the MHRA are currently in discussion on the information on the safe use of bed rails for users, carers and staff with responsibility for the provision, prescription, use, maintenance and fitting of bed rails.

References

Boocock, M., Weyman, A. and Mcilroy, R. 2002, *Bedside safety rails: An ergonomics evaluation.* HSL Internal reports ERG/02/05

Boocock, M., Weyman, A. and Mcilroy, R. 2006, *Bedside safety rails: assessment of strength requirements and the appropriateness of current designs* Ergonomics 49 (7), 631–650

BS EN 1970:2000. 2000, *Adjustable beds for disabled persons – requirements and test methods, ISBN: 0580360369*

BS EN 60601-2-38:1997. 1997, *Medical electrical equipment – part 2-38 Particular requirements or the safety of electrically operated hospital beds, ISBN: 0580273008*

FDA. 2006, *Guidance for Industry and FDA staff. Hospital bed System Dimensional and Assessment Guidance to Reduce Entrapment*

Hoffman, SB., Powell-Cope, G., MacClellan, L and Bero, K. 2003, *Bed safe, a bed safety project for frail older adults.* Journal of Gerontological Nursing 29 (11), 34–42

HSE. 2007, SIM 07/*2007/06 Bed rail risk management*

HSE. 2008, *Welsh nursing homes inspected.* www.hse.gov.uk/press/2008/w01708.htm

MHRA. 2004, MDA/2004/014 Alert *Replacement mattresses for hospital beds used in hospitals and the community* www.mhra.gov.uk/home/idcplg?IdcService=GET_FILE...

MHRA. 2006, *DB 2006(06) Safe Use of Bedrails http://www.mhra.gov.uk/Publications/Safetyguidance/DeviceBulletins/CON2025348*

NPSA. 2007, *Bedrails – Reviewing the Evidence, A Systematic Literature Review* www.npsa.nhs.uk

PeopleSize software, Open Ergonomics Ltd 1993-2008. www.openerg.com

Pheasant, S. and Haslegrave, C. 2006. *Bodyspace Anthropometry, ergonomics and the design of work.* Taylor and Francis, 3rd edition

Todd, J.F., Ruhl, C.E. and Gross, T.P. 1997, *Injury and death associated with hospital bed side rails: reports to US food and drug administration from 1985-1995.* American Journal of Public Health, 87 (10), 1675–1677

Veterans Health Administration Warning System. 2001, *Patient Safety Alert; Bedrail Entrapment.* www.va.gov/NCPS/alerts/BedEntrap.doc

MUSCULOSKELETAL DISORDER RISK MANAGEMENT PERFORMANCE IN SONOGRAPHY WORK IN THE NHS

Simon C. Monnington

Human Factors and Ergonomics, Corporate Specialist Division, Health and Safety Executive, Government Buildings, Ty Glas, Cardiff, CF14 5SH

Healthcare professionals undertaking diagnostic imaging work using ultrasound equipment have a high prevalence of musculoskeletal disorders (MSD). There is a useful amount of published information on MSD risk management specific to sonography. However, it is not clear how this information translates into practice in the UK. This paper describes a project to examine MSD risk management performance across a range of sonography operations at UK NHS Trusts. Fourteen NHS trusts undertaking sonography work were visited across the UK by Specialist Ergonomics Inspectors from the Health and Safety Executive. Awareness of MSD issues was very high but action to identify, prioritise and tackle the risks was considered poor. MSD ill-health is a key professional issue among sonographers and this paper suggests that further management resource is required to help mitigate the situation.

Introduction

Diagnostic imaging using ultrasound equipment is an effective and useful tool for a range of clinicians in the National Health Service (NHS). Ultrasound uses high frequency sound waves to produce a real time image of the internal body. These images are used to check function, look for abnormalities and take measurements and stills for the most part non-invasively. Three main professions do ultrasound work: Radiographers work in obstetrics, gynaecology and general ultrasound; Cardiac Technicians undertake echocardiography (the heart); and Vascular Technicians investigate blood vessels using ultrasound. Midwives can undertake aspects of the obstetric work (e.g. dating scans etc). Ultrasound equipment is also deployed in musculoskeletal investigations. Medical practitioners can also use ultrasound machines. Sonography work involves an operator guiding a handheld transducer against the patient's body while simultaneously monitoring a screen and using a keyboard to control/record the image.

Musculoskeletal disorders (MSD) in sonographers

Several studies have identified aspects of sonography work that were associated with MSD. These included: one-sided static working position; prolonged pinch gripping of the ultrasound transducer; abduction at the shoulder during transducer placement; and insufficient recovery time (Vanderpool *et al.*, 1993; Magnavita *et al.*, 1999; Schoenfeld *et al.*, 1999; and Village and Trask, 2007). Most recently, Dodd-Hughes (2008) found evidence that work organization (psychosocial) factors (such as demand, control, support and role) as well as physical factors were associated with musculoskeletal problems in a sample of UK sonographers. Musculoskeletal disorder prevalence rates in excess of 80% have been reported for sonographers in the literature (Smith *et al.*, 1997; Magnavita *et al.*, 1999 and Dodd-Hughes, 2008). Worryingly, Brown and Baker (2004) found that around 20% of sonographers reportedly left the profession due to persistent discomfort.

Key information on managing MSD risks in sonographers

The Society of Diagnostic Medical Sonography (SDMS, 2003) developed consensus guidance on industry standards for the prevention of work related MSDs in sonographers. This process involved a number of key stakeholders including equipment manufacturers. The guidance described a range of useful ergonomics measures concerning:

- Equipment – ultrasound machine, table, chair, monitor etc
- Work organisation – workload, scheduling and work area
- Professional issues – adoption of best practice and training

The Society of Radiographers (2003 and 2007) in the UK produced useful documentation to help trusts, departments and individuals to tackle the risks of MSD that appear to be linked to sonography work. The Health and Safety Executive (HSE) has produced guidance material that is not specific but highly applicable to the management of MSD risks in sonography. For example, HSG 60 – upper limbs disorders and L26 – the display screen equipment (DSE) regulations.

The above indicates there is a useful amount of information available on the management of MSD risks associated with sonography. It is not clear how far this has been used in practice Consequently, a project was undertaken to examine MSD risk management performance in sonography within the NHS.

Project approach

Checklist development

Pilot visits were carried out to obtain relevant information on sonography tasks and issues that surround the sonographer within their workplace. This information was used in conjunction with current and available guidance to develop the checklists for use by the participating Specialist Inspectors (Human Factors and Ergonomics). Two checklists were developed, one aimed at the NHS Trust and the other for the individual departments. The Trust checklist contained items to prompt the inspector to consider aspects such as Trust awareness of MSD issues, occupational health provision and the approach used to monitoring and investigate cases. The department checklist considered the number, range and duration of scans undertaken; risk assessments; and controls measures implemented.

Project briefing

A team briefing was undertaken to ensure that members of the inspection team were fully aware of the issues surrounding sonography, the aims of the visits and what needed to be delivered. A briefing pack was issued that contained relevant information: a literature review concerning MSD in sonography; the checklists; industry good practice guidance; and useful information from manufacturers and suppliers.

Project inspections

An opportunistic sampling approach was used to select trusts for inspection. Local HSE inspectors were contacted and potential candidates for inspection were identified. Each visit used the checklists as a basis for the inspection to promote consistency of approach. Checklists were completed for each trust and the departments visited. Each trust visited was presented with the specific findings and subsequent recommendations for action. The completed checklists were returned to the project leader for analysis.

Inspection findings

Thirty-two departments undertaking sonography work were inspected from 14 NHS Trusts across the UK. All trusts conducted obstetric and general radiology sonography. Seven (50%) trusts operated an echo cardiology service and 12 (85%) operated a stand alone vascular service. The trusts were located UK wide (4 in Wales, 2 in Scotland and 8 in England). The sample was not selected at random but was it considered to be reasonably representative of sonography in the UK.

Awareness of MSD risks

Awareness of sonographers being a group of workers with a particular MSD issue was exceptionally high in the departments and trusts visited. In 94% of departments visited managers were aware of individuals having experienced or currently experiencing MSD problems attributed to their work. In at least 4 trusts visited awareness of the MSD issue in sonography at the trust level was stimulated by employee litigation claims. As a result, Inspectors were not required to persuade managers and staff that MSDs were a real concern requiring attention.

An overwhelming majority of departments (96%) were aware of some form of guidance on the topic. The most commonly cited guidance was from the Society of Radiographers (SoR) and HSE's guidance on the Display Screen Equipment

Regulations (L26). The SoR has done an effective job of raising levels of awareness but this may well have been limited to sonography modalities operated and run by Radiographers. Echo and vascular teams were less aware of the SoR guidance. In addition, guidance and information on MSD provided by their respective professional bodies appeared to be much less comprehensive than that supplied by the SoR. It is important to note that there is a great deal of commonality between sonography modalities in terms of the range and depth of MSD risks present in their professions. Hence, all modalities can make effective use of the range of existing guidance and should do so.

A small opportunistic sample of sonographers in each department was asked about the musculoskeletal problems they had experienced. In common with the extensive literature supporting a high prevalence of MSD in sonography staff the interviews identified that nearly 90% had experienced some type of musculoskeletal discomfort during their career. Around half of the interviewees reported the discomfort they experienced as 'severe'. Almost 30% of those experiencing problems had also taken time off work as a result. There was strong agreement between the findings of the sonographer interviews and department manager awareness of MSD problems in their unit.

Occupational health provision

All of the trusts reportedly had systems in place for prompt access to occupational health support. However, it was unclear why, given that so many sonographers experienced musculoskeletal ill-health, occupational health support was not being utilised more often than it was? In some cases time pressure and workload may have been enough to deter sonographers seeking assistance. There were signs that some staff considered pain and discomfort to be 'part of the job'. Most Trusts investigated MSD incident reports with support from the BackCare Advisor and/or Occupational Health (OH). Triggering the involvement of these professionals appeared to take a mostly reactive route. It was also found that 85% of trusts' investigation process included identifying remedial actions. This suggested that most trusts recognised that part of the investigation process was to identify and implement interventions that will reduce the risk of injury occurring again.

MSD Monitoring

Effective and informative health monitoring for MSD was not widely observed at the trusts/departments visited. In fact, MSD monitoring was identified in just 35% of the Trusts visited and these were usually passive and reactive in nature.

Scanning workload

Sonographers can be expected to undertake up to 35 scans per day. Most departments seemed to operate in the range of 12 to 20 scans per sonographer per day. The number of scans did vary with department function and the mix of full and part time staff. The quickest scans can take as little as 5 minutes to complete but some abnormality,

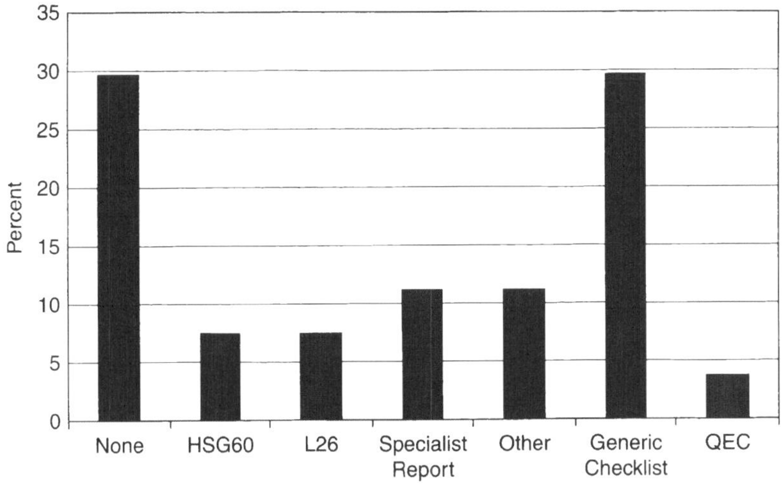

Figure 1. The distribution of assessment methods used in sonography departments.

abdominal, echo and vascular scans can take between 25–50 mins to complete. Most general ultrasound departments tended to work on a 15 to 20 minute slot for scans when booking appointments. It seemed challenging for a general or obstetric department to work effectively using a standard expected slot time given the wide range of scan times reported. Sonographers often commented that their work load encroached on their planned breaks. Some suggested that their workload was such that they were left with insufficient time for a proper lunch break. Extensions to the working day were frequently mentioned and at some departments it appeared to be common practice for extra patients to be added to the, already full, work list. There were suggestions that sonographers have been instructed to increase the number of patients they see in a day and/or take part in additional weekend or evening clinics without consultation or thorough assessment of the risks that this may pose.

MSD risk assessment

Seventy percent of trusts had undertaken some risk assessment for sonography work with respect to MSD (see Figure 1). Only 26% of the assessments were found to be suitable and sufficient in that they thoroughly examined the range of MSD risks involved. Adequacy of the assessment methods used was a concern at 90% of trusts with any MSD assessment. For example, the application of generic risk x probability style matrix assessments was commonly seen as was using a manual handling assessment approach (e.g. L23 manual handling checklist). The trusts that utilised in-house professionals such as OH physios or Backcare Advisors produced high quality assessments. Only 35% of assessments appeared to identify potentially useful risk reduction and control methods. Given that industry better practice guidance has been around for several years and was generally well known in sonography departments this was quite disappointing. However, it was often found that assessments did not document control measures.

Thorough assessment of the MSD risks is required at each department. They need to take account of the specific mix of risks associated with the sonography task, equipment used, scanning room, sonography staff, patients and work organisation. Checklist methods as are described in HSG60 and L26 (DSE regs) provide a useful basis. Some professionals (e.g. ergonomists or physiotherapists) have used quantitative MSD assessment tools like the Quick Exposure Checklist (QEC) to assess sonography risks. Involvement of these individuals would seem to be a very effective approach to take given the level of sonographer ill-health in the UK.

Risk reduction and control

Despite the poor quality of the risk assessments most departments (approx 90%) had made attempts to introduce MSD risk controls. The process of identifying control measures was usually not well documented or evaluated. Given the lack of assessment focus it was unsurprising that MSD issues and symptoms persisted despite some controls being introduced. Internal cooperation between departments that use sonography was poor in most cases as almost 75% showed no evidence of cooperation. Posture was the main focus for risk reduction and control and this overwhelmingly concerned the provision of seating.

The project illustrated that there are many ways to reduce / control MSDs in sonography. Selection of appropriate controls needs due consideration at department and trust level and for best effect needs to be underpinned by thorough risk assessment and worker involvement. It would seem most likely that a department would need to select a range of controls to reduce the risks to the lowest level reasonably practicable. It is essential that the controls deal with work organisation (e.g. work lists and slot timings) as well as operator posture and working conditions. Controls introduced to tackle the work organisational aspects were less apparent. However, it is these measures designed to tackle the psychosocial aspects of the work that seemed to offer untapped potential to help mediate the MSD problems in sonography. The findings of Dodd-Hughes (2008) seemed to corroborate this opinion.

Fundamental changes in scanning practice should be thoroughly examined. For example, adoption of methods to increase variation such as placing the patient in different positions ('Stockport position' in echocardiography) and using the left hand to scan (for some of the time). Changes to working practices can be difficult to introduce unless staff have the time and resource to develop the new skills. The involvement of individuals with ergonomics competence would seem to have a useful potential in terms of helping trusts and departments question the fundamentals of how their sonography departments operate.

Worker involvement

Information on worker involvement in the process of tackling MSDs was gathered (see Table 1). The majority of departments sought the involvement of their staff in

Table 1. Worker involvement (%) in departments when tackling MSD.

	No	Yes
Worker views are sought and valued	23	77
Actions to tackle MSD have been taken via worker involvement routes	50	50

health and safety matters (including ergonomics related issues). Around half the departments visited had introduced controls to reduce the MSD risks via worker involvement routes. However, this mainly concerned trialling seating options. The direction and effect of some of the worker involvement projects could have been better especially given the psychosocial risks observed. Departments that involved competent professionals (e.g. Ergonomists, Occupational Health Physiotherapists or Backcare Advisors) seemed to better facilitate the worker involvement.

The generally poor state of department risk assessments may have made it more difficult to identify and introduce controls – if the risks are not clearly understood how can methods to reduce them be identified? Direction of worker involvement appeared more effective when it was based on the risks of the job. Worker involvement is particularly important in terms of the scanning workload and this might be expected to help reduce the psychosocial burden.

Sonography workstation

A total of 59 exam rooms were inspected during the project. Conditions varied a great deal within and between departments.

The average age of the UltraSound(US) machines seen was 3.6 years (range 4 months to 10 years). Overall, newer models of US machine appeared to offer considerable improvements. They were found to be easier to manoeuvre and adjust, possess greater adjustability, lighter transducers and improved software compared to machines obtained 5+ years previously.

Most patient couches were pendant operated electric couches capable of height adjustment at the very least. Vascular departments tended to be equipped with couches capable of elevating the patient to an angle of 45 degrees or more for investigations of the legs. Echocardiography rooms sometimes used purpose designed 'echo' couches. These had handles to help patients adopt the right position and cut outs so access to the appropriate area was less of a hindrance. Patients are sometimes scanned while they are in bed or in a wheelchair. Sonographers reported that scanning a patient in the bed or in a wheelchair can be more awkward and demanding compared to scans on a purpose designed couch.

The room environments were also examined: 33% of rooms had dimmer switches, 37% were equipped with air conditioning and 24% were considered by the inspectors to be undersized. It was found that some aspect of basic room specification was an issue in approximately 80% of the consultation rooms visited during the project. Conversely, 20% of consultation rooms were found to be adopting better practice in terms of their specification for sonography work.

Education and training

In almost two-thirds of visits discussion with staff and managers led the inspector to conclude that discipline training did not include MSD risk awareness and coaching that was specific to sonography. At 20% of departments manufacturers of US equipment had provided department staff with some special training and this may have included effective operation to reduce MSD risks. Efforts to integrate the health and safety messages with the practicalities of getting the work done to the quality standard did not appear to be commonplace.

Patients with a high body mass index

The issue of scanning obese patients was raised in nearly every visit. Sonographers have often reported that when scanning deeper tissues the obese patient can require greater transducer pressure and reaching or stretching postures. Scan quality can be difficult to guarantee with obese patients and staff may persevere with the procedure and still end up with unsatisfactory image quality. Reducing unnecessary transducer time should be a key priority and the issue extends beyond the obese patient issues. Management systems such as an extension of bariatric patient policies combined with working procedures could help to take account of some of these difficulties. Some departments have found it effective to use paired scanning whereby two sonographers cooperate to complete the scan – 1 on the transducer and 1 on the control panel.

Other DSE work

During the visits poor DSE arrangements were often observed in the consultation rooms. There would seem to be useful scope for improvements here as this would have a beneficial influence over a sonographer's overall exposure to MSD risks. Once targeted, the trusts usually had the necessary DSE risk management system to tackle these risks appropriately.

Conclusions

Inspections of 14 NHS trusts in the UK were undertaken to examine how effectively MSD risks were managed in sonography work. Overall, performance was found to be much less effective than it should be and is cause for concern because of the high levels of MSD associated with sonography professionals and the large amount of industry specific guidance in existence. The main issues were:

- Risk assessments generally fell short of being considered suitable and sufficient
- Despite awareness of the MSD issue being very high action to tackle the problems tended to be quite superficial

- Risk reduction and controls in a few cases were diverse but generally not comprehensive enough to reach the lowest level reasonably practicable standard
- Workload was generally high in sonography and work organisational/ psychosocial factors appear to be very influential with respect to MSD issues
- In most cases sonographer training was not specific to MSD risks in sonography and how to conduct best practice scanning
- Occupational Health provision was generally good although often it tended not to be utilised in a proactive way for sonographers
- MSD health monitoring was rare even for individuals with established MSD problems
- Generally the arrangements for sonography work are less than optimal. This has the potential to affect human performance and increase the risk of clinical error

It is recommended that trusts adopt the HSG60 seven stage approach for managing MSDs in sonography and across other specialist departments. Efforts seemed to be most pressing in the area of risk assessment and control. NHS trusts should provide the necessary resources to help all departments tackle their MSD risks as effectively as possible.

Efforts should be coordinated across the various sonography modalities at a trust. Risk reduction should target work organisation. Increased sonographer control should be considered as well as efforts to balance workload to enable rest breaks to be taken. Sensible changes to the physical environment like the introduction of slave monitors can be effective. The considerable potential of introducing fundamental changes in sonography practice such as scanning with both hands and making effective use of voice activated systems should be a priority.

References

Brown, G. and Baker. J. 2004, Work-related musculoskeletal disorders in sonographers. *Journal of Diagnostic Medical Sonography*, 20, 85–93

Dodd-Hughes, K. 2008, *An Evaluation of Musculoskeletal Discomfort Experienced by Sonographers*, CSD HF Team report

Health and Safety Executive. 2002, *Upper Limb Disorders in the Workplace*, HSG60, HSE Books

Health and Safety Executive. 2002, *Work with Display Screen Equipment (L26)*, HSE Books

Magnavita, N. Bevilacqua, L. Mirk, P. Fileni A. and Castellino, N. 1999, Work-related musculoskeletal complaints in sonologists, *Journal of Occupational and Environmental Medicine*, 41, 981–988

Russo, A. Murphy, C. Lessoway V. and Berkowitz, J. 2002, The prevalence of musculoskeletal symptoms among British Columbia sonographers, *Applied Ergonomics*, 33, 385–393

Schoenfeld, A. Goverman, J. Weiss, D. and Meizner, I. 1999, Transducer user syndrome: and occupational hazard of the ultrasonographer, European *Journal of Ultrasound*, 10, 41–45

Society of Diagnostic Medical Sonography. 2003, *Industry Standards for the Prevention of Work-Related Musculoskeletal Disorders in Sonography*, SDMS, Texas

Society of Radiographers. 2002, *The Causes of Musculoskeletal Injury Amongst Sonographers*, SoR, London

Society of Radiographers. 2007, *Prevention of Work-Related Musculoskeletal Disorders in Sonography*, SoR, London

Smith, A. J. Wolf, G. Xie and M. Smith. 1997, Musculoskeletal pain in cardiac ultrasonographers: results of a random survey, *Journal of the American Society of Echocardiography*, 10, 357–362

Vanderpool, H. E. Friis, B. Smith, and K. Harms. 1993, Prevalence of carpal tunnel syndrome and other work related musculoskeletal problems in cardiac sonographers. *Journal of Occupational Medicine*, 35, 604–610

Village, J. and Trask, C. 2007, Ergonomic analysis of postural and muscular loads to diagnostic sonographers, *International Journal of Industrial Ergonomics*, 37, 781–789

THE DESIGN AND APPLICATION OF AMDAT, A HEURISTIC TOOL TO ASSESS ADOLESCENT USER NEEDS OF MEDICAL DEVICES

Alexandra R. Lang, Jennifer L. Martin[1], Sarah Sharples[1,2] & John A. Crowe[1,3]

[1]*MATCH (Multidisciplinary Assessment of Technology Centre for Healthcare)*
[2]*Human Factors Research Group, Faculty of Engineering, University of Nottingham, NG7 2RD, UK*
[3]*Electrical Systems & Applied Optics Research Division, Faculty of Engineering, University of Nottingham, NG7 2RD, UK*

Adolescents are currently overlooked in medical device development and the industry is in need of tools that target this specific user group. This paper describes the development of a new heuristics tool, AMDAT (Adolescent Medical Device Assessment Tool) for the evaluation of medical devices for use by adolescents. The development stages of the tool are presented, followed by a hypothetical example application of the tool.

Introduction

Adolescence *"a period of personal development during which a young person must establish a personal sense of individual identity and feelings of self worth which include an alteration of his or her image, adaptation to more intellectual abilities, adjustments to society's demand for behavioral maturity, internalizing a personal value system and preparing for adult roles".*

(Ingersoll, 1989)

The stage of adolescence, characterised by more changes and development than any other outside of infancy (Elliott & Feldman, 1990), has often been overlooked by academia. Medicine and healthcare research have sometimes provided a cursory nod to this audience, for example Moore describes how "only lip service is paid to the notion of giving children a voice in the development of health promotion intervention designs for them" (1998). With the exception of popular issues of health behaviours related to smoking, drinking and mental health it is still an under-researched stage of life.

With regard to medical device design and manufacture, whilst there is often significant consultation with adult and child user groups during design and development, teenage users are generally 'shoehorned' into using devices made for one or other of these groups. This is reflected by the lack of published ergonomics research on this topic. It is likely, therefore, that not all of the needs of adolescent users of medical equipment are currently being met, which may be resulting in low compliance with treatment regimens (Cameron, 1993; Fielding & Duff, 1999; Suris *et al.*, 2004). This is supported by a recent study by Lang *et al.* (2009) which asked a sample of adolescents to evaluate a variety of medical devices used by patients of the same age. The participants reported that current device design did not meet their specific needs, with particular criticism aimed at the aesthetics and social acceptability of the devices; issues which are likely to dis-incentive adolescents from using them in public and when in company. We know from the experiences of other research studies (Alderson, 2007; Lang *et al.*, 2009) that adolescent participants for research studies are difficult to access. Two reasons for this are the practical and ethical difficulties of accessing adolescents and the lack of appropriate tools for effectively engaging them in the design process.

The ethical guidelines surrounding the concept of 'capacity to consent' are numerous and conflicting. The UK General Medical Council (GMC, 2007) states that for clinical research there is no consensus on the age of consent for children's involvement in medical research and that "it depends more on a young person's ability to understand than their age…a young person under 16 may have the capacity to consent, depending on their maturity and their ability to understand what is involved". Whilst for non clinical research it is more difficult to involve adolescents, due to the fact that parental consent is required for research participants until they reach the age of majority; 18 years old in England and 16 years old in Scotland (BMA, 2001).

Researchers must also choose the topics they include in adolescent research carefully, balancing the need for sensitivity when dealing with personal issues that participants may not wish to discuss whilst at the same time respecting them and their advancing adulthood (Santelli *et al.*, 2003). Research should acknowledge that many adolescents will already be responsible for managing their own healthcare needs and therefore it is imperative that these are taken into account when designing equipment for them. To achieve this, the medical device industry requires appropriate methods and tools for involving adolescents in device development.

There are many tools for the elicitation of users' views available to the medical device industry, see Figure 1. These include traditional methods such as focus groups and interviews and more novel ergonomics methods such as contextual inquiry and cognitive walkthrough are also beginning to be used (Martin *et al.*, 2006). However the applicability of such methods to the study of adolescent views may be limited. For example observational research methods may be inappropriate for ethical reasons whereas focus groups and interviews may fail to effectively engage adolescents in the topic under investigation. There is a need therefore, for a novel approach of bringing adolescents into the design process which is rigorous as well as being practical for device developers to perform.

This paper proposes that a Heuristic Evaluation tool designed from data derived from adolescents and their proxies would provide a focussed method for considering these young people during the design process.

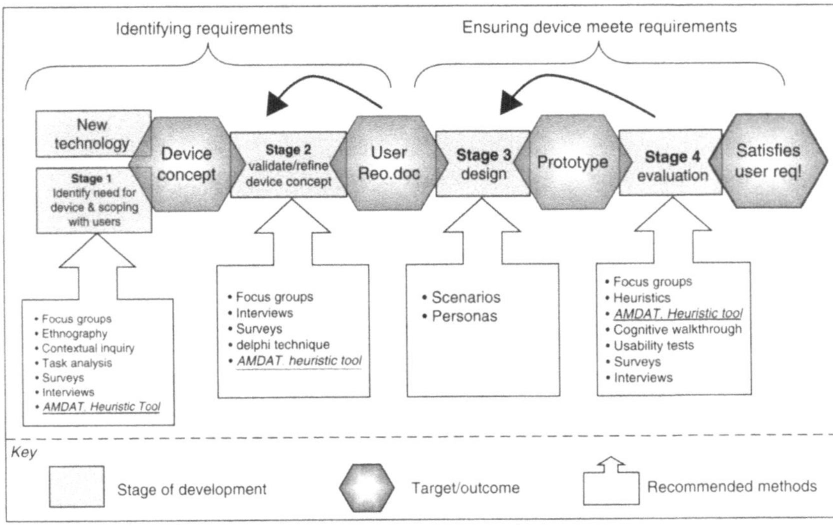

Figure 1. Medical Device Design Process.

Heuristic Evaluation is a discount (shortened) form of usability testing (Nielson, 1994), which uses evaluators to test a device rather than real end users and therefore may be a useful tool when real users cannot be involved (Zhang *et al.*, 2003). It is most useful in the early stages of evaluation to identify major usability problems with a device by assessing its compliance with a list of heuristics, principles or 'rules of thumb' that have been identified as important in meeting the device users' requirements. A list of heuristics against which medical devices designed for use by adolescents could be assessed would be a useful addition to the methods available to developers. It could allow them to conduct 'in-house' early evaluations of initial prototype designs to identify major issues before performing full user evaluations with real-adolescent users on a later, more refined prototype and could also provide benchmark scores to inform the evaluation of future devices.

This paper describes the development of a new heuristics tool, AMDAT (Adolescent Medical Device Assessment Tool) for the evaluation of medical devices for use by adolescents. The following stages of development are presented:

1. Generation of items for inclusion within the heuristics tool
2. Development of categories to group heuristics
3. Validation of the tool by administering items to experts

The final section of this paper presents a hypothetical example of the application of AMDAT to two sample user populations – adolescents and parents of adolescents – who might be asked to evaluate a medical device. The purpose of this was to demonstrate the way in which AMDAT can be used to identify differences between the views and opinions of adolescents compared with a proxy.

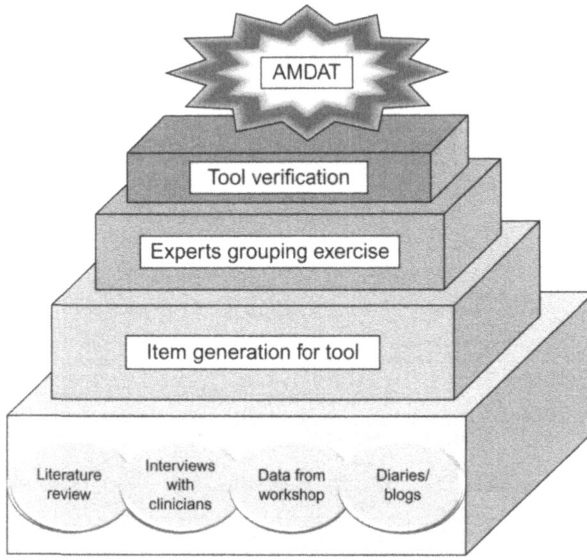

Figure 2. Developmental stages of AMDAT.

Development of AMDAT

A systematic approach was taken to develop AMDAT which was similar to that used by Wagenaar *et al.* (1994) in the development of the TRIPOD safety strategy, whereby a number of individual processes contributed to the generation of items and the development of heuristic categories, in a bottom-up manner (Figure 2)

Item Generation

To ensure that the items included in the AMDAT heuristic tool represented the breadth and depth of issues affecting adolescent use of medical devices, information from a number of different sources and different types of users was collected and considered.

Literature Review – A detailed review of the academic and grey literature on the subject of adolescents, medicine and device use was conducted. Literature was included from a range of disciplines including medicine, ergonomics, design, psychology and social science. 86 items were produced from this source.

Expert Interviews – A variety of different experts were consulted about the issues affecting adolescent use of medical devices including a variety of healthcare professionals (nurses, doctors, paediatric and adult physiotherapists and counsellors) and ergonomists; these contributed to 61 of the items.

Table 1. Heuristic Category Development 1st Stage Titles of Heuristic Categories Generated by Human Factors Professionals.

Researcher 1	Researcher 2
Everyday Life, Time, Portability	Social Acceptance
Aesthetics and Materials	Transport and Portability
Cognitive	Fit in with Life
Self Consciousness and Emotion	Aesthetics
Context and Environment	Encourage regular and correct use
Maintenance and Hygiene	Ease of use
Ease of Use and Comfort	Maintenance and Calibration
Anthropometry	Inclusive Design
Compliance, Behaviour and Technique	Customisable
Feedback and Interaction	Learning and Training
Social Acceptance and Relationships	Safety

Workshops with adolescents'- 4 workshops were conducted in secondary schools with a sample of 71 healthy adolescent participants (aged 14 to 18). They were presented with a variety of medical devices designed for use by adolescents and their opinions, concerns and suggestions collected.188 items were derived from this exercise.

Diaries and Blogs – Information from these grey sources provided an insight to the experiences of adolescents with long term health conditions. This contributed to 5 individual items.

This process resulted in the generation of 256 individual items. These were presented in question format, whereby the respondent would answer yes or no. The rationale for the items being presented as a question, with yes/no answers rather than a Likert Rating Scale was based on the techniques used in the development of TRIPOD by Wagenaar *et al.* (1994). They found that Yes/No question statements were a more practical and efficient mechanism for obtaining responses from participants.

Development of Heuristic Categories

The items were inspected by two human factors experts (1st and 2nd authors) who carried out analysis of the items generating categories and classifying the items into these (Table 1).

The results of this were discussed by the authors and any areas of repetition or disagreement examined. The individual heuristic items were also inspected and scored as either positive or negative according to their yes/no response. This process resulted in six final heuristic categories with definitions.

Portability and Practicality – Factors which either facilitate or impede the use of the device on an everyday basis so that it does or does not 'fit' in with the user's life and the many contexts and environments that this may include.

Social Acceptance – Factors which either encourage either engagement or alienation with the device; including elements such as relationships, society and acceptance or intolerance.
Maintenance – Involves the continual upkeep of the device to ensure correct working order. This may involve processes additional to the medical task which the device is designed for.
Ease of Use – Elements of design which enables or hinders the user to use of the device easily and with minimum complaint or inconvenience. This will include aspects of comfort and of cognitive ease of use as well as physical attributes.
Aesthetics – This element of design consists of the artistic and material qualities of the device; contributing to a user's perception of device attractiveness and visual satisfaction.
Adherence to regular and correct use – Factors of the device which either help to promote or to maintain the correct routine of device use. This includes aspects which prevent or deter the user from using the correct technique or that lead to low compliance

Originally there was a seventh heuristic: 'safety, accuracy and error prevention'. However it was decided that evaluating the safety of medical devices requires formal user testing and risk assessment and was, therefore, outside of the remit of the AMDAT tool.

Tool Verification

To initially test the AMDAT items and categories 18 lay participants completed a survey, ranking in order which, if any, of the heuristic categories each of the items was linked to. As the list of items requiring checking was long (256 items) it was anticipated that some participants would not fill out the questionnaires completely. To therefore avoid a situation where only the items at the start were classified, five versions of the survey were created.

Following this verification process a number of items were removed or re-worded resulting in a total of 240 individual items. The number of items in each category ranged from 240 to 38. 'Adherence to regular and correct use' is the category with the most items, This is consistent with the literature, which suggests that adolescents are often dissatisfied with the aesthetics and social acceptability of their devices which can result in low compliance (Lang *et al.*, 2009).

This section provides an example of how the AMDAT tool could be applied to the evaluation of a medical device by two different types of users: in this case an adolescent user group and a group which is commonly used as a proxy for adolescents: their parents.

1. **Selection of items.** For each individual heuristic survey, a random sample of items will be used; this is based on the technique used by Wagenaar *et al.* (1994) who state that this "enables the construction of many parallel tests that do not

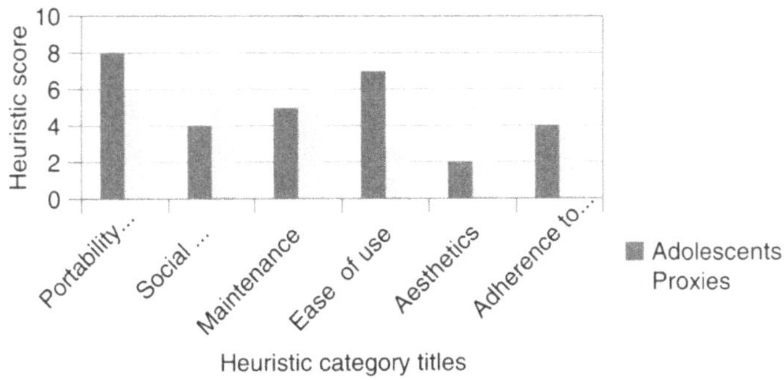

Figure 3. Examples of Adolescent and proxy AMDAT scores of an asthma inhaler.

have items in common". The number of items selected per category can vary according to the needs of the individual project but ideally should be no more than 12 so as not to place an excessive burden on respondents.

2. **Description of device**. The device is presented or described to the participant, using visual prompts where available. Information should be standardised to ensure that respondents have equal knowledge and experience of the device under investigation.
3. **Administration of tool**. Participants are presented with a list of item statements. The statements are not grouped into categories and are presented in a random order. Participants are instructed that they should carefully read and consider each statement in relation to the device and then answer yes or no in response to each statement.
4. **Scoring**. A score of 1 is assigned to an answer that indicates a positive aspect of the device. For some items, reverse scoring is applied, so an answer of 'no' would be assigned a score of 1, for others a response of 'yes' results in a score of 1. For each category, the maximum score (i.e. the most positive view of the device) will be equivalent to the number of items in that category.
5. **Analysis**. The scores can then be used to identify common and conflicting views both across and between different populations. Parametric statistics may be used, as the scores can be considered to be interval data.

Figure 3 demonstrates an example output of AMDAT and how the heuristic category scores provide easily interpretable visual outputs. This form of analysis will be a useful resource to developers, clearly presenting the levels of satisfaction between respective participant groups. This example shows that adolescents have rated the device lower with regard to social acceptance, aesthetics and maintenance that the parent proxy, suggesting that design development in these areas should be considered. The scores for portability and practicality and ease of use however indicate that adolescent requirements are more satisfactorily met.

Discussion

The aim of AMDAT is to provide a tool to aid the elicitation of adolescent user needs early on in the development of new medical devices. Adolescent requirements are frequently overlooked in the development of devices aimed at this age group and one reason for this is the lack of tools for engaging with this important user group, particularly at early stages of design evaluation.

A key challenge for this project was ensuring that the AMDAT tool covered the wide range of issues that are relevant to adolescent medical device users. To ensure this was achieved a number of adolescents and adults that work closely with adolescents were directly involved with the research. This information, combined with the other sources of data which fed into the tool, should ensure that the content, vocabulary and terms used in AMDAT are not only appropriate and relevant to this user group but will also be understandable by them.

Another challenge for a project such as this is developing a tool which is comprehensive enough to enable it to be applied to the evaluation of as wide a range of devices as possible, whilst allowing for some degree of customisation to fit the specific needs of any one device. AMDAT deals with this by allowing different scoring systems to be used. For examples, measureable integer scores can be calculated and may be useful for comparing different designs of devices that have similar purposes. Additionally, the scores from the different categories can be used to direct developers to aspects of device design which need further refinement or to conduct comparisons of existing devices to identify good and bad design features.

Another application of AMDAT could be to investigate whether the perceptions of medical device users systematically differ. It is often assumed that the proxies used to gather adolescent views such as clinicians and parents have different priorities and may underestimate the importance of issues such as social acceptance and aesthetics. AMDAT will allow issues such as this to be explored more formally than what has been done so far.

One feature which could improve the utility of AMDAT is a weighting feature to identify the priorities between the heuristic categories for different user groups. For examples if the heuristic group priorities were known for adolescent users, parents and clinicians then the responses given from these different respondents could be weighted so that their scores more accurately represented their views.

Developing the AMDAT to this stage has required extensive work, however extensive testing is now required and this is currently underway with the aim of identifying shortcomings and refining the tool. The practicality of the tool for the medical device industry must be tested, particularly because of the large number of items that have been generated. In practice will developers want to spend time producing individual surveys with randomly selected items? This may provide useful information, but when the time and budget pressures of design development require quick evaluations in design development, can a tool such as AMDAT feasibly fit into this?

There is also the issue of what developers do with the information from AMDAT, the tool will only identify issues that need further work, it will not provide the design solutions. It is important to stress that, as with traditional heuristics, this tool is not designed to be used in isolation but in combination with others methods as part of a general user-centred design approach to device development. It is intended to be used as an initial method for the evaluation of early prototypes; it should not replace full user-testing with an appropriate sample of real end users or appropriate risk analysis procedures.

The inclusion of adolescent needs in medical device design is long overdue. With the aid of this tool and the continued progression of this research it is hoped that a better awareness of this user group can be achieved and by addressing their needs in a systematic way; through the use of AMDAT, adolescent user requirements can be better satisfied with the aim of improving treatment adherence and better long term health.

Acknowledgements

The authors acknowledge support of this work through the MATCH Programme (EPSRC Grant GR/S29874/01), although the views expressed are entirely their own.

We would also like to acknowledge and thank the staff of the Paediatric Respiratory Team at the QMC Hospital Nottingham. We would also like to thank for their contributions, Zara Hooley of Helium Learning, Dr. Sue Cobb of the University of Nottingham, School of Mechanical, Materials and Manufacturing Engineering Ethical Review Committee and Prof. Elizabeth Murphy of The University of Leicester.

References

Alderson, P. 2007. Competent children? Minors' consent to health care treatment and research. *Social Science & Medicine* 65(11): 2272–2283

British Medical Association (BMA). 2001. *Consent, Rights and Choices in Healthcare for Children and Young People.* (London, UK. BMJ Books)

Cameron, K. T. 1993. Treatment compliance in childhood and adolescence. *Cancer* 71(S10): 3441–3449

Elliott, G. R. & Feldman, S. S. 1990. *Capturing the adolescent experience. At the threshold: The developing adolescent I.* 1–13. (Cambridge, MA. Harvard University Press)

Fielding, D. & Duff, A. 1999. Compliance with treatment protocols: interventions for children with chronic illness. *Arch Dis Child* 80(2): 196–200

General Medical Council (GMC). 2007. 0–18 years: guidance for all doctors. (London, UK. General Medical Council)

Ingersoll, G.M. 1989. *Adolescents*. (Englewood Cliffs, NJ. Prentice Hall)

Lang, A.R., Martin, J.L., Sharples. S. & Crowe, J.A. 2009. *Enabling adolescents to participate in the design and improvement of medical devices*. International Ergonomics Association, 17th World Congress on Ergonomics. Beijing, China. 9th–14th August 2009

Martin, J. L., Murphy, E., Crowe, J. A., & Norris, B. J. 2006. Capturing user requirements in medical device development: the role of ergonomics. *Physiological Measurement*. 27: R49–R62

Moore, H. L. & Kindness, L. 1998. *Establishing a research agenda for the health and wellbeing of children and young people in the context of health promotion. Promoting the health of children and young people. Setting a research agenda*. London, Health Education Authority (HEA)

Nielsen, J. 1994. *Usability Engineering*. (San Diego. Academic Press) pp. 115–148

Suris, J. C., Michaud, P. A. & Viner, R. 2004. The adolescent with a chronic condition. Part I: developmental issues. *Arch Dis Child* 89(10): 938–942

Santelli, J. S., Smith Rogers, A., Rosenfeld, W.D. *et al*. 2003. Guidelines for Adolescent Health Research: A position paper of the Society for Adolescent Medicine. *Journal of Adolescent Health* 33(5): 396–409

Wagenaar, W. A., Groeneweg, J. & Hudson, P.T.W. 1994. Promoting safety in the oil industry. *Ergonomics*. 37. No 12. 1999–2013

Zhang, J., Johnson, T. R., Patel, V. L., Paige, D. L., Kubose, T. 2003. Using usability heuristics to evaluate patient safety of medical devices, *J. Biomed. Inf.* 36, pp. 23–30

IMPLEMENTING THE WHO SURGICAL SAFETY CHECKLIST AT A UK HOSPITAL

K. Bowling[1], G. Moon[1], F. Austin[1] & F. Watts[2]

[1]*Human Engineering Ltd, London, UK*
[2]*National Patient Safety Agency, London, UK*

In 2009, the National Patient Safety Agency (NPSA) issued an alert to UK NHS surgical organisations requiring the launch of the World Health Organisation (WHO) Surgical Safety Checklist (SSC). This Checklist is a tool which aims to reduce clinical safety risks associated with surgical procedures. In collaboration with the NPSA, Human Engineering undertook a programme of work to support a UK Hospital (H1) to prepare for this implementation. It was understood that successful SSC application would require significant process and behaviour change. A user centred design approach incorporating observations and interviews was used to identify barriers, develop persona characteristics of user groups and process map existing pathways so that an implementation plan could be agreed. This paper outlines significant learning points and the recommendations of developing an implementation plan for the SSC in a UK hospital.

Introduction

In England and Wales in 2007 the NPSA received 129,419 surgical incident reports from thc 8 million surgery procedures that were conducted. Of these over 1,000 resulted in severe harm and 271 resulted in the death of the patient. Although it is unclear if all these incidents were preventable, the term 'never events' has been applied to serious, largely preventable patient safety incidents that should not have occurred if the available preventative measures are implemented. 'Never events' are not a symptom of the NHS, they are the result of human failure in the high risk medical practices which includes the arena of surgery, where complex operations requiring the skills of a team of people allow human weaknesses in memory, attention, team working and communication to have disastrous consequences. This problem is mirrored around the globe to such an extent that in June 2008, the WHO launched the 'Safe Surgery Saves Lives' campaign, to reduce the number of surgical deaths across the world. As part of this initiative the WHO developed the Surgical Safety Checklist (SSC) which aimed to reduce the risk of human error in a surgical setting. The National Reporting and Learning Service (NRLS) in collaboration with an expert reference group adapted the checklist for use in England and Wales, which the NPSA issued as a Safety Patient Alert in January 2009. This alert required organisations to ensure an executive and a clinical lead are identified in order to implement the surgical safety checklist within the organisation; ensure the checklist is completed for every patient undergoing a surgical procedure (including local anaesthesia) and; ensure that the use of the checklist is entered in the clinical notes or electronic record by a registered member of the team.

Early indication from research conducted by Haynes *et al.* (2009) in a year long, international pilot study suggests that implementing the checklist successfully has positive outcomes for patients. For example, introduction of the checklist showed a reduction of nearly a third in inpatient complications and a staggering reduction of deaths of over 40%. However, with one human factors issue dealt with, another became apparent. The success of the SSC will be directly related to its successful implementation. As noted by Mahajan and Gemmell (2009), there will be a number of human factors issues that contribute to the practical and productive use of this tool such as strong visible leadership, knowledge of the tool and motivating staff in its use, as well as the inevitable adaptation and/or alignment of processes and procedures, which vary slightly in every institution.

Following a UK pilot at a London surgical centre, the SSC was modified for a UK setting. The next stage was to roll out the SSC across UK surgical centres, which included H1. The modernisation team at H1 had already done much to implement NPSA alerts and actions, including an audit of compliance with the NPSA Correct Site Surgery Checklist. Specifically, H1 had already established a team of clinical leads to co-ordinate the roll out of the Checklist across its hospital theatres. Human Engineering supported the next stage of planning and implementation, covering two key focus areas:

- The integration of the SSC into existing surgical pathways (including into the patient pathway documentation).
- Alignment of surgical team behaviour with the requirements of the SSC, and overcoming any barriers to achieving this.

In addition to the above aims, Human Engineering considered three other factors when designing the approach. The first was finding the time, capability and resource to implement the WHO SSC in an effective manner. The second was dealing with negativity and scepticism of surgical staff towards the SSC. The third was implementing the SSC in the 'right' way, to overcome barriers and minimise risk.

Method

It was understood that successful SSC implementation would require significant process and behaviour change. The project methodology focused on the use of observations of existing surgical procedures and interviews with a range of different surgical staff. This user-centred design process put a focus on understanding the characteristics, requirements and barriers associated with the end-user. The end goal was to develop recommendations for the integration of the SSC in a way that is usable, practical, minimises the risk of human error, and overcomes any specific barriers identified. The approach taken by Human Engineering was therefore based on four key concepts:

- To focus on the requirements and characteristics of people who will be expected to use the SSC (taking a user-centred approach)
- To fully understand the barriers that will prevent effective application and take up of the SSC.
- To integrate the checklist in a way that fits best with the existing process and ways of working.
- To carefully plan and execute implementation, making use of a phased approach.

To gather the necessary information the following methods were used:

Interviews

A series of site visits took place over a three week period. Interviews were completed with staff involved in the surgical pathway, from a range of disciplines and specialisms. The sample consisted of a Theatre Matron, 2 Theatre Managing Nurses, 2 Consultant Anaesthetists, 2 Anaesthetists, 3 Consultant Surgeons (orthopaedics, gynae-oncology, and general surgery) a Managing Nurse Day Surgery (pre-operative holding area) 2 Operating Department Practitioners (ODPs), 3 Healthcare Assistants, a Risk Management Sister, and a Modernisation Project Nurse. A proforma was used to structure these interviews which included questions examining knowledge of the SSC and its potential benefit, as well as comparisons with current procedures and open ended questions designed to get feedback from the participants on implementation issues.

Observations

Observations of the existing day surgical pathway and procedures were completed over three days. This included an observation of the 'patient journey' through day surgery, from arrival at reception through to recovery. The interaction between staff and the patient, as well as any existing checks performed, were noted. In addition, the perioperative process was observed repeatedly across several theatres, covering three different surgeons, three different types of anaesthesia, and three different surgical procedures. This included the observation of 3 mastectomies (general anaesthetic), 2 hip injections (local anaesthetic) and a hip replacement (spinal anaesthetic). The observations focused on the day surgery preoperative process due to the useful observations that could be gained from the quick turnaround of patients.

Documentation review

In addition to interviews and observations existing day surgical pathway documentation was assessed against the contents of the SSC, and any duplications were recorded.

The data gathered were used in variety of ways to meet the end goal of providing practical recommendations. First the information collected allowed a process map specific to H1to be constructed so that process barriers could be identified. The interview data and observations were used to develop persona typologies. These are a way of organising end user characteristics and requirements, which reflect existing attitudes, opinions, and concerns of end user groups. This information was then applied to guide decisions about the implementation of the WHO SSC to suit preferences and ways of working.

The persona results, along with interview and observation data was used to identify other barriers to the implementation of the SSC with particular reference to attitudes and knowledge barriers, and culture and organisational barriers.

Results

Personae development

Four key personae were identified when examining the factors of goals and priorities; understanding of the SSC; experience; attitudes; working environment; and specific barriers or facilitators to using the SSC. The personae do not relate to specific individuals. They are an amalgamation of characteristics derived from a series of interviews and observations.

These personae were:

- **The advocate:** Sees that many benefits can be derived from the use of the SSC – "The SSC would be a great mechanism for patient safety as well as improving communications and teamwork". At its extreme, this group included 'evangelists'.
- **The tentative supporter:** Supportive of the theory behind the SSC, but sceptical about the way that it is implemented – "The SSC will be good if we ever get it implemented- however we have seen initiatives like this fail before…".
- **The cynic:** Very sceptical of the SSC – "this is a political tool- we do not need this. Experienced theatre staff are seasoned professionals who rarely make errors. They already know that they must check and do it already".
- **The efficiency-focused:** Agrees with the SSC concept but is concerned about the impact on patient pathways – "I am concerned about the extra paperwork, and the impact on process efficiency, particularly that associated with 'time out'".

A common observation in similar change programmes is that seemingly initially supportive and engaged individuals could become compliant but minimalist performers when it comes to execution. This is hard to predict at this stage, however it would be beneficial to employ specific measures during training to reduce the likelihood of this.

The identification of barriers to the implementation of the SSC

The results indicated that barriers to implementation fitted into three key categories – (a) process, (b) attitude and knowledge, and (c) culture and organisation.

Process barriers

Process barriers are related to existing surgical systems and processes that may perpetuate behaviours that hamper the progress of adopting the SSC, or may conflict with the use of the SSC with a detrimental result. The first was a common viewpoint that an efficient, sustainable day surgery process had been established, which included patient turnaround times. Key priorities were to minimise turnaround time and to reach national pathway targets (i.e. the time between treatment decision and completion of the procedure). There was concern that the use of the SSC would negatively impact this carefully designed and refined process and make it unsustainable. In particular there was a perception that the SSC would take too long, would create a delay in the surgical process, and would thus impact targets and turnaround time.

There was also concern that the SSC conflicts with existing checks. It was noted that the pathway already has a series of formal checks. This implementation barrier concerns the perception that (a) the SSC adds unnecessary 'repeats' of these checks, and (b) at the final stage, the SSC check is 'too late'.

There was concern that the SSC requires a change in working practice, in terms of the formalisation and documentation of checks that are currently nonexistent, informal or verbal. The perception was that this would mean an increase in paperwork. A barrier to change was therefore the fear that this paperwork will take attention away from practical tasks at busy and critical points of the surgical process (e.g. immediately prior to incision). A general comment was that overall paperwork is disliked and is seen as a chore.

The final process barrier identified was the potential impact on the patient experience by asking repeated questions through the pathway (particularly those concerning confirmation of name, procedure, and site). There is the concern that this will act as a barrier to some members of staff performing these checks with the patient at Sign In.

Attitude and knowledge barriers

Attitude and knowledge barriers related to any SSC requirement for skills, behaviours and abilities that were different from those required in the past. In general, there was evidence of a general lack of understanding of the psychology and human factors reasoning behind the SSC, which could undermine the motivation of people to attempt to change behaviours.

There was a tendency for some individuals to overestimate their own, or their team's skills, or the quality of existing hospital processes. This could lead to an underestimation of the risk of an error occurring. The perception that human error 'does not apply', would naturally lead to the attitude that the SSC is irrelevant. If staff believe that the SSC is unnecessary, then it will be very difficult to establish any sort of momentum for change. Unfortunately however this tendency to overestimate was recognised only through subjective observation and was not examined by any statistical means.

There was a lack of understanding of the range of SSC benefits and these were superseded by the issues raised. These benefits include:

- Empowerment of more junior members of the team through providing them with an opportunity to voice opinions and concerns.
- Increased communication and information sharing in the theatre.
- Potentially the reduction of some repetitive checks by several members of staff.
- Increased collaborative decision making between theatre staff.

Finally, one of the major impacts to staff would be the requirement to verbalise information to a theatre audience that currently may be internalised, or may occur in an informal, one-to-one fashion. This will require the development of new skills and attitudes relating to theatre behaviour. Some staff members expressed the concern that they would not get heard or listened to in noisy environments, and many theatres were also observed to be relatively quiet and non-vocal places.

Culture and organisational barriers

These barriers related to specific factors associated with the existing culture and organisational systems that may hamper adoption of the checklist, e.g. use of language, leadership, and communication. A few issues were found regarding language used in the SSC and how this influenced individuals' perception of the tool. The SSC has been developed following a process of global pilot studies and trials to promote consistency across countries as much as possible. The language of the SSC is therefore designed from a global perspective. This has also been evaluated in a UK context at St Mary's Hospital in London. However, there are some aspects of SSC language that may not suit each individual hospital's culture and environment. Many of these points relate to specific SSC items, for example, what exactly the term 'medicine check' means. Another example of a mis-match in terminology is the use of 'Sign in', 'Time Out' and 'Sign Out'. On first appraisal of the SSC, these terms were viewed as quite 'Americanised' to some individuals interviewed. 'Time Out' for instance has immediate negative connotations concerning lean processing and time efficiency.

Effective implementation of the SSC relies on a democratic culture of communication, whereby all team members listen and value the input of other staff, regardless of seniority or experience. It is also important that more senior members of the team provide junior members with the respect and confidence to provide views and opinions. The existing theatre culture observed is currently a relatively 'quiet' environment in a verbal/conversational sense. There is a great reliance on non-verbal communication and the assumption that all team members 'already know what they are doing'.

Recommendations

Recommendations made as part of this project and the resulting full implementation plan will not be of interest to the majority of readers as they are specific to the processes and users at H1. Therefore more general learning points and recommendation are outlined below that should give the reader some insight into developing plan for a typical SSC implementation.

Recommended integration approach

The main practical considerations when integrating the SSC into existing surgical pathways and patient documentation, in a way that overcomes implementation barriers were:

1. Where in the day surgery process the SSC items are completed.
2. Who takes responsibility for (a) verbally leading the SSC, and (b) controlling the documentation?
3. How the SSC will be integrated into documentation.
4. How the completion of the SSC will be recorded (i.e. by ticks or signatures).

A useful approach was to evaluate the above options against the following criteria:

- Existing process: Does the option have a close fit to existing ways of working and the current patient pathway? There will be less of a change impact if the SSC can be aligned with existing ways of working where possible.
- Staff preference: What was the feedback during interviews and observations? Is this likely to be a well-received approach? If not, why not?
- Theatre performance: What impact would this option have on theatre performance in terms of patient turnaround, target pathways, procedure time?
- Compliance: Is this option in line with the guidance provided by the WHO and the NPSA? Does it fulfil the implementation criteria? There is little point in implementing the SSC if it is not compliant and does not bring the correct benefits.

The main general finding that was unearthed using the above approach was that to achieve compliance with the WHO it was recommended that the SSC was implemented in its current one-page format. However, a recommendation was to add an adapted pre-operative checklist to the other side of the SSC, making it a two-sided document for maximum usability during the pathway. Side one would contain an adapted pre-operative checklist for completion in the pre-operative holding areas by the escort, and side two would contain the WHO checklist. This recommendation would allow current forms and checks to be consolidated into this document reducing duplication and formalising a number of verbal checks.

Recommended implementation approach

A workshop involving the implementation team was supported by the NPSA and Human Engineering. This workshop served to feedback the results of the observations, and described the key barriers that were to be overcome. The implementation team then collectively developed a specific implementation plan that would suit the needs of the hospital. This included the development of phases of the project with timelines, taking into consideration piloting the SSC.

Conclusions drawn from this workshop and from human factors good practice indicated a number of implementation principles and communications approaches that could be used to suit the specific requirements of H1. However, these are largely applicable to any SSC implementation.

The implementation principles included:

Involvement of users throughout: During this initiative a number of surgical staff had been involved and consulted during the development implementation plan. However, many others could not be included so it is important to consult a wide range of staff throughout the implementation.
Feedback and communication: Opportunities should be provided for the provision of feedback and issues from the end users. The implementation team should create an open approach that is focused on listening to feedback and improving where necessary, as well as communicating the reasons behind any decisions made.
Consider the barriers raised including those in the personae: These should be dealt with one by one through careful planning.
Use trials in a limited number of theatres to explore the options.
Co-ordination of implementation effort: The implementation should follow a structured, considered implementation approach which is controlled and co-ordinated by a central implementation team. This should include the message that existing processes should be adhered to until a designated launch date.
Planned and controlled communications: Try to prevent misleading communications coming from sources other than the implementation team.
Focus on getting it right: The implementation plan should allow enough time to get things right, and should not be rushed.
Consider a phased approach: For instance rolled out over theatres in stages, or implementing one part of the SSC at a time (e.g. Sign In first).

Communications approach

One of the key advantages of the approach taken for this project was the potential for the implementation team to tailor a communications strategy specifically for the surgical team at H1. Specifically it led to a development of frequently asked questions sheet which can be used by leaders to reassure staff but can also act as a training guide.

Key communication messages that will be important to circulate amongst surgical teams were also identified. Many of these specific messages were based on the barriers identified. An indication of the type of messages that are more generally applicable are outlined below:

- Surgery should be a team effort. Each member of the team contributes and has valid opinions and viewpoints in helping to reduce error and maximise patient safety.
- A team that works effectively together to engage their knowledge and abilities on behalf of a patient can avert a considerable proportion of life-threatening complications (Gawande, 2009).
- The implementation plan for the SSC and the fact that it must be implemented
- The benefits of the SSC (and not just to reduce human error- the SSC will bring other benefits such as improved communications and empowerment).

- Anyone can make errors. Errors are not usually the fault of one individual but a culmination of multiple failings in the system. The SSC helps to address this.
- What documentation the SSC replaces so that time savings are communicated.
- The SSC is based on a real need to reduce human error in surgery, incidents statistics.

Finally, the approach allowed training requirements to be identified which fall under four broad topics:

- Basics in the WHO Checklist and why it is required
- Understanding the new pathway
- Verbalising the Checklist
- Teamwork & leadership styles in the theatre

Conclusions

Since the implementation plan is currently underway implementation results cannot be confirmed. Therefore there is no evidence that the above approach is the 'best' way of implementing the SSC and there are factors that inevitably will arise at other medical institutions that were not found in this research. What this project has helped identify is that with a comprehensive human factors assessment that considers the task, the people and the process in relation to current and new ways of working it is likely that any implementation will fall short of the desired objective. In the wider context, using independent staff skilled in behaviour change and human factors analysis means that investigations can be carried out in manner that surgical staff are more likely to perceive as fair, and therefore real issues can be raised and considered in the implementation plan.

References

Gawande, A., 2009. The *Team is Key: Endorsement of the WHO Surgical Safety Checklist*. www.who.int/patientsafety/safesurgery/en/

Haynes, A.B, Weiser, T. G., Berry, W. R., Lipsitz, S. R., Breizat, A. S., Dellinger, P., Herbosa, T., Joseph, S., Kibatala, P. L., Lapitan, M. C. M., Merry, A.F., Moorthy, K., Reznick, R. K., Taylor, B., and Gawande, A. A., 2009. A Surgical Safety Checklist to Reduce Morbidity and Mortality in a Global Population. *New England Journal of Medicine*, 360; 5.

Mahajan, R. P. & Gemmell, L. 2009. Saving Lives in Surgery. *The Ergonomist*, 471: 12–13

National Patient Safety Agency, 2009. *Patient Safety Alert.* Available here: http://www.nrls.npsa.nhs.uk/resources/type/alerts/?entryid45=59860

National Patient Safety Agency and the Royal College of Surgeons, 2005. *Patient Safety Alert: Correct Site Surgery.* www.npsa.nhs.uk.

World Health Organisation, 2009. *Surgical Safety Checklist Implementation Manual*. www.who.int/patientsafety/safesurgery/en/

KEY TEAM BEHAVIORS DURING IN SITU CRITICAL EVENTS USING BEHAVIORAL MARKERS

W. Riley[1], S.E. Davis[2], K.K. Miller[2], H.E. Hansen[1], F. Sainfort[1] & R. Sweet[1]

[1]*University of Minnesota, USA*
[2]*Fairview Health Services, USA*

Effective team communication is essential to ensure patient safety, yet little is known regarding team behavior in clinical settings during critical events. This study examines differences in key team behaviors between physicians and nurses during simulated critical events. Perinatal staff participated in 11 in-situ simulation trials over a six-month period involving 8 MDs, 17 RNs, and 20 other staff. Videotapes of simulations were evaluated and scored to assess five categories of communication behaviors over different stages of OB critical events scenarios. Results indicate that physician and nurse professionals performed differently on key behavioral markers during critical events, and such performance is associated with different phases of team formation and reformation. Implications for patient safety are discussed.

Introduction

There is growing recognition regarding the importance of non-technical skills in health care to ensure patient safety. For example, communication failure is associated with over 70 percent of sentinel events, underscoring the necessity of effective team communication. However, very little evidence is available to indicate non-technical skill performance of rapidly formed health care teams based on behavioral markers developed specifically for health care. This study examines differences in key team behaviors among team members during simulated critical events. The purpose is to determine whether there are differences between physicians and nurses in key non-technical skills during critical events based on grounded theory about how health care teams differ from other types of teams during specific stages in critical events.

Background

Non-technical skills in health care are the cognitive and interpersonal skills necessary to ensure safe patient care which supplement clinical and technical skills (Yule *et al.*, 2007, Paisley *et al.*, 2001). While cognitive skills (e.g., memory, decision making, perception) are impaired by fatigue, stress and interruptions, interpersonal skills (e.g., leadership, teamwork, communication) are lacking due to an absence in professional training curricula. Consequently, different strategies are required to ensure patient safety and achieve high reliability (Miller *et al.*, 2009). Safeguards to compensate for deterioration of cognitive skills need countermeasures and forcing functions, while strategies to improve interpersonal skills need training and practice. While non-technical skills are critical for good medical practice, they are typically not addressed explicitly in normal training. The need to train and assess non-technical skills is growing, yet these activities must be based on properly developed frameworks and validated measurement tools (Fletcher *et al.*, 2003).

Teams in health care, however, differ in fundamental ways from teams in other settings such as the military, sports teams and teams in business settings (Riley *et al.*, 2008, Miller *et al.*, 2009). Many types of healthcare teams do not have stable membership or constant leadership, and team members rarely train together (Riley *et al.*, 2009). Much of the theory regarding team behavior in health care is imported from other fields and not based on how health care teams actually behave in clinical settings during critical events. There is a need to examine differences in key interpersonal behaviors among various health care team members. Previous findings indicate that rapidly formed health care teams exhibit sequential patterns of team formation and reformation during critical events (Riley *et al.*, 2008). In this study, we explore potential differences between nurses and physicians in non-technical skills during different phases of critical events.

Methods

Study Setting and Design

We used an *in-situ* simulation strategy with a prototype behavioral marker system to assess non-technical team skills of health care providers during critical events. Perinatal staff participated in 11 *in-situ* simulation trials over a six-month period in a medium-sized suburban community hospital perinatal unit in the Midwest in the United States. All staff on the labor and delivery unit and operating suite was eligible to participate with 47% of the clinical staff taking part in the experiment. The professional categories included 8 MDs, 17 RNs, and 20 others from various departments in the hospital. A team of experts scored simulation videotapes on four categories of non-technical skills during two stages of labor and delivery critical events scenarios.

We selected two stages of team formation within a critical event where non-technical skills of nurses and physicians are most prominent. In the first stage – identification of a clinical crisis by the nurse (RN) – the nurse leadership is the critical team factor whereas in the second stage – declaration of an emergency by the physician (MD) – physician leadership is the critical team factor. Each stage was reviewed and scored on four categories: 1) situational awareness, 2) shared mental models, 3) leadership and teamwork, and 4) closed loop communication which includes Situation-Background-Assessment-Recommendation-Response (SBAR) communication). This study is part of a larger project to assess and improve team performance during critical events. The university and hospital Institutional Review Boards approved the protocol for the entire project.

Behavioral Markers

Behavioral markers are observable non-technical skill behaviors to measure performance (Thomas *et al.*, 2004, Fletcher *et al.*, 2003) that contribute to team effectiveness within a work environment (Carthey *et al.*, 2003). *Behavioral markers systems* are rating scales based on taxonomies used to identify observable, non-technical behaviors that measure performance (Flin *et al.*, 2008). Behavioral markers are derived from empirical data and developed into structured skill taxonomies, combined with a rating scale. Behavioral markers are used as frameworks for assessment, training and research to develop and test countermeasures.

Behavioral makers have been used in the medical domain starting with anesthesia (Howard *et al.*, 1992, Greaves and Grant 2000, Sexton *et al.*, 1998, Fletcher *et al.*, 2002, Fletcher *et al.*, 2003) and emergency medicine (Small *et al.*, 1999). They can be identified by focus groups with clinical experts (Thomas *et al.*, 2004, Yule *et al.*, 2007, Fletcher *et al.*, 2003) or by frequency of behavioral misses in a laboratory setting (Riley *et al.*, 2008, Davis *et al.*, 2008). In our broader study, we developed a weighted interdisciplinary team behavior (WITBI) index to identify and measure the behavioral markers. Our behavioral markers are based on an event set methodology, which require the desired behavioral markers to be performed (Hamman 2004). While errors and threats management are important components of behavioral marker systems in safety critical industries (Helmreich and Foushee 1993, Carthey 2003), we developed scenarios to incorporate conditions (setting and clinical symptoms), triggers (sudden change in clinical condition which must be correctly recognized by providers) and distracters (interruptions and threats to patient safety).

Behavioral markers have been created using content experts to develop lists of behaviors that are important for safer patient care (Gaba *et al.*, 1998, Carthey *et al.*, 2003, Thomas *et al.*, 2004, Fletcher *et al.*, 2003, Yule *et al.*, 2007). Rather than create lists of teamwork behaviors based on expert opinion, we created the markers based on empirical observation using four stages: 1) literature review to identify team failures associated with patient safety failures, 2) application of key behaviors from the AHRQ TeamSTEPPS team training program (AHRQ, 2009), 3) a pilot test with 27 focus group debriefings to identify learning lines based on the simulation experience, and 4) categorization of behavioral markers based on observed frequency in the actual findings. In this study, we used four categories of behavioral markers associated with effective team functioning during two stages (stages 2 and 3 selected from six stages identified in the broader study) of a critical event as shown in Table 1. Each category is comprised of 4 to 5 behavioral elements.

Data Collection

Video recording produces reliable information on team performance (Mackenzie *et al.*, 1996) and is useful in extensive drill down of behavioral markers by raters. For this study, we video-recorded 11 *in-situ* simulation sessions of experienced health care teams in their actual work setting. Video recording was used for two purposes 1) as a discussion aid with the teams during debriefings, and 2) to allow images to be stored for viewing at a later time, and for multiple reviews and analyses of

Table 1. Behavioral Markers Categories and Examples.

Human Factor Category	Stage 2 (RN) Examples of Behavioral Elements	Stage 3 (MD) Examples of Behavioral Elements
Situational Awareness Conscious, mindful observation of one's own environment or recognition of patient condition	**6 Elements** • RN verbalizes changes in maternal or fetal condition • RN minimizes distractions from critical tasks	**4 Elements** • MD verbally acknowledges RN's information. • MD is timely in verbalizing assessment of deteriorating maternal and fetal condition
Closed Loop Communication Communication to a specific person that is acknowledged by the receiver and then affirmed by the sender	**3 Elements** • RN (primary or secondary) acknowledges receipt of information, instruction, requests • SBAR (Situation, Background, Assessment, Recommendation, Response)	**4 Elements** • MD checks for understanding with mother, spouse/partner • Questions posed to elicit clarification by MD to RN • SBAR (Situation, Background, Assessment, Recommendation, Response)
Shared Mental Model A team trait characterized by an articulated common understanding of the problem and/or the plan. Everyone "being in the same movie"	**4 Elements** • RN establishes direct communication with MD, requests his/her presence to evaluate mother's condition • RN engages in clear, concise, timely exchanges of information with mother and family about fetal condition and treatment plans	**4 Elements** • MD and RN verbalize essential information about mother's condition, including urgency and agreed plan of care. • MD calls Code C-section promptly and clearly to notify OR personnel to assemble for emergency
Leadership Explicit handoff of responsibility or providing direction from one team member to another. Often occurs when team reforms or mission changes	**4 Elements** • RN maintains nursing leadership role specifies roles, assigns, coordinates tasks or transfers leadership to another RN • New team members introduced and integrated	**5 Elements** • MD establishes and maintains medical leadership role • MD verbalizes instructions, information and likely intervention

each event. Eleven *in-situ* simulations were video recorded and analyzed by a team of expert reviewers using a behavioral marker tool developed for measuring non-technical skills during two stages of critical events scenarios to isolate the physician and nurse behaviors.

We used a form divided into stages of team formation and reformation sections. While both observability and frequency (Thomas *et al.*, 2004) could be rated with two separate scales, numerous activities occur simultaneously by many persons

Table 2. Measurement Index Composition.

	Stage 2 # of Items	Stage 3 # of Items
Situational Awareness	6	4
Shared mental Model	3	4
Leadership and Team	4	4
Closed Loop Communication	4	5
Total	17	17

during critical events. Thus, instead of recording cumulative behaviors separately, which would not have been reliable, and because we were examining both individual and team behaviors, we combined both scales into a three-point scale as follows:

0 = Behavior Present ≤49% of the time
1 = Behavior Present 50–89% of the time
2 = Behavior Present ≥90% of the time

Table 2 shows the structure of the instrument and the number of items measured for each of the four categories for each stage.

Scale Construction

We weighted the four behavioral markers by ranking their importance within each stage. The importance of each behavioral marker varies according to the stage. For example, situational awareness is more important than closed loop communication when identifying a serious change in a patient's condition, whereas closed loop communication is more important than situational awareness when conducting a leadership transfer between the two stages. Three clinical experts (MD/OB, RN clinician and RN researcher) rank-ordered the behavioral markers within stage 2 and stage 3. Rank-reciprocal weights were assigned to the markers accordingly. A similar process was used for the items within each behavioral marker categories. Item weights were used to compute a weighted additive score for each marker. Marker weights were used to compute a weighted additive score for each stage.

Results

Inter-Rater Reliability

Inter-rater reliability is of particular concern when creating behavioral markers for health care (Fletcher *et al.*, 2003). Inter-rater reliability was assessed using the Kappa statistics. Each item in the instrument is rated as an ordinal variable with three values (0, 1 and 2). We arranged the pairs of responses for each item in a 2-way contingency table with a pair of independent raters agreement on a given item if both assign the same rating to it. For such variables, three measures of concordance can be calculated: percentage agreement, Cohen's kappa, and weighted

Table 3. Percentage Agreement and Kappa Statistics.

	Percent Agreement	Kappa	Weighted Kappa
Stages			
Stage 2	72.7%	0.300	0.436
Stage 3	71.1%	0.421	0.470

kappa (Liebetrau 1983). To interpret the strength of agreement, we followed the guidelines proposed by Landis and Koch (1977) and Altman (1991), so that a kappa value between 0 and .20 indicates "slight" agreement; a value between .21 and .40 indicates "fair" agreement; a value between .41 and .60 indicates "moderate" agreement, a value between .61 and .80 indicates "substantial" agreement, and a value over .81 indicates "almost perfect" agreement. It should be noted that both percent and Kappa values are needed for full understanding of the strength of the relationship (Viera and Garrett 2005).

Baseline videos in a previous phase of the study were used to assess inter-rater reliability of each category of behavioral markers. Three raters independently rated all behavioral markers, organized in four categories within the two stages. Kappa values ranged from 0.69 for closed loop communication to 0.7 for situational awareness, thus indicating a substantial level of agreement among raters in their observations (Riley *et al.*, 2008). In the videos used for this analysis, two expert raters observed and analyzed the videos. The percentage agreement, the Cohen's kappa and the weighted kappa are reported in Table 3 for each stage.

Thus, even with a smaller number of raters and observations, Kappa coefficients indicate fair to moderate agreement and high levels of percent agreements. While the lowest Kappa of 0.30 (indicating "fair agreement") is obtained for the stage 1, the percentage agreement is still high at 72.7%. As indicated by Viera and Garrett (2005), this occurs when the prevalence of some ratings is low compared to the prevalence of other ratings. Thus, overall, the interrater reliability was satisfactory. Final ratings for each item were obtained via joint review of the video and a consensus between the two raters.

Nurses versus Physicians

The findings in Table 4 show the weighted interdisciplinary team behavior index (WITBI) trends for the four behavioral markers over the *in-situ* simulation sessions. It also shows the total performance of each clinician. Stage 2 consists solely of the nurse as the team leader and the behavioral markers represent non-technical behaviors by nursing staff. Stage 3 consists of the OB physician as the expected dominant team leader and the behavioral markers represent non-technical behaviors by the physician.

The magnitude of team performance on the scale we used indicates that the behaviors observed and scored are present between 50–90% of the time on average for both physicians and nurses. Physicians ranged from a high of 1.76 (leadership and team) to a low of 0.92 (closed loop communication), while nurses ranged from a high of

Table 4. Results – Means (Standard Deviations).

	Stage 2 Nurses	Stage 3 Physicians
Situational Awareness	1.63 (.52)	1.59 (.43)
Shared mental Model	1.61 (.32)	1.49 (.66)
Leadership and Team	1.40 (.64)	1.76 (.40)
Closed Loop Communication	1.07 (.70)	0.92 (.60)
Overall Score	1.48 (.48)	1.27 (.42)

1.63 (situational awareness) to a low of 1.07 (closed loop communication). We also observed moderate to strong correlations in both behavioral markers and aggregate performance between nurses and physicians (correlation coefficients range from 0.41 to 0.73) indicating parallel performance between nurses and physicians across the simulation exercises. Physicians scored significantly higher than nurses on leadership and team (at the .05 level of significance). Nurses scored higher than physicians on closed loop communication and shared mental model. While this difference is not statistically different with the 11 videos, it becomes significant (at the .05 level of significance) when stage 2 data from the 11th video is removed (a clear outlier in the data set). There is no difference between physicians and nurses on situational awareness. With regard to overall score, nurses perform slightly higher than physicians in the aggregate behavioral marker (at the .10 level of significance).

Discussion and Conclusion

These findings indicate both similar and different patterns between physicians and nurses in key behavioral markers during critical events conducted with *in-situ* simulation exercises. While nurses tend to perform better on closed loop communication and shared mental models, the physicians perform better in team leadership. Overall, aggregate scores were higher for nurses than physicians and performance across simulation exercises was strongly correlated between nurses and physicians. These findings indicate differences in behavioral markers in non-technical skills and a range in competence for both physicians and nurses in the same skills. The findings also suggest that neither physicians nor nurses consistently exhibit high reliable performance in communication and teamwork skills during critical events. This gives credence to the need for training of health care professionals in key teamwork skills. Physicians, nurses and other health professionals perform differently on key behavioral markers during critical events, and such performance is associated with different phases of team formation and reformation. While team members have different roles, certain key behaviors are the responsibility of all roles. Given the complexity and the dynamic nature of the healthcare world, it seems imperative that such critical training be incorporated into safety plans of any institution.

While there are several limitations with this study relating to 1) the fidelity of *in-situ* simulation, 2) the sole reliance on observation, and 3) the use of different elements for physicians and nurses in assessing the behavioral markers, the results have important implications for patient safety. This study is unique in that simulations were all performed at the worksite in the actual operating room and suggests that experiential exercises of this type can be critical in changing team behavior of practicing clinicians. The findings also suggest an important role for applied *in-situ* simulation techniques as an effective assessment tool as well as an effective training tool. It should be recognized, however, that structured observational methods can be more suited to some healthcare domains than others. For example, settings with frequent unpredictable changes, movement of staff around a large area over time, and with numerous triggers and distractions may be less conducive to such approach (Wear 2000). However, this study confirms that behavioral markers can be effectively used as frameworks for assessment, training and research to develop and test countermeasures as suggested by Fletcher *et al.* (2003). Finally, the study demonstrates that non-technical markers need to be considered to measure performance in critical crises to provide training and system design for a set or error countermeasures in health care (Pizzi *et al.*, 2001; Howard Gaba *et al.*, 1992; Sexton *et al.*, 1998; Small *et al.*, 1999).

References

Agency for Healthcare Research and Quality, 2007. TeamSTEPPS™: Team Strategies and Tools to Enhance Performance and Patient Safety. Available online at http://teamstepps.ahrq.gov/index.htm

Altman D.G., 1991. *Practical Statistics for Medical Research.* London: Chapman & Hall

Carthey, J., 2003. The role of structured observational research in healthcare, *Qual Saf Health Care*, 12, 13–16

Davis S., *et al.*, 2008. Failure Modes and Effects Analysis Based on *In-situ* Simulations: A Methodology to Improve Understanding of Risks and Failures. In *Advances in Patient Safety: New Directions and Administrative Approaches.* Agency for Healthcare Research and Quality

Fletcher, G.C.L., *et al.*, 2002. The role of non-technical skills in anaesthesia: a review of current literature. *British Journal of Anaesthesia*, 88, 418–429

Fletcher, G., *et al.*, 2003. Anesthetists Non-Technical Skill (ANTS): Evaluation of a behavioral marker system, *British Journal of Anesthesia*, 90, 580–588

Flin, R., O'Connor, P., and Crichton, M., 2008. *Safety at the sharp end: a guide to non-technical skills.* Burlington VT: Ashgate

Gaba, D.M. *et al.*, 1998. Assessment of clinical performance during simulated crises using both technical and behavioral ratings, *Anesthesiology*, 89, 8–18

Greaves, J.D. and Grant, J., 2000. Watching anaesthetists work: using the professional judgement of consultants to assess the developing clinical competence of trainees, *British Journal of Anaesthesia*, 84, 525–533

Hamman, W.R., 2004. The complexity of team training: what we have learned from aviation and its applications to medicine. *Qual Saf Health Care.* 13 Suppl 1: p. i72–9

Helmreich, R.L. and Foushee, H.C., 1993. Why crew resource management? Empirical and theoretical bases of human factors training in aviation. Cockpit resource management. In Wiener, E.L. (ed.), Kanki, B.G. (ed.), Helmreich, R.L. (ed.) *Cockpit resource management*. (Academic Press, San Diego), 3–45

Howard, S.K., *et al.*, 1992. Anesthesia crisis resource management training: teaching anesthesiologists to handle critical incidents, *Aviat Space Environ Med*, 63, 763–70

Landis, J.R., and Koch, G.G., 1977. An Application of Hierarchical Kappa-type Statistics in the Assessment of Majority Agreement among Multiple Observers, *Biometrics*, 33, 363–374

Liebetrau, A.M., 1983. *Measures of Association: Quantitative Applications in the Social Sciences*. London: Sage Publications

Mackenzie, C.F., *et al.*, 1996. Comparison of Self-Reporting of Deficiencies in Airway Management with Video Analyses of Actual Performance, *Human Factors*, 38, 623–635

Miller, K.K., *et al.*, 2008. *In-situ* simulation: A method of experiential learning to promote safety and team behavior. *Journal of Perinatal and Neonatal Nursing* Vol 22, No 2, 105–113

Miller, K.K., Riley, W., and Davis, S., 2009. Identifying Key nursing and team behaviors to achieve high reliability. *Journal of Nursing Management*, 17: 247–255

Paisley, A.M., Baldwin, P.J., and Paterson-Brown, S., 2001. Validity of surgical simulation for the assessment of operative skill, *British Journal of Surgery*, 88, 1525–32

Pizzi, L.T., Goldfarb, N.I., and Nash, D.B., 2001. Promoting a Culture of Safety. In *Making Health Care Safer: A Critical Analysis of Patient Safety Practices*, Evidence Report/Technology Assessment Number 43, (Agency for Healthcare Quality and Research), 447–457

Riley, W., *et al.*, 2008. The Nature, Characteristics and Patterns of Perinatal Critical Events Teams. *In* Advances in Patient Safety: New Directions and Administrative Approaches. Agency for Healthcare Research and Quality

Riley, W., 2009. High Reliability and Implications for Nursing Leaders. *Journal of Nursing Management*, 17, 238–246

Sexton, J.B., Thomas E.J., and Helmreich, R.L., 2000. Error, stress, and teamwork in medicine and aviation: cross sectional surveys. *BMJ*, 320, 745–749

Small, S.D., *et al.*, 1999. Demonstration of High-fidelity Simulation Team Training for Emergency Medicine. *Academic Emergency Medicine*, 6, 312–323

Thomas, E.J., Sexton, J.B., and Helmreich, R.L., 2004. Translating teamwork behaviors from aviation to healthcare: development of behavioral markers for neonatal resuscitation, *Quality and Safety in Health Care*, 13, i57–i64

Viera, A.J. and Garrett, J.M., 2005. Understanding Interobserver Agreement: The Kappa Statistic, *Family Medicine*, 37, 360–3

Wear, D., 2000. Toward Qualitative Understandings of Health Phenomena. *Qualitative Health Research*, 10, 277–283

Yule, S., *et al.*, 2007. Surgeons' Non-technical Skills in the Operating Room: Reliability Testing of the NOTSS Behavior Rating System, *World Journal of Surgery*, 32, 548–556

METHODS AND TOOLS

AUTOMATED LINK ANALYSIS USING RADIO FREQUENCY IDENTIFICATION (RFID)

N.J. Mansfield, J.J. Evans, C. Andrews, A.P. Lancaster, I.R. Ransome, Z. Moti, L. Li, K. Kniveton, Y. Zhao & S. Hignett

Department of Ergonomics (Human Sciences), Loughborough University, Loughborough, Leicestershire, LE11 3TU, UK

Being able to accurately record the interactions which take place within any environment is beneficial for understanding human behaviour in a wide range of industries. Link Analysis (LA) is a standard technique that analyses the interactions between people and items/locations in an environment. Traditional pen and paper methods of LA can be cumbersome and time consuming. This paper details a way to automate the recordings of interactions between a human and their current environment by using radio frequency identification (RFID) tags and a subject-mounted receiver. Using the results from the system, it is possible to instantly create conventional Link Analysis diagrams and tables, reducing the time and resources required for data collection and analysis. The system has been developed in partnership with the Healthcare Ergonomics and Patient Safety Unit (HEPSU) at Loughborough University, with initial focus being on monitoring paramedics, patients and environment interactions within an ambulance; however, the technologies and the analyser system are not limited to use within this particular field.

Introduction

Link Analysis (LA) is an established tool to assist in the understanding of dynamic interactions in an environment. In LA, systems are composed of links and elements. Elements comprise items in the system which are drawn in a logical pattern, often representing the physical layout of the system being studied. Links are drawn between elements in order to record the sequence of activity. When a LA is complete, a diagram is produced showing all links between elements, thus providing information on activity patterns. The LA might represent activity for an entire task or for a fixed period of time. If elements have many links between them it indicates that they are often used in close chronological proximity; it might therefore improve efficiency if they are located in close spatial proximity. Thus LA is a valuable tool for optimisation of workplace ergonomics. LA has been used to understand work patterns of healthcare professionals (e.g. Ferreira and Hignett, 2005; Jones *et al.*, 2009).

When performing Link Analyses the interactions have been recorded manually, with researchers observing paramedics, in real or simulated emergency situations.

The interactions between the patient and the ambulance equipment and consumables were then recorded on paper. This is a labour intensive process where the researchers have to carefully observe the paramedics and try and record every single interaction. This is sometimes augmented by the use of video recording. In a confined environment such as an ambulance, there is limited space for the observer or their cameras, and there could be ethical and privacy issues related to direct observation. It would therefore be beneficial to automate part of the LA procedure such that the observer does not need to be physically present during data collection.

This paper describes a development of LA which uses Radio Frequency Identification (RFID) tags to instrument an ambulance and a wrist-mounted RFID detector. This system allows for automated recording of the LA links between the tags (elements). A LabVIEW front end system is utilised as the computer based interface which records the links and automatically draws the LA diagram and produces an interaction matrix instantly. The system is wireless, using Bluetooth technology, thus allowing healthcare professionals minimal disturbance from the system.

Principals of RFID

The goal of any RFID system is to collect information and data automatically, rapidly and accurately. There are currently no defined standards for the architecture and infrastructure of such systems but there are three main components that are part of RFID systems (Graafstra, 2006):

- RFID Tags are placed on objects and can be unobtrusive. They contain at least a chip with a unique identification number and antenna which can be used to identify the host object. Tags can be active (which require constant power) or passive which are powered inductively.
- The RFID Reader includes a radio antenna and is capable of at least reading the tag identification number but can sometimes also write data to the tag. Tags and readers must operate at the same radio frequency; the frequency influences the read range (distance between reader and tag to obtain a 'read'), antenna size and power consumption.
- The Data Processing System (Server) collects and processes the data and controls the read/write operations of the RFID Reader. The Server can be integrated with the Reader or can use a wired or wireless data connection. The Server must include data correction algorithms to clean duplicates or mis-reads and to buffer data if multiple tags are in range of the Reader.

In order to read a tag the Server commands the RFID Reader to make a read (Figure 1). The RFID Reader then sends a read request. Any RFID Tags within range will respond with their ID number and this number is then relayed to the Server. For an automated system, the Server will request reads at regular intervals (e.g. 10 times per second).

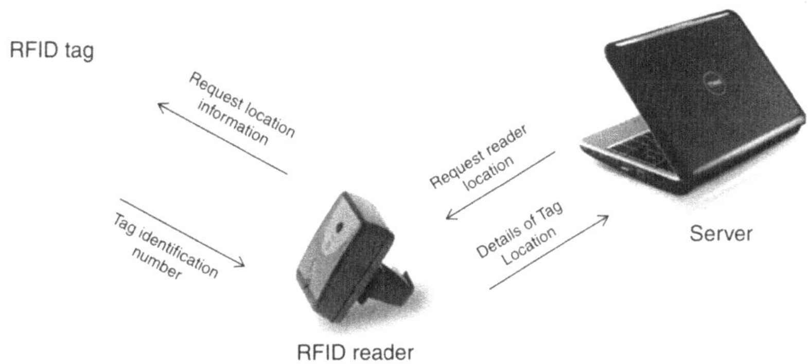

Figure 1. Example of RFID system hardware.

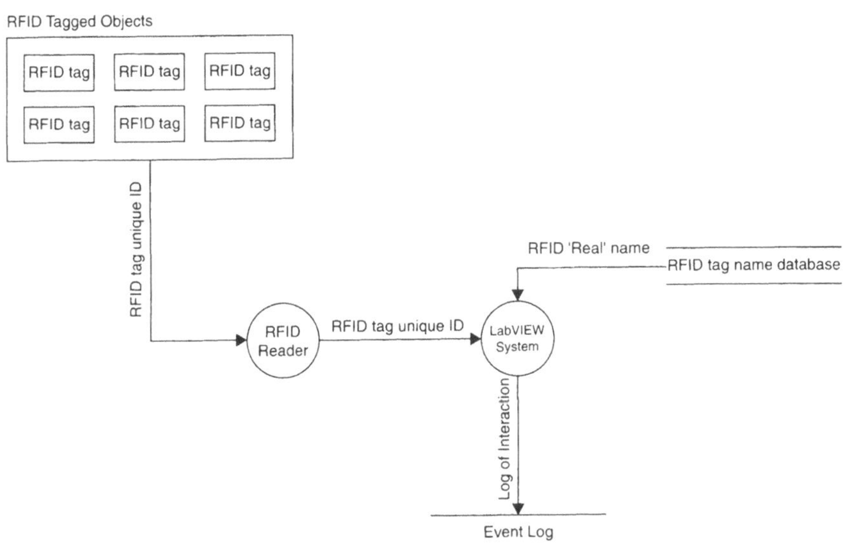

Figure 2. System architecture.

RFID Link Analysis system overview

For the RFID Link Analysis system, the RFID tags are placed on objects which the user intends to monitor. In the system architecture diagram (Figure 2) this is represented by the RFID Tags entity. A Bluetooth RFID reader is mounted on each wrist of up to three subjects (i.e. six readers). If the reader is moved into the readable range of any of the tags, the system is alerted of this event and it is logged into a database. Once the required amount of data have been collected the user can select the desired output from options including raw data tables, link analysis diagrams and basic interaction diagrams.

The RFID tags used in the system are passive tags so require no power source. Every tag has a unique identity number which can be read by a reading device. Currently the system utilises low-cost wipe-clean credit card size tags; these are 54 × 85 mm and 1mm in thickness. These cards can easily be placed on cupboards, equipment and on test mannequins. Smaller RFID tags are available which could be incorporated into the patient stretcher in order to monitor patient-crew interactions in real-life situations. Each RFID tag has an ideal read range of around 110 mm and for large scale items, multiple tags can be used. The software allows for tag clustering where multiple tags represent one element in the LA. This is stored in the RFID Tag Name Database shown in Figure 2. If close to metallic objects, read ranges can be reduced to about 55 mm and so care must be taken to ensure appropriate placement or shielding should be used. There is no limit to the number of tags that can be used with the system, therefore complex tag arrays can be constructed providing many more elements than are possible in conventional Link Analysis. Although the tags themselves have a wipe clean surface, are reusable and have an indefinite life, if there are infection control concerns, they can be considered disposable, as each tag only costs around £1.50 and are available off the shelf.

The RFID readers are 40 × 55 × 20 mm and can be fixed into a wrist strap. The readers contain a small rechargeable battery which lasts for around 5 hours. They communicate with the Server through standard Bluetooth technology, as already used in ambulances. Tags must be sited to be in ideal positions relative to the Reader. For example, tags on the handle of a cupboard door should be mounted beneath the handle such that they are within the read range of the wrist-mounted reader while the door is being opened.

RFID Link Analysis data analysis

The LA software must be programmed to incorporate each tag and reader. The software uses an image as a background for the data analysis (Figure 3). Each tag is programmed into the system by scanning with the reader to obtain the tag ID at which time the user is prompted to select a location for the tag on the image.

The current system records each read in a data file. If a user maintains their position close to a tag then the tag will be read at each read request which occurs 10 times per second. In the data analysis phase, the raw data file is cleaned to remove duplicate reads and leaves a results file recording the time of read-start and read-end for each link. This allows for a traditional LA diagram or interaction matrix to be drawn but with additional information such as the direction of the link and the precise timing of interactions.

The two main benefits of the RFID Link Analysis system are the speed of obtaining results and the possibility of monitoring up to six Readers simultaneously. The data analysis phase of the LA process is very rapid. One hour of observational data takes about two minutes to analyse and to produce the final Link Diagram using a basic laptop computer. Quicker processing can be achieved on more powerful computers.

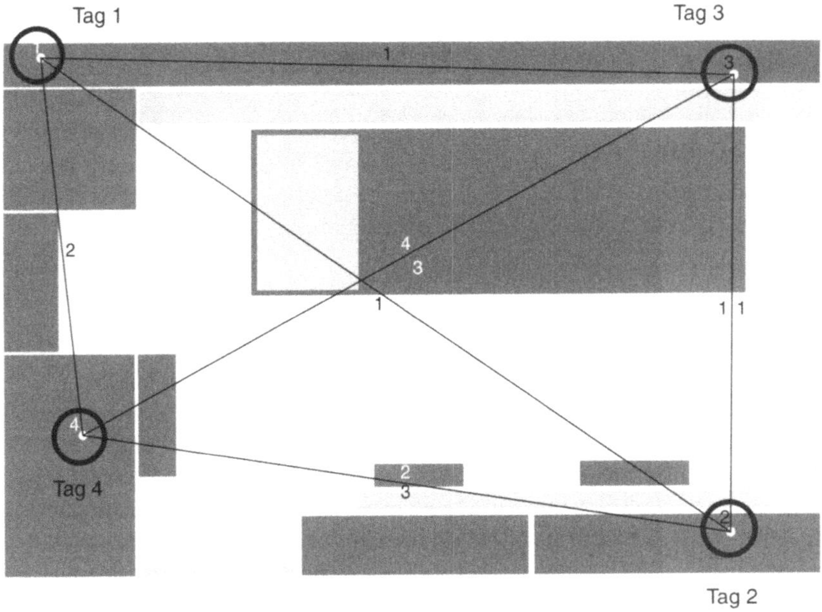

Figure 3. Example basic LA output showing location of four tags, links between each tag and diagrammatic representation of ambulance interior.

The output can be saved as a graphic image or as a data file. These files can include results for the complete data set or subsets for each subject.

Future development

Although this system has been developed within the healthcare environment, it will be generally applicable to many work environments. Furthermore, the reduction of the need for the observer to be physically present in the workspace has potential benefits in terms of privacy and space restrictions.

Readers could be mounted on ankles instead of wrists. This would mean that the position of a worker could be tracked via tags mounted into the floor of a workspace and a reader mounted onto their ankle. It would be feasible to manufacture a floor covering incorporating a large array of tags which could be installed into a test facility. This would allow for movement to be monitored in terms of body position rather than interaction with elements.

The current limit of six readers is a design limitation to ensure that the level of complexity in the analysis was manageable. If additional data are required, this could be achieved using a second array of readers paired with a second Server PC whilst still using the same array of RFID tags. With this approach, the number of subjects being observed may be unlimited.

To date, the system has reached full Beta test stage and includes hardware, software and data analysis packages, and a user manual. It has been used by novice users and LA experts. It has been used and demonstrated functioning in an ambulance under controlled laboratory conditions. Further research is needed to verify that it is reliable under full field test conditions.

RFID technologies could be implemented in different ways to perform LA and other data collection tasks in the field of ergonomics.

Summary

Historically human interaction monitoring using Link Analysis has taken place using direct observation or some method of video capture; videoing a scenario from multiple angles and focusing on the direct monitoring of the people and their behaviour. The new RFID method takes longer to set-up than the traditional pen and paper method, but analysis is almost instantaneous. The level of detail which the system provides is far greater than that obtained using the traditional method in both detail and breadth, and the simplicity and speed of the system offers new possibilities to the field of human activity monitoring.

References

Graafstra, A. 2006, *RFID Toys*, First Edition (Wiley Press, UK)

Ferreira J. and Hignett S. 2005, Reviewing ambulance design for clinical efficiency and paramedic safety, *Applied Ergonomics*, 36, 97–105

Jones, A., Hignett, S., Benger, J. (2009) Mobile Pods: Technology to Support the Delivery of Community-Based Urgent Care. *Proceedings of the 17th Triennial Congress of the International Ergonomics Association* Beijing, China, 9–14 August 2009

OCCUPATIONAL HEALTH AND SAFETY

UNIVERSITY STUDENTS' VIEWS OF THE DESIGN AND USE OF THEIR STUDY ENVIRONMENT

Rachel Benedyk

*UCL Interaction Centre, MPEB, University College London, London WC1E 6BT, UK**

University students are characterised as a user group as independent, multi-national, peripatetic, social and skewed to younger adult age group. They use dedicated study environments halls of residence, lecture theatres, classrooms, campus facilities, intensively, and appear to be unaware or uncaring of ergonomic risks in many cases. However, the university provides and designs these environments for the purpose of student study; who should take responsibility for reduction of student risk, the students or the university? This paper presents some data from a recent survey of students who use laptop computers, illustrating the attitudes students have to their study environment and to their own role, and that of the university, in moderating such risks.

Introduction

It is sometimes said that university students are the most frequently studied group of experimental subjects, mostly because they are available on site at universities where research is carried out. In this study, however, we focus on university students for their own sake, as a user group with particular characteristics, attitudes and behaviours. Design for this group of consumers is of economic significance; in the UK alone there are 2.3 million students attending university campuses, and the UK universities budget approaches £8bn pa. (Universities UK, 2008)

Students attending campus-based courses as opposed to distance-learning study, make use of dedicated facilities and resources designed and supplied by the university. These roughly fall into estates provision buildings, spaces, security, transport etc,, teaching facilities provision lecture theatres, laboratories, classrooms, libraries, equipment etc, and living facilities provision cafes, bars, healthcare, sports, halls of residence etc,. Almost all these facilities and their infrastructure and management are supplied complete by the university; the student simply arrives, and makes use of them. Is the student by implication a passive consumer then? And do the university authorities take due account of the needs of this user group? What happens when student needs are not met – are students capable of taking some individual responsibility for their work and live environment? How well do they do this?

The challenge to an arriving student, planning for a stay of perhaps several years, is to find or create optimal living and working conditions for their study period.

* Contact: r.benedyk@ucl.ac.uk

They will need to adapt to the generic university facilities provision, and select what facilities to use and how. They may alter or amend things for their own preferences and/or supply or design facilities provision that the university lacks, but in doing so they must identify their own personal needs that are not being met. This implies that they do have some ability to assess the fitness of their work and live environment. This study examines this premise, as demonstrated in a recent ergonomic study of students' use of laptop computers at university (Benedyk *et al.*, 2007).

Characteristics of students as a user group

The majority of students using university campuses, whether home or international, are living independently from their families, at least during term-time. They are taking decisions for themselves, some of them for the first time. Many may arrive well-supplied with parental advice, but on campus the main sources of guidance will be their peer-group. University advisory staff may be available; however, young people are not known for seeking advice in general from those who don't share their values (Twyman *et al.*, 2008).

Despite growing opportunities for mature students' tertiary study, a majority of students are young, mostly 18–23 years old (Universities UK, 2008). This skew to a younger adult age group is relevant to behaviour, in particular risk-taking (Sharland, 2005). Young adults often tend to understand risk poorly, and care about its consequences less. They exhibit a higher tendency to take personal risks, both in sensation-seeking and in ignoring danger. Unintentional injury is the largest cause of death in this age-group. (Thomas *et al.*, 2007)

Students lead rather peripatetic lives, moving constantly between varying, short-lived settings for living, working, travel and leisure. Even while studying on campus, they move around between teaching rooms, libraries and study areas. They often carry all they need with them – books, laptops, sports gear, etc.

The student population is multi-national and there is a wide range of cultural expectations. A large proportion are very social, seeking stimulation and leading active social lives that may leave them fatigued during work time. In addition, many students are short of money and need to take part-time jobs during their student years; these activities too can contribute to fatigue.

In summary, the student population is generally independent and prone to risk-taking, adaptable to varying settings, culturally varied, highly social, often fatigued, and used to making do with limited resources. Students are susceptible to peer influence and in the main, unsupervised. How do these factors affect students' attitudes to risks in their main setting – their study environment?

The Student Laptop Study

Students are expected to use individual computers at university level; however, frequently there are insufficient desktop computers supplied on the campus and in halls

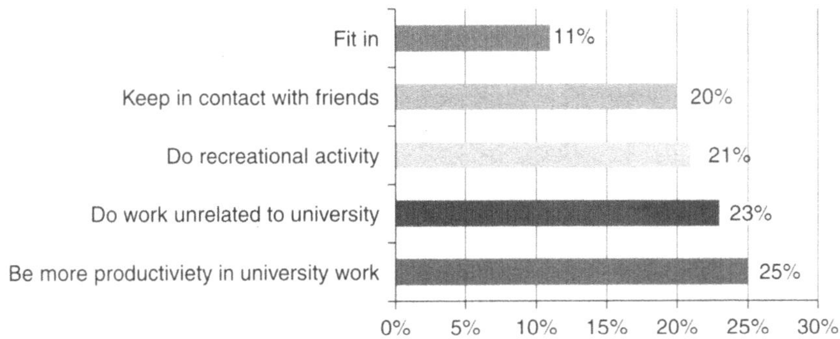

Figure 1. Student motives for bringing a laptop to university.

of residence. Increasingly, therefore, students bring their own laptop computers to university. This allows them to work in varying settings classroom, library, hall of residence, to carry their work with them, and to make use of wireless networks, as well as to maintain social contacts and for entertainment (see Figure 1).

Laptops are considered high risk for intensive use under the DSE Regulations 2003, not only because their configuration imposes postural strain and high visual load, but also because of the variety and unpredictability of their environments of use. (Benedyk *et al.*, 2007)

A large ergonomic survey of university student use of laptops was recently carried out at a city centre university in the UK (Benedyk *et al.*, 2007). 650 students responded to the intranet-based survey, 46% male, 54% female, both undergraduate and postgraduate. Small incentives were offered to attract respondents. Respondents' identities were not known to the researchers. There was extensive follow-up to key respondents, and a parallel survey canvassed staff attitudes to student laptop use.

Ergonomic risks by students

86% of the student respondents have used a laptop computer recently; most use it every day. Asked to explain why, the most prevalent reason (61%) was portability: the ease with which work on the laptop can be transported around campus or back to a home base outside London or outside the UK (see Figure 2).

On average, students use their computer for 5.24 hours per day (SD 3.14), of which around half the time is while at university. This period of working on a laptop would be considered extensive in a DSE risk assessment (HSE, 2003). They bring it to university on an average 10.5 days per month, and often carry it around all day when they do, because there is a severe shortage of secure lockers on campus. About half the respondents admitted it was difficult to transport their laptop, for two major reasons: laptop weight (67% of respondents) and fear of crime (49%). Although the main advantage of the laptop is its portability, in fact laptops can present a significant manual handling load if carried around extensively in this way.

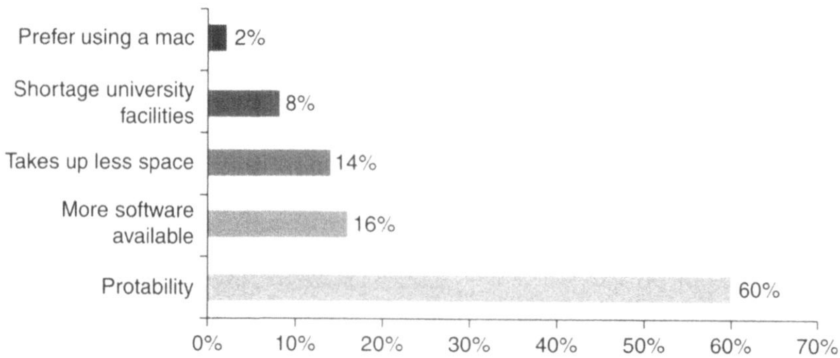

Figure 2. Reasons why students use a laptop rather than a desktop pc.

The vast majority of students never use a docking station nor link the laptop to a bigger screen or separate keyboard or external input device; these are ergonomic adjustments recommended in the DSE regulations (Heasman and Brook, 2000). Nearly all the laptops used have no inherent ability to separate screen from keyboard, nor to swivel the screen.

Many students report frequently using a laptop on their lap (42%), lying on their front (10%), or while in bed (29%) (see Figure 3). These configurations are probably benign for brief periods, but for extensive use such as reported by students, they are not likely to provide good support nor maintain a healthy posture.

In addition, students report that the university study environment often provides poor work conditions and unsuitable furniture for laptop use both on campus and in halls of residence, such as non-adjustable chairs, desks too small to hold a desktop computer, and inconvenient siting of power points; these, too, contribute to poor user postures.

These data are evidence of students taking ergonomic risks, in both overuse and poor use of a laptop, which can cause postural and visual strain. Many aspects are contrary to the DSE Regulations (Heasman and Brook, 2000). There is also potential manual handling risk (HSE, 1998). It is less than surprising then, that 57% of the students reported having experienced pains in the body as a result of using their laptop, 7% of them a lot of the time (see Figure 4). However, 20% also said that they would not take a work break, even when in pain! Such breaks are regarded by ergonomists as essential in moderating postural strain. (Pheasant and Haslegrave, 2006, page 181).

Almost no students agreed they would seek treatment or advice for the pain, either from a medical service or from the university computer service. They had little perception that such advice might help prevent their discomfort getting worse in the future, even though most of them understood that the pain originated from their use of the laptop. Very few students suggested that they might give up laptop use: their motives to continue to use the laptop seemed too strong, and their understanding of the ergonomic risks too skimpy. In interviews, when the ergonomic risks were

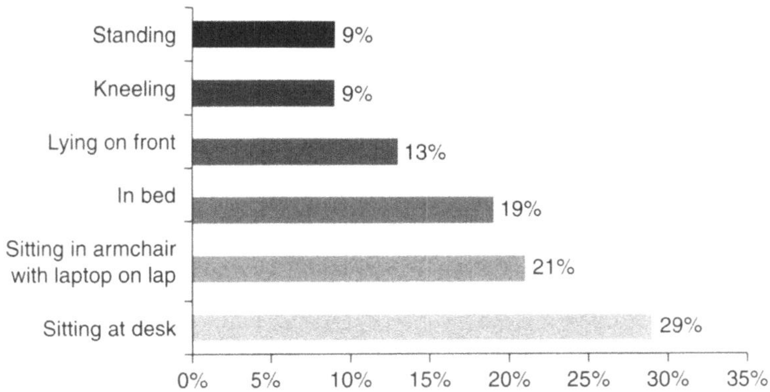

Figure 3. Postures adopted by students using laptops.

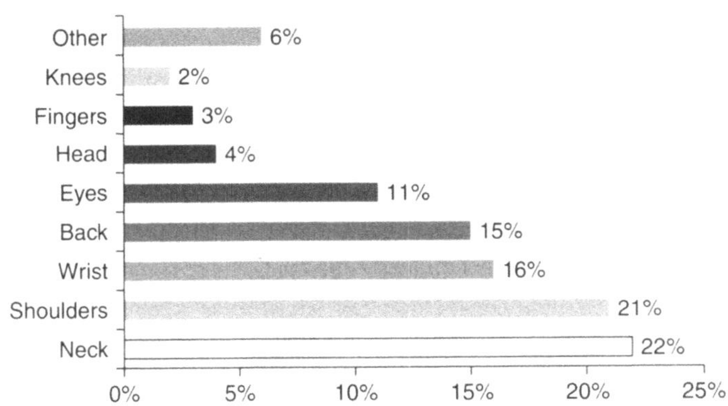

Figure 4. Areas of pain reported by students using laptops.

explained clearly, students still often gave pragmatic reasons for not giving up or limiting use of their laptop, such as lack of desktop machine access, or frequency of work deadlines.

Risk – whose responsibility?

The majority of students surveyed have no knowledge of the potential benefit of using positive ergonomic adjustments or ergonomic furniture, nor of the risks taken by working for such lengths of time in such stressful postures. Moreover they have had no guidance from the University about laptop use; the university officially supports only desktop computers for students, and there is no technical nor medical support supplied for those using laptops. Students receive induction courses and on-line information about the hazards of computer use, but only as related to desktops.

Most students are not under contract of employment to the university, and are therefore regarded as 'self-employed' for the purposes of risk assessment for computer use. Nevertheless, the university has a duty of care to them as visitors, under the Health and Safety at Work Act. With no official technical support for laptops, the university seems to discount laptop use in its care plan. The expectation is that students will assess and moderate their own risk. However, most students have no knowledge of health and safety risk assessments, and little understanding of the potential ergonomic risks of laptop use. Add in the inherent factors of young people embracing risky behaviour, and their reluctance to ask advice in general, and the picture looks bleak.

Concerns from students

The survey data did show two areas where students showed good understanding and expressed concern. The first area is crime prevention; students are very aware that laptops in a city are a frequent target for thieves and muggers, and they often take careful steps to avoid being mugged by disguising the laptop in an ordinary rucksack, and by guarding it at all times on campus. The potential theft is one related to the value, both monetary and convenience, of the computer; the perception of risk here is therefore well understood.

The second area of student concern relates to facilities provision by the university for laptop users. A variety of comments criticised the university for, for example: the lack of power points in teaching and library areas for students to plug in their laptops; the lack of secure lockers in which students could keep laptops during the day; the lack of table space in study areas on which to rest a laptop; the lack of technical support for laptop users; and the patchiness of the wireless network.

These concerns were expressed in terms entirely of inconvenience and resentment, not in terms of ergonomic risk. For example: students wanted more power sockets merely to deal with competitive demand among students, not to relieve poor user postures resulting from difficult siting of sockets; and students wanted lockers so that they wouldn't have to guard their laptops all the time, not because it would relieve their manual handling load. The perception of ergonomic risk here is far less well understood.

In selected follow-up interviews to students, this premise was generally borne out. Students admitted that they used laptops for long periods in poor postures without recognition of this as a potential risk to health. Even among students who complained of aches and pains following laptop use, most did not take consequent action to adjust their mode, environment or position of use. However, they often did adjust their working time periods, simply because the onset of pain would remind them to stop.

Interestingly, some students held the university responsible for the student needing a laptop at all, for example because there were insufficient desktop computers provided, or because the work-desks in the hall of residence were too narrow to support a desktop computer. These students were originally reluctant to bring a laptop to university, sometimes because of the risk of crime, but also because of the inconvenience and expense; they had later felt they had no choice but to bring one.

Students in university settings – the laptop example

In this case study we observed that students using laptops generally do not undergo risk assessments, for themselves or by the university. They demonstrate over-intensive use under DSE Regulations, with risks of postural and visual strains and discomfort. Their environment of usage is varied and unpredictable, with poor support furniture.

These students were often unaware and/or uncaring of ergonomic risks – they placed the responsibility external to themselves – and they are thus a vulnerable user set. Yet they have strong motivations to continue this behaviour, because it holds value to them by way of social and work benefits. On campus there is no strong guidance from authoritative sources to steer them otherwise, and in line with Twyman, 2008, they mostly do not seek such sources out.

Moderation of risks

It would appear that in certain situations, students are able to take responsibility for moderation of their risk. For example, in taking action to prevent being mugged for their laptop, students recognise the risk and its consequences and judge it sufficiently relevant to them. They also have the knowledge of what action to take.

We suggest however, that there are many situations where other factors intervene to prevent student moderation of risk. The student may be ignorant of the risk or its potential consequences, or unaware of actions that could be taken to moderate it. For example, many students did not know to improve their posture or to take breaks in order to prevent pains when using the laptop. However, it is also likely that the student is uncaring of the risk, both as a function of their attitude to life in line with Sharland, 2005, as we saw earlier, and because the benefits to them outweigh the risks. Students report that it is useful to bring a laptop to university for both work and leisure reasons, and some even bring it for reasons of style. Cultural approbation evidently plays a significant role.

The university can play some part in guiding young people in the risks and their moderation eg a leaflet on laptop use has now been issued – downloadable from http://ucl.ac.uk/uclic/people/r.benedyk/laptops.pdf, but there is also a critical role in optimising facilities provision that would help prevent the occurrence of much of the risk. Where universities do not set the objectives for student laptop use, they may not see the need to provide support and facilities. However, students will continue to bring laptops to university – there is too much benefit to them for this trend to reverse – and this needs to be recognised and supported appropriately.

This survey found a concerning lack of ergonomic knowledge by students, in line with previous studies (Noack, 2003). Given the young age of students at university, physical problems could be introduced early and stored up, contributing to health problems in later life (Harris and Straker, 2000). We question the universities' current approach to reducing risk for students in their care, and suggest musculoskeletal problems among such groups are likely to increase unless this policy changes. It also falls upon educators at school level to ensure young people understand more about risks and their prevention before leaving for university in order to equip them better to manage their own risks.

We conclude that many students are not taking responsibility for their exposure to risk; both ignorance and lack of concern are playing a part. The optimal route to shaping such young people's behaviour in favour of risk moderation has yet to be determined. It may be over-optimistic, therefore, to expect young people to assess their own environment and to take design decisions supportive of their well-being.

References

Benedyk, R., Harder, A., Hussain, R., Ioannidis, A. and Perera, M. 2007, The ergonomic impact of student use of laptops in university settings. *Contemporary Ergonomics 2007*, ed P. Bust, pp. 229–234, Taylor and Francis, London

Harris, C. & Straker, L. 2000, Survey of physical ergonomics issues associated with school childrens' use of laptop computers. *International Journal of Industrial Ergonomics* 263, 337–346

Heasman, T. and Brooks, A. 2000, Health and safety of portable display screen equipment. HSE Contract Research Report 304/2000, HMSO

HSE 1998, Manual Handling Operations Regulations 1992. Guidance on the Regulations L23, 2nd Edition. HSE Books

HSE 2003, Health and Safety Display Screen, Regulations 1992, 2nd Edition. HSE Books

Noack, K. 2003, College student computer use and ergonomics, *MSc Thesis* North Carolina State University. Accessible at www.lib.ncsu.edu/theses, Site accessed March 2007

Pheasant, S. and Haslegrave, C. 2006, Bodyspace 3rd Edition, Taylor and Francis, London

Sharland, E. 2006, Young people, risk taking and risk making: Some thoughts for social work. *British Journal of Social Work*, 362, 247–265; doi:10.1093/bjsw/bch254

Thomas, J., Kavanagh, J., Tucker, H., Burchett, H., Tripney, J., Oakley, A. 2007, Accidental injury, risk-taking behaviour and the social circumstances in which young people live: a systematic review. London: EPPI-Centre, Social Science Research Unit, Institute of Education, University of London

Twyman, M., Harvey, N. and Harries, C. 2008, Trust in motives, trust in competence: Separate factors determining the effectiveness of risk communication. *Judgement and Decision-Making*, Vol 3 No 1 pp 111–120

Universities UK 2008, http://www.universitiesuk.ac.uk/ Site accessed May 20.

This work was supported by the Stephen Pheasant Memorial Fund

VALIDATION OF A MULTI-FACTORIAL MODEL FOR PREDICTING VEHICLE DISCOMFORT

N.J. Mansfield, S.J. MacMull, G. Singla, H. Capstick, J. Mackrill & A.N. Rimell

Department of Ergonomics (Human Sciences), Loughborough University, Loughborough, Leicestershire, LE11 3TU, UK

Combining results from studies on long-term sitting discomfort by Porter et al., and vibration-induced discomfort by Ebe et al., a multi-factorial model for overall driver sitting discomfort was proposed in 2005. This model includes static factors, dynamic factors and temporal factors. Experimental testing of the model showed that the original 2005 version needed to be extended to include an interaction effect which represents how the presence of vibration accelerates the onset of driver discomfort in a non-linear manner. Further field and laboratory studies have validated the updated model. The model demonstrates that only considering vibration or sitting duration alone is inadequate for prediction of overall long-term discomfort.

Introduction

Drivers and passengers can experience discomfort in the vehicles in which they are travelling. Factors affecting the discomfort could include the physical shape of the seat, the fit of the seat to the occupant, the materials which the seat is made from (which will dictate its stiffness), the thermal environment to which the occupant is exposed, exposure to whole-body vibration and/or mechanical shocks, opportunity for changing postures (thus changing the muscle groups which are supporting the body weight) and the length of time the occupant has been sitting in the seat.

Ebe and Griffin (2000a,b) proposed a qualitative and quantitative model of overall seat discomfort. The approach of their model was to separate discomfort into 'static' seat factors and 'dynamic' seat factors. Static factors focused on seat stiffness, but these could also include any factor which does not change in response to the dynamic environment of the vehicle. Dynamic factors are those which are related to the whole-body vibration caused by the vehicle environment. The concept of Ebe's model is that when vibration is absent, the total seat discomfort is dictated by the static factors, but if there is a vibration input to the seat, the total seat discomfort is comprised a combination of static factors and dynamic factors. Thus, as the vibration magnitude increases, the relative importance of the dynamic increases accordingly. The limitation of Ebe's model is that it does not take the duration of vibration exposure/sitting into account.

For some vehicle environments, the typical magnitude of the vibration entering the seat is low. This could be due to smooth road surfaces and high quality vehicle suspension. In this case, the dynamic seat factors are less important than the static seat factors and therefore improved comfort could be achieved by optimising the static factors (such as seat stiffness, fit/adjustments, seat shape). However, should the seat be exposed to high vibration magnitudes, it might not perform very well, resulting in a poor 'ride' for the occupant. An example of this type of case is a luxury car. Other vehicle environments expose the seat to high magnitudes of vibration. This could be due to a requirement to drive off-road or on rough roads. In this case, the dynamic seat factors can be as (or more) important than the static seat factors and therefore improved comfort would require optimisation of both static and the dynamic seat factors. However, when the vehicle is driving on smooth surfaces, the static properties of the seat might not be optimal.

It is well known that sitting in the same posture for extended periods of time will lead to discomfort. It is recommended that sitting comfort trials for car seats have a duration of at least 2 hours to differentiate between seat designs (Gyi and Porter, 1999). At the beginning of a sitting trial, Porter et al. (2003) showed that although some seats are considered uncomfortable by about 1/3 of participants after 15 minutes, others are initially considered comfortable but become uncomfortable after about one hour.

If seat discomfort is assessed under 'showroom' conditions, evaluations are short and do not involve exposure to whole-body vibration. Both of these factors have been shown to be important by previous researchers. Porter et al. have shown that short-term evaluations whilst driving are likely to be inadequate, as they fail to consider the effects of fatigue and long-term sitting; Ebe and Griffin have shown that evaluations which do not include vibration are likely to be inadequate, as they fail to consider the effects of whole-body vibration exposure. An extended conceptual version of Ebe's model has previously been suggested (Figure 1; Mansfield, 2005). This paper reports results from two laboratory trials and one field trial which serve to validate the model with quantitative data.

Methods

Study 1 – laboratory study of long term discomfort in a car seat

Study 1 was an independent measures design and used three groups of 12 subjects. Each subject was required to sit in a car seat for 90 minutes. Group V0 was exposed

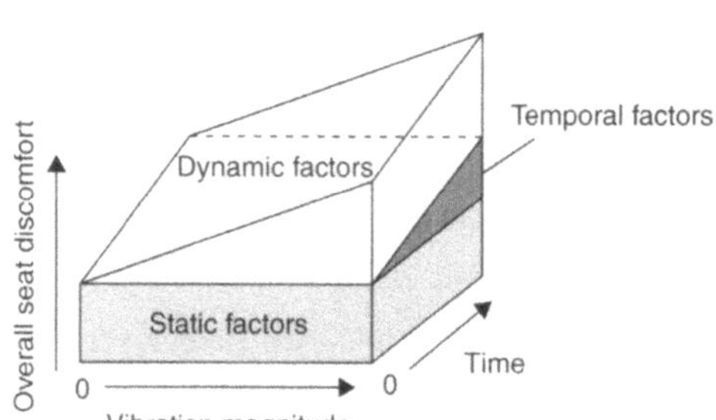

Figure 1. A model of vehicle seat discomfort incorporating static, dynamic and temporal factors (from Mansfield, 2005).

to no vibration; Group V1 was exposed to 80 minutes of random tri-axial vibration with a magnitude of 0.3 m/s^2 r.m.s. in each axis; Group V2 were exposed to 80 minutes of random tri-axial vibration with a magnitude of 0.6 m/s^2 r.m.s. in each axis. Groups V1 and V2 were required to remain seated with no vibration exposure for the 10 minutes after cessation of vibration.

Study 2 – laboratory study of long term discomfort in a passenger train seat

Study 2 was a repeated measures design and used one group of 12 subjects who were each exposed to vibration stimuli V0, V1 and V2. Subjects sat in a train passengers' seat for 60 minutes with 50 minutes of vibration (V1 and V2).

Study 3 – field study of long term discomfort in a car

Study 3 was a repeated measures design and used 16 subjects who were each required to travel on a 70 minutes journey by car (magnitudes 0.15, 0.16, 0.25 x-, y-, z-axes, weighted) and experience a simulation of the same journey with no motion in the same vehicle. The simulation comprised a projection of the field of view and replay of recorded sounds in the vehicle.

In each of the studies, subjects were required to complete Porter's seat discomfort questionnaire at 10-minute intervals (Figure 2). The second study was shorter than the first to ensure total time in the laboratory was acceptable for the participants (who attended on three separate occasions); the duration of the third study was dictated by the route.

Results and model development

Groups V0 were exposed to no vibration but were required to sit in the vehicle seat for extended periods. This paralleled the experiments previously reported by Porter et al. Each of the subjects in these groups showed a general trend of increasing in discomfort as time accrued, and this is shown in the mean scores (Figure 3). Thus findings previously reported by Porter et al. are validated. In all cases there was a significant main effect of time on discomfort ($p < 0.0005$) and vibration magnitude for studies 1 and 2 ($p < 0.0005$).

For each of the studies the discomfort had a general trend of increasing with vibration magnitude. The only exception occurred for Study 1 which used an independent measures design. Being independent measures it was not possible to track the response of individuals as vibration magnitude varied. The crossing of mean discomfort lines at $T < 20$ minutes is likely to be due to the different populations within each group. These data therefore support the models suggested by Ebe.

The model shown in Figure 1 predicts a linear increase in discomfort with the addition of vibration. However, results as illustrated in Figure 3 show that the presence of vibration not only increases the discomfort at $T = 0$ but the magnitude of the increase in discomfort escalates with duration. This indicates that vibration both increases discomfort and accelerates fatigue. A model of overall discomfort therefore not only requires static, dynamic, and fatigue components but also an interaction effect. This can be represented by:

$$\Psi = s_s + f_t t + d_v a + i_{tv} ta$$

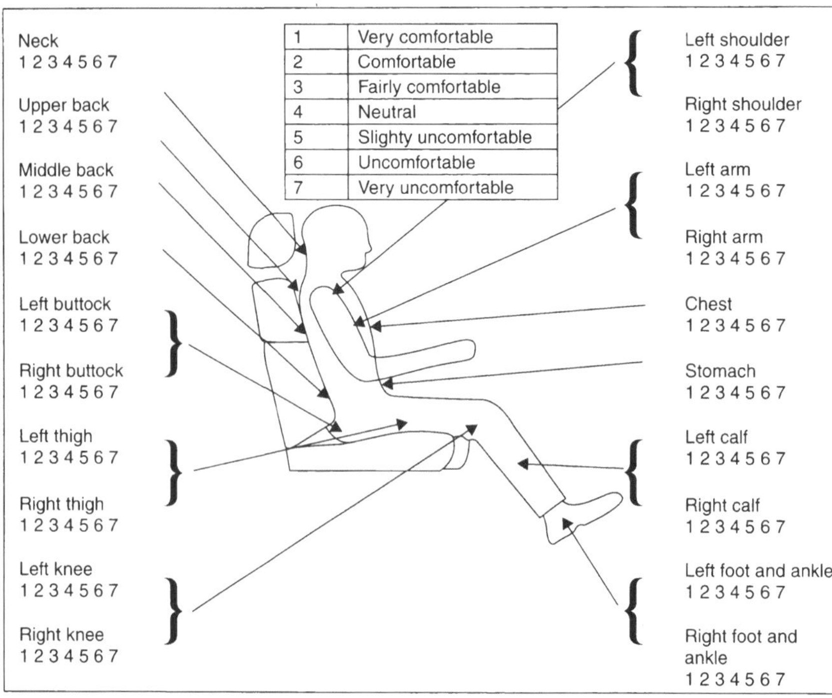

Figure 2. Porter's seven point comfort rating scale, as used in the experiments.

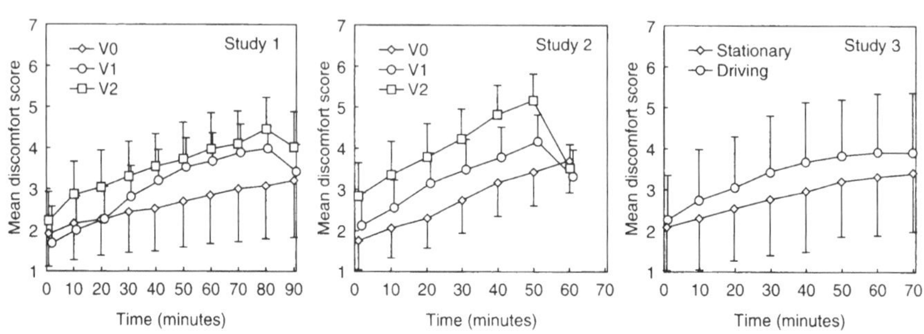

Figure 3. Mean discomfort scores for the 3 studies (jitter added on time axis to improve clarity).

where Ψ is the mean discomfort, s_s is a static discomfort constant, f_t is a fatigue constant, d_v is a vibration constant, i_{tv} is an interaction effect, t is the time in minutes and a is the r.s.s. acceleration. For this model the variables are not designed to correspond to any biological system but variables can be clustered into those that do not change over time 'static'; those that are affected by vibration 'vibration'; those that do change over time 'fatigue'. The 'interaction' is used in order to compensate for non-linearity.

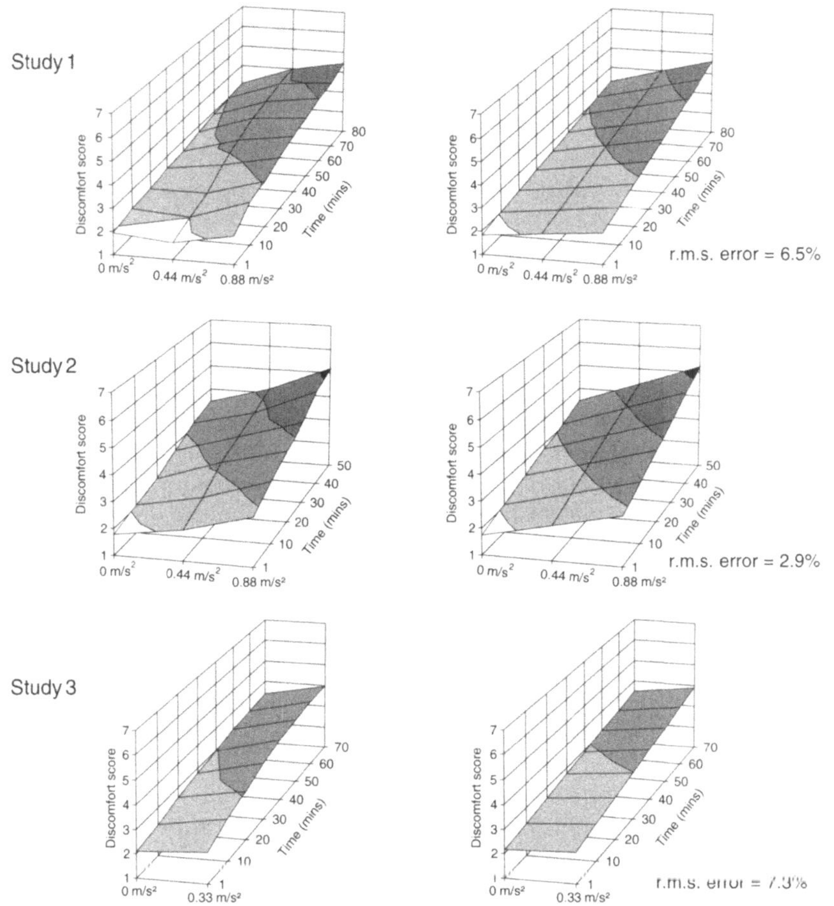

Figure 4. Mean discomfort scores for the three studies (left column) and fitted data (right column) according to the four-parameter model.

Fitting parameters to the model provides a good fit to the experimental data (Figure 4). Static discomfort s_S was best for the train seat in Study 2 (1.72 Ψ) and worst for the vehicle seats in Study 3 (2.16 Ψ). f_t was similar for the two automobiles (0.018 and 0.019 Ψ/min) but greater for the train (0.035 Ψ/min) indicating that discomfort onset was more rapid for the train seat. d_v was again similar for the two automobiles (0.57 and 0.45 $\Psi s^2/m$) and poorer for the train (1.31 $\Psi s^2/m$). The interaction constant, i_{tv}, was similar for both Study 1 and Study 2 (0.012 and 0.014 $\Psi s^2/m$ min). Three levels of vibration are required to determine i_{tv} and therefore it was not possible to calculate the constant for Study 3. 0.01 was used as an estimate based on Study 1 and Study 2. r.m.s. errors for the fitted data were between 2.9 and 7.3%. As would be expected, the best model could be fitted for the repeated-measures laboratory study.

Variables contained in the model illustrate the limitations of commonly used methods for seat assessment. Discomfort from short term static test data corresponds to s_S. For a case of a 60 minute exposure to vibration at 1.0 m/s^2 r.s.s. this would only account for 44%, 29% and 50% of the overall discomfort based on the fitted models from the three studies. Short term tests with vibration correspond to $s_s + d_v a$; this would correspond to 57%, 51% and 60% of the overall long term discomfort respectively. Long term static tests correspond to $s_s + f_t\, t$; this would correspond to 69%, 64% and 76% of the overall discomfort respectively. Thus, omitting vibration from the assessment would omit between 24% and 36% of the discomfort contributors; omitting fatigue from the assessment would omit between 40% and 49% of the discomfort contributors.

The studies reported in this paper show good internal agreement and are coherent with previous studies of discomfort from vibration, and discomfort from long-term sitting. They were designed to use continuous vibration stimuli and therefore results, and the model, cannot be directly generalised to scenarios where vibration is intermittent of varies in magnitude. Further work is required to validate and develop the model for intermittent signals, and to include the recovery time following cessation of vibration exposure.

Conclusions

The paper describes a multi-factorial model for predicting discomfort, validated through two laboratory and one field trial. The model demonstrates that only considering vibration or sitting duration alone is inadequate for prediction of discomfort, but that improvements in prediction can be obtained through consideration of static, dynamic, fatigue and interaction effects.

Acknowledgement

Study 1 was supported by EU Asia-Link ASIE/2005/111000 CIRCIS (Collaboration in Research, and Development of New Curriculum in Sound and Vibration).

References

Ebe, K. and Griffin, M.J. 2000a, Qualitative models of seat discomfort including static and dynamic factors. *Ergonomics*, 43, 771–790

Ebe, K. and Griffin, M.J. 2000b, Quantitative prediction of overall seat discomfort. *Ergonomics*, 43, 791–806

Gyi, D.E. and Porter, J.M. 1999, Interface pressure and the prediction of car seat discomfort. *Applied Ergonomics*, 30, 99–107

Mansfield, N.J. 2005, *Human response to vibration*. Boca Raton; London: CRC Press

Porter, J.M., Gyi, D.E., and Tait, H.A. 2003, Interface pressure data and the prediction of driver discomfort in road trials. *Applied Ergonomics*, 34, 207–214

A TOOL TO MEASURE THE SUCCESS OF PATIENT HANDLING INTERVENTIONS ACROSS THE EUROPEAN UNION

Mike Fray & Sue Hignett

Healthcare Ergonomics and Patient Safety Unit
Department of Ergonomics, Loughborough University, UK

Patient handling intervention strategies are many and varied. The focus of interventions has primarily been on the health, safety and welfare of care givers. Data from 4 EU focus groups and 2 world-wide expert panels were used to evaluate whether other types of outcomes were perceived as having relative importance. Qualitative and quantitative analysis showed that organisational and patient outcomes were also highly rated by the participants. The data had good agreement between the 4 different EU sources (Kendall's Concordance significant at 0.005) and the 12 highest rated measures were considered eligible for inclusion in further study. In parallel, a wide ranging analysis of patient handling intervention literature was considered to evaluate the qualities of each individual study. Using the 12 most important outcomes from the initial study and the most appropriate and accessible measurement tools from the literature analysis, the Intervention Evaluation Tool (IET) is proposed. The IET is a single set of measurements that can be used for evaluating all organisational and individual patient handling interventions in healthcare.

Introduction

Patient handling is a known cause of musculoskeletal risk for healthcare staff. A range of ergonomic and other approaches have been used to reduce the effects of these tasks, e.g. risk assessment and management, training, equipment provision, culture change. Comparing the effectiveness of these interventions has been difficult due to the different outcome measures used to evaluate success. Fray and Hignett (2006) found that published patient handling studies used staff outcomes in 77% and represented patient outcomes in less than 8% of the investigations. Recent systematic reviews have concentrated on the specific measures of musculoskeletal disorders (MSD) in healthcare staff using the highest level of scientific data (Bos *et al.*, 2006, Amick *et al.*, 2006, Martimo *et al.*, 2008) and deduce there is little high quality evidence available and little proven benefit on the rate of MSDs. In comparison Hignett *et al.* (2003) used an inclusive methodology and a quality assessment system to allow a wider range of information to be accessed and included.

The literature does include many different methods for measuring outcomes from patient handling interventions. At present it is very difficult to conduct a meaningful comparison between different styles of interventions or different methods of measuring outcomes. This paper will discuss the relative values of the wider range of outcome measures used to evaluate patient handling interventions and describe the development of an 'inclusive' evaluation tool. If a wide application evaluation tool can be developed then it would prove useful to report successful patient handling interventions and guide organisations to a more directed and streamlined approach of future investment to improve their services.

Experts and practitioners from four European Union (EU) countries participated to add to the content validity and strengthen the evaluation tool for the use across all members of the European Panel of Patient Handling Ergonomics group (EPPHE).

The overall aim of this study is to develop an intervention evaluation tool (IET) that allows the comparison of different types of interventions on a single score system using a range of outcomes. The two stages of this process are a) identify which outcomes are preferred by patient handling practitioners, b) develop a tool that measures all the preferred outcomes in a single calculation. This study is part of a longer term research partnership investigating patient handling interventions sponsored by Arjo-Huntleigh ab and the EPPHE group.

Methods

Development of the Intervention Evaluation Tool (IET)

The first aim, to create identify the most important outcomes was achieved using focus groups across 4 EU countries. The 4 countries included in the study were selected using the following criteria:

- Ability to access a range of suitable participants.
- Support delivered by a key facilitator to recruit and organise the focus group.
- Achieve a geographical and demographic spread across the EU.
- A mix of levels within the actions taken to implement/answer the EC directive on manual handling in healthcare (Hignett *et al.*, 2007).

The countries selected for the focus groups were UK, Finland, Italy and Portugal.

The structure for the focus group was based on a model by Higgins (1994), known as the 'Nominal Group Technique'. This allows participants to individually record their own thoughts based on a question set. The method was tested at 2 UK and 2 international pilot sessions. Recruitment for the focus group facilitators took place through the EPPHE group network. Several key stages were used to improve the between-groups validity. Advice was circulated to the focus group facilitators. The following items were included in the guidance:

- The documentation was translated into Finnish / Italian / Portuguese (in the UK) and sent to the facilitator to checking for translation and content errors.
- Standards for recruiting participants for the group.
- Guidance for the focus group room, facilities and timetable.

As it was essential to standardise the translation and cross checking of feedback from the focus groups, the following process was used:

- A translator was supplied to translate the participant feedback information.
- A whispering interpreter was provided to report on the group in real time.
- The discussion groups were transcribed in Finnish/Italian/Portugese. The transcriptions were translated into English.
- The whispering interpreter was taped and transcribed in English.
- The transcriptions from the discussion groups (Finnish/Italian/Portugese) and whispering interpreter (English) were compared for differences.

Specific instructions were developed to assist the EU facilitators. The EU facilitators all participated in the international pilot studies to learn the process and format. The Principal Investigator (MF) was present at all the EU groups to assist with the standardisation of the process and with the development of the discussion group check list of topics for discussion in collaboration with the EU facilitator and the interpreter. Each focus group was centred around a scenario describing a patient care centre. The participants were invited to give advice to the centre and they were asked which outcomes would they like to measure.

The relative importance of the outcomes was considered within each of the individual and homogenous groups. The following simple analytical style was used:

a) The initial recordings of the preferred outcomes were scored on the content.
b) A computerised qualitative analysis package (NVivo) was used to identify themes and content from the focus group discussion. This will be reported in a future paper.
c) The ranked priority lists created by the participants at the close of the focus group interview were scored and ranked on a 5 point scale.

Corrections were made for unequal group sizes before comparisons were made between groups. The ranking scores were added to give a group preference list. All group lists were compared to check for similarities and differences. All the groups were accumulated to give an overall list of the preferred outcomes.

Literature Analysis

To achieve the second aim of selecting methods for measuring each of the preferred outcomes a detailed analysis of published patient handling studies was completed. Studies were collected using the search strategy used for Hignett *et al.* (2003) was extended to December 2008. 752 additional papers were assessed against the inclusion criteria. The total number of papers included in the analysis was 343. Each paper was analysed by two independent researchers and the following data were recorded:

- Design of the study
- Characteristics of the intervention

- Quality Rating (QR; Downs and Black, 1998)
- Level of outcome measure (Robson, 2007)
- Ranking of outcome (12 factors from EU study)
- Practitioner rating (from Hignett *et al.*, 2003)

Certain factors were used when analysing these outcomes against the different categories. Robson (2007) level 3 data are mostly organisational outcomes and level 2 data are mostly staff and patient outcomes. Level 1 outcomes are measures of the intervention and are not included in this analysis. The full comparative data found in this analysis will be presented in a future paper.

Results

Four EU focus groups and 2 worldwide expert panels were completed (n = 44, 9 countries were represented).

Development of the Intervention Evaluation Tool (IET)

The results from the focus groups and the individual scores were analysed for content and theme (Table 1). 210 outcome qualities were recorded in the focus group discussions. The outcome qualities were grouped and compared to give a complex definition for each theme. The translated material was returned to the Finnish/Italian/Portuguese focus group facilitator to check for errors in language and translation. The three sets of qualitative and quantitative scores were combined to identify the most highly valued outcomes. The 12 most highly rated outcomes were worthy of further inclusion.

The ranked scores recorded at the end of the focus groups were then combined against the thematic definitions to give the ranked list for each country and in total (Table 2). Themes that scored less than 5 in any countries combined scores were removed. The same 12 outcomes were seen as most important in each of the 4 countries.

Statistical Analysis There is similarity between the 4 EU sets of rankings as the same highly ranked outcomes (safety culture, compliance and MSD measures) are seen in all countries as are the lower ranked outcomes of finance and patient related

Table 1. Number of different recorded outcomes.

Beneficiary	Outcomes included	Outcomes included in rankings
Organisational	65	13
Staff	57	14
Patient	40	7
Task	30	3
Others	18	1
Totals	**210**	**38**

measures. It was therefore more appropriate to conduct an analysis for association rather than difference. Kendall's Measure of Concordance was performed using the correction factor for tied ranks and $W = 27.66$ ($N = 12$, df 11, $k = 4$) is significant at the 0.005 level and indicates close agreement between the four European groups.

The literature analysis examined all the methods used to measure outcomes in the included studies ($n = 343$). All papers with a QR of >50% were included. Table 3 shows the number of methods used for each outcome.

The outcome measurement methods were assessed using the following inclusion criteria:

- Level of academic quality of the study (QR rating >50%).
- Evidence of peer reviewed validation studies for the method.

Table 2. Ranked themes for individual and combined EU countries.

	Outcome theme	Italy	Portugal	Finland	UK	Total
Organisation	Accident numbers	8	3	11	6	6
	Absence or staff health	3	8	2	4	4
	Financial	12	12	7	10	12
	Safety Culture	2	1	1	2	1
Staff	MSD measures	1	5	8	1	2
	MSD Exposure measures	12	9	5	12	10
	Competence, compliance	4	2	4	6	3
	Psychological well-being	10	7	9	4	7
Patient	Patient injuries	8	12	11	9	11
	Patient perception	8	10	11	8	9
	Patient condition	6	7	6	11	8
	Quality of care	5	4	3	7	5

Table 3. Number of methods used to measure outcomes.

Preferred outcome	No. methods included
1 Safety Culture	5
2 MSD measures	45
3 Competence Compliance	21
4 Absence or staff health	19
5 Quality of care	1
6 Accident numbers	2
7 Psychological well being	8
8 Patient condition	1
9 Patient perception	26
10 MSD exposure measures	170
11 Patient injuries	0
12 Financial	10

Table 4. The measure and sources of the IET.

Preferred outcome	Method for collection	Source paper
1 Safety Culture	Organisational audit of safety systems (PHOQS)	Hignett (2005)
2 MSD measures	MSD level in staff (Nordic Questionnaire)	Dickinson (1992)
3 Competence/Compliance	Observational checklist (DiNO)	Johnsson (2004)
4 Absence or staff health	Standard absence per work population (OSHA)	Charney (1997)
5 Quality of care	Ward and patient survey to evaluate care quality	Nelson (2008)
6 Accident numbers	Accident numbers and non-reporting ratios	Menckel (1997)
7 Psychological well being	3 part worker for satisfaction and well being (Bigos)	Evanoff (1999)
8 Patient condition	Patient survey to evaluate clinical needs	Nelson (2008)
9 Patient perception	Survey for comfort, security, fear etc	Kjellberg (2004)
10 MSD exposure measures	Workload based on patient handling tasks	Knibbe (1999)
11 Patient injuries	Measure for detrimental effects of poor handling	No source
12 Financial	Calculation of costs versus investment	Chokar (2005)

- Previously used to score a peer reviewed intervention trial.
- Most frequently used measurement devices.
- Complexity of the data collection in healthcare.

The IET incorporated the 12 most preferred outcomes (Table 2) and the most suitable method for measuring each of those outcomes (above). Some of the methods chosen (Table 4) were closely related to known peer reviewed tools and studies (1, 2, 3, 4, 7, 9). But others required careful consideration of a range of tools (5, 6, 10, 12). The patient outcomes were poorly represented in the literature review and needed new methods of measurement to be devised.

Discussion

This study has reviewed a wide variety of data from literature and empirical sources. The process of measuring and comparing different types of patient handling interventions has been addressed with the development of the Intervention Evaluation Tool. Every effort has been made to draw the content from studies and measurement methods that have either a good academic score or have proven validation. Some outcome areas were poorly represented in patient handling studies, in particular those relating to patient conditions and quality of care. The IET has undergone several peer review evaluations and has been translated into a further 3 EU languages to allow for further evaluation.

The IET is created to collect a comprehensive set of data from a ward or unit and calculate 12 individual section scores and an overall score to show the effectiveness of the management processes for patient handling. It can be used as a before and after intervention comparison or to compare different types of interventions in similar settings. The IET has been developed to include two distinct forms. Firstly there is a guidebook for managers which outlines the structure of the IET and clearly shows the calculation and scoring process. Secondly there is a data collection format and a series of data collection forms to aid the process. The data collection consists of only 4 sections; a management survey for workload and staff structure, a safety culture audit, transfer observations for 25% of patients, and a questionnaire survey for 50% of staff and 25% of patients. The calculation of the IET scores for the separate and combined scores has been developed into an excel spreadsheet. The following evaluation programme for the IET outlined below has been completed:

1. UK pilot in acute hospital, redefine model and guidance.
2. Expert panel of stakeholders in UK patient handling field.
3. Use of IET in acute hospital (UK).
4. User focus group and feedback (UK).
5. EU translation and back translation of IET.
6. Use IET in Finland / Italy / Portugal in acute hospitals.
7. User feedback groups in EU countries.
8. EU expert focus group (EPPHE) for final evaluation panel.

The detailed investigation of the different intervention studies and the focus groups has developed a clearer picture of the outcomes that are valued among the patient handling specialists in healthcare. There has been a move towards a more organisational and behavioural focus. The measures of safety culture and competence/compliance have featured highly alongside the traditionally high ranking MSD and sickness absence. This shift of perspective may suggest that most patient handling specialists consider the physical risks are manageable with the equipment/engineering solutions that are available in the marketplace and that the future developments are to be focussed on delivering a more compliant organisation.

Full sets of data have now been collected in 4 EU countries and this has allowed further work to be completed on the language and format of the tool and the data collection strategy. At present a simple 1–13 score has been assigned to the 12 section scores to give a total IET score (87). The weightings and calculation structure will need to be part of the ongoing evaluation and validity process that is recommended. The initial results are encouraging and allow for differentiation between different management systems. Concern was raised with early peer review regarding the time for collecting full sets of data. If a local facilitator can distribute the staff survey and prepare the access for the transfer observations, the time on the ward does not exceed one half day. Complications of missing data and lack of observation do increase data collection time. Procedural instructions are being developed as part of the evaluation phase.

Much work needs to be completed to develop and validate this proposed tool. If the IET proves to be a usable and efficient measurement tool then it will be possible to identify the strengths and weaknesses in an organisation from the individual rating scores in the 12 sections and an overall performance score for patient handling interventions. This will allow future interventions to be designed with specific outcomes and gains for the participating organisation, giving the opportunity for more directed interventions to enable best return on financial investment.

Acknowledgements

Thanks in particular go to the members of the EPPHE panel that have assisted in the study and all participants in the workshops.

References

Amick, B., Tullar, J., Brweer, S., Irvine, E., Mahood, Q., Pompeii, L., Wang, A., Van Eerd, D., Gimeno, D., Evanoff, B. (2006). Interventions in health-care settings to protect musculoskeletal health: a systematic review. Toronto: Institute for Work and Health, 2006

Bos, E.H., Krol, B., Van Der Star, A., Groothof, J.W., (2006), The effect of occupational interventions on reduction of musculoskeletal symptoms in the nursing profession. *Ergonomics* 49, 7, 706–723

Charney, W. (1997). The lift team method for reducing back injuries. *AAOHN.* Vol. 45, No. 6, 300–304

Chhokar, R., Engst, C., Miller, A., Robinson, D., Tate, R., Yassi, A. (2005). The three-year economic benefits of a ceiling lift intervention aimed to reduce healthcare worker injuries. *Applied ergonomics.* 2005 Mar; 36(2): 223–9

Dickinson, C., et al, (1992). Questionnaire development: an examination of the Nordic Musculoskeletal Questionnaire. *Appl Ergon*. Vol 23 No 3 June 1992.

Downs, S.H., Black, N. (1998). The feasibility of creating a checklist for the assessment of methodological quality of both randomised and non-randomised studies of healthcare interventions. *Journal of Epidemiological Community Health* 52: 377–84

Evanoff, B., Bohr, P., Wolf, L. (1999). Effects of a participatory ergonomics team among hospital orderlies. *Am. J. Industrial Medicine* 35:358–365

Fray, M., Hignett, S. (2006) An Evaluation of Outcome Measures in Manual Handling Interventions in Healthcare. In Pikaar, R.N., Konigsveld, E.A.P., Settels, P.J.M. (Eds.) *Proceedings of the XVth Triennial Congress of the International Ergonomics Association, Meeting Diversity in Ergonomics* 11–14 July 2006, Maastricht, Netherlands

Higgins, J.M. (1994). *101 Creative Problem Solving Techniques*. Florida: New Management Publishing Co

Hignett, S. (2003). Intervention strategies to reduce musculoskeletal injuries associated with handling patients: a systematic review. *Occup Environ Med;*60:e6

Hignett, S., Crumpton, E., Alexander, P., Ruszala, S., Fray, M., Fletcher, B. (2003). *Evidence based patient handling- Interventions, tasks and equipment.* London Routledge

Hignett, S., and Crumpton, E. (2005). Development of a patient handling assessment tool. *Int. JTR* April 2005, Vol 12, No 4 178–181

Hignett, S., Fray, M., Rossi, M. A., Tamminen-Peter, L., Hermann, S., Lomi, C., Dockrell, S., Cotrim, T., Cantineau, J. B., Johnsson, C. (2007). Implementation of the Manual Handing Directive in the Healthcare Industry in the European Union for Patient Handling tasks. *International Journal of Industrial Ergonomics* 37, 415–423

Johhnsson, C., Kjellberg, K., Kjellberg, A., Lagerstrom, M., (2004). A direct obser-vaion instrument for assessment of nurses' patient transfer technique (DINO). *Appl Ergon.* 35 (2004) 591–601

Kjellberg, K., Lagerstrom, M., Hagberg, M., (2004). Patient safety and confort during ransfers in relation to nurses' work technique. *J. of Adv. Nursing,* 47(3), 251–259

Knibbe, J., and Friele, R. (1999). The use of logs to assess exposure to manual handling of patients, illustrated in an intervention study in home care nursing. *IJIE*, 24 (1999) 445–454

Martimo, K.P., Verbeek, J., Karppinen, J., Furlan, A.D., Takala, E.P., Kuijer, P., Jauhianen, M., Viikari-Juntura, E. (2008). Effect of training and lifting equipment for preventing back pain in lifting and handling: systematic review. *BMJ* doi: 10.1136/bmj.39463.418380.BE

Nelson, A., Collins, J., Siddarthan, K., Matz, M., Waters, T. (2008). Link between safe patient handling and patient outcomes in long term care. *Rehabilitation Nursing*, Vol 33, No 1 Jan 2008

Robson, L.S., Clarke, J.A., Cullen, K., Bielecky, A., Severin, C., Bigelow, P.L., Irvin, E., Culyer, A., Mahood, Q. (2007). The effectiveness of occupational safety management system interventions: A systematic review . *Safety Science 45 (2007) 329–353*

THE HUMAN FOOTPRINT – SCOPING HFI FOR GENERIC TECHNOLOGY OR SYSTEMS OF SYSTEMS

John Harrison

Hu-Tech Associates, UK

The Human Footprint (of a system or new technology) is the totality of its impact on human activity. It covers all tasks that are changed, influenced or created by its presence. The paper proposes a graphical representation of the footprint, as a tool to support exploratory discussion with stakeholders when considering the human aspects of the introduction of a new technology to a complex socio-technical system, or systems of systems. The diagram starts with the item in question, and around it develops layers of 'task spaces' that will be directly or indirectly affected by its presence. An example is given of how the approach was used by a Ministry of Defence funded study where the technique was initially applied, an exploration of the training and selection issues associated with the widespread introduction of UAVs (unmanned air vehicles) into military operation. This example is used to elucidate the key features of the approach, and to discuss its apparent strengths and weaknesses.

Introduction

The key Human Factors questions to answer about any introduction of technological artefacts within a socio-technical system are:

- Who is affected?
- How are they affected?
- What is the result?
- Does it matter?

In systems engineering terms the first question defines the effective system boundary. With more complex systems, the answer becomes far from trivial, especially when several systems are inter-related (the so-called systems of systems).

Early in a project, in exploratory discussions with stakeholders, it is common to use some sort of framework to provide coherence while permitting open-ended or discursive responses, typically a semi-structured set of questions. Some sort of representational 'model' of the system can also help to ground the discussion in a common view of what is being discussed. Such a model could be built from scratch in a workshop, or when interviewing stakeholders individually, it could be seeded using prior knowledge, with each stakeholder invited to recognise, challenge, change or elaborate it. The initial straw man version may be used with all stakeholders, or the model may be evolved using the insights of successive stakeholders.

The model should not be too complex, and should be easy to interpret with only basic explanation. It should contain enough detail to provide a coherent framework, but be general enough not to close off lines of thought. It should be easy to change (in response to suggestions). Such models are often presented as diagrams, structured lists, or a combination of the two. A diagram should provide a good 'intuitive' mapping onto the problem space. For example: a tree can be useful when the essence of the problem space is a hierarchy or decomposition; a net can be useful when the essence is about links or relationships; a timeline can be useful when the essence is about sequence; a Venn diagram can be useful when the essence of the problem space is overlap; a map (plan view) can be useful when the essence of the problem is spatial; and so on. Other forms are possible.

The model must show, or at least hint at, the sort of detail that stakeholders will need to discuss, but it should be coarse enough to permit the whole problem space to be encapsulated in a diagram of reasonable complexity. Fitting comfortably on a single sheet of paper is a useful aim, with the added benefit that participants can draw and scribble on it, which can be a more liberating form of interaction than asking for changes to be made on screen.

The Human Footprint

The term 'footprint' is now widely used to refer to the total (direct and indirect) impact of something (often a technology or activity) on an area of concern. Thus the carbon footprint of a product represents all CO_2 emitted to the atmosphere during its manufacture, use and disposal. By analogy, the 'human footprint' represents the total impact of a deployed item (technology, process or operational concept) on an organisation and the work of people within it.

A diagram of the human footprint seeks to represent all significant areas of work within an organisation or a set of related organisations that are influenced by the presence of the item (a technology, a process or an operational concept). The diagram uses a succession of 'layers' of 'task spaces', which contain groups of activities related to the item and the people involved with them. The layers represent progressively less direct interaction, for example:

- Direct hands-on interaction – People who are physically present with the equipment and do things to it. Their presence in the work system may be because of the equipment, or they may be there anyway, and its presence changes how they work (by adding, removing or altering what they do). (For example they may set the equipment up.)
- Direct hands-off interaction – People who are not physically present with the equipment, but whose tasks still interact directly with it, either as well as or

instead of other tasks they would do if it were not present. (For example, they may operate the equipment remotely.)

- Indirect interaction – People who do not interact directly with the equipment (even remotely) but whose tasks exist because of its presence. (For example, they may supervise and direct equipment operators.)
- Implicit (consequential) interaction – At this most remote level of impact, people do not interact in any meaningful way 'with' the technology, and in principle their jobs are neutral to the technology, but in practice their jobs are either enabled by it, for example using information or artefacts that rely on it, or in some other way affected by it. For example, people who order goods via the Internet do not overtly interact with the logistics system that stores, distributes and delivers the goods, but the nature of the logistics system could have a transforming effect on how they work, and the extent to which they use the system.

Example application

The Human Footprint concept was developed and used during a study for the UK Ministry of Defence (MoD) into the training and selection issues associated with widespread introduction of UAVs (Unmanned Air Vehicles) into military operation (Harrison & Kendal, 2008). At the time of the study, several UAVs were already in use and the technology was expected to be deployed much more widely in the future. Current UAVs range from small hand-launched craft, manually controlled within visual range, to craft the size of a small airliner controlled via satellite link by a team in another continent.

The study was tasked to look beyond current or planned programmes to issues that would be faced with widespread deployment of many UAV types across all three armed services, each of which has a different operational context and a distinct approach to manpower selection, training and career development. The study was specifically intended to draw together these different needs and perspectives, in a tri-service perspective.

The broad scope required input from many stakeholders, with diverse experience and perspectives. Capturing the diversity was important, but the different strands had to be related in a coherent way. Some sort of framework was needed as a backdrop for discussion. It had to be generic to cater for all the differences, and it had to encourage exploration of the key questions about how the technology would affect people's work, and how far through the organisation this would spread. And in order to be usable it had to be relatively simple.

An Operation Sequence Diagram was offered from a UAV project, which clearly showed the interplay between different parts of the organisation. This approach was rejected because it could not be generalised. The diagram offered was tightly linked to a specific organisation and a specific class of UAV. An alternative way was needed to represent the way that functions affected by the UAV would map

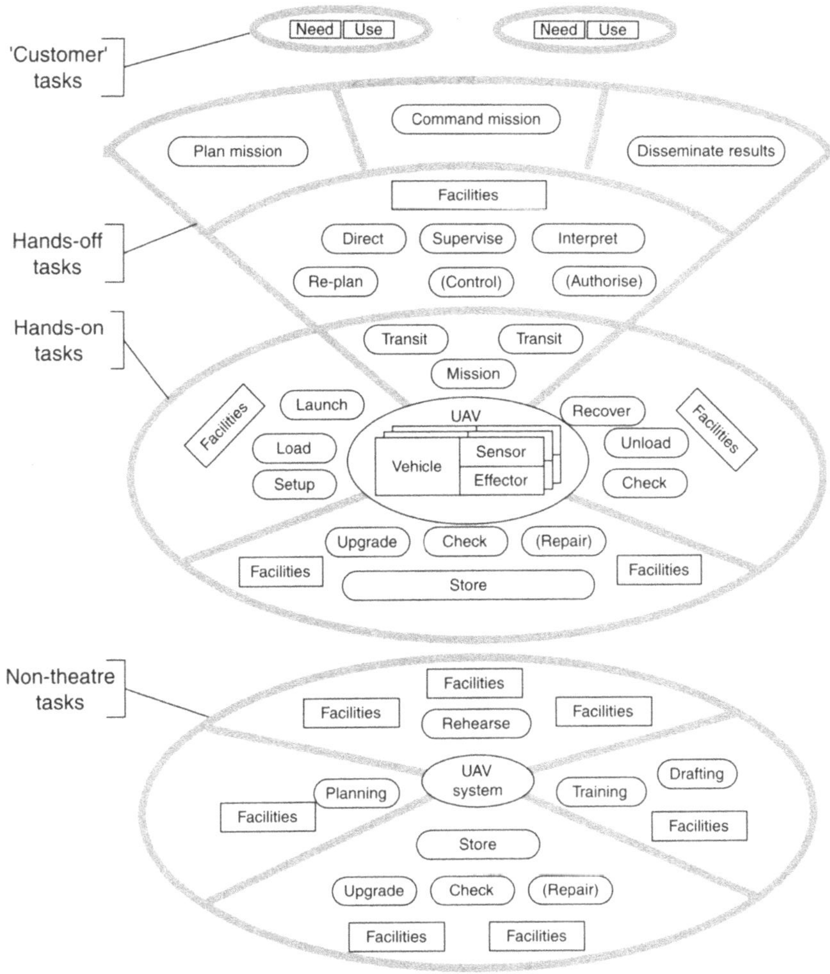

Figure 1. Human footprint diagram for a UAV.

onto people in various organisations. The alternative that emerged was the human footprint diagram, shown slightly simplified in Figure 1.

In the middle of the diagram is an ellipse containing a generic UAV, metaphorically dropped into the organisational pond. It was described simply as 'vehicle + sensors + effectors'. (This is the generic term for something that produces an effect. It might be some sort of weapon, but could equally be a device like a jammer or distractor, intended to disrupt radar or communications.) The rest of the diagram showed the human task spaces through which its ripples would spread.

The first layer of 'hands-on' task spaces reflected the cyclic nature of UAV operation, i.e.: before flight, during flight, after flight, and 'between flights'. Each task space contained the names of a few generic activities, which could (and in most cases would) involve people to some degree. For example pre-flight included 'setup', 'load' and 'launch'. Even if an action occurs autonomously (for example some vehicles are able to take off automatically) there are still likely to be some human initiation and checking tasks before it is allowed to do so. The 'between flights' task space included 'storage', 'repair', 'upgrade' and 'checking'. The 'in-flight' task space included activity (the mission itself, plus transit to and from the mission) but by definition it is empty in terms of hands-on human tasks (because the vehicle is unmanned).

The layer of 'hands-off' task spaces in the diagram was confined to this in-flight phase (which was empty of hands-on tasks). The task mix here can vary quite a lot with vehicle capability, but generically it includes some level of control, re-planning and authorisation.

The next layer (indirect interaction) included mission planning, mission command and dissemination of results. Each was given a separate task space to cater for the fact that they might be performed in different parts of the organisation. The distinction between this command level and the more direct control level reflects the separation between deciding what must be done (and allocating resources to it) and deciding how it is to be done and then doing it. The inclusion of dissemination at this level also reflects this distinction. Whereas at the lower level the objective is to fly the mission, at this higher level the objective is to deliver the results from it, for example surveillance data.

The task spaces at the implicit (consequential) level were drawn detached from those clustered round the UAV, reflecting the fact that they are quite detached. They relate to jobs that exist despite the presence of UAVs. If UAVs did not exist, they would get some sort of information from some other source. The reason their jobs are affected by the presence of UAVs is that the volume, quality and timeliness of information from a UAV can far exceed what they can get from other sources, for example in the extreme case they could be relying on information from men concealed on the ground with binoculars. In the Figure 1 these most indirect task spaces are designated 'customer', and the tasks within them are described completely generically in terms of having a need and using a result.

The addition of the lower part of Figure 1 reflects the two-state nature of military life (for people and equipment) where part of the time is spent 'in theatre' and part is 'back at base'. While there is a natural focus on operational activity performed 'in theatre', with its cycle of mission and inter-mission activity, any whole life Human Factors Integration (HFI) perspective must embrace both phases, especially when considering domains like training and selection, which were the subject of the UAV study. Thus the footprint diagram included a second, simpler cluster of task spaces centred round the presence of the UAV when not in theatre. This included task spaces for 'rehearsal' and 'storage', which broadly mirrored their operational counterparts, and two extra task spaces. One related to planning (longer term resource planning, etc). The other covered tasks related to manpower provision, i.e. drafting and training.

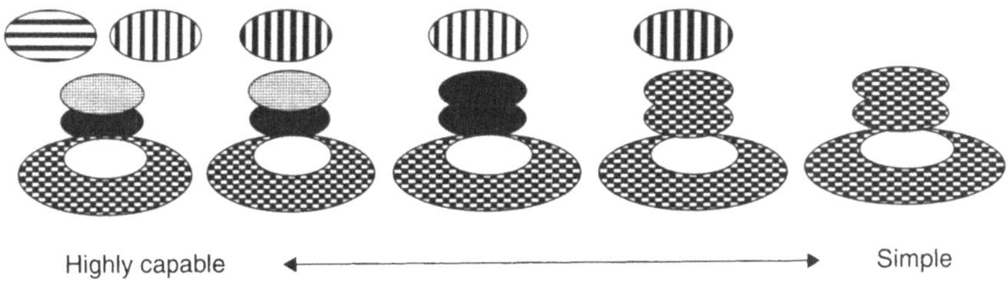

Figure 2. Organisations mapped onto different types of UAV.

During the project, stakeholders were able to relate to the diagram, and it served its purpose as a reference framework from which to discuss the activities surrounding both current and potential future UAVs. The diagram evolved from an initial minimal form to a final more comprehensive form, in response to stakeholder-suggested changes and additions.

Derivative forms of the diagram

The UAV study used an even more simplified version of the Footprint Diagram later in the project to provide a graphical summary of the diversity of the human impact of different types of UAV, see Figure 2. It shows 'thumbnail' versions of the 'in theatre' part of the footprint diagram for five different UAV systems ranging from highly capable to very simple. In each thumbnail, the different 'layers' were preserved, but all the detail within them was suppressed. (The shape discrepancy in the middle layers – separate ellipses rather than sectors centred on the main ellipse below – was an artefact of the tools used to generate the diagrams.) Different shading was used to show where different organisational groups are involved in different task spaces. (The UAV itself is unshaded in the diagram, because it is unmanned.)

For the most capable system (left) the different shading shows that each task space is occupied by a different organisation. Ground handling, in-flight control and mission direction are all separate, and there are multiple 'customers', in different organisations. In contrast, the simple system (right) is operated and managed by a single (quite small) organisation for its own purposes, with no separate external 'customer'.

Discussion

No directly comparable approach is known. The contextual process of using semi structured interviews to elicit information from stakeholders is well developed, and is described for example by Stanton et al (Stanton, et al, 2005) in terms of questions. The use during discussion of models that represent the area of concern seems a natural adjunct of the questions, and perhaps does not merit comment. Such diagrams will be specific to the topic being explored, and as noted above, flow diagrams, tree diagrams, geographical maps, etc are all natural ways to think about some sorts of problem.

The approach used here, a course-grained representation of activity associated with a particular artefact or technology, organised in terms of cycles of activity and degree of directness or indirectness of interaction, appears to be novel. Stanton et al (Stanton, et al, 2005) review and describe over 80 methods and techniques organised into 11 different groups. Many use diagrammatic representations, and several are relevant to systems that involve different groups of people, for example Operation Sequence Diagrams (OSD) and Comms Usage Diagrams (CUD), but these are finer grained descriptions.

Descriptive granularity

The Footprint Diagram uses a very coarse-grained representation, quite different from those used in design (such as the OSD that was offered and rejected as unsuitable by the UAV study). It does not make explicit each interaction between organisations, or every sequential dependency, but just the broad groups of activity associated with phases in a cycle of use. A coarse grain is necessary when trying to capture a large scope (in the UAV case, anything that might affect training and selection) within a relatively simple diagram.

A coarse grain can also be used to focus on generic features, as in this case. The activity groups (set up, load, launch, etc) are generic across all classes of UAV, even though the details of how the activities are conducted may vary dramatically. For example, launching a micro UAV may entail little more than throwing it into the air, whereas launching a UAV at the other end of the scale could entail pre-flight procedures, runway and air traffic clearance, and handover between local 'pilots' responsible for take-off and remote 'pilots' responsible for the mission.

There is of course much detail hidden in each task space, not represented explicitly, some of which will be germane to the issues being explored. For example within the pre-flight task space there could be two separate organisations involved, one responsible for certifying that the vehicle as fit for flight, and the other responsible for preparing it for its mission. From a safety perspective, the division of responsibility between them will be important. Such issues should emerge from stakeholder discussions, of the questions that the Footprint Diagram is intended to support. Further exploration of individual cases may require more explicit representations, for example task hierarchies, Operational Sequence Diagrams, and so on.

Time is also coarse grained, with broad phases of activity (pre-flight, in-flight, post-flight, etc.) that are generic to any UAV, even if the detail that is glossed over would vary with organisational processes and/or UAV type. Stepping back to a longer timescale should also encourage a broader perspective. In the UAV example this was evidenced by closure of the between-flight gap to provide a complete cycle, and also by the inclusion of the out-of-theatre, as well as in-theatre activities.

Traditional approaches to systems sometimes treat 'maintenance' and/or 'support' as separate, disjoint activities, rather than as a continuum of system life, and can hence under-play the risks associated with these aspects of the total system.

The weakness of a course grained model is that it can only be used to prompt discussion of what lies in each task space, and can't be used to record the detail. The (more detailed) discussion thus prompted must either resort to supplementary models or be conducted without visual aids. Supplementary models could be expansions of individual task spaces to show more detail, or they could use other notations (task trees, flow diagrams, etc.)

Layering

The four layers (direct hands-on, direct hands-off, indirect and implicit/consequential) fit the UAV domain but might need some adaptation when considering the impact of other technologies. For example with 'soft' technologies, like 'intelligent' embedded decision aids, the distinction between hands-on and hands-off interaction would be harder to define.

The outermost layer (implicit/consequential interaction) might not exist in all cases, as was shown with the 'simple' UAV in Figure 2. However, it seems sensible to retain it within the approach, for use as a prompt, while accepting that it might be empty.

'Facilities'

The primary purpose of the task spaces in the diagram is to represent activities (notably those that may be performed by people). Most spaces in Figure 1 also include 'Facilities'. This makes explicit the fact that the activities, and the people performing them, may be associated with other tools and equipment, as well as with the object under consideration (in this case the UAV).

It might be argued that this is redundant, because stakeholders knowledgeable about the activities implicit in headlines like 'load' and 'set up' would know about any associated equipment and tools. They might consider the facilities to be implicit in the activities, and thus not even mention them in discussion. For this reason, their inclusion in the diagram seems appropriate, since the aim of early discussions with stakeholder is to transfer insights from them to the analyst, which includes making explicit things that might otherwise be implicit.

Should facilities be represented with a granularity match that of the associated activities? They aren't in Figure 1 (the grain is even coarser) but this seems appropriate, given the desire for a simple diagram. In fact the labels used in Figure 1 are slightly simplified from the originals (to make the diagram more compact here). For example 'facilities' in the pre-flight task space were 'launch facilities'. In that respect, the simplified version is more generic than the original, since facilities would no doubt be used for load and set-up activities, as well as launch.

Clustering

The UAV diagram in Figure 1 goes beyond the basic idea of concentric cycles of activity. It adds a separate cluster to represent out-of-theatre activities, and multiple 'islands' in the customer layer. The in-theatre v out-of-theatre distinction is domain specific, and may or may not have parallels in other domains.

The customer 'islands' could more purely have been shown as a stack of segments one behind the other, to avoid breaking the underlying layout convention. Would that have been more helpful to stakeholders? If this hybridisation of the layout is useful, then it seems likely that application in other domains may lead to the use of more clusters, for example where a complementary set of technologies is embedded within different organisations, with the aim of influencing their interaction with each other as part of a whole system. An example is the set of multiple changes to the European Air Traffic Management System that are being explored by the SESAR consortium (SESAR 2008).

Conclusion

The human footprint diagram complements rather than replaces existing approaches. It can help to visualise the impact of a new technology on the work of different organisations in a complex socio-technical system. It can handle generality, and was found useful in one project. Use in other domains requiring a generic representation seems possible, but may require some domain specific adaptation.

References

Harrison, J.A., Kendal, R. 2008 *Tri-Service Training, WP2: Training for Unmanned Systems*, (Haldane–Spearman Consortium)

SESAR Consortium, 2008 *SESAR Masterplan D5* (SESAR Consortium)

Stanton, N.A., Salmon, P.M., Walker, G.H., Baber, C., Jenkins, D.P. 2005 *Human Factors Methods – A Practical Guide for Engineering and Design* (Ashgate)

WORK-RELATED LOWER LIMB MUSCULOSKELETAL DISORDERS – A REVIEW OF THE LITERATURE

Olanrewaju O. Okunribido[1] & David Lewis[2]

[1]*Health and Safety Laboratory, Harpur Hill, Buxton, UK*
[2]*Health and Safety Executive, Liverpool, UK*

This work was undertaken to determine the current knowledge base concerning work-related lower limb musculoskeletal disorders and injuries (LLD). Results showed that LLD are a problem in many workplaces and they tend to be associated with conditions in other areas of the body. There is appreciable evidence for kneeling/squatting, climbing stairs or ladders, heavy lifting, walking/standing, and slips and trips hazards as causal risk factors for LLD; the evidence is less clear for jumps from height, driving and sitting. Workplace redesign/modification initiatives, implementation of protective equipment and participatory programs appear to be the most suitable interventions. Further work is suggested to determine whether persons who suffer back pain are also likely to suffer LLD and vice versa, and to determine the suitability of prevention strategies proposed for conditions in other regions of the body against LLD.

Introduction

Musculoskeletal disorders (MSD) are a major cause of work-related disability and account for absence from the workplace in many occupational groups. However, relative to work-related MSD of the upper extremities and lower back area, much less attention has been given to the epidemiology of lower limb musculoskeletal disorders and injuries (LLD) (D'Souza *et al.*, 2005). Furthermore, regulations and guidance that have been provided to date are largely targeted at the musculoskeletal health of workers' backs and upper limbs. Therefore, their suitability for application to the musculoskeletal health of workers' lower limbs is questionable. LLD are distinct from MSDs affecting the back, the neck and the upper limbs, in that they can often give rise to greater degrees of immobility and can degrade the quality of life substantially (Bruchal, 1995; Lohmander *et al.*, 2004). This work was undertaken to determine the current knowledge base on the nature and extent of work-related LLD that are suffered by workers as well as the key risk factors and to identify suitable strategies for risk control and injury prevention.

Methodology

Studies included in this review were selected from searches of electronic databases, i.e. MEDLINE, OSHROM, and Ergonomics Abstracts on-line, using key words (lower limb, MSD, leg, extremity, injury, disorders, musculoskeletal, pain, interventions, etc.). The search was restricted to studies published after 1990 and based on the following criteria: they examined the prevalence, nature and work-relatedness of the disorders; they evaluated strategies for risk control or injury prevention, and they involved worker populations. Studies that focused primarily on prognosis or clinical treatment and management rather than prevalence or prevention were excluded. The majority of the studies reviewed was in English and published in peer-reviewed journals. This was a qualitative literature review as the studies were too heterogeneous for meta-analysis.

Results

Nature of LLD

Like disorders of the upper limb and axial skeleton (neck and trunk), LLD involve the muscles, tendons or nerves, ligaments and other tissues, and they are generally manifested by inflammation, pain, discomfort or tingling. Both acute and overuse disorders/injuries were identified for populations of industrial workers, as well as populations of athletes and military personnel (Bennell and Crossley, 1996; Rudzki, 1997; Akesson *et al.*, 1999). However, acute injuries tend to be more common among athletes and military personnel than among industrial workers (Olsen *et al.*, 2005). Hip osteoarthritis (OA); knee OA, knee bursitis, meniscal lesions/tears; stress fracture/reaction injury and venous disorders/varicose veins of the lower legs were the most commonly reported overuse injuries among industrial workers. Table 1 presents a list of LLD conditions that have been associated with occupational populations in the literature (industrial workers, athletes and military personnel).

Burden of LLD

The prevalence of LLD varied widely and seemed to depend on the type of study and/or the population surveyed. Thus, studies based on a general population or archival data (Chen *et al.*, 2006; Smith *et al.*, 2006; Jones *et al.*, 1998), report prevalences between 10% and 30%; those based on specific worker populations (Jensen and Kofoed, 2002; Forde *et al.*, 2005; Gallis, 2006) report higher prevalence rates of between 20% and 60%. The reports of LLD symptoms were often not independent of reports of symptoms in other areas of the body and, more than one site of pain was often mentioned. The prevalences for symptoms and disorders of the knee were generally higher than those for symptoms and disorders of other regions of the lower extremity.

Various consequences are associated with prevalence of LLD (Welch *et al.*, 1999; Hildebrandt *et al.*, 2000). These tend to depend on the particular disorder suffered, the number of joints affected and the job demand on the worker, but may also depend on how concerned the worker is about their condition, i.e., their tolerance level

Table 1. LLD conditions that have been associated with occupational populations.

Variable	LLD conditions according to the regions of the lower extremity		
	Hip/thigh	Knee/lower leg	Ankle/foot
Overuse Injuries	Osteoarthritis (OA) Piriformis syndrome Trochanteritis Hamstring strains Sacroiliac pain	Beat knee/Hyperkeratosis Bursitis Meniscal lesions Osteoarthritis (OA) Patellofemoral pain syndrome Pre-patellar tendonitis Shin splints Infra-patellar tendonitis Stress fractures	Achilles tendonitis Blisters Foot corns Halux valgus (bunions) Hammer toes Pes traverse planus Plantar fasciitis Sprained ankle Stress fractures Varicose veins Venous disorders
Acute Injuries		Meniscal tear	Ankle fractures Metatarsal fractures

(Cromie *et al.*, 2000). Days of restricted duty, sick leave and days receiving hospital treatment were the most commonly identified consequences, but ability to carry out normal work tasks and recreational activities may also be affected, particularly when the condition is chronic. Compared with workers who suffer conditions of the upper extremities and torso, workers with LLD are more likely to have symptoms beyond two months and require time off work (Jarvholm *et al.*, 2008). Compared with workers in light jobs (involving sedentary tasks with little or no manual handling of loads), those in heavy jobs (often involving heavy or physically demanding tasks) who suffer a LLD condition, particularly a knee condition, are more likely to seek medical treatment and require hospitalization.

Risk factors for injury

Studies concerning the risk factors for injury (Andersen *et al.*, 2007; Choobineh *et al.*, 2006; Roelen *et al.*, 2008) suggested diverse factors in three broad categories: occupational (physical) risk factors, personal (and demographic) risk factors and psychosocial risk factors. Many of the factors are not specific to the lower extremities, as they have also been associated with disorders of the upper extremities and the trunk. For LLD, the studies generally indicate that the most likely causal agents are physical work-related factors, particularly, the following:

- Kneeling/squatting,
- Climbing stairs or ladders,
- Heavy lifting,
- Walking/standing

- Slips and trips hazards.
- Jumps from height
- Driving, particularly continuously for more than four hours
- Sitting, in an awkward position or for long durations at a time (>2 hours), particularly for hip pain. However, sitting, as a work posture, has often been investigated comparatively with standing, and generally showed a negative association with lower limb (knee and ankle/foot) pain. This indicates it is a preventive rather than a causative factor, when compared with standing.

Personal and psychosocial factors that were most often associated with increased risk are: previous injury, physical condition (no activity outside work), obesity, gender, lack of job satisfaction, low level of control over work, little or no social support from colleagues and no support from supervisors.

Based on the observations from the literature, Figure 1 provides an overview for the more common lower limb conditions and associated risk factors.

Risk reduction and control

Studies on risk reduction and control e.g. Brooker *et al.* (2001), Baisden *et al.* (2009), Minns *et al.* (2004), Shaffer and Uhl (2006), and Carrivick (2005), generally discuss three groups of interventions: workplace redesign/modification initiatives, implementation of protective equipment, and socio-organisational regimes and training.

Workplace redesign/modification interventions, e.g. Jensen and Friche (2008) and Brooker *et al.* (2001) can help prevent/control LLD in the workplace, particularly when the worker is encouraged to adopt optimal work positions/postures and to exert reduced levels of force in the tasks. There are also real cost benefits to be gained (Nicholson *et al.*, 2006).

Two groups of protective equipment are differentiated: Personal protective equipment and handling assist devices.

- Personal protective equipment: Hip protectors and knee pads, shoe insoles, limb supports and anti fatigue matting have most often been reported on for lower limb injuries, though the efforts to provide protection devices for the hip have largely been in relation to falls of the elderly.
- Handling assist devices: The usual selection (trolleys, carts, etc.), may be used to try to miminise lower limb injuries. Portable adjustable tables, lifting aids, extended handle bars and other simple low-cost tools may also be introduced to enable performance of operations at better working heights.

The influence of workplace policies, such as exercise regimes and participatory initiatives, as interventions for preventing injury and controlling risks has been reported on appreciably (Chen *et al.*, 2007). The role of regulations and codes of practice has also received some consideration. van der Molen (2007)

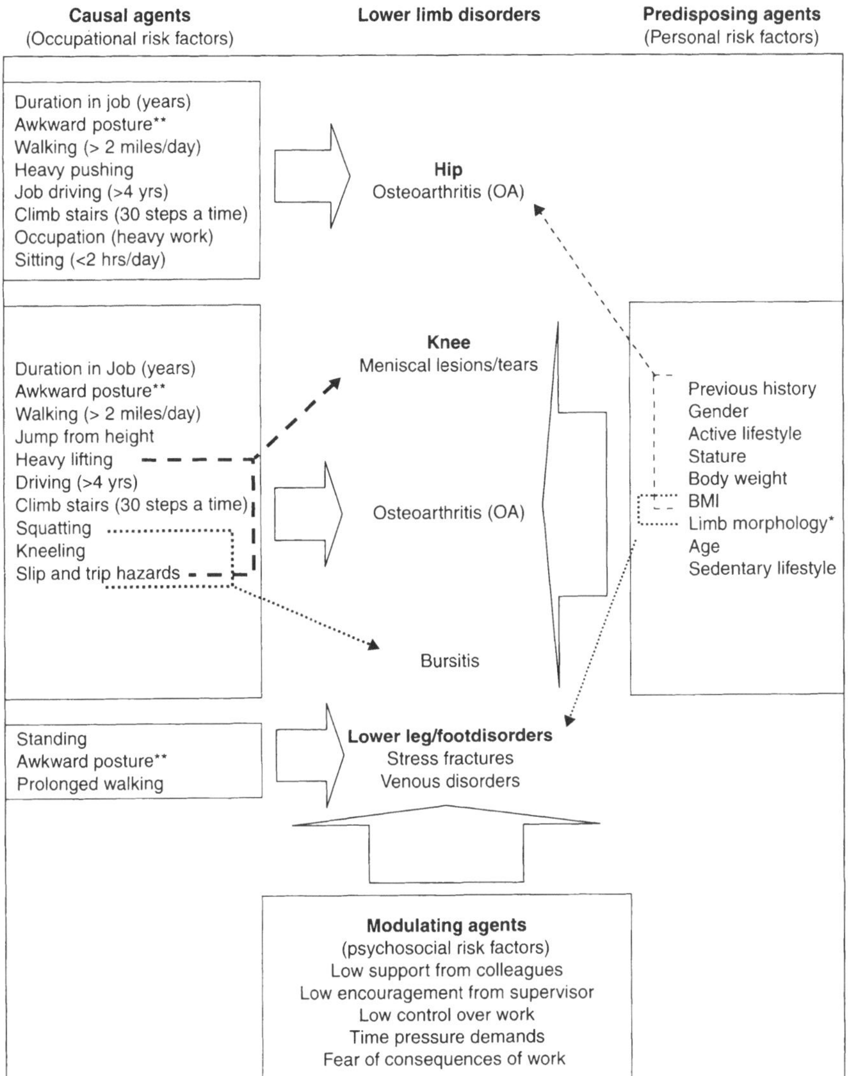

*Anatomical alignment, this is suggested to impact on the biomechanical stresses imposed on joints of the limb during activity.
**Awkward posture, e.g., constrained/twisted positions of the limbs

Figure 1. A framework of LLDs and the risk factors.

assessed the effects of three types of regulatory interventions for preventing injuries among workers at construction sites: regulations, safety campaigns and drug-free workplace programs. They concluded that the vast majority of technical, Human Factors and organisational interventions, which are recommended by standard texts, consultants, and safety courses, have not been adequately evaluated.

Despite the efforts to understand MSDs and several interventions developed to control the risks and prevent their occurrence; there are still doubts about the efficacy of many of the control strategies. Notwithstanding, the studies reviewed suggested that workplace redesign/modification initiatives and implementation of protective equipment are the most suited interventions for control of LLD risks, especially when they are applied in combination.

Discussion

The literature reviewed confirmed that LLD are a problem in many workplaces and that there are consequences of the injuries for society, the economy and industry in terms of lost work time, medical treatment and hospitalisation, decreased work ability of employees and decreased quality of life.

In these regards, Figure 1 indicates first, that the risk factors for LLD conditions are not specific to any region of the limb or type of condition; secondly, that more risk factors are associated with knee conditions than for either hip or lower leg/ankle conditions. These observations could be due to a number of factors, such as study design. Indeed, majority of the studies that investigate risk factors alongside prevalence or incidence of injury applied cross-sectional and case-control study designs, and defined the conditions in terms of effects such as perceived fatigue, pain and complaints. Only a few studies applied prospective study design methodologies. Also, the conclusions about the associations of the risk factors with symptom/pain were based largely on uni-variate analyses and often without investigation of dose-response effects. Another major drawback of the literature is the manner in which risk factors are assessed. This was often through self-reporting and usually focused on only one of three dimensions required to completely quantify exposure: the amplitude of the exposure, the duration and the repetition. Self-reporting has several drawbacks. Sometimes, a spurious association may occur if those with injury recall the past exposures more completely than controls (those without). There is also the question of common beliefs in different occupational groups about the occupational hazards experienced by their group, which could influence self-reporting. Finally, Anderson *et al.* (2007) identified that, even though the temporal relationship between measures may fulfill an acceptable standard procedure, the self-reporting of physical exposure could easily be flawed by the subjects' pain status. The suggestion that there are many more risk factors for knee conditions than for hip and lower leg/foot conditions and for knee OA than for other specific knee conditions, may be attributed to the fact that many more of the studies concerned the knees as the region of injury and OA.

Despite the heterogeneity of the methodologies and the associated shortcomings, a number of the risk factors repeatedly emerged as significantly associated with pain (or other symptom of LLD) and in some, also indicated dose-response relationships. This was particularly true for physical and personal factors, many of which have been directly implicated in the aetiology of MSDs. Prevention of LLD in the workplace therefore requires the creative use of control strategies to balance the physical and psychosocial demands with the characteristics of the individual(s). The indications were that workplace redesign/modification interventions, implementation of protective equipment and participatory programs might be most suited for control of the risks and prevention of LLD. However, there was also evidence that an intervention suited for one type of disorder may not protect against another. In addition, it was not clear whether personal protective devices provided protection against all types of disorders or only a subset, those with a similar aetiology. Furthermore, there were doubts about the effectiveness of organisational interventions, such as guidance, regulations and safety campaigns for encouraging employers and workers to comply with safety measures.

Conclusions

Lower limb musculoskeletal disorders and injuries, particularly knee conditions, are a problem in many workplaces and they tend to be associated with conditions in other areas of the body. There is appreciable evidence for kneeling/squatting, climbing stairs or ladders, heavy lifting, walking/standing, and slips and trips hazards as causal risk factors for LLD; the evidence is less clear for jumps from height, driving and sitting. Workplace redesign/modification initiatives, implementation of protective equipment and participatory programs appear to be the most suitable interventions for controlling the risks in the workplace. Further work is required to:

- Clarify the inter-relationships between injury/pain at different regions of the body, i.e., to determine whether persons who suffer back pain are also likely to suffer pain in the lower extremity and vice versa, and to determine whether the relationships are dependent on the type of injury suffered or not;
- Provide more objective measures of workplace ergonomics risk exposures;
- Determine suitability of prevention strategies proposed for conditions in other regions of the body (back/upper limbs) against LLD; and,
- Identify organizational strategies other than regulation that would aid increased awareness of the problems in workplaces, and encourage the commitment of employers.

References

Akesson, I., Johnsson, B., Rylander, L., Moritz, U. and Skerfving, S. 1999, Musculoskeletal disorders among female dental personnel – clinical examination and a 5-year follow-up study of symptoms, *International Archives of Occupational and Environmental Health*, 72, 395–403

Andersen, J.H., Haahr, J.P. and Frost, P. 2007, Risk factors for more severe regional musculoskeletal symptoms: a two-year prospective study of a general working population, *Arthritis and Rheumatism*, 56, 1355–1364

Baisden, J., Carragee, E.J., Resnick, D.K., Shaffer, W.O. and Loeser, J.D. 2009, Surgery for Low Back Pain: A Review of the Evidence for an American Pain Society Clinical Practice Guideline, *Spine*, 34, 1094–1109

Bennell, K.L. and Crossley, K. 1996, Musculoskeletal injuries in track and field: incidence, distribution and risk factors, *Australian Journal of Science and Medicine in Sport*, 28, 69–75

Brooker, A.S., Cole, D.C., Hogg-Johnson, S., Smith, J. and Frank, J.W. 2001, Modified work: prevalence and characteristics in a sample of workers with soft-tissue injuries, *Journal of Occupational and Environmental Medicine*, 43, 276–284

Bruchal, L.C.H. 1995, Occupational Knee Disorders: An Overview. In A.C. Bittner and P.C. Champney (eds.) *Advances in Industrial Ergonomics and Safety VII*, (Taylor & Francis, London), 89–93

Carrivick, P.J., Lee, A.H., Yau, K.K. and Stevenson, M.R. 2005, Evaluating the effectiveness of a participatory ergonomics approach in reducing the risk and severity of injuries from manual handling, *Ergonomics*, 48, 907–914

Chen, J.C., Linnan, L., Callahan, L.F., Yelin, E.H., Renner, J.B. and Jordan, J.M. 2007, Workplace policies and prevalence of knee osteoarthritis: the Johnston County Osteoarthritis Project, *Occupational and Environmental Medicine*, 64, 798–805

Chen, Y., McDonald, J.C. and Cherry, N.M. 2006, Incidence and suspected cause of work-related musculoskeletal disorders, United Kingdom, 1996–2001, *Occupational Medicine*, 56, 406–413

Choobineh, A., Rajaeefard, A. and Neghab, M. 2006, Association between perceived demands and musculoskeletal disorders among hospital nurses of Shiraz University of Medical Sciences: a questionnaire survey, *International Journal of Occupational Safety and Ergonomics*, 12, 409–416

Cromie, J.E., Robertson, V.J. and Best, M.O. 2000, Work-related musculoskeletal disorders in physical therapists: prevalence, severity, risks, and responses, *Physical Therapy*, 80, 336–351

D'Souza, J.C., Franzblau, A. and Werner, R.A. 2005, Review of Epidemiologic Studies on Occupational Factors and Lower Extremity Musculoskeletal and Vascular Disorders and Symptoms, *Journal of Occupational Rehabilitation*, 15, 129–165

Forde, M.S., Punnett, L. and Wegman, D.H. 2005, Prevalence of musculoskeletal disorders in union ironworkers, *Journal of Occupational and Environmental Hygiene*, 2, 203–212

Gallis, C. 2006, Work-related prevalence of musculoskeletal symptoms among Greek forest workers, *International Journal of Industrial Ergonomics*, 36, 731–736

Hildebrandt, V.H., Bongers, P.M., Dul, J., Van Dijk, F.J.H. and Kemper, H.C.G. 2000, The relationship between leisure time, physical activities and musculoskeletal symptoms and disability in worker populations, *International Archives of Occupational and Environmental Health*, 73, 507–518

Jarvholm, B., From, C., Lewold, S., Malchau, H. and Vingard, E. 2008, Incidence of surgically treated osteoarthritis in the hip and knee in male construction workers, *Occupational and Environmental Medicine*, 65, 275–278

Jensen, L.K. and Friche, C. 2008, Effects of training to implement new working methods to reduce knee strain in floor layers. A two-year follow-up, *Occupational and Environmental Medicine*, 65, 20–27

Jensen, L.K. and Kofoed, L.B. 2002, Musculoskeletal disorders among floor layers: is prevention possible?, *Applied Occupational and Environmental Hygiene*, 17, 797–806

Jones, J.R., Hodgson, J.T., Clegg, T.A. and Elliott, R.C. 1998, *Self-reported work-related illness in 1995: Results from a household survey*, (HSE Books, Sudbury, Suffolk)

Lohmander, L.S., Ostenberg, M., Englund, M. and Roos, J. 2004, High prevalence of knee osteoarthritis, pain and functional limitations in female soccer players twelve years after anterior crutiate ligament injury, *Arthritis and Rheumatism*, 50, 3145–3152

Minns, J., Dodd, C., Gardner, R., Bamford, J. and Nabhani, F. 2004, Assessing the safety and effectiveness of hip protectors, *Nursing Standard*, 18, 33–38

Nicholson, A., Smith, C., Mitchell, A. and Wright, R. 2006, *Cost benefit studies that support tackling musculoskeletal disorders*, (HSE Books, Sudbury, Suffolk), RR 491

Olsen, O.E., Myklebust, G., Engebretsen, L., Holme, I. and Bahr, R. 2005, Exercises to prevent lower limb injuries in youth sports: cluster randomised controlled trial, *BMJ.*, 330, 449

Roelen, C.A., Schreuder, K.J., Koopmans, P.C. and Groothoff, J.W. 2008, Perceived job demands relate to self-reported health complaints, *Occupational Medicine*, 58, 58–63

Rudzki, S.J.1997, Injuries in Australian Army recruits. Part I: Decreased incidence and severity of injury seen with reduced running distance, *Military Medicine*, 162, 472–476

Shaffer, S.W. and Uhl, T.L. 2006, Preventing and treating lower extremity stress reactions and fractures in adults, *Journal of Athletic Training*, 41, 466–469

Smith, D., Otsuka, Y. and Kaida, K. 2006, Musculoskeletal disorders and their adverse sequelae among Japanese workers, *Journal of Occupational Health and Safety – Australia and New Zealand*, 22, 45–49

van der Molen, H.F., Lehtola, M.M., Lappalainen, J., Hoonakker, P.L., Hsiao, H., Haslam, R., Hale, A.R. and Verbeek, J. 2007, Interventions for preventing injuries in the construction industry, *Cochrane Database Systematic Reviews* CD006251

Welch, L.S., Hunting, K.L. and Nessel-Stephens, L. 1999, Chronic symptoms in construction workers treated for musculoskeletal injuries, *American Journal of Industrial Medicine*, 36, 532–540

RISK FACTORS FOR MUSCULOSKELETAL DISORDERS IN CALL CENTRE WORKERS: A HOLISTIC APPROACH

Gideon de Wet

Cope Ergonomics

Aims: The aim of the study is to determine the prevalence and possible interactions of a variety of risk factors for developing Musculoskeletal Disorders (MSDs) in a customer services call centre environment and make recommendations for change where appropriate. Risk factors for MSDs identified in the literature included psychosocial, physical and physiological risk factors.

Methods: Focus groups were held to incorporate the knowledge of all the stakeholders of the Display Screen Equipment (DSE) risk assessment process in order to contribute to the content of an online survey that was circulated to DSE users. The survey assessed physical conditioning and activity levels, psychosocial risk factors, DSE knowledge and behaviours. The 'Workstyle Short Form' was used to assess psychosocial risk factors. Exercise frequency, chair adjustment awareness and Body Mass Index (BMI) were also assessed. The data were analysed to investigate any significant interactions/correlations between the risk categories and reports of pain during working.

Results: Operators who experienced pain that they related to their work had significantly higher mean Workstyle scores than those who did not report any work associated discomfort (28 vs. 19; $n = 110$, $p = 0.007$). Operators with no pain tended to exercise more frequently and be more aware of the adjustments available on their chairs. No correlation was found between pain and age, gender or BMI.

Conclusion: DSE risk assessment programmes need to include a check for psychosocial risk factors and cannot rely on traditional musculoskeletal risk factors, ergonomic equipment provision or physical conditioning alone. Operators need to be aware of the ergonomic adjustments available on their seating in order to gain any benefit from these features. Physical exercise performed more than 4 times per week may also help to reduce the risk for developing MSDs.

Introduction

The prevention of Musculoskeletal Disorders (MSDs) and stress-related illnesses are some of the greatest challenges facing Governments and private organisations in all developed countries (Rantanen, 1999). Commercial organisations have to combat these conditions as they can cause severe financial losses in terms of sickness absence, medical treatment and litigation costs. The annual economic cost of sickness absence and worklessness associated with working-age ill-health in the UK is estimated to be over £100 billion. This is greater than the current annual budget for the NHS and equivalent to the entire GDP of Portugal (Black, 2008). MSDs are some of the largest contributors to this overall cost (HSE website statistics, 2008). MSDs in the workplace are a significant problem for all types of workplaces. In the United Kingdom, around 1.1 million people suffered from MSDs caused or made worse by work, in 2001/02 (HSE, 2008). Stress related absences are also a major problem and similarly are on the increase, claiming 13.8 million lost work days in 2006/7 (HSE website statistics 2008). In industries where the primary working tasks are performed using Display Screen Equipment (DSE) – a computer at a desk, with or without a telephone, the typical MSDs encountered include Work Related Upper Extremity Disorders (WRUEDs), Low Back Pain (LBP), as well as a variety of stress related problems such as anxiety, insomnia and heart disease (Karasek, 1979). These conditions have proven to be very expensive for employers to eliminate and manage once they have arisen. Preventive programmes are therefore the first choice in dealing with these costly conditions. In the call centre working environment, preventative programmes rely heavily on the process of completing a DSE risk assessment and the provision of basic training on how to interact with computer equipment in order to reduce the risk of developing MSDs. The scope of the topics covered in these programmes generally follows the schedule of Regulations as set out by the HSE. The topic of this paper is to highlight important risk factors that are perhaps not addressed thoroughly enough in the current guidance. There is a growing body of evidence that also indicates a link between psychological and psychosocial triggers and the risk of developing MSD associated with work, with various theoretical models having been postulated.

The biopsychosocial model suggests that increased workload produces physiological responses (e.g. muscle tension, increased cortisol and catecholamine responses) that may either "overload" or "underload" workers through monotonous or repetitive work tasks (Frankenhauser, 1991, Lundberg, 2002). Balance theory (Carayon *et al.*, 1999) suggests that job stress results from imbalance in various aspects of the work system and this produces changes in multiple psychological, psychobiological and behavioural processes that influence health. The Bio-behavioral model attempts to explain the effect of stress by focusing on the physiological responses to stress or increased workload. This includes changes in several different body systems. Recent developments of this theory have mapped out the most likely pathways/links between the physiological changes and the stressors in the environment (Feuerstein, 2002). Most of these theoretical models are not contradictory but rather complementary in nature. The existence of such a variety of different models is further evidence to the complex nature of the interaction between the various risk factors. This study does not propose a new model, but rather draws from the strengths of many of these designs in order to address the risk factors which are most likely to affect the workforce.

Research Design

A review of the relevant literature made it clear that the three main categories of risk (Physical, Physiological and Psychosocial) would all have to be considered in any research intervention in order to build a Holistic model for MSD risk reduction. Figure 1 gives a summary of the variables identified from the literature review. In order to gain a better understanding of the issues that may exist, focus groups were held at several of the call centres. A set list of topics was used to guide the stakeholder attendees to discuss the various issues with the existing DSE risk management process. The output of the focus groups was transcribed and the emergent themes were used to guide the survey design. These focus groups are not discussed in detail here. For reasons of ease, speed and cost, it was decided that an online survey would be used to gather operator data. Questions were asked in each of the three main risk categories. These categories do overlap and the three different classifications should not be viewed as exclusive labels for each of the risk factors. Figure 2 gives a schematic of the variables tested in the three main fields and how they contribute to understanding the overall individual MSD risk profile. In the category of physiological risk factors, the factors found identified in the literature were age, gender, BMI (inferred from asking height and weight), smoking status and level of physical activity (defined as number session of vigorous physical activity lasting 30 minutes or more). The results from the focus groups suggested that an important risk factors in relation to the physical environment included operator awareness of the adjustability of their equipment. This was addressed by asking questions about the adjustability of their seats; who they could talk to if they had any questions; how they used their mouse and keyboard and if they used any supports for their wrist during mouse or keyboard tasks.

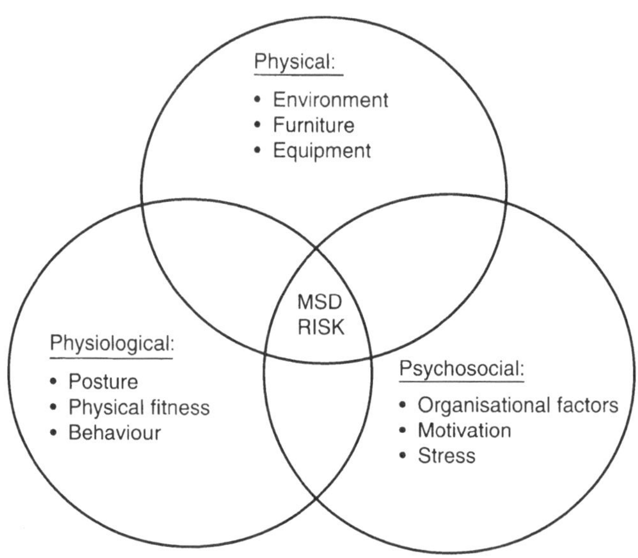

Figure 1. Illustration of the interactive nature of various risk factors.

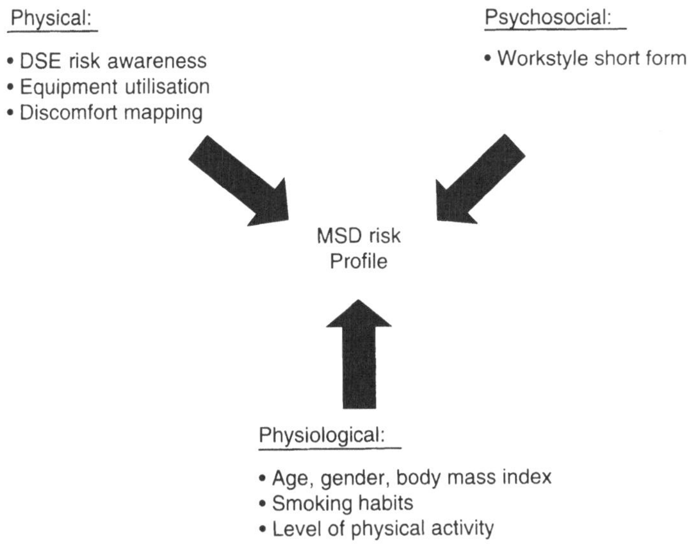

Figure 2. Survey content.

Psychosocial risk factors were covered by using the 'Workstyle Short Form' (Feuerstein, 2006). This recently developed questionnaire shows promise in addressing most of the major categories of psychosocial risk on a Likert scale in a relatively short set of questions, making it easier to use in commercial settings where 'down time' due to research is costly. The 'Workstyle Short Form' describes a threshold test score of 28 or more as increased or high risk of developing work related MSDs. The methodology of how this score is calculated is not discussed in detail here.

Results

The initial research design was based on access to a sample size of 5,000 operators. Unfortunately, due to operational limitations, this sample was reduced to 447 at a very late stage of the research programme, rendering many of the originally envisaged statistical analyses impractical.

Response rates

An initial pilot test on operators (similar to those in the test sample) provided positive feedback and no readability issues were reported. Two different sets of call centre operators were tested: One call centre dealt with internal customers (other departments of the same organisation) and another dealt with external customers (the general public). For the call centre for external customers (n = 150), 69 individuals started the survey, but only 31 complete data sets were obtained: a response rate of 20%. The high drop-out rate was reportedly related to network problems: operators found the web page did not respond, or it took too long to download. In the internal call centre (n = 297), 90 individuals started the survey and 68 completed

Table 1. Summary of two call centers.

	Internal Customers: (n = 68)	External customers: (n = 31)
Age in years (Mean)	34 SD = 8.4	34 SD = 10.3
Males	43	9
Females	25	21
BMI (Mean)	24.0 SD = 6.3	24.5 SD = 4.5
Exercise frequency sessions per week (Mean)	3.9	3.9
'Workstyle' score (Mean)	23.6	26.4
Chair adjustment awareness (Mean)	6.9	6.7

the survey, equivalent to a 23% response rate. In both cases the response rates were average for online surveys (Hamilton, 2003). The two call centres were initially assessed individually due to existing evidence that indicates that operators who deal with external customers or the general public are at greater risk of reporting lower levels of job satisfaction and higher stress levels than their back office counterparts (Ferreira, 2002).The groups were compared using an independent *t* test to check for any significant differences between the two call centres for BMI, age, gender and the presence of pain. Mann-Whitney tests were performed for the non parametric data items: exercise frequency, 'Workstyle' scores and chair adjustment awareness. No differences were found, making it possible to combine the data sets for further statistical analysis.

Age, gender and BMI

The two call centres are profiled in Table 1. Respondents were not required to provide their weight and height and 3 data sets did not include this information. In two cases the height and weight data provided was clearly incorrect. This resulted in slightly reduced number of values (2 incorrect + 3 missing) for BMI calculations. One person from another location also completed the survey due to an emailing error, so this data was also excluded from the analysis.

Pain and No pain

Operators were asked if they were suffering from aches and pains that they specifically attributed to their work. The response to this question was used to classify operators into two groups – Pain and No Pain. The data sets of the two call centres were combined in order to allow for more powerful analyses on the basis of reports of pain following the testing mentioned above ensured that there were no differences between the groups. The findings are summarised in Table 2. The combined data set was divided into two groups, Pain and No pain and independent *t* tests were performed on the two groups for the same categories mentioned above. No differences were found between the two groups for age, gender or BMI.

Table 2. Pain vs. No Pain: Characteristics of participants reporting pain and no pain: Mean values and Standard Deviations.

	Age	BMI	Exercise frequency (30 minutes session per week)	Workstyle score	Chair adjustment awareness
Pain	32.9 (SD = 8.3)	25.4 (SD = 3.6)	3.7	28.0	6.5
No pain	35.0 (SD = 9.7)	23.3 (SD = 7.8)	4.2	19.6	7.2

Table 3. Workstyle scores for Pain and No pain.

	Mean	Median	SD	Highest score	Lowest score
Pain	28.0	26.5	14.5	62	6
No pain	19.6	18.0	12.8	50	0

Workstyle, exercise and chair adjustment

Mann-Whitney tests on the two groups reporting Pain and No Pain revealed that there was a statistically significant difference ($p = 0.007$) for Workstyle scores between the Pain (mean = 28) and No pain (mean = 19.6) groups, with the Pain group reporting a higher Workstyle score. The mean score for the Pain group was also very close to the score indicated as the threshold for high risk by the authors of the Feuerstein 'Workstyle Short Form' (2006). See Table 3 for a summary of the Workstyle scores.

The differences in exercise frequency ($p = 0.051$) and chair adjustment awareness ($p = 0.052$) between the Pain and No Pain groups were just outside of statistical significance.

Additional findings

More than 80% of operators in both call centres indicated that they used their keyboard raised up on its feet/legs. No correlation between keyboard height and pain was found. Too few data sets were obtained to analyse the effectiveness of wrist supports for keyboards and mice and comparison to incidence of pain. The body discomfort map also provided insufficient data for analysis. Neck pain was the most commonly reported problem (30%), followed by shoulder pain (20%). Low back pain (18%) and wrist pain (17%). Discomfort was mostly reported in multiple body parts for the same individuals: 56 individuals reported 129 areas of discomfort between them, an average of at least 2 body parts per operator affected. The analysis of different types of work tasks did not show any statistically significant relationships to pain, but reports of high frequency of talking on the phone and using the mouse revealed a trend ('trend' defined as $0.05 < p < 0.1$) toward developing pain: $p = 0.066$ for both tasks.

Discussion

Several operators reported network problems in trying to access the survey – this highlights the impact of known network infrastructure shortcomings for this site, which has also reportedly caused increased frustration for operators – a well known job stressor (Smith and Carayon, 1996).Of all the known work related MSD triggers tested, the high/positive Workstyle score proved to be the most accurate indicator for work attributed MSDs. The lower correlations for the physical and physiological risk could be explained by the existing risk reduction programmes that are controlling the other possible triggers effectively. The results must therefore be interpreted with caution. The high workstyle scores could perhaps be seen as an indication of residual risk not yet effectively controlled by current interventions. The physical fitness and level and 'Ergonomic awareness' also appeared to play an important role, although not statistically significant in this limited sample. Insufficient data was gathered to reliably check for correlations between the Workstyle answers and other risk categories. Whilst the mean workstyle score did correlate with the prescribed positive test value, the spread of test scores appear to be very wide and clearly the questionnaire did not positively detect all the persons reporting pain. This may be due to the fact that their discomfort is not caused by psychosocial factors, but rather by risks in the other two categories. Clearly this tool is not foolproof and additional validation is required.

Conclusion

The results of this survey among call centre operators appear to indicate that psychosocial risk factors for developing MSDs as measured by the 'Workstyle Short Form' (Feuerstein, 2006) appear to be associated with work related musculoskeletal discomfort related to psychosocial arousal. A trend was also noted for operators that were aware of the adjustments available to them on their chairs to be less likely to report discomfort, as well as those who were more physically active (exercise $\geq$4 per week). A larger sample may have revealed significant relationships for these factors also. This study suggests that even in environments where highly adjustable seating and other high quality equipment is provided, operators may still be at higher risk of developing work aggravated MSDs due to their personal levels of psychosocial arousal and their perceptions of the workplace. These issues are not addressed in great detail in the DSE risk assessment programmes currently used in most organisations. More guidance, training and support may be required to assist operators in dealing with the psychosocial risk factors encountered at work and train operators to adjust their equipment properly, rather than spending the same money on more adjustable seating. The 'Workstyle Short Form' scores also appeared to match the initial validation data used to develop the tool closely; this suggests that the tool can be used in the United Kingdom workforce and appears to provide a reliable test tool for the presence of MSDs aggravated by work related psychosocial risk factors. More studies will be required to further improve the validity of the tool.

References

Carayon, P., Smith, M.J. Haims, M.C. Work organisation, job stress and work related musculoskeletal disorders. *Human factors*. Vol.44:644–663

Ferreira, M., Saldiva, P.H.N. 2002. Computer-telephone interactive tasks: predictors of musculoskeletal disorders according to work analysis and workers perception. Applied Ergonomics. Vol.33:147–153

Feuerstein, M. 2007. Workstyle: It's probably time we address both biomechanical and biobehavioural approaches? *Pain*. Vol.132, p.3–4

Feuerstein, M. Nicholas R. A. Development of a short form of the Workstyle measure. *Occupational Medicine* 2006; 56:94–99 Published online 15 December 2005 doi:10.1093/occmed/kqi197

Feuerstein, M. 2002. Biobehavioural Mechanisms of Work related upper extremity disorders: A new agenda for research and practice. *American journal of Industrial Medicine*. Vol.41:293–297

Feuerstein, M., Shaw, W.S., Nicholas, R. A., Huang, G. 2004. From confounders to suspected risk factors: psychosocial factors and work related upper extremity disorders. *Journal of Electromyography and Kinesiology*. Vol.14:171–178

Frankenhauser, M. The psychophysiology of workload, stress and health: Comparison between the sexes. *Annals of Behavioural Medicine*. Vol.13: 197–204

Hamilton, M.B. 2003. Online Survey Response Rates and Times: Background and Guidance for Industry. http://www.supersurvey.com. (Tercent Incorporated)

Health and Safety Executive, 1992, 2002. Work with display screen equipment: Health and Safety (Display Screen Equipment) Regulations 1992 as amended by the Health and Safety (Miscellaneous Amendments) Regulations 2002. HSE Books

Health and Safety Executive. 2002. Upper limb Disorders in the workplace – HSG60. HSE Books

Health and Safety Executive Website: www.hse.gov.uk. Accessed 23 05 2008. Key search terms: MSD, call centre, WRULD DSE

Nicholas, R. Feuerstein, M. Suchday, S. 2005. Workstyle and Upper Extremity Symptoms: a Biobehavioural perspective. *Journal of Occupational and Environmental Medicine* 47:352–361

Rantanen, J. 1999. Challenges for Occupational Health from Work in the Information Society. *American journal of Industrial Medicine*. Suppl.1:1–6

Smith, M.J. Carayon, P. 1996. Work organisation, stress and cumulative trauma disorders. In: Moon, S.D. and Sauter, S.L. (editors). *Beyond biomechanics: Psychosocial aspects of musculoskeletal disorders in office work*. Taylor & Francis, Bristol, 23–42

Waddell, G. Burton, K. 2000. Occupational health guidelines for the management of low back pain at work: evidence review. *Occupational Medicine*. Vol.51 (2), 124–135. Oxford University Press

ERGONOMICS AND OCCUPATIONAL 'STRESS' – WHERE DO WE STAND

A.J. Baird

University of Derby, UK

Occupational 'stress' is a relatively recent health concern, but one which is reported to affect thousands of people and place a heavy financial burden on organisations and the economy as a whole. It is an ill defined issue however, having been described as both a stimulus and a response and as both a state and a process. If we are not to repeat the mistakes made with the management of MSDs, then a clear understanding of the theory and evidence associated with this complex and emotive subject is required. Actions in relation to stress must be driven by theory/evidence and not by the desire to replicate existing processes or to produce packaged interventions that are easy to sell.

Background

'Stress' appears to be something of a modern plague, apparently affecting thousands of individuals and costing the UK economy billions. The term 'stress' is widely used in lay language and it is not unusual to see newspaper headlines like this one from June 2009 – "Stress costs more working days to be lost than 1970s strikes" – or for such articles to contain comments like "but today a more modern problem – work-related stress – is proving more costly than industrial disputes did back then, say researchers". The figure of 13.5 million lost days used in this report relates to Health & Safety Executive Statistics (HSE 2008) for days lost due to "stress, depression or anxiety". This rather alarming statistic explains the level of priority given to this topic. Ergonomists are likely to become increasingly involved as employers seek to address this health problem, so this discussion paper attempts to raise key issues and questions in the field and work towards a way forward for the discipline.

From reading the previously mentioned newspaper article (which represents fairly typical reporting of the data) it would be easy to conclude this is a work problem. Such a conclusion is supported by the fact that the term "work-related stress" is used alongside another commonly referenced statistic, from the 'Bristol Study' (Smith *et al.*, 2000), reported as "as many as one in five workers – or around five million – are believed to suffer from work-related stress". The World Health Organisation (WHO) have defined work-related stress as "the response people may have when presented with work demands and pressures that are not matched to their knowledge and abilities and which challenge their ability to cope" (Leka *et al.*, 2003), so it could easily be concluded that large numbers of people are being made ill by their work and that those 13.5 million days relate to problems in the workplace.

However, in the same newspaper as the article mentioned above, the previous month had seen an article with the title "Parents find family holidays more stressful than work" which suggested that "two thirds of parents admitted that they think going on holiday with the kids is more stressful than being at work". The current economic situation has also prompted numerous articles on 'stress' associated with the 'credit crunch'. It is clear therefore, that 'stress', whatever it may be, is not something solely associated with work and the WHO definition is unhelpful. Just as with an issue like MSDs, problems outside the workplace may well impact upon an individual's ability to work, so we need to be very careful with the use of the term 'work-related stress'. This isn't simply an issue of semantics, as again, just as with MSDs, appreciating the origin of problems facilitates the provision of appropriate interventions. Arguably, the term 'work relevant', introduced in HSE's research report on upper limb disorders (Burton *et al.*, 2008) may be just as suitable for mental health problems.

The debate over whether 'stress' should be used as a clinical diagnosis has raged for the best part of a century, if not longer. After World War 1, a War Office Committee looked into the issue of shell shock and the term's use, concluding that it was wholly misleading to use it as a diagnosis and recommended that "the disorders hitherto included under this heading being designated by the recognized medical terms for these conditions" (Symonds 1943) – this 'shell shock argument is equally applicable to what we now call 'stress'. We will come on to definitions later, but at this point it is vital to recognise that no academic/theoretical or 'professional' (i.e. recognised guidelines) definition would suggest that 'stress' is an illness. Stress is a psychological construct – a set of models that can help us understand how and why ill health might arise in certain circumstances, but it does not represent ill health in itself. There remains a huge stigma associated with mental health issues (Horton 2007) which may make it appear preferable for a label of 'stress' to be given rather than one of 'depression', but this is entirely inappropriate and does not help with providing treatment and support to the individual. We also risk losing sight of the huge problems societies around the globe face in relation to mental illness. About 14% of the global burden of disease can be attributed to neuropsychiatric disorders (Prince *et al.*, 2007) but it would be wrong to view these problems only in terms of 'stress'. By 2030, it is predicted that the single biggest burden of all health conditions will be related to depression (Saxena 2009), but whilst work relevant, this will not simply reflect a 'workplace problem'.

In 2006, the HSE published a research report (RR449) (Cox *et al.*, 2006) which looked to define a 'case' of 'work-related stress'. Whilst such a definition would no doubt be attractive to personal injury lawyers and could be argued to be beneficial in relation to epidemiological research, as stress is not an illness, it does not make logical sense to define a case of it. Admittedly, one of the main problems in stress research is the collection of meaningful, comparable data, but great care needs to be taken not to confuse this already complex issue further – if 'pragmatism' causes greater confusion, it does nothing to move things forward. Perhaps unsurprisingly, the conclusion of this research was that no single and universal case definition was possible. This failure was largely attributed to the complex nature of stress itself, but also to the realisation that key stakeholders were clear that they did not wish stress to be seen as an 'illness', but rather as a 'process'. We should conclude therefore, that trying to conceptualise stress as a disease is not only inappropriate and confusing, but ultimately futile. On the HSE website, it is made clear that "stress is not an illness – it is a state", but it is questionable that the general population recognise this from the information they receive and grouping 'stress' with anxiety and depression in terms of occupational health statistics certainly confuses the issue.

Development of the construct

The field of stress has shown marked development over the course of almost 100 years, but it does not reflect a single, linear, evolutionary process. Essentially there are differing perspectives, held by different types of researcher and studies are still undertaken using markedly different paradigms. In terms of professional practice, the major developments have occurred in the last 20 years with the first HSE report on the subject published in 1993.

Models and theories

It is beyond the scope of this paper to provide a detailed review of the field, but it is important to recognise key developments and perspectives in as much as they relate to operational definitions and approaches. Broadly speaking there are three major perspectives – engineering, physiological and psychological (Cox 1993).

In engineering environments, the term 'stress' has a quite specific meaning – engineers building a bridge will ascertain the 'load' it will bear, and from that calculate stresses and strains associated with particular components. An analogy can be made with the elastic properties of materials and their ultimate failure point – a certain amount of stress may produce a distortion that would disappear once the load was removed, but a higher level of stress could produce plastic (non-reversible) changes. Ultimately people, like a mechanical structure, can fail – to use a common expression, they can be 'pushed beyond breaking point'.

From an engineering perspective, 'stress' is all about the stimulus. When studying air crew in WWII, Air Commodore Symonds made it clear that "stress is that which happens to the man, not what happens in him; it is a set of causes not a set of symptoms" (Symonds 1943). This is the exact opposite of the perspective taken by psychologists like Richard Lazarus who are primarily concerned with what happens 'in' the person, particularly with respect to how they perceive the situation. Whilst these simplistic perspectives on stress are no longer prevalent in health fields, they can still be seen in the area of human performance, albeit it is more likely that stress will be seen as a response to a stimulus termed the 'stressor' (Kavanagh 2005).

When exposed to fear, shock, or general anxiety, physiological changes take place which affect the body's function. This is perhaps most famously described by Hans Selye with his 'General Adaptation Syndrome' (GAS) produced in the 1950s (Selye 1956). For Selye, stress was "a state manifested by a specific syndrome which consists of all the non-specific changes within the biologic system" triggered by aversive or noxious stimuli. So, stress could be seen as a generalised and nonspecific physiological response. More recently it has been shown that Selye's description is not supported by the evidence and in fact responses do tend to be specific (Jovanovich 2006).

In practice, these models are better suited to more acute and/or severe responses and do not explain the typical presentation of 'workplace stress' as we may see it. Another analogy with MSDs can be made here – simplistic biomedical models fail to explain the presentation of MSDs we see on a day-to-day basis (Baird 2008), but they do work quite effectively in cases of trauma (i.e. more 'extreme' cases). Proponents of these models point to the fact that measures of indicators such as cortisol provide a more objective measure of a stress response. However, it can be argued that physiological changes are not just of interest in their own right but also in as much as they influence behaviour and cognition and the relationship between chemical indicator and eventual behaviour is a far from simple one. These models do not figure in operational definitions, but it is still very common to hear media reports on stress which make reference to a 'fight or flight' response.

Within the psychological literature, it can be said that there are two 'strands' in relation to stress – those who see stress as an 'interaction' and those who view it as a 'transaction'. The difference is not simply one of philosophical standpoint, but to some extent one of pragmatism. Interactional theories are simpler and easier to operationalise. The two 'classic' interaction theories are Person-Environment Fit and Demand-Control.

Person-Environment Fit Theory (French *et al.*, 1982) – suggests that 'stress' occurs when there is a mismatch between the individual and their environment. In ergonomics terms, we can say that there is a mismatch between the demands of the job/environment and the capabilities of the person. Away from a traditional ergonomics perspective, the mismatch may also relate to an individual's needs and aspirations (which also links to Effort-Reward imbalance models). In their classic study, French, Caplan, and Harrison demonstrated that a mismatch was indeed associated with psychological, physical and biological 'strains'. Subsequent studies have produced similar results and it appears that mismatches associated with an individual's needs are the strongest predictor of 'strain'. However, many of these studies have been criticised for methodological shortcomings, and recent work suggests that the relationship may well be complex and curvilinear rather than a simple linear interaction.

Demand-Control Model (Karasek, 1979) – this differentiates between two basic dimensions of work place factors – the demands of the job and the worker's level of 'decision latitude'. Job demands relate to the physical and psychological demands experienced by the individual. Decision latitude refers to the worker's decision authority and skill discretion (i.e. the control they can exert). The Demand-Control model combines the two dimension of job demands and decision latitude in a 2×2 matrix. Karasek suggests that a combination of high demands and low decision latitude in the 'high strain' jobs is most detrimental for people's health and well-being. It is suggested that a combination of high demands and high decision latitude in the 'active jobs' will produce little harm for the individual, which indicates that high decision latitude attenuates the negative effects of high demands. This model was subsequently developed to include a third axis – 'Support'.

These models have an innate 'rightness' about them and it isn't difficult to think of examples where one of these models would fit. Indeed, the highly influential case of Walker v Northumberland County Council – the first civil action in the UK in relation to stress in 1995 – could easily be described in terms of Demand-Control. John Walker was a social worker who flagged concern over his case load, but could exert no control over it and was not offered any support from his superiors. Equally, when Thelma Conway sued her employer (Worcestershire County Council), it was basically on the grounds of Person-Environment Fit as she was thrust into a management role despite her protestations and with no effective training to allow her to develop the competencies required for the job. However, these models are not without their critics and whilst working well in some situations they still fail to adequately explain all of the 'stress' problems we see in the workplace. As such, it can be argued that in terms of an explanation of what we see, if not ease of application, the more complex transactional models are more effective – i.e. as with bio-psychosocial models of MSDs, we mustn't see the individual as a passive component in the stress process.

True transactional models of stress are the antithesis of engineering models – the focus is not so much on the environment, but on how the individual reacts. Fundamental to transactional models are the concepts of 'appraisal' and 'coping'. Appraisal is described as the evaluative process that gives meaning to the person-environment transaction. Initially conceptualised as a single process, it was later broken down into two stages. Primary appraisal therefore draws attention to the presence of a problem, whereas secondary appraisal defines what the individual can do about it. The idea that 'coping' would be important when studying stress may appear self evident, but when Lazarus first began researching the subject he found very little enthusiasm for it. He did, however, feel it was central to the issue of stress and indeed made the comment "in my view, stress itself as a concept pales in significance for adaptation compared with coping".

Lazarus defines coping as "constantly changing cognitive and behavioural efforts to manage specific external and/or internal demands that are appraised as taxing or exceeding the resources of a person" (Lazarus & Folkman 1984). Initially Lazarus and other researchers had thought of coping as a trait or style, but in the 90s it became increasingly conceptualised as a process. For Lazarus and those championing a transactional perspective, stress is very much an individual experience – a process driven by cognitive appraisals. In the 90s Lazarus advocated research into appraisal and coping based on an intra-individual analysis of the stress phenomenon, but whilst some research was done in this way, the drive as the 90s unfolded was to analyse stress as an organisational problem. Unfortunately, Lazarus' views and transactional models in general, do not sit comfortably in the field of occupational safety & health. Stress as an occupational health problem was increasingly conceptualised as an organisational issue, requiring organisational interventions, to be addressed using a traditional risk assessment approach.

Operational definitions

To understand the nature of interventions proposed by practitioners, it is essential to appreciate the context of the definitions they work with. In the US, NIOSH define 'job stress' as the "harmful physical and emotional responses that occur when the requirements of the job do not match the capabilities, resources, or needs of the worker" – this is very much a person-environment fit definition, with the emphasis on problems with the design of jobs.

In Canada, 'workplace stress' is defined as "the harmful physical and emotional responses that can happen when there is a conflict between job demands on the employee and the amount of control an employee has over meeting these demands" – this is a demand-control definition and again, another interactional perspective

Worksafe in Australia define 'stress' rather than 'work stress' and use the WHO definition of stress as "the reaction people may have when presented with demands and pressures that are not matched to their knowledge and abilities and which challenge their ability to cope" which extends the Person-Environment Fit Theory to consider coping and is thus more of a transactional perspective. However, in their explanation they pick out issues of support and control and don't mention appraisal.

In the UK, the HSE's first guidance document on stress – HS(G) 116 (HSE 1995)– used the definition: "Stress is the reaction people have to excessive pressures or other types of demand placed upon them. It arises when they worry that they can't cope." – this then is the closest definition to a transactional model as it addresses appraisal (worry) and coping. To be a true transactional definition however, it would need to allude to the fact that 'excessive pressures' are not the key issue so much as pressures that are perceived to be excessive (beyond thc individual's perceived ability to cope).

However, the HSE have since amended their definition and guidance released in 2001 - HSG218 - defined stress as "the adverse reaction people have to excessive pressures or other types of demand placed upon them" (HSE 2001). This remains the HSE's current definition. So, from arguably the closest definition to transactional models, the UK definition has become closer to an engineering definition. From a consideration of appraisal and coping, the definition has moved to one akin to a stressor-strain approach. Whilst this might reflect the move through the 90s to see stress as very much an organisational problem and the desire to have this problem addressed in the same way as any other health & safety problem, this is not a good definition and it doesn't reflect current scientific thinking in terms of mechanisms of harm. It is akin to trying to deal with MSDs without considering psychosocial factors.

How can we conceptualise and 'operationalise' transactional models

It is vital that disciplines learn from past mistakes and in the case of Ergonomics, we must ensure we do not repeat the mistakes associated with 'traditional' approaches to the MSD problem. Simple biomechanical/biomedical approaches to MSDs had face validity, were easy to describe and easy to 'sell'. Nevertheless, they failed. Inappropriate approaches led to a "20th Century medical disaster" (Waddell 2004). A similar mismanagement of mental health issues must be avoided at all costs.

The first step in avoiding such a disaster is the 'operationalisation' of transactional models – i.e. we must provide not just a model, but a way of using it and explaining it. As a profession, ergonomists tend to consider themselves as 'systems thinkers' and a systems perspective is useful when conceptualising stress. It has been argued that organisational 'culture' can be treated as an emergent property of the organisation as a system (Cox & Thomson 1999) and it is useful to consider stress as just such an 'emergent property' - a function of the interaction of many factors rather than a simple consequence of an individual issue. Stress therefore, like MSDs, should be seen as a multi-factorial problem.

Whilst appreciating a systems-based context, a simple way to present transactional models is by extending a concept commonly used in ergonomics – the idea of balance. Ergonomics is often seen as the science of matching the demands of the job to the capability of the worker. Similarly, in its advice on stress, HSE show scales describing a balance between "demands & pressures" and "skills & ability" (HSE 2009). To take these interactional ideas and create a model more appropriate to transactional theories, it must be recognised that far from being simple scales, the situation is complicated by a pivot point that is not fixed. Effectively the pivot point can move as a function of the individual's perception of the scale of the threat faced and their perceived ability to cope with such a threat i.e. the pivot point can be conceptualised as 'appraisal' – see Figure 1.

It follows therefore that balance cannot be achieved through looking at 'stressors' or 'coping resources' alone. Such a statement is not controversial in itself, but it does have significant implications for the way in which we address this health problem. In particular, it precludes a 'Five Steps', hazard-spotting approach (HSE 2007) to managing risks associated with stress.

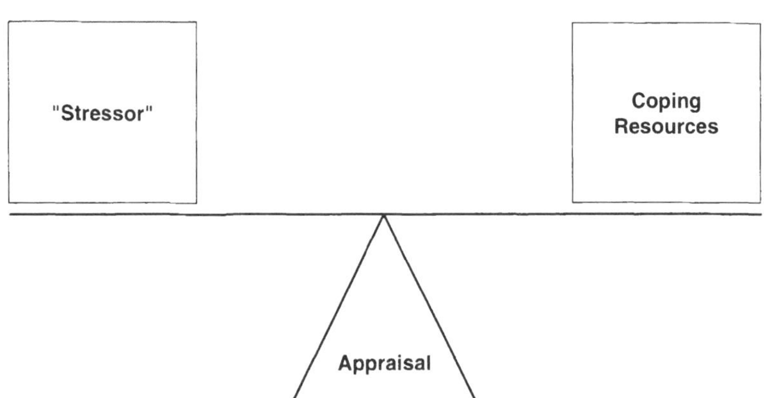

Figure 1. Stress as a function of balance from a transactional perspective.

What should Ergonomists be doing?

There are two key areas to consider – how we talk about stress and how we attempt to manage it. The first issue may seem relatively trivial, but is far from it – if we are to address the issue, we must take control of the terminology. If we are to use the HSE's definition, then stress is inherently 'bad' and there can be no such thing as an 'acceptable level' of stress. We must clearly distinguish between that which may increase demands on an individual (pressure) and the negative/harmful outcome that may (but equally may not) occur. Being short-staffed, under-resourced, facing a major deadline, etc. should not be confused with stress, even if they do have the potential to contribute to a problem. As the Appeal Court ruling stated "there are no occupations which should be regarded as intrinsically dangerous to mental health" (the case of 'Hatton v Sutherland' and Barber v Somerset) – just because something looks bad doesn't mean that it is. We must also be very wary of the term 'caused' – if we recognise stress to be an emergent property and/or a multi-factorial problem, then no one issue can be said to have caused stress.

In Canada, the Canadian Centre for Occupational Health & Safety, guidance talks about some stress being a normal part of life. Arguably such a stance has merit and an analogy could be made with pain – a life without stress is about as likely as a life without pain. It may surprise some people that a resource pack, endorsed by HSE in 1998 (Doherty & Tyson 1998) had a case study in which the guidance started with the line "stress is a normal part of life". This makes for messy definitions and confusion, but stress, like pain, *is* to some extent a normal part of life.

Which leads to the question of how stress should best be managed? At the highest level, the answer is quite simple – we need to understand people and manage them well. Turning such a noble goal into action however, is less than straightforward. A traditional, 'hazard-spotting' approach simply does not make logical sense. One cannot look for stress in the same way as one would look for trip hazards, noxious chemicals or excessive noise. In addition, HSE's own research casts much doubt over the measures being used to assess 'psychosocial hazards' (Rick *et al.*, 2001).

In 2004 the HSE introduced their "Management Standards". The six areas highlighted are undoubtedly vital to the successful running of an organisation, but are arguably of general importance and not just an issue in relation to stress. Whilst these issues are unquestionable, that does not mean that they translate well to headings on a risk assessment form. If assessing anything, it should be organisational systems and not 'psychosocial hazards'. This may make those working in health & safety a little uncomfortable, but recent research coming out of HSE raises an issue with which they are very familiar – competence. Yarker *et al.* (2007 & 2008) set out to define management competencies in relation to stress and it is telling that one of the conclusions was that "stress management is a part of normal general management activities; and that there is no single behaviour needed for effective stress management, so managers need to think about using a complementary set of behaviours". Essentially, it is about managing people well. Part of effective management is also the production of effective systems – organisations will typically have a wealth of data which could indicate a developing problem (from IT complaints to non-specific sickness absence) and it must be ensured that such data is acted upon. Just as has been argued for MSDs (Baird 2008), the key is to ensure that chronic, disabling conditions are avoided – complete avoidance is unrealistic.

However, whilst the above may sit with the 'organisational problem requiring an organisational solution' mindset, we must not lose sight of the first element of our overall aim – to understand people. Ultimately, stress has no meaning outside of the individual and as with pain, we are dealing with individual experience. This is clearly recognised by HSE as each of the Management Standards areas is associated with the phrase "systems are in place locally to respond to any individual concerns". Actions at the individual level are often essential (particularly in relation to competence) and if addressing appraisal, they are effectively primary prevention activities. It is quite appropriate to aim for hardy and resilient individuals.

Conclusions & way forward

'Stress' is something that has been defined as both stimulus and response and as both a state and a process. It is therefore no surprise that the area is confused. Practitioners and academics alike must recognise however that there is a theoretical basis behind the construct and that there are strengths and weaknesses to all of the models available. In extreme cases, simple models may appear effective, but just as with MSDs, if we want to make sense of problems as they present in the workplace, the more complex, psychology-based theories are required. The best of these are transactional theories which have appraisal and coping at their core. Trying to deal with stress without considering appraisal is akin to trying to address MSDs without considering psychosocial factors. The discipline of ergonomics must not preside over another health 'disaster' as it did with MSDs. To ensure that we do not, we need clarity of definition and action. Those definitions and actions must be based on the best available models and evidence and not on that which may appear easiest to explain, package and sell.

References

Baird (2008) Teaching about Musculoskeletal Disorders – Are we Barking up the Wrong Tree? In Bust, P. (ed) Contemporary Ergonomics 2008, Taylor and Francis, London

Burton, A., Kendall, N., Pearce, B., Birrell, L. and Bainbridge, L. (2008) Management of upper limb disorders and the biopsychosocial model. RR596. HSE Books

Cox, T. (1993) Stress Research and Stress Management: Putting Theory to Work. HSE Contract Research Report 61/1993. HSE Books

Cox, T. and Thomson, L. (1999) Organizational healthiness, work-related stress and employee health. In: Coping, health and organisations. Eds Dewe, P., Leiter, M. and Cox, T. Taylor & Francis

Cox, T., Griffiths, A. and Houdmont, J. (2006) Defining a case of work-related stress. HSE Research Report 449. HSE Books

Doherty, T. and Tyson, S. (1998) Mental well-being in the workplace – A resource pack for management training and development. HMSO

French, J., Caplan, R., & Harrison, R.V. (1982). *The mechanisms of job stress & strain*. New York: Wiley.

Horton, R (2007) Launching a new movement for mental health. *The Lancet*, 370, 806

HSE (1995) Stress at Work: A guide for employers. HS(G)116. HMSO

HSE (2001) Managing the causes of work-related stress. HSG118. HSE Books

HSE (2007) Managing the causes of work-related stress. HSG118 2nd Ed. HSE Books

HSE (2008) Self-reported work-related illness and workplace injuries in 2007/08: Results from the Labour Force Survey. HSE Books

HSE (2009) What is Stress. HSE Stress pages, downloaded 22/9/09 http://www.hse.gov.uk/stress/furtheradvice/whatisstress.htm

Jovanovich, J. (2006) Theoretical Approaches to Problem of Occupational Stress. ACTA FAC MED NAISS, 23 (3): 163–169

Karasek, R. A. (1979). Job demands, job decision latitude, and mental strain: Implications for job redesign. *Administrative Science Quarterly*, *24*, 285–307

Kavannah, J. (2005) Stress and Performance A Review of the Literature and Its Applicability to the Military. RAND Technical Report available on line at: http://www.rand.org/pubs/technical_reports/TR192/#

Lazarus, R. and Folkman S. (1984) *Stress, appraisal and coping*. New York: Springer, p 141

Leka, S. Griffiths, A. and Cox, T. (2002) Work organisation and stress: systematic problem approaches for employers, managers and trade union representatives. Protecting workers health series No. 3. World Health Organisation

Prince, M., Patl, V., Saxena, S., Maselko, J., Philips, M. and Rahman A. (2007) No health without mental health. *The Lancet*, 370, 859–877

Rick, J., Briner, R., Daniels, K., Perryman, S. and Guppy, A. (2001) A critical review of psychosocial hazard measures. Contract Research Report 356/2001. HSE Books

Saxena (2009). Mental Health: a global challenge. Downloaded (22 September 2009) from http://www.bbc.co.uk/worldservice/news/2009/09/090902_mental_health_wt_sl.shtml

Selye, H. (1956) The Stress of Life. New York:McGraw Hill

Smith, A., Johal, S., Wadsworth, E., Smith, G. and Peters, T. (2000) The Scale of Occupational Stress – The Bristol Health and Safety at Work Study. HSE Contract Research Report 265/2000. HSE Books

Symonds, C. P. (1943) The Human Response to Flying Stress – Lecture 1: Neurosis in Flying Personnel. British Medical Journal. Dec 4, 703–706

Waddell G. (2004). The Back Pain Revolution. London: Churchill Livingstone

Yarker, J., Lewis, R., Donaldson-Feilder, E. and Flaxman, P. (2007) Management competencies for preventing and reducing stress at work. RR553 HSE Books

Yarker, J., Lewis, R. and Donaldson-Feilder, E (2008) Management competencies for preventing and reducing stress at work – Phase 2. RR633 HSE Books

PSYCHOLOGY & COGNITIVE

LESSONS FROM STOCKWELL: HOW A COMBINED APPROACH TO ACCIDENT ANALYSIS CAN INFORM SYSTEM DESIGN

Daniel P. Jenkins[1], Paul M. Salmon[2], Neville A. Stanton[3], Guy H. Walker[4] & Laura Rafferty[3]

[1]*Sociotechnic Solutions Ltd, St Albans, UK*
[2]*Monash University, Australia*
[3]*University of Southampton, UK*
[4]*Heriot Watt University, UK*

This paper takes a fresh look at how human factors' modelling techniques can inform the design of systems and organisations in the light of accidents, errors, or undesirable outcomes. Considering incidents at both an organisational and individual level, the approach models the flow of information and orders within and system and considers the impact on task performance. The shooting of Jean Charles de Menezes at Stockwell Underground station is used as a case study to illustrate the technique. Using an approach based upon AcciMaps, the salient findings of a number of published reports are summarised in a single representation. The model captures both latent conditions and active failures, ranging from component failures to overarching bureaucratic or political issues. The paper also considers how the information available to those at the 'sharp-end' has the potential to impact perception and task performance. Based upon these findings, changes to the organisational structure are proposed, to provide those at the sharp end with the information, strategic understanding and authority to keep pace with dynamic situations.

Introduction

Retrospective analysis of situations in which systems cross the boundary of safe operation is one of the most powerful, and commonly used, approaches for improving safety. Not only do they highlight specific cases in which the system has become unsafe at a component level, they can also expose underlying weaknesses at a bureaucratic level. Reason's (1990) 'Swiss cheese model' serves as a powerful metaphor for introducing this concept. The model highlights how latent conditions at different organisational levels can set the conditions for an accident to occur. While the concept of considering accidents at multiple levels of the organisation is becoming prevalent in the scientific community, a culture of individual culpability, and blame, is still prevalent in our society. In the follow-up to accidents, particularly in cases where human life is lost, calls are frequently made to identify individuals involved and hold them accountable. As Dekker (2007) discusses, the Newtonian view of causality, which assumes symmetry between cause and effect, can lead to the last person performing an action being held responsible for the catastrophic consequences. As this paper will show, the relationship between cause and effect within complex sociotechnical systems is rarely straightforward.

According to Sheridan (2008), the standard remedies for human error are: design the system so it is simple and easy to use, training, active warnings that anticipate a system state that will likely lead to error, and restricting the operator's exposure to opportunities for error. A typical response to error prevention is to introduce 'defences in depth'. Ostensibly, developing 'defences in depth' involves modifying the system at higher organisational levels in an attempt to ensure that subordinates operate within an identified safe boundary of operation (e.g. by adding in supervisory controls or legislation). An assumption is made that limiting the actions of individuals to those that are known to be safe will prevent error. While this approach may prove effective in predictable, tightly constrained, deterministic, environments, it is less suitable for application in complex sociotechnical systems. In complex systems, workers' environments, objectives and priorities are likely to be highly dynamic. Workers are required to adapt to meet their objectives, balancing the conflicting constraints of productivity and efficiency. In many cases, the abundance of legislation and set-procedures, put in to increase safety, results in workers' discretionary space being limited beyond the point that allows them to function effectively. Furthermore, there is often a mismatch between the level of accountability placed on an individual and their authority.

While Reason's (1990) 'Swiss-cheese' metaphor, has undeniably raised awareness of the importance of considering error causality at a number of organisation levels. This linear, metaphor hides the inherent complexity in large sociotechnical systems. Others, such as Helmreich (1990), and O'Hare (2000), favour 'causal webs', rather than a linear sequence, as they expose the complexity of the system. One such approach, based on a similar theoretical standing as Reason's (1990) model, is the use of AcciMaps (Rasmussen, 1997; Rasmussen & Svedung, 2000; Svedung & Rasmussen, 2002). Rather than identifying and apportioning blame, the AcciMap is used to identify and represent the causal flow of events and the planning, management and regulatory bodies that may have contributed to the scenario, with a view to improving system design and safety (Svedung & Rasmussen, 2002). As Woo & Vicente (2003) point out, "accidents in complex socio-technical systems do not usually occur because of an unusual action or an entirely new, one-time threat to safety. Instead, catastrophes result from a combination of a systematically induced migration in work practices and an odd critical event that ends up revealing the degradation in safety that had been occurring all the while".

This paper originates from a larger study on sociotechnical systems. An example incident is used to illustrate the approach, where an innocent man, Jean Charles de Menezes (JCdM), was killed after being mistaken for a suicide-bomber. While superficially the death JCdM can be directly linked to the officer's decisions to fire the fatal shots, further examination reveals that their actions were predated by a number of latent, error-causing conditions. This paper is predominantly based on the findings of the Independent Police Complaints Commission (IPCC, 2007) and the Metropolitan Police Authority (MPA, 2008). The reports summarises these findings in a single representation, modelling their interdependencies and the causal flow. The paper also draws additional conclusions, questioning the organisational structure and the authority permitted to those at the sharp end.

The case study

The fateful events in Stockwell, on the 22nd July 2005, need little introduction. The city of London was still grieving the loss of 52 innocent civilians, killed in suicide bombings earlier in the month (the 7th of July). A manhunt was on for the perpetrators of the previous day's attempted bombings. A gym membership card, found with one of the failed devices, connected Hussain Osman and the address '21 Scotia Road' to the attacks, additional intelligence confirmed Osman as a credible threat. An operation was mounted at the address to apprehend Osman as he left the flat's communal entrance. At 09.33 hrs a man, allegedly bearing a resemblance to Osman, left the flat. Officers followed him on his 33-minute journey to Stockwell Tube Station. Two minutes after he entered the station, members of the Metropolitan Police Service's (MPS) specialist firearms department (CO19) entered the underground station with orders to 'stop' a suspected suicide-bomber. Surveillance officers directed them towards the suspect. Moments later, two of the CO19 officers approached the man and between them fired seven shots into his head and one into his shoulder from close range. This man was later found to be Jean Charles de Menezes (JCdM), a completely innocent Brazilian national.

The organisation involved, the Metropolitan Police Service (MPS), failed to balance its responsibilities of preserving the life of the public and officers, with the lives of suspects. While the situation can, unquestionably, be defined as complex, dynamic, and safety critical, procedures to tackle suicide bombers, defined by the codename Kratos, have been in existence since 2003. Thus, the pertinent question is how was this outcome able to happen, and what systemic changes can be made to reduce the chances of it happening again.

Results

The AcciMap for the Stockwell shooting incident is presented in Figure 1. The events presented within the AcciMap are coded based on when, in the overall chain of events, they occurred. The AcciMap is comprised of six main levels. There were various failures at each of the six levels that contributed to the shooting incident. Had JCdM been identified or challenged earlier, in a more controlled manner, it is possible that his death could have been avoided [1]. The system crossed the boundary of safe operation as soon as a suspected suicide-bomber was allowed to board public transport. It is evident that the direction of the Gold Commander 'to stop all residents of the Scotia Road flat', briefed five-hours earlier, was not carried out [2]. The operation was poorly coordinated; the firearms officers assigned to the operation were not in a position to challenge JCdM until he had boarded the train at Stockwell. The suspect was allowed to board a bus, even though a bus had been a target of a bomb earlier that month. Furthermore, the firearms officers arriving

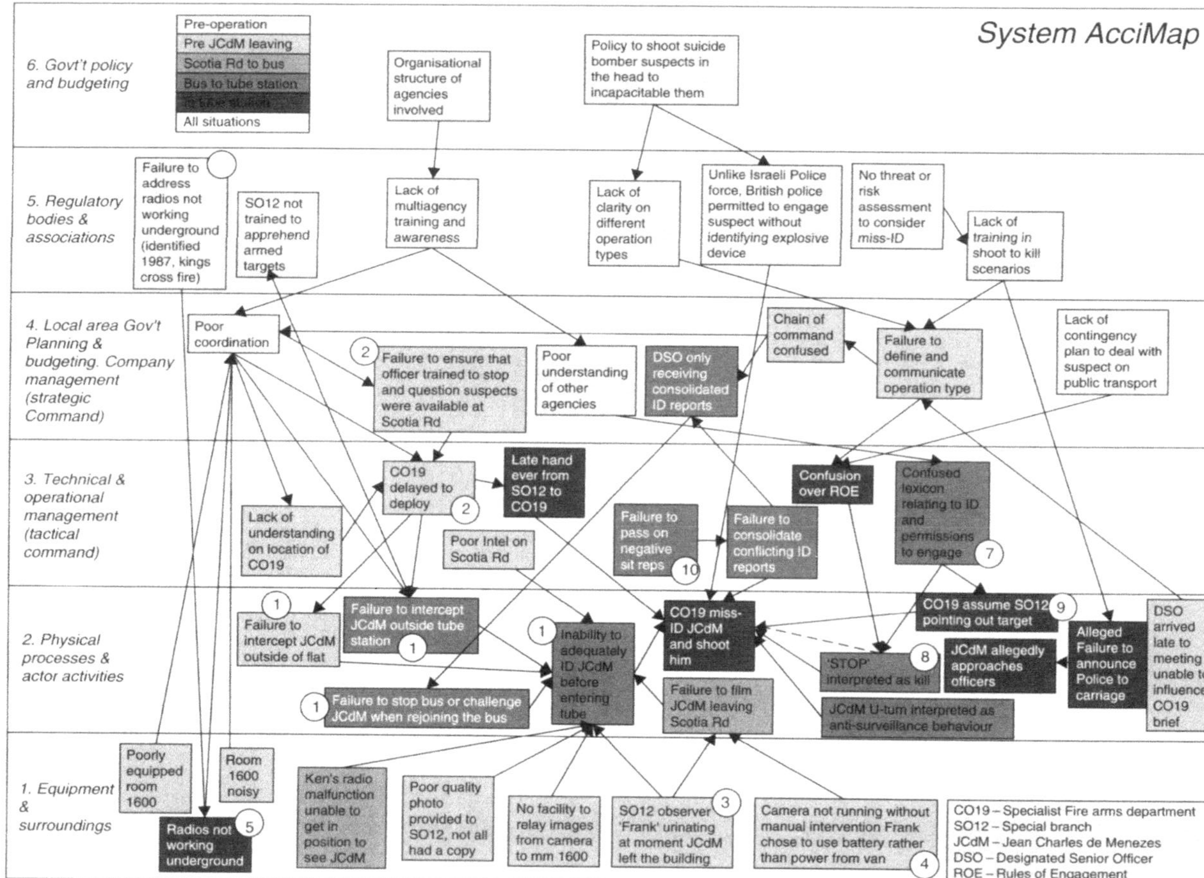

Figure 1. AcciMap (numbers in circles relate to the commentary in text).

on the scene had received limited information on the unfolding events. Had they been better integrated into the identification process, they may have had a greater understanding of the unfolding situation.

The model clearly highlights that failing were present at all of the identified organisational levels. For example, at the lowest level, the observer 'Frank', positioned in a van right outside the flat's communal door, failed to capture an image of JCdM. His 'call of nature' preventing him from switching the video camera on [3]. Had Frank connected the camera to the van's power source [4], or had there been a second observer in the van, vital information could have been captured that may have led to the identification of JCdM. This information could have been compared with a photo of Osman in the van or passed to others to inspect.

A number of factors clearly predated the operation, in this case at the higher organisational levels. Many of these failing relate to the systems ability to support the passage of unambiguous information. A failure to equip officers with radios capable of working underground [5] (identified by the Kings Cross fire over 17 years before [6]) meant that CO19 officers had no contact with their superiors after entering the station. The organisation appears to have lacked a well-written and well-trained lexicon relating to orders and rules of engagement [7]. According to the IPCC report, the *'stop'* command was not clear [8]. Also, the comment from the surveillance officer to the firearms officers, *'There he is'*, could also be misunderstood as, *'There the bomber is'*, rather than, *'There the suspect is'* [9]. A mismatch in understanding between the different levels of command is also evident. Information from officers on the ground was consolidated without considering the negative identifications [10]. This undoubtedly affected the perception of those in the distributed command centre. A more comprehensive discussion of Figure 1 can be found in Jenkins *et al.* (in press).

Discussion

The analysis highlights the importance of effective communication between team members. Salient points include the explicit, communication of well-written orders and the communication of situation assessments within risk adverse settings. Modelling the same the scenario on Rasmussen's (1974) decision ladder (see Jenkins *et al.*, under review) highlights that the type and format of information delivered to the operator at the sharp-end, can influence their perception of the situation, and in turn, their actions. Thus, the role of orders in complex, highly dynamic, environments needs to be considered. In situations where significant ambiguity exists, superiors should take great care in delivering information. A lexicon needs to be developed that clearly distinguishes information (that should be used to refine perception) from orders (that should be acted upon without question). The briefings provided to the CO19 officers also have great potential to influence the schemata selected (see Jenkins *et al.*, under review). While schemata will be continually updated by subsequent information, in high stress situations, officers may reject contradictory information to focus on information that confirms their existing schema. Training is also likely to significantly affect the development of schemata. Therefore, training packages need to be carefully developed to consider how the information presented will be interpreted, and how this interpretation may be used to generate and modify schemata. A greater emphasis on identification in training, possibly encouraged by changes to Standard Operating procedures (SOPs) or rules of engagement (ROE), is likely to influence the schemata adopted. Any change to the way orders are delivered, training, SOPs or ROE need to be carefully considered within a broader context. However, it may be possible to adopt a safeguard, such as that required in Israel, that would forced the officers to 'control' the suspect until they were certain he was carrying a device.

Besides collating the findings of the IPCC into a single graphical representation, the approach also reveals, and raises additional considerations. Both papers (Jenkins *et al.*, in press; under review) question the suitability of the MPS's organisational structure to support rapid-paced operations, particularly where reliable identification of a suspect is not possible. The organisation structure in place on the 22nd July required a designated senior officer to make the final, difficult, call relating to a challenge on the suspect. The structure can therefore be defined as hierarchical. Information on the suspect was collected by a surveillance team on the ground, this was then passed to the distributed decision making team located at New Scotland Yard. The decision-making team used this information to decide upon a course of action and briefed this as orders to the firearms team. The effective passage of information, up and down the command chain, is therefore vital with this kind of distributed hierarchical structure. From a detailed analysis of the information available to those on the ground and the information passed to the distributed decision makers (modelled on the decision ladder, see Jenkins *et al.*, under review), it is apparent that the system was unable to support the passage of complex system states in the time provided. Incompatible system states were generated by each of the three distributed teams. Furthermore, as technical constraints, at the time of operation, prevented communications underground leaving the final call to Scotland Yard is fundamentally flawed.

As highlighted by Jenkins *et al.* (under review), to match the fast pace of the operational environment, a flatter, horizontally integrated, structure is proposed, in which greater autonomy is granted to the units on the ground. To make informed decisions, those on the ground require up-to-date information on the situation, along with an understanding of strategic goals. If the organisational structure is unable to support the passage of information within the time provided, it may be appropriate that the roles of surveillance and tackling the suspect be combined. Not only does the decision maker need to be provided with information on the unfolding situation and the strategic aims, they also need to be granted a level of authority permitting them to make decisions within an appropriately sized discretionary space. Currently, there appears to be a mismatch between the authority provided to the officers and their level of responsibility. Within dynamic sociotechnical systems, the rules, laws and instructions put in place to safeguard the system often become unworkable because of shifting environmental conditions. In order to maintain other objectives, and to allow the system to function, workers are required to violate these formal rules.

According to Rasmussen (1997), the implication of this is that following an accident it will be easy to find someone involved in the dynamic flow of events that has violated a formal rule by following established practice, and who is, therefore, likely to be exposed to punishment. However, these decisions should not be considered in isolation. When these, on-the-surface, irrational decisions to violate a rule are considered in terms of the context and the previous decisions and actions at higher levels, they can often be found to be rationalised. The actions taken by some of the officers involved, to cover up their behaviour suggest a fear of blame culture. Questions are raised within the IPCC report (2007) over attempts by the officers involved to pervert the course of justice by falsely claiming that there were shouts of 'armed police' (IPCC, 2007 p136) as they entered the carriage. In addition, allegation have been made that the shooters' statements were falsified to indicate that JCdM was wearing a 'bulky denim jacket (IPCC, 2007 p137). Perhaps the elements of the officers' statements found by the jury to be untrue (the finding that, Officers did not shout 'armed police' before shooting, and that JCdM did not walk towards armed officers) were amended to make JCdM appear to be non-compliant, so their decision to act was concordant with the briefed rules of engagement.

The concept of redesigning an organisation to meet dynamic environmental changes is not unique to this domain. The majority of manufacturing systems have now moved from long-term planning to flexible production. Drucker (1988; cited by Rasmussen *et al.*, 1994) observed that design, production planning, and production could no longer take place as a slow, serial, fragmented process. Rasmussen *et al.* (1994) point out that, the transition to a concurrent system is required, along with the decentralisation of decision-making and advanced communication systems. Thus the system's ability to match the fast pace of the environment is dependant on the distribution of competence and decision making authority across the systems. Not only is the competence and authority required, but the availability of up-to-date information and the criteria with which to base judgements. In this case, information on the suspect and an understanding of the balance strategic goals.

While a flat organisational structure may alleviate many of the identified issues related to communication, it is acknowledged that providing local autonomy may introduce other potential errors. Thus, it seems apparent that to mitigate the level of risk, the organisational structure should be dynamically adjusted to fit the operational requirements. According to Rasmussen *et al.* (1994), centralised, hierarchical planning meets the requirements of longer term, up-fount, strategic planning while the fast response required for operation planning call for more horizontal integration. Thus, a number of predefined organisational structures are proposed for the MPS, the choice of structure informed dynamically by the operation type and unfolding events. In situations where a hierarchical structure is unable to match the pace of the operation, a predefined structure can be implemented, allowing the same individuals to collect information, make a decision and challenge the suspect. While it is not appropriate to specify these structures in detail without further study, it is proposed that the current hierarchical structure used by the MPS is most applicable for pre-planned events (falling under the existing codename, operation Clydesdale). A new, flat structure, granting greater autonomy to those at the scene is more applicable for spontaneous operations (falling under the existing codename, operation Andromeda).

By limiting the organisational structures to two or three variants (possibly in line with current operational classifications, Andromeda, Beach, or Clydesdale), it is postulated that it would be possible to develop clear 'rules of engagement' and train them. Operations could be classified at initial briefings, allowing a common understanding of roles and responsibilities. Furthermore, in dynamic situations, it may also be possible to adjust the organisational structure, in real time, in response to unfolding, or presupposed events. An example of this would include a hierarchical structure in place at the start of the operation. However, this could be briefed to reconfigure to a flat structure if a suspect attempts to enter the public transport network; granting autonomy to the units on the ground to respond to the threat.

The AcciMap approach has been found to be useful as an accident analysis approach for complex sociotechnical systems. The approach takes a holistic, exploratory look, at the events and their interdependencies, creating a web of conditions to active failures. The approach is comprehensive and questions the apportioning of blame to individual operators on the front line. Further, the links between failures at each of the different levels allow the development of appropriate countermeasures aimed at the system and not the individuals (e.g. retraining, reprisals). When used in isolation, the approach lacks the depth found in analyses focusing on individual levels; however, as in this case, when combined with an exploration of the processes used by individuals to develop situational understanding (see Jenkins *et al.*, under review, for a comprehensive description) greater insights can be gained. An approach is presented that, holistically, considers how the organisations can be structured to provide those at the sharp-end with appropriate levels of direction, information, and flexibility to respond to dynamic, emergent, events.

The approach is not without its limitations. The problem of hindsight affects all retrospective accident analysis approaches; Dekker (2002), for example, points out that hindsight can potentially lead to oversimplified causality and counterfactual reasoning. Further, the approach is entirely dependent on the provision of accurate and reliable data for the incident in question, some of which may stretch back years from the incident itself. Such data may not be available for many incidents. However, due to its domain-specific, activity-independent nature, it is contended that the approach can be applied pro-actively to a wide range of sociotechnical domains. While adding validation to the analysis components, the approach taken in this paper provides an example of how multiple techniques (AcciMaps, Schema theory, and Rasmussen's Decision Ladder) can be combined to deliver a result that is greater than the sum of its parts.

References

Dekker, S. 2002, *Reconstructing human contributions to accidents: the new view on human error and performance.* Journal of Safety Research, 33, 371–385

Dekker, S. 2007, *Just Culture: Balancing Safety and Accountability*, (Ashgate, Aldershot, England)

Drucker, P.F. 1988, *The Coming of the New Organization*, Harvard Business Review, January-February, pp 45–53

Helmreich, R. 1990, *Human factors aspects of the Air Ontario crash at Dryden*, Ontario, in Final Report of the Commission of Inquiry in to the Air Ontario Crash at Dryden, Ontario (Ottowa: Minister of Supply and Services)

Independent police complaints commission (IPCC). 2007, Stockwell One: *Investigation into the shooting of Jean Charles de Menezes at Stockwell underground station on 22 July 2005.* (Published by the Independent Police Complaints Commission on 8 November 2007)

Jenkins, D.P., Salmon, P.M., Stanton, N.A. & Walker, G.H. in press, *A systemic approach to accident analysis: a case study of the Stockwell shooting*, (Ergonomics)

Jenkins, D.P., Salmon, P.M., Stanton, N.A. Walker, G.H. & Rafferty, L. under review, *What could they have been thinking? A case study of the Stockwell shooting. (Ergonomics)*

Jenkins, D.P., Stanton, N.A., Salmon, P.M. & Walker, G. H. 2009. *Cognitive Work Analysis: Coping with Complexity*, (Ashgate, Aldershot, England)

Metropolitan Police Authority (MPA). 2008, *MPA Stockwell Scrutiny.* (Available at http://www.mpa.gov.uk/downloads/reports/stockwellscrutiny.pdf accessed 23/01/2009)

Metropolitan Police Service (MPS). 2008, *Suspected Suicide Bombers – Operation Kratos.* (Available at http://www.met.police.uk/docs/kratos_briefing.pdf Accessed 12/02/2009)

O'Hare, D. 2000, *The 'Wheel of Misfortune': a taxonomic approach to human factors in accident investigation and analysis in aviation and other complex systems.* Ergonomics, 43, (12), 2001–2019

Rasmussen, J. 1974. *The human data processor as a system component: Bits and pieces of a model* (Report No. Risø-M-1722 Roskilde, Denmark: Danish Atomic Energy Commission)

Rasmussen, J. 1997, *Risk management in a dynamic society: A modelling problem.* Safety Science, 27(2/3), 183–213

Rasmussen, J., Pejtersen, A. & Goodstein, L.P. 1994, *Cognitive systems engineering.* (New York: Wiley)

Rasmussen, J. & Svedung, J. 2000, *Proactive risk Management in a Dynamic Society.* Karlstad, Sweden. Räddningsverket; Swedish Rescue Services Agency. (Available at www.raddningsverket.se/Shopping/pdf/16252.pdf accessed 12/01/2009)

Reason, J. 1990, *Human Error,* (New York, Cambridge University Press.)

Svedung, I. & Rasmussen, J. 2002, *Graphic representation of accident scenarios: mapping system structure and the causation of accidents.* Safety Science, 40, 397–417

Sheridan, T.B. 2008, *Risk, Human Error, and System Resilience: Fundamental Ideas.* Human Factors 50, (3), 418–426

Woo, D.M, Vicente, K.J. 2003, Sociotechnical Systems, Risk Management, and Public Health: Comparing the North Battleford and Walterton Outbreaks. Reliability Engineering and System Safety, 2003, 80: 253–269

TOWARDS AN INTEGRATED SOCIOTECHNICAL FRAMEWORK FOR SITUATION AWARENESS AND DECISION-MAKING FOR EMERGENCY CARE

Harris Abd Hamid[1,2] & Patrick Waterson[1]

[1] *Loughborough University, UK*
[2] *International Islamic University Malaysia, Malaysia*

Emergency Care Practitioner (ECP) is a work role in the Ambulance Services in the UK that was introduced to provide an alternative emergency care within the community. In order to further understand the cognitive aspects of the work role, a study was carried out to identify factors influencing situation awareness and decision-making among ECPs in an Ambulance Service in England. Two ECPs participated in observations (40 hours of ride-outs) and semi-structured interviews. Analysis of the interview reveals factors that can be categorised into a number of systems-related levels of analysis (macro-micro). The outcomes from the study form the basis of an integrated sociotechnical system framework for emergency care with a focus on situation awareness and decision-making.

Introduction

The last few years have seen a huge growth in research focusing on human factors and patient safety. Within this growing body of literature, human factors methods and techniques, and the systems approach in particular, have been used in order to gain a better understanding of human error and decision-making and how these relate to patient safety. A number of specific domains within healthcare have been examined in detail including surgery, anaesthesia, the operation of emergency departments, and the work of ambulance dispatchers. In this paper, we focus on situation awareness (SA) and its relationship with decision-making within the domain of pre-hospital care, specifically the work of advanced ambulance paramedics or Emergency Care Practitioners (ECPs). In the next section, we review research that has examined the relationship between situation awareness and decision-making and their application within healthcare contexts.

Situation awareness (SA) and decision-making

Endsley (1988, p. 97) defined SA as: "*...the perception of the elements in the environment within a volume of time and space, the comprehension of their meaning, and the projection of their status in the near future*". An appropriate level of SA is seen as a requirement for decision-making which in turn affect performance

(Endsley *et al.*, 2003). Decision-making in complex and dynamic work systems requires regular updating of SA. The Naturalistic Decision Making (NDM) approach is well suited for studying the relationship between SA and decision-making (Klein *et al.*, 1993) in applied, dynamic and complex work environments. The Recognition-Primed Decision (RPD) model (Klein *et al.*, 1993), for example, describes how people make decisions quickly by matching the current situation to a pattern they have already learned. Within the RPD model, SA is seen as an important prerequisite in making a decision.

Situation awareness and decision-making within emergency care

Emergency care can be divided into a number of stages, decision-making points (e.g., the decision to transport a patient to hospital as compared to discharging the patient once treatment has been administered by ambulance personnel), types of roles and technological support. Figure 1 shows a simplified pathway for emergency care alongside example of the types of decision-making made by the various personnel associated with stages in the pathway.

The first part of the emergency care pathway in Figure 1 (call taking and dispatch) has been investigated in detail by Blandford and Wong (2004). Their work focused on SA and decision-making among Emergency Medical Dispatch (EMD) operators in the control room of London Ambulance Services. Wong and Blandford (2001) made use of what they termed the "Integrated Decision Model", or IDM (Wong, 2000), to review the decision concepts revealed in studies of ambulance dispatch. The IDM is built upon existing NDM models like the RPD. It describes decision making by highlighting the goal, priorities and constraints, decision process, and SA.

Other work has examined in detail the dynamic and fluctuating nature of emergency care. Anders *et al.* (2006) for example, observed an emergency department within a hospital as the department coped with the challenge of limited resources (e.g., staff shortages) and uncertainties (e.g., quiet and busy periods of activity). They

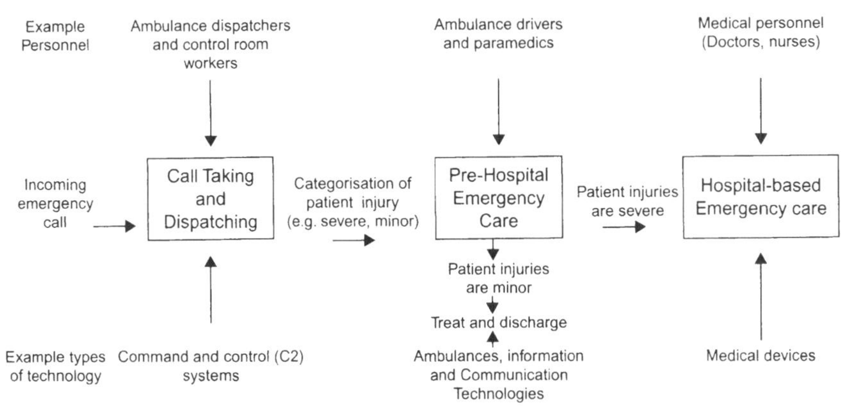

Figure 1. Stages of emergency care pathway.

interpreted their findings as they related to decision-making using a resilience engineering perspective. In general, the emergency department copes with limited resources by reconfiguring spare resources from other departments in the hospital. In some cases this involved allowing paramedics to take over patient care for less critical patients and using the available resources (i.e., doctors and nurses) to attend to more critical cases.

Study aims and objectives

The work of Blandford and Wong (2004; Wong and Blandford, 2001) and Anders *et al.* (2006) provide researchers with an impression of SA and decision-making as it applies to the first (call taking and dispatch) and third parts (hospital-based emergency care) of the pathway described in Figure 1. The main aim of the current paper is to fill in the missing part of the model, namely the process of SA and decision-making as it applies to pre-hospital emergency care. In order to do this we describe a study examining the requirements for SA and decision-making alongside an account of the types of constraints that apply to these processes within this context. A second aim of the paper is to use the findings from the case study to present an outline of an integrated sociotechnical framework for emergency care.

Methods of study

Participants

The advanced paramedics, or ECPs, nominated by the Director of Operations, participated in this study. They are both males aged 39 and 48 with three and one year experience as ECP respectively. Three Triage Nurses working in the ambulance control room also participated in this study.

Procedures

The study adopted a participative observation approach. The observations were performed during the participants' normal working hours. SA related issues were noted and categorised using Endsley's (1988) 3-level classification of SA. The first author followed two ECPs during their regular shift for two shifts each for a total of 40 hours. To increase the probability of attending a variety of patient cases, both day and night shifts were covered during weekdays and weekends.

In addition, two observation sessions on separate days were carried out with three Triage Nurses each lasting four hours. The ECPs were also interviewed using a semi-structured interview schedule, along with one of the Triage Nurses. The primary aim of these interviews was to gain an overview of the operation of the control room and the communication with a focus on the ECP. The findings from the observation and interview sessions were subsequently used to develop a communication map of the role of the ECP within the Ambulance Services. This map was then validated by a subject matter expert (the Director of Operations for the Ambulance Service).

Findings

Communication between the control room and the Emergency Care Practitioners

Incoming emergency ("999") telephone calls are filtered by Control Room staff and information systems (two medical information systems (MIS1 and MIS2 in Figure 2) and the Advanced Medical Priority Dispatch System (AMPDS) before assigned to ECPs. Calls are taken and information about the patients are gathered to make a decision about the patients' clinical needs. Using the triage system, calls may be passed on to the ECPs by the Triage Nurses. During regular work time, colleagues (ambulance technicians, paramedics) who require assistance may contact the ECPs for their advanced knowledge. In cases where the patient died, the police had to be notified. Additionally the ECP may liaise with other medical services like walk-in health centre, general practitioner's (GP) clinic, District Nurses, and Social Workers.

Situation awareness, decision-making and other factors influencing the work of the Emergency Care Practitioners

Examples of issues that are relevant to SA are presented in Table 1. These issues represent threats to SA. Based on the observations and interviews, five factors were identified as influencing the work of the ECPs. These factors cover both micro and macro levels of the system. The SA issues and five factors are presented in this section.

(1) Targets, levels of resources

National and local standards are set for the ECPs. These standards seem to affect the work assigned to the ECPs. For example, one ECP explains the reason for him being assigned non-ECP type job.

Table 1. Examples of threats to SA by SA levels.

SA Levels	Observation	Inference
Level 1: Perception	The MIS1 screen in the ECP's vehicle displayed truncated postcode	ECP cannot perceive complete information
Level 2: Comprehension	ECP reported that a patient's ECG output is different from what was expected based on pulse pattern	ECP cannot comprehend the contradictory symptoms the patient presented
Level 3: Projection	ECP reported that he wish he know whether the patients are all right after taken by the ambulance to the A&E	ECP cannot make accurate projection of patients' future health condition

> *"There is an immense pressure at the moment to respond to calls within a certain time and so you're often called in."* (ECP2)

Resources allocated for technology also influence the effectiveness of ECPs. The ECP thought that investment should be made with regards to the communication technology (especially the MIS1 and satellite navigation system).

> *"The system here is, for some reason, .. they have not fully invested in it. ... If they want us to respond immediately, (they) should really provide the equipment to be able to achieve that."* (ECP2)

(2) Quality and nature of information presented to the ECP

The Control Room acts as a gatekeeper for the tasks assigned to the ECPs. An issue that rises from such set up is the quality of the information relayed to the ECP. One ECP commented that the patient information from the control room (as displayed on the MIS1 screen) does not help much in assessing a patient's condition.

> *"we're not really getting the appropriate information unless I then can manage to speak to the nurse who took the call."* (ECP1)

The control room staff's situation assessment did not seem to transfer well via the electronic communication technology. In some cases, the transfer of information occurred electronically only, and not at a personal level. In these cases, very limited amount of information had been passed to the ECPs.

> *"sometimes they don't ring you. They just send you a job...I've gone to job where I hadn't spoken to them at all."* (ECP2)

A threat to Level 1 SA mentioned in Table 1 is due to the limited screen area of the Management Information System (MIS1) hardware installed in the ECP's vehicle. The incomplete information increased the time required to input the correct information into the satellite navigation system (which is separate from the MIS1).

(3) Patient-related influences

At a micro level within the system, information originating from the interaction with the patient (e.g. checking pulse pattern) affects SA as indicated in Table 1. Additionally, the patients themselves, who are viewed as the main source of information, may not be able to convey the information well.

> *"Sometime (the patients) are frightened. They don't know how to explain their condition."* (ECP2)

(4) Training and knowledge

The decision making process of the ECPs is influenced by experience and expectation. One ECP reported having more confidence in leaving the patients at home as a result of ECP training. Knowledge about local geography is also pertinent as evident in the problem regarding the incomplete postcode. The information displayed on the MIS1 screen does not lead to a satisfactory understanding (Level 2 SA) of the geographical location of the patient. The ECP who has just started working in the new area was not familiar, and had not been trained, with the street names.

"*.. you pressed that you're responding and you're moving and actually you're not because I .. don't know the area that well, and you're spending time putting it in the SatNav* [Satellite Navigation], *which takes time.*" (ECP2)

"*If you have a rough idea of an area you can be heading in the direction while the SatNav's firing itself up...rather than if you were to put me in a completely different city or area.*" (ECP1)

(5) Job design: feedback and skill variety

The example for level 3 SA in Table 1 shows how SA is related to a job design issue. The ECP reported that a feedback for a completed job would be beneficial. The feedback is important for the development of better or more accurate SA in the future.

"*[I] always try to reflect and think ... was that the best plan of action or have I put them in the hospital for no real reason and ...quite often there is no real feedback.*" (ECP2)

Another job design issue is skill variety. The opportunity to work on different clinical cases is regarded as valuable. Thus, the ECP did not mind being asked to perform non-ECP jobs.

"*I mean my skills got better. I've been to (emergency) calls and actually been able to deal with them and involved ECP skills as well.*" (ECP2)

"*Because we are so specific in what we're doing, we're not really performing the full gamut of ECP tasks. Working in different clinical areas and being exposed to different clinical things on a regular basis (will) stop the skill erosion*" (ECP1)

Towards an integrated socio-technical framework for emergency pre-hospital care

This study revealed the relevance of SA as applicable in decision-making in the works of ECPs. The findings point to a range of influences spanning various levels within the sociotechnical system. An appropriate way of capturing and visualising these is to first use a framework that enables the examination of the various levels. The IDM (Wong, 2000) is a suitable model to examine some of the micro issues with a cognitive engineering orientation. Meanwhile, at the other end of the care pathway, the resilience engineering orientation, represented by the model proposed by Wears *et al.* (2008), is a suitable candidate. The two orientations are combined in an integrative manner (Figure 2) to complement each other with safety being a common thread linking them together.

Using the communication map, the Ambulance Services is divided into dispatch team and responders in the field. This distinction is meant to highlight the different nature and scale of SA and decision-making taking place in both areas. The IDM appended to the ECP is based on the IDM reported by Wong (2000). Further evidence is required to populate the modified IDM as it relates to ECPs.

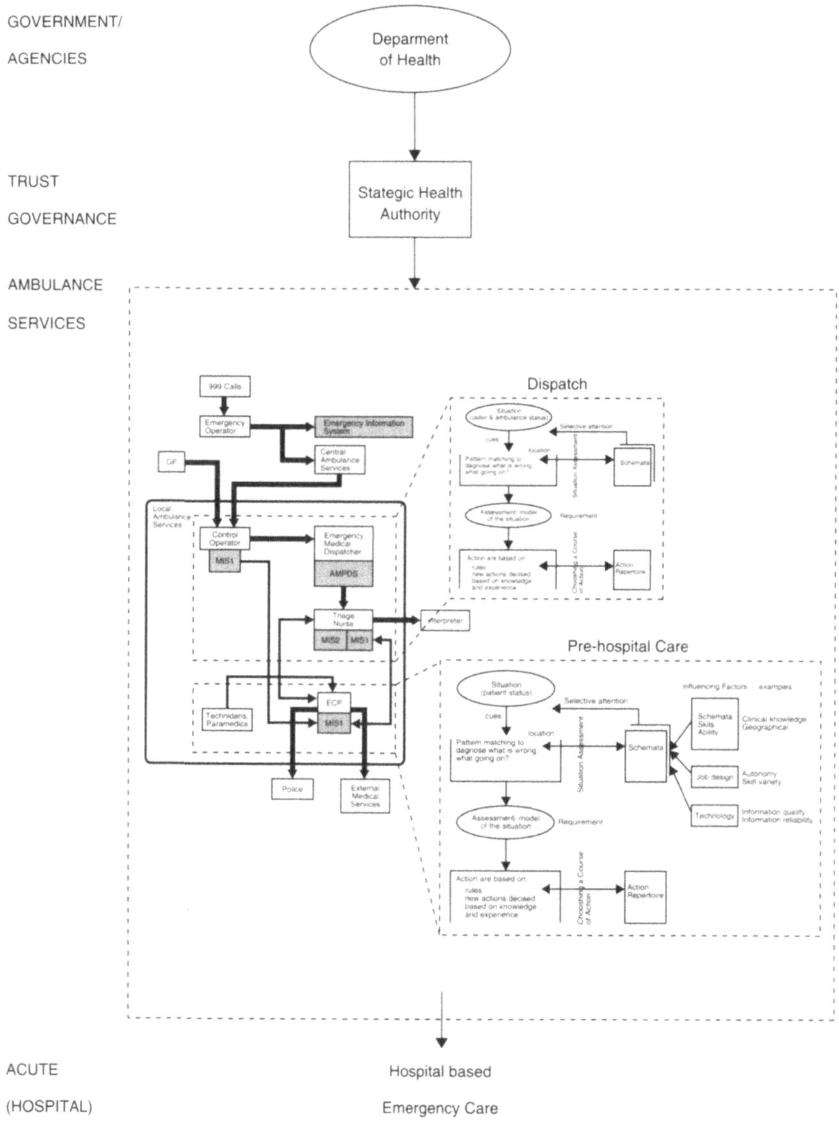

Figure 2. An integrated sociotechnical framework of SA and decision-making in emergency care.

Conclusion and implications

This paper described the development of an integrated sociotechnical system framework for SA and decision-making among ECPs. This framework will be used to investigate SA from a systems ergonomics perspective. Realising the limited number of participants in this study, this framework should be tested and validated with a larger sample size. Nonetheless, the proposed framework can be a good starting point as it is also based on existing literature. As a research framework, it can be helpful to investigate issues in emergency pre-hospital care in a more comprehensive manner. The linkages across the system elements can be examined.

These linkages can be used to suggest job design changes. As mentioned by the participants, task variety is deemed as important. However, these aspects of the job are influenced by how the ambulance services deploy the ECPs – either mainly as solo responders treating non-life threatening cases, or as part of the pool of resources for all emergency cases. This in turn has to be aligned with the targets that the Ambulance Services has to achieve. If the existing resources are insufficient, the ECPs might be deployed to cover the shortage and meet the time-to-respond target.

For example, the framework can be tested by comparing SA and decision-making in different organisations to examine the macro level influences. A record of 999 calls received, the resources assigned, and the outcome of the patients would provide a systems perspective of the work performed by the ECPs. A part of the future work also includes getting an alternate way of validating the communication map. A different managerial staff will be approached to provide a map that will be compared against the existing one.

For the Ambulance Services, the test and validation of the framework would provide an empirical evidence for developing the ECP role. By focusing on SA and decision-making, individual factors can be identified, as well as organisational factors and job design issues. For example, the mastery of geographical knowledge can be improved to supplement the satellite navigation system. The information system devices themselves should be designed better to support the ECPs' knowledge. Thus, the ECP role can be developed in a more holistic way. Consequently, interventions for improving patient safety can also be scoped better. This would be a way for providing better emergency care in the community.

Acknowledgements

We thank the Ministry of Higher Education Malaysia, and the International Islamic University Malaysia for financial support to the first author, and Dave Whiting of East Midlands Ambulance Services for his support to this project.

References

Anders, S, Woods, D.D., Wears, R.L., Perry, S.J., and Patterson, E. 2006, Limits on adaptation: modeling resilience and brittleness in hospital emergency departments. *Paper presented at the 2nd Symposium on Resilience Engineering*, Juan-les-Pins, Paris, France, November 8–10

Blandford, A. and Wong, W. 2004, Situation awareness in emergency medical dispatch. *International Journal of Human–Computer Studies*, 61(4), 421–452

Cooper, S., Barrett, B., Black, S., Evans, C., Real, C., Williams, S., et al. 2004, The emerging role of the emergency care practitioner. *Emergency Medicine Journal*, 21(5), 614–618

Endsley, M.R. 1988, Design and evaluation for situation awareness enhancement. *Proceedings of the Human Factors Society 32nd Annual Meeting*, (Santa Monica, CA: Human Factors Society), 97–101

Endsley, M. R. and Robertson, M. M. 2000, Training for situation awareness in individuals and teams. In M. R. Endsley, & D. J. Garland (Eds.), *Situation Awareness Analysis and Measurement*, (Erlbaum, London), 349–365

Klein, G.A., Orasanu, J., Calderwood, R. and Zsambok, C.E. (Eds.). 1993, *Decision making in Action: Models and Methods,* (Ablex Publishing Corporation, Norwood, New Jersey)

Wears, R.L., Perry, S.J., Anders, S., and Woods, D.D. 2008, Resilience in the emergency department. In E. Hollnagel, C.P. Nemeth, and S. Dekker (Eds.), *Resilience Engineering Perspectives, Volume 1: Remaining Sensitive to the Possibility of Failure*, (Ashgate, Aldershot, England), 193–210

Wong, W. (2000), The Integrated Decision Model in emergency dispatch management and its implications for design. *Australian Journal of Information Systems*, 2(7), 95–107

Wong, W. and Blandford, A. 2001, *Naturalistic Decision Making in Emergency Ambulance Command and Control* (Discussion Paper Series No 2001/11, (Department of Information Science, University of Otago, Dunedin)

WHEN ARE YOU TOO TIRED TO BE SAFE? EXPLORING THE CONSTRUCTION OF A FATIGUE INDEX IN ATM

N. Cebola[1] & A. Kilner[2]

[1]*Instituto Universitário de Lisboa*
[2]*EUROCONTROL*

Human beings in general are very poor at assessing their state. In safety critical industries not being able to assess your status has safety implications. In Air Traffic Management (ATM) fatigue is a major factor that can affect controller's performance and safe conduct. This paper explores an approach to developing a prototypical Fatigue Index Tool (FIT) to assess factors relevant to fatigue and their impact on Air Traffic Controllers (ATCOs). Preliminary research has evaluated sources of literature and identified key common factors in shift workers related to fatigue and its countermeasures. A review of the factors identified was undertaken by controllers at EUROCONTROL's EXPERIMENTAL CENTRE (EEC) and a further validation exercise with operational staff at the Air Navigation Service Provider (ANSP) "CRNA NORD (Athis Mons)".

Introduction

The associated risk with mistakes in the ATM environment makes it imperative to understand fatigue and tiredness so that levels of safety may be maintained and accidents prevented. Fatigue is regularly reported as a factor that reduces the capabilities of ATCOs and also as a factor in incidents (Calvaresi-Barr 2009). The purpose of this research is to explore the feasibility of developing and presenting a Fatigue Index Tool (FIT) and provide countermeasures for individual use. The FIT concept was for a tool to be used by anyone during a shift, and was made simple so that no experts were required for its application. In researching the tool, a systematic review was conducted on factors in published literature to identify the factors related to fatigue. Two sets of interviews were also conducted: at the EEC and secondly at a French ANSP.

Method

A systematic review limited to English language publications was conducted via Google Scholar, Science Direct, and Scopus search engines. Its aim was to compile a list of factors regarding fatigue that have been shown to be relevant in ATM and in other shift working industries and a set of recovery factors used by workers when dealing with fatigue. The search terms used included among others "shift work",

"work hour", "sleep*", "fatigue", "cumulative fatigue", "time on task", "shift length", "duty length", "time of day", "rest break", "rest period", "nap*", "ATC", and "Air traffic controller". The main criterion for the selection of articles was the relevance to fatigue.

Causes and consequences of fatigue – The literature

The review produced a list of factors that were assessed in the development of FIT. These factors are "sleep debt", "sleep loss", "sleep quality", "recovery time between shifts", "shift start time", "time on duty", "circadian rhythms", "shift length", "workload", "breaks", plus "naps", "stress", and "boredom". All these factors were assessed and implemented within FIT where appropriate. As is usual with constructs in Human Factors, there is no consensus on a single definition of fatigue either by the public or academia. In FIT the Åkerstedt (1996) definition was followed in which fatigue is described as the tiredness and sleepiness that results from insufficient sleep, extended number of waking hours and circadian rhythms. Fatigue consequences on the other hand have more agreement and Neri, Dinges, & Rosekind, (1997), and RSSB's feeling tired work (2004) agree on physiological and cognitive impairments, mood degradation, and sleepiness. Kaida *et al.* (2006) have shown that there is a good correlation between subjective fatigue ratings and psychomotor vigilance task (PVT) lapses ($r = 0.56$) – a common performance test. Gillberg, Kecklund, & Åkerstedt (1994), state that the PVT and the Visual Analogue Scale, both measuring performance, have shown correlations with the KSS, e.g. Åkerstedt (1996) found mean intraindividual correlations up to $r = 0.79$. Van Dongen *et al.* (2003) found a correlation between accumulated wakefulness and lapses in the PVT with the model explaining 83% of the variance.

Interviews 1 – Ex-controllers

Based on the literature review a structured interview checklist was created to assess relevant factors and seven air traffic controllers at the EEC were interviewed. Of the seven controllers interviewed, six were male, there were 4 different nationalities, all had worked in civil control but two had experience of military centres with an average of 15.5 years experience. Controllers were asked about factors they had experience that related to sleepiness, workload, stress and recovery. Secondly, they were asked to extrapolate from the factors identified in the literature to the broader concept of fatigue in the ATC environment. The results substantiated the factors found in the literature as well as the set of countermeasures identified both in ATC studies and in other industries.

Interviews 2 – Controllers in the wild

The findings from the interviews were used to develop an embryonic version of FIT. A second set of interviews was undertaken to gather information on how to develop a fatigue assessment tool in a practical and useful form for ATCOs. Exploring the deployment method, interviews were conducted at the ANSP "CRNA NORD

(Athis Mons)" over a period of three shifts (morning, day and night shifts), rostered controllers were asked to participate in the interviews. 15 controllers (11 women, 4 men) from three teams were interviewed. Controllers wanted a shorter tool, of pen and paper format that was very easy and quick to use. Finally, the recommendations provided by FIT must be practical and useful. Controllers believe they are aware of fatigue, and consider fatigue management as being personal. The use of naps was reported extensively and ATCOs expressed doubts on the usefulness of FIT.

The Quality of Sleep scale was assessed based on the original sleep scale developed by Smith, Allen, & Wadsworth (2006, p. 32). Analysis showed that the sleep quality results had a very strong correlation to a single question on the scale – "how would you rate your quality of sleep?" ($r = 0.90$, $\rho <= .01$). It was, therefore, decided to use this single question to assess sleep quality.

Is it sensible – What are the components?

In reviewing the literature, it is clear that the KSS is the most commonly used scale. The KSS reliability with objective measures, namely Electroencephalographic (EEG) and the Accumulated Time with Sleepiness (ATS) measures has been extensively tested and its accuracy confirmed (Åkerstedt, & Gillberg, 1990), (Kaida *et al.*, 2006). The KSS was therefore used as a basis to assess the data to be used in FIT. Data was first converted into KSS and used to calculate the fatigue score, then converted into a five point scale for ease of consideration by controllers which provide us with some confidence that FIT is doing what it was intended to do. Although derived independently, FIT uses the same three components system identified by Spencer *et al.* (2006) to assess fatigue. These are the cumulative fatigue, the shift fatigue, and the recovery factors.

Cumulative fatigue

The cumulative fatigue component is an isolated factor that relates to the amount and quality of sleep attained during a roster. Quantity and quality of sleep contribute equally to fatigue (Neri *et al.*, 1997), (Rosekind, Gander, Connell, & Co, 2001). The same literature suggests that extended wakefulness is a better fit in modelling fatigue than sleep debt, and that sleeping is the only way to effectively return fatigue to baseline levels. The cumulative component of fatigue is, therefore, not influenced by breaks but only by naps. Furthermore it has been showed that taking a nap has no strong effect on the recovery impact of sleep (Matsumoto, & Harada, 1994), (Sallinen Härmä, Åkerstedt, Rosa, & Lillqvist, 1998).

Wakefulness and Sleep Quality

Van Dongen, *et al.* (2003) assessed chronic sleep restrictions in laboratory conditions with participants receiving one of three sleep doses (4 h, 6 h, or 8 h) over 14 days and a group that experienced a 3 day complete sleep restriction. Results showed that the impact of sleep loss and sleep debt on performance relates more closely to the cumulative duration of wakefulness in excess – 15.84 h. Assessing vigilance, the research showed a linear build-up of psychomotor vigilance performance impairment in all four experimental conditions allowing the relation between PVT and wakefulness to be derived. It was necessary now to derive the relationship between wakefulness and KSS.

The relationship was inferred by reviewing the data information available for the validation of the KKS conducted by Kaida *et al.* (2006). In their research they divided the KSS-J (Japanese version of the KSS) scores into bins (1–3, 4–5, 6, 7 and 8–9) "since all individuals did not provide ratings at all levels of the KSS-J (i.e. from 1 to 9)" (pp. 1577). Using their results it was possible to reverse engineer the KSS score related to the cumulative component of fatigue. This is a little irregular but this inference was necessary as none of the papers reviewed in the production of FIT (over 100) showed directly the relationship, as a coefficient between wakefulness, fatigue or any other sub scale under consideration. The analysis showed that after approximately 10 h of wakefulness a KSS score of 1–3 is achieved, after 14 h a score of 4–5, after 18 h a score of 6, after 22 h a score of 7, and after 34 h a score of 8–9.

Rosekind, Gander, Connell, & Co, 2001), in two studies found that "sleep quality and sleep quantity account for approximately equal amounts of variance in sleepiness" (p. 593). Åkerstedt *et al.* (2004) in a study with 5720 participants in the Stockholm area found a correlation of 0.84 between fatigue and sleep disturbances where they conclude that "disturbed sleep has to be considered in the aetiology of mental fatigue" (p. 432).

Recovering from long shifts and extended shift work

Consistent indications of the time taken to recover can be derived from Tucker, Smith, Macdonald, & Folkard (1999), Belenky *et al.* (2003), Åkerstedt, Kecklund, Gillberg, Lowden, & Axelsson (2000). The findings indicate that two days off between shifts for rest are sufficient to return fatigue levels to baseline values. Working shifts that disrupt the circadian rhythms however may need up to 4 days to recover. It is assumed in FIT that ATCOs within the roster system will have enough time to recover.

Shift work

Shift fatigue is composed of the fatigue levels at the start time of the duty (cumulative fatigue), the time of day throughout the duty period (circadian rhythm) the amount of time on duty and workload. Spencer *et al.* (2006) established "that the effect of these factors is approximately additive" (p. 51). Each is explored in more detail below.

Shift Work – Shift Start time, Time on duty, and Circadian Rhythms

McGuffog *et al.* (2004) assessing fatigue in train drivers found the time of day the shift started was the most important factor contributing to fatigue at the "start" of the shift and that the reason behind it was the reduced hours of sleep. The highest rating of fatigue occurred in the early morning: the earlier the start the shorter the amount of hours slept. FIT, instead of using the "average" results found in the literature, uses cumulative fatigue results for a shift starting between 18.00 and 22.00 as the indicator of fatigue at shift start. These figures ensure that the calculation of fatigue include the loss of sleep that McGuffog *et al.* (2004) found to be the major influence behind the fatigue levels at the shift start time.

Spencer *et al.* (2006) when putting together the HSE fatigue shift work index found that fatigue increases 0.14 per hour on the KSS, the same found by McGuffog et al. (2004).

Spencer *et al.* (2006), analysing data from train drivers and aircrew on long and short-hauls found that fatigue was represented by a sinusoid (circadian effect) which peaked at 05.15. This data to replicate the variation on fatigue based on each hour of the day. In this feasibility test of the prototype FIT the values of time of day in fatigue were averaged to represent four shift start periods – morning (06 h to 09 h), day (09 h to 12 h), afternoon (12 h to 17 h), and night (18 h – 22 h). This was a necessary step in order to ensure the reduced size of the tool and keep data entry simple for users. The bounds of the start time should be examined further so as to appropriately account for the anecdotal evidence of higher fatigue levels arising during "graveyard" shift.

Workload

Rogers *et al.* (1999) in the first HSE fatigue index review found information in a study with ATCOs that a high workload period only impacts fatigue after periods of work longer that 120 min ($\rho < 0.001$). Within many European control centres, periods of work longer that 120 min are prevented by legislation. From Rogers' review it might be inferred that controllers do not become fatigued at work as a result of workload.

Recovery

The final component of fatigue accounted for in FIT is that of recovery, comprising breaks and naps. Research suggests that naps have an impact both on shift fatigue and cumulative fatigue. Breaks only have an impact on shift fatigue.

Recovery – Breaks and Naps

Gillberg, Kecklund, Göransson, & Åkerstedt (2003) collected data regarding the impact of breaks on fatigue both during the day and night shifts on power plant operators. Their results show that the positive effect of breaks was seen 20 min after the break but not 40 min after, with no significant differences found between the night and the day shifts. This suggests a positive impact of breaks but one that has a time limit. Neri *et al.* (2002) in a flight simulator with 42 crew members investigated the effectiveness of five seven-minute breaks during the course of a six-hour night time flight which commenced at 02:00. Similar results were found with a reduction in fatigue up to 25 min after a break but not after 40 min. Given the similarity of the findings in the two papers, but with no additional evidence to suggest which was most appropriate, an initial instantiation of the impact of breaks has been embedded in FIT using an average of the impact of the data from these studies.

Rosekind *et al.* (1995) in a series of studies with long haul flight crews concluded that micro-sleeps associated with physiological sleepiness occurred twice as much in groups not allowed to nap (40 min) than in groups that were. When assessing short vs. long naps in Della Rocco,(2005), had 60 ATCOs complete a four-day protocol where they worked three early morning shifts (0700–1500) followed by a rapid rotation to the midnight shift (2300–0700). They found a stronger perceived effect of long naps (120 min) over short naps (45 min). The use of subjective measures (Stanford sleepiness scale) showed a similar 50% difference in fatigue ratings much the same as Rosekind *et al.* (1995).

Based on this information the effect of naps on FIT is calculated as reducing the fatigue score to half after taking a nap.

Using FIT

This short description is intended to show how simple it is to calculate a Fatigue Score.

- Step 1: estimates sleep debt. The number of hours of sleep is entered into the Table 1 for each day (adding them up). This is compared to the baseline score of optimum sleep for each day within a shift cycle.
- Step 2: calculate the difference between the actual number of hours and the optimum sleep hours and read across for an initial fatigue score.
- Step 3: using table 2 accounts for the quality of sleep; select the column that has your initial fatigue score. Rate the quality of your sleep over the previous nights (1 is good, 5 is bad), and read off a second index score.
- Step 4: Using table 3 accounts for time on shift; read down the left hand column to find your second index score, read across the table to the column reflecting time into the shift.
- Step 5: This is the basic fatigue index score. It has been colour coded for simplicity, green through to red, green being not fatigued, and red being very fatigued.

It is now possible to bias this score to account for napping, workload and breaks by deducting and adding points to the score and moving between the different coloured zones of the tool. FIT is presented in Figure 1.

Cumulative component

Day	Hours of sleep	Baseline	Difference	Score
1		8h10	> 10h	1
2		16h19	> 10h	2
3		24h28	> 18h	3
4		32h38	> 22h	4
5		40h48	> 30h	5
6		48h57	> 30h	5
7		57h07	> 30h	5
8		65h16	> 30h	5

Length (columns) / Quality (rows)

Sleep	1	2	3	4	5
1	1	1.5	2	2.5	3
2	1.5	2	2.5	3	3.5
3	2	2.5	3	3.5	4
4	2.5	3	3.5	4	4.5
5	3	3.5	4	4.5	5

Remember to add up your nap time to your total of hours slept.

Shift work component

Night Shift 18h00 to 22h00

Cumulative score	2h	4h	6h	8h	10h	12h
1	1.38	1.60	2.19	2.50	2.57	2.72
1.5	1.38	2.30	2.70	3.00	3.17	3.22
2	2.38	2.80	3.20	3.50	3.67	3.72
2.5	2.38	3.30	3.70	4.00	4.17	4.22
3	3.38	3.80	4.20	4.50	4.67	4.75
3.5	3.38	4.30	4.70	5.00	5.17	5.22
4	4.38	4.80	5.20	5.50	5.67	5.72
4.5	4.38	5.30	5.70	6.00	6.17	6.22
5	5.38	5.80	6.20	6.50	6.67	6.72

After napping

Cumulative score	2h	4h	6h	8h	10h	12h
1	1.00	1.00	1.10	1.25	1.34	1.36
1.5	1.00	1.15	1.95	1.50	1.59	1.61
2	1.19	1.40	1.60	1.75	1.84	1.96
2.5	1.44	1.65	1.85	2.00	2.09	2.11
3	1.69	1.90	2.10	2.25	2.34	2.36
3.5	1.94	2.15	2.35	2.80	2.59	2.61
4	2.19	2.40	2.60	2.75	2.84	2.86
4.5	2.44	2.65	2.85	3.00	3.09	3.11
5	5.38	2.90	3.10	3.25	3.34	3.36

For every 2 breaks (30 min or more) reduce a point in your fatigue rating. One point = one color

In high workload add one point per 2 hours.

XXXXXXXX
Alert
XXXXXXXX
XXXXXXXX
Externely sleepy, fighting sleep

Recommendations

Cumulative fatigue – The only way to counteract cumulative fatigue is to incease the quantity and/or quality of sleep

- Napping – the effects are stronger closer to the circadian low points – between 02 h and 05 h.
- Avoid caffeine or other stimulants 4 to 6 h before bedtime and limit your dose to around 300 ml a day.
- Don't stop for a drink after work. You may feel relax at first but alcohol disturbs sleep.
- Try to get around 8h30 of sleep a day. Continuously sleep would be better.
- Create a sleep conductive environment before going to bed.
- Fresh air and a room temperature of around 18 degrees will give you the best sleeping conditions
- Soaking in hot water before retiring to bed can ease the transition into a deeper sleep.
- Some people find that a milky drink or light carbohydrate snack promotes sleep.
- Light is a powerful resynchronizer so make you room dark if sleeping during the day.

Shift fatigue – Represents the increase in fatigue during your shift and can be fought in several ways

- Start a shift with a meal of proteins to increase alertness, finish with carbohydrates to facilitate sleep.
- Avoid large meals (more than 600 calories) before a during your shift as they can induc sleepiness.
- Naps during the mid afternoon (14 h-17 h) and early morning hours (02 h-08 h) are more effetive
- Caffeine has beneficial effects on cognition, particularly in those who are sleep-deprived.
- Use it close to circadian low point for a stronger effect – 14 h-15 h and 03 h-05 h.
- Chewing a piece of gum can relieve sleepiness
- Stretch reglarly during the shift to improve blood flow restricted by sitting for long periods
- Scoring a high **orange** or a **red** point can only be alleviated with sleep

Figure 1. FIT.

What to do if you score "FATIGUED"

A great number of recommendations regarding combating fatigue have been put forward through reviews on fatigue (e.g. Turner, & Stone, 2004; Neri *et al.*, 1997). The set of recommendations is, however, somewhat limited and all of them inevitably revolve around the same suggestions. These are firstly how to improve sleep, and secondly measures that can be applied at work to counteract the effects of fatigue. A set of the most practical of these recommendations was chosen for FIT and is presented (Figure 1).

Conclusions

FIT provides an embryonic tool that encompasses many of the factors considered as relevant when assessing fatigue. Further development was limited both by time, and by the lack of traceable information within the literature. A great number of academic papers have been published in the area of fatigue, but few have provided detailed information on the sources of data and relationships that have been expressed between the different components they establish as contributing to fatigue. This has resulted in an examination and piecing together from different sources to understand:

- Contributors to fatigue
- The relationship between sub factors of fatigue
- The mathematical relationships between sub factors to produce a fatigue score
- The relationship of the fatigue score to perceived fatigue

Approximations and reasoned arguments have been incorporated into the development of FIT to account for the lack of clarity in the literature, and to ensure it is clear where FIT is based on clear guidance, and where the authors have made interpretations based on best available evidence. FIT is the results of an initial feasibility study into whether it would be possible to create a fatigue assessment tool for the domain of ATC. It shows that it is possible to create a tool that assesses fatigue in accordance with the principles documented in academic literature. However we have insufficient confidence at this stage of the quantitative relationships expressed within the tool used to actually calculate fatigue. The study has been successful in so far as:

- A simple to use tool – less than a minute – has been created
- It is paper based
- It accounts for all the major factors identified within the literature
- It provides recommendations to resolve high fatigue scores.

FIT is tool primarily developed for ATCOs allowing them to assess their own level of fatigue as an awareness warning at any point before or throughout a shift and take action against it. However, if information is made available to supervisors these can gain awareness on fatigue levels on their team and plan rosters accordingly, and finally management can use it to gain knowledge on the possibility of a fatigue problem and decide on further measures to minimize it.

References

Åkerstedt, T. (1996). Subjective and objective sleepiness in the active individual. International Journal of Neuroscience 1990; 52, pp. 29–37

Åkerstedt, T., & Gillberg, M. (1990). Subjective and objective sleepiness in the active individual. *International Journal of Neuroscience*, 52, pp. 29–37
Åkerstedt, T., Kecklund, G., Gillberg, M., Lowden, A., & Axelsson, J. (2000). Sleepiness and days of recovery. *Transportation Research Part F: Traffic Psychology and Behaviour*, 3, pp. 251–261
Åkerstedt, T., Knutsson, A., Westerholm, P., Theorell, T., Alfredsson, L., & Kecklund, G. (2004). Mental fatigue, work and sleep. *Journal of Psychosomatic Research*, 57, pp. 427–433
Calvaresi-Barr, A. (2009). Report on ARC potential Fatigue Factors. FAA report Number AV-2009-065
Defence Evaluation and Research Agency (DERA), Centre for Human Sciences (CHS). (1999). *Validation and development of a method for assessing the risks arising from mental fatigue*. Farnborough, Hampshire, UK: Rogers, A.S., Spence, M.B., Stone, B.M.
EUROCONTROL EXPERIMENTAL CENTRE: European Organization for the Safety of Air Navigation. (2006). *Low Vigilance in Air Traffic Management*. Bretigny-Sur-Orge, France: Gordon, R., Straussberger, S.
Gillberg, M., Kecklund, G., & Akerstedt, T. (1994). Relations between performance and subjective rating of sleepiness during a night awake. *Sleep*, 17(3), pp. 236–241
Gillberg, M., Kecklund, G., Göransson, B., & Åkerstedt, T. (2003). Operator performance and signs of sleepiness during day and night work in a simulated thermal power plant. *International Journal of Industrial Ergonomics*, 31, pp. 101–109
Health & Safety Executive. (2006). *The development of a fatigue/risk index for shift workers*. Hampshire, UK: Spencer, M.B., Robertson, K.A., & Folkard, S.
Kaida, K., Takahashi, M., Åkerstedt, T., Nakata, A., Otsuka, Y., Haratani, T., & Fukasawa, K. (2006). Validation of the Karolinska Sleepiness Scale against performance and EEG variables. *Clinical Neurophysiology*, 117, pp. 1574–1581
Maritime and Coastguard Agency, Research Report 464. (2006). *Seafarer fatigue: the Cardiff research programme*. Cardiff, United Kingdom: Smith, A.P., Allen, P.H., & Wadsworth, E.J.K.
Matsumoto, K., Harada, M. (1994). The effect of night-time naps on recovery from fatigue following night work. Ergonomics, 37(5), pp. 899–907
National Aeronautics and Space Administration, Ames Research Center. (1997). *Sustained Carriers Operations: Sleep Loss, Performance, and Fatigue Countermeasures*. California, USA: Neri, D. F., Dinges, D.F. & Rosekind, M.R.
National Aeronautics and Space Administration, Ames Research Center. (2001). *Crew Factors in Flight Operations X: Alertness Management in Flight Operations Education Module*. California, USA: Rosekind, M.R., Gander, P. H., Connell, L.J., & Co, E.L.
Rail Safety and Standards Board. (2004). *T059 Human factors study of fatigue and shift work Appendix 1: Working patterns of train drivers – Implications for fatigue and safety*. London, UK: McGuffog, A., Spencer, M., Turner, C., Stone, B.
Rail Safety and Standards Board. (2004). *T059 Human factors study of fatigue and shift work Appendix 2: Review of Coping Strategies*. London, UK: Turner, C., & Stone, B.

Rosekind, M.R., Smith, R.M., Miller, D.L. Co, E.L., Gregory, K.B., Webbon, L.L., Gander, P.H., Lebacqz, J.V. (1995). Alertness management: strategic naps in operational settings. *Journal of Sleep Research*. 4, Suppl. 2, pp. 62–66

Sallinen, M., Härmä, M., Åkerstedt, T., Rosa, R., & Lillqvist, O. (1998). Promoting alertness with a short nap during a night shift. *Journal of Sleep Research*, 7(4), pp. 240–247

Tucker, P., Smith, L., Macdonald, I & Folkard, S. (1999). Distribution of rest days in 12 hour shift systems: impacts on health, wellbeing, and on shift alertness. *Occupational and Environmental Medicine*, 56, pp. 206–214

U.S. Department of Transportation Federal Aviation Administration. (2000). *The Effects of Napping on Night Shift Performance*. Oklahoma City, USA: Della Rocco, P.S., Comperatore, C., Caldwell, L., & Cruz, C.

Van Dongen, H.P.A., Maislin, G., Mullington, J.M., & Dinges, D.F. (2003). The Cumulative Cost of Additional Wakefulness: Dose-Response Effects on Neurobehavioral Functions and Sleep Physiology from Chronic Sleep Restriction and Total Sleep Deprivation. *Sleep*, 26(2), pp. 117–126

FLUIDITY OF PERSONAL BOUNDARIES IN ONLINE SOCIAL MEDIA: IMPLICATIONS FOR NETIQUETTE

Damian Ballam & Chris Fullwood

Wolverhampton Internet and Technology Society, Psychology Department, University of Wolverhampton, UK

Etiquette serves the dual functions of consolidating group membership and restricting access to outsiders. Experiences of offline and online etiquette among a university student population were explored via five focus group discussions. Thematic analysis suggested conceptualising etiquette violations as transgressions of personal space. Online applications – particularly social networking sites – problematise territorial boundaries, as well as ownership and control over information important to self-definition and -presentation. Online disinhibition and violations of netiquette may result from the ambiguity and fluidity of self/other boundaries online. Future phases of the research programme will explore the influence of both individual differences and the features of online media interfaces on the emergence of new netiquette rules.

Introduction

Etiquette can be broadly defined as a set of social norms or conventions, which aim to facilitate smooth, comfortable social interactions and are based upon principles which honour and respect the equal rights of all people (Fox, 2007; Morgan, 2001; Post, 2004). However, from an historical perspective, etiquette would seem to serve the opposing functions of both including and excluding people from social groups and norms; of offering both a code of conduct that facilitates safe and comfortable social interaction, as well as being a means of maintaining structures of social inequality (Wouters, 1995; Arditi, 1999; Elias, 1994). Etiquette can also be seen as codes of conduct that are deeply embedded in, and are an embodiment of, the social milieu in which they exist. They are "the proverbial glue that holds society together" (Post, 2004, p. 6). Further, Elias (1994), charting 'the civilising process' in Europe from the 16th to the 20th centuries, discusses the interrelationships between increasing social interdependencies, the increasing role of self- and social-awareness and self-control in social behaviour, and the development of the superego as an integral constituent of personality.

Etiquette thus changes as the structure of society changes. Consequently, etiquette guides over the 19th and 20th centuries shifted from a treatise format (offering conduct advice for all situations based around a unified conceptual core) to a reference book format (offering quite different conduct advice for different social contexts). Arditi (1999) argues that "[a] person is no longer called to be a coherent being, to be consistent from one situation to the next, but can become a different being in relation to each center." (p. 36). Thus, one is expected to display sensitivity to different social expectations of different social contexts.

The recent rapid expansion of Internet-based communication has led to a number of challenges to everyday notions of etiquette. The rules that govern social interaction online would seem to be very different to those that direct our offline behaviour. Social, nonverbal cues are often lacking in what has until recently been a solely text-based communication medium. Moreover, individuals sometimes behave very differently online and this may be a consequence of a number of factors that make Internet communication unique. For example, it is easier to remain anonymous online and therefore disassociate oneself from ones online activity. This 'online disinhibition effect' would seem to affect individuals differently. For instance, whereas some people may be more likely to be rude/aggressive towards others (toxic disinhibition), others are more polite, helpful, and more likely to disclose personal information online than face-to-face (benign disinhibition) (Suler, 2004). A further consideration centres on the difficulty in considering online interactions homogenously, as varying online applications necessitate specific styles of interaction. Indeed, Herring and Paolillo (2006) suggest that there are common conventions and shared purposes within specific online communities and these factors shape the interaction style. Therefore, just as one would expect differing rules of etiquette in various offline social scenarios, the same may be true of the online world.

Changes in social behaviour in cue-limited online settings clearly have implications for etiquette, and particular guidelines for online conduct (netiquette) have been developed. For example, Cerf (1994), Shea (1994), Hambridge (1995), and more recently Strawbridge (2006) have published general netiquette guidelines. In addition, the more recent guides to general etiquette have included material relating to electronic communication (Morgan, 2001; Post, 2004; Fox, 2007). These guidelines have necessarily changed in character as the nature of the medium has changed from text-based to multi-media environments. Early concerns included dealing with possible ambiguities in a reduced-cue medium of asynchronous text-only communication, clarity and brevity of messages (respecting others' connection time and costs), and concerns about bandwidth usage. Only very recently have there been attempts to advise people regarding new applications such as social networking sites (SNS) (Rocker-Gladen, 2007; Martin, 2008).

It is unclear how aware general Internet users are of existing netiquette guidelines, or indeed how they actually use different online applications – especially instant messaging (IM) and social networking sites – on a day-to-day basis. There is also little in the way of research which has explored problems with etiquette and social conduct in new online media, where technology and the affordances it provides seem to be changing rapidly. New social applications afford ways of interacting (for example, by publicly ranking or even 'auctioning' one's friends online) that existing face-to-face social norms cannot accommodate. Research regarding perceptions and use of such applications is thus useful for both users and site designers, to facilitate smoother and more positive social interactions online through carefully conceived and designed interfaces.

The present study represents a qualitative exploration of the ways in which general Internet users use online communication media, and what they consider acceptable and unacceptable behaviour online. Focus group interviews were conducted to discuss a wide range of attitudes and experiences of online communication, to uncover what counts as good and bad netiquette among current Internet users. Given that everyday etiquette varies between different social contexts, and given the variety of methods of online communication, the aim of the present analysis was to look beyond specific behaviours that count as generally 'good' or 'bad' netiquette to participants' reasons and motivations for treating others in certain ways. What, in other words, determines the decisions people make online regarding their treatment of others?

Methods

Participants

A total of 25 individuals (21 females, 4 males) took part in the study. Participants were all undergraduate students, recruited opportunistically through internal university email system and a participant pool credit system.

Design and Procedure

The study utilised semi-structured focus group interview methods, with groups of up to six participants. The researcher sought to approach the topic with a minimum of bias or theoretical input to the discussions, to allow the participants to speak for themselves about the issues they found important. Participants were thus seen as the experts on the topic. Sessions were designed to be largely participant led, around a broad structure of three main sections: what counts as good and bad etiquette in everyday life?; what are the parallels between these everyday etiquette concerns and online communication?; and what specific issues arise with the use of particular online applications (e-mail, IM and SNS)?

Analysis

Digital audio recordings were made of all focus group discussions, which were subsequently transcribed verbatim. Data were analysed qualitatively according to Braun and Clarke's (2006) guidelines for thematic analysis. This approach was chosen for its flexibility and freedom from strict theoretical underpinnings. A bottom-up data-driven approach was taken to the data. Themes were identified and developed by grouping commonly coded (descriptive) elements of the data corpus to provide a more interpretive overview. The process of analysis was recursive. At every stage of abstraction – from audio files to transcripts, from transcripts to coding, and from coding to the development of themes – the preliminary findings and interpretations were referred back to previous levels, to ensure that they accurately summarised and reflected what participants had reported.

Analysis was conducted with the aid of NVivo 8 qualitative analysis software, which was used for primary and secondary coding. Initial coding involved tagging any words or phrases the researcher found interesting or relevant to the general research question of how people treat each other, and what possible reasons there could be for this. The NVivo 8 software allows researchers to code any word or section of text, which can then viewed with varying degrees of surrounding transcript as context, allowing great flexibility when re-reading coded extracts. After producing an extensive and inclusive list of codes, the extracts were re-read to ensure that the code labels were appropriate and representative of the data. The next level of abstraction involved organizing the list of codes into meaningful groups. Extracts within each group were again compared to ensure they formed a coherent, meaningful set. Finally, these groups themselves were summarized and collated under a small number of interpretive themes. Themes were identified which not only gave what the researcher considered to be a meaningful, organised account of the data, but also which were related to each other, around a central concept. The following report is necessarily selective in terms of the themes represented, yet it offers a meaningful interpretation of the important concerns expressed by these participants, in relation to etiquette and social conduct online.

Findings and Discussion

General netiquette rules

As a group of general Internet users, the present participants were largely unaware of existing netiquette rules or conventions. The 'rules' of netiquette discussed were more idiosyncratic –focusing on aspects of online communication that personally offended participants rather than any overarching set of rules or norms. The way one behaves online was seen to depend very much upon the context of communication, upon which "bit of the Internet you happen to be dwelling on at any given point". The attitudes towards what constitutes good and bad etiquette, acceptable and unacceptable behaviour online, varied from group to group and participant to participant. Online behavior was often characterised as a "free for all", and involved complex negotiation and often negative exchanges, in part because the applications do not make clear and visible their codes of conduct, or basic rules for novices (e.g., that writing in capitals is the equivalent of shouting).

When considering concepts underlying the *reasons* for their attitudes, most etiquette-related decisions could be found rooted in a notion of *boundaries*. Participants discussed boundaries and barriers in a number of different ways; both physically, in terms of "personal space" and "territory", and socially, in terms of a "familiar circle". In both cases, being outside of the boundaries made participants more self-aware, and more concerned with making good impressions. Boundaries were also discussed in terms of a "line" in any social situation. As one participant noted, "I think every member of a group or a social interaction has to be aware of that line and when one person views it as very very differently to other people then, that's when you can have a breakdown in etiquette."

Etiquette infringements were seen largely as either violations of one's personal space (e.g., playing loud music or sitting too close on public transport), or being involuntarily drawn into another's personal space (e.g., by witnessing personal arguments in public places, or by others' inappropriate self-disclosure). Online breaches of etiquette were seen to result from more complex notions of boundaries and spaces in online contexts. The inner and the outer are less clearly defined, and controlling access and gaining insight into others' inner spaces of meaning are more complex online than face-to-face. Findings are discussed in relation to three broad themes. First, the ways in which one's personal space is more ambiguous and fluid online. Second, the ways in which people maintain the boundaries between self and other online. Third, the ways in which seeking knowledge and providing information to others are used to facilitate smoother interactions which are unique to online communication.

The nature of boundaries: Social distance and proximity online

Online environments, on a simple level, were discussed as providing an additional boundary, a physical barrier, the exclusion of nonverbal cues between interactants. This barrier serves to increase social distance and facilitate behaviours that are different or less inhibited than offline behaviours. This heightened *physical* barrier allows for the breakdown of *social* barriers or reserves. "People feel that they can say exactly what they want to say without having any repercussions.... so, yeah, it sort of breaks down a lot of barriers really". The notion that one is braver in one's own space, one's own territory,seems to be important in influencing the extent of self-disclosure, and indeed of the kind of tone one is able/willing to adopt with others. Many participants discussed online behaviour in terms compatible with Suler's (2004) disinhibition effect, both with strangers and with those who are close. "I think it's like, you're in your own house, you're in your own space, you feel more, relaxed." Disinhibition experienced in online environments differs markedly from everyday life in an apparently increased complexity in defining the boundaries. At once, someone engaging in online communication may be within a familiar and private *physical* environment and also in a publicly visible *online* space. In a sense, one is simultaneously within and without one's own territory. On the one hand, one can act with the comfort, security and sense of familiarity (and sometimes the casuality and complacency) afforded by one's own physical surroundings, and also display affront towards unwanted online communication which, because it appears and is accessed within the home environment, may be perceived as an invasion or violation of that personal space.

A prime example of such a fluid, paradoxical boundary can be found in Social Networking Site profiles. On the one hand, one's profile is one's own space ("it's more, customisable on MySpace, you can make it more your own, well, it's *called* MySpace, you can make it more your own space, really!"). On the other hand, it is publicly available – either in a limited sense in which it is available to one's network of friends, or in a broad sense in which, without privacy settings in place, it is accessible to anyone online. One's profile will contain information, messages and files uploaded by the profile owner, and those uploaded or added by other people (such as tagged pictures). Disclosure of personal information in such a medium, which is both personal and impersonal, can become problematic. It can feel appropriate to send because one is within one's own physical space; however, it can feel inappropriate to receive because it places the reader in a position of feeling obligated to respond, and having to make difficult decisions about how best to do so. "It makes people more un-at-ease." In posting such information, which will appear on others' profiles (i.e., in *their* own personal spaces), one makes personal information not only public, but also personal to the recipients.

Using the boundaries: keeping a distance and presenting a face

Etiquette plays a major role in maintaining the boundaries between self and other, in increasing, decreasing or maintaining social distances (e.g., Wouters, 1995; Arditi, 1999). In relation to online communication, participants particularly discussed appearing offline and tagging pictures for these purposes.

In synchronous (i.e. environments in which messages are received instantaneously) online settings, especially in IM and SNS chat facilities, participants often reported using the option to appear offline to control who had access to them in their personal territory, and to select to whom they revealed their presence. It was however acknowledged that such behaviour was "quite rude of me" or "a bit creepy". Appearing online/available in a synchronous application also means that in a sense one is both online and present in one's own home, and also simultaneously visible to one's other contacts, in *their* own homes. Social presence (especially in a format which is usually one-to-one online communication) can be awkward and ambiguous. It creates a sense of obligation to offer a greeting and to interact with others by nature of the apparent intimacy of the presence. However, because of the limited nature of the medium, one cannot be sure whether the other person is truly present and available, or is talking to someone else. The desire to be sociable and the desire not to intrude can create conflicts in a way that face-to-face social encounters tend not to: "Sometimes when people are online you feel the need to say hello to them, even if you haven't got anything to say to them, just think 'oh, they can see that I'm online, I'd better be polite and say hello."' One participant in particular mentioned using the "appear offline" status as a courtesy to others, because "they might not wanna sign on when you're on."

SNS allow users to add tags of friends' names to pictures, creating a two-way link between profiles. In Facebook for example, the tag acts as a hyperlink to the friend's profile, while the tagged picture itself appears not only in that person's news feed, but also under a link titled "View photos of [name]". Participants expressed mixed opinions on tagging. They function to decrease social distance, to act as physical manifestations of the links between people, and sometimes "the worst thing you can do is not tag somebody in a photograph", raising questions such as "are we not friends anymore?" Tagging may also be undesirable, however, especially if the pictures are not flattering. It is something which can contribute to creating an impression, over which the person themselves has no immediate control. Some participants reported removing tags, either because they did not want the pictures to appear on their profiles, or because the pictures allowed a gateway via which strangers can see and contact them. Removing tags is a way of maintaining the boundaries, of restricting access to unwanted individuals, maintaining one's privacy, and also maintaining an acceptable impression of oneself.

Asynchronous media (such as e-mail) in particular afford a temporal boundary by which participants discussed having "time to edit yourself", to present a more considered, favourable response, and thus to be more polite and more helpful (supporting Walther's, 1996, hyperpersonal model of online communication). For some participants, however, this offers too much space to craft online messages and consider how the recipient will interpret the message. "Sometimes I find it harder that you have the choice of editing it, because you write something and then you think 'Oh no, they'll take that as a nasty comment', and then you write it again, and then you think 'Oh no, but they might take that as being too nice' or something."

Compensating for boundaries: Seeking, and granting knowledge

Online media were largely seen as limited, cue-restricted forms of communication. The interfaces act as a barrier to the visual, nonverbal kinds of knowledge that are normally available and usable in everyday communicative contexts. Without such cues, communication can be "very cold and clinical", and when written without emotional cues in the text (such as emoticons or expressive punctuation), "It just, it comes across as what it is – written text, on a screen, – rather than someone talking to you". This can lead to misinterpretation and confusion regarding other's intentions. Even in the cue-richer online applications such as SNS, one's "perception of the person is narrower". Online communications are "hard to read", and "you can make excuses for them more because there might actually be a reason for it."

In online communication, participants reported using language differently, depending on the intended recipient, the purposes of the message, and the medium used. "You adapt your language, you probably use *more* words, *more* language to say the same thing than you would face-to-face". People also compensate for the lack of nonverbal cues by intensifying the things they write online, or saying "more outrageous things" Participants reported deliberately using language online to create interpersonal distance, via formality of style, wording and punctuation, and the use of short, non-expressive sentences. They also deliberately used language to create closeness/decrease interpersonal distance, through the use of informal style, casual spelling (e.g., "Helloooooo"), expressive punctuation, and the use of emoticons to "put a *face* onto, online settings, to sort of make it more personal."

Knowledge and interpretive cues are lacking not only in overt textual messages, but also in the meanings of others' behaviours. For example, taking longer to reply to an online message may be misinterpreted for ignoring or being disinterested in what another is saying. Most participants agreed that access to knowledge of other's intentions is lacking online, and that "there could be any excuse" for apparently rude or ignorant online behaviour. This state of obliviousness, as well as leading to negative interpretations can also lead interactants to "cut people more slack online". Participants frequently reported being more understanding when others had to quickly leave an online conversation in ways that would seem rude in everyday life, such as failing to say goodbye.

Many participants discussed how the usual cues we use to compartmentalize and stereotype people are missing online, meaning that interlocutors have to be more cautious and conservative in their judgments because important cues are missing. Good etiquette requires not jumping to (negative) conclusions about the motivations behind other's behaviour online. Indeed, it was considered the single most important rule of netiquette by participants in one group. This is in contrast with the way some participants described the *lack of understanding* shown in everyday situations, where people make judgments about others based on the kinds of nonverbal cues that are missing in online contexts. People "judge by what they see", based on others' physical appearance, "if you don't understand it, or you don't dress like them as such, you w-, you do think that your lifestyles are different and that kind of rolls over to your personality, to your behaviour, and even possibly manners."

One mediating factor is the degree to which interactants know each other in real life. Knowing someone well, having them in one's "inner circle", results in less concern that they will take one's comments negatively and a more relaxed attitude towards politeness and the efforts that goes into communications. It can also result in less concern and ambiguity about how to interpret others. It may, therefore, be the case that communication in anchored relationships online (i.e. environments in which interactants are likely to have had some form of prior face to face contact) necessitate different approaches to etiquette than communication with strangers (which may be more common in chat rooms for example).

Participants referred to sometimes elaborate or covert means of seeking out information about people, to compensate for the lack of nonverbal cues. Commonly, participants reported reading others' public profiles, comments and conversations. Some participants also reported using the applications themselves to covertly discover information to explain other's behavior, such as checking whether a particular e-mail had been read, discovering who was blocking them, and trying to counter the software's measures to block information on SNS.

Conclusions and recommendations

Participants proved to be creative and often idiosyncratic in using online media to express themselves, to understand others, to allow people into their territories and to keep others out. The design of such applications needs to facilitate not only more open and expressive communication, but also to allow others to protect their own boundaries and privacy. With SNS, tagging pictures caused many disputes among friends. For users, it would seem prudent to know in advance whether friends would wish to be tagged before uploading pictures. For site designers, it would seem prudent to ensure, in the same way that friend requests must be confirmed by both parties, that tags are acceptable to everyone concerned before going live.

With SNS the process of friending offers only a single category into which all contacts are grouped. In offline communication, people interact and present themselves differently to close friends, acquaintances, colleagues, and family members, all of whom may be "friends" in one's online social network. Regarding privacy concerns, participants' advice to "just be careful who you add as a friend" seems sensible. And for site designers, the present participants would have appreciated a means of segregating 'friends' to simplify their concerns over conflicting modes of self-expression and self-presentation online, in spaces which are simultaneously personal and public to a diverse audience.

SNS profiles in particular and also IM applications (with lists of online/offline contacts, profile pictures and status messages) provide spaces which are simultaneously personal and shared. These elements are important to their design and functionality, though the present participants – particularly those with less experience of online communication – reported confusion and ambiguity within these unique environments and would appreciate more obvious structure and guidance when they first start using these applications.

References

Arditi, J. (1999). Etiquette books, discourse and the deployment of an order of things. *Theory, Culture and Society, 16*, 25–84.
Braun, V., & Clarke, V. (2006). Using thematic analysis in psychology. *Qualitative Research in Psychology, 3*, 77–101.
Cerf, V. (14th August, 1994). Draft v0.1 guidelines for conduct on and use of Internet. Accessed 14th May 2008 from http://www.isoc.org/internet/conduct/cerf-Aug-draft.html
Elias, N. (1994). *The civilizing process.* Oxford: Blackwell
Fox, S. (2007). *Etiquette for dummies.* (2nd ed.). John Wiley & Sons
Hambridge, S. (October 1995). RFC 1855: Netiquette guidelines. Accessed 13th May 2008 from http://tools.ietf.org/html/rfc1855
Herring, S.C., and Paolillo, J.C. (2006). Gender and genre variation in weblogs. *Journal of Sociolinguistics, 10*(4), 439–459
Martin, N. (13th June, 2008). Debrett's guide to online etiquette. Daily Telegraph, London UK, p. 27
Morgan, J. (2001). *Debrett's new guide to etiquette and modern manners.* New York: Thomas Dunne books
Post, P. (2004). *Emily Post's etiquette.* Harper-Collins World
Rocker-Gladen, N. (24th June, 2007). MySpace etiquette tips: 10 great netiquette rules for social networking. Accessed from http://social-networking-tagging.suite101.com/article. cfm/myspace_etiquette_tips
Shea, V. (1994). *Netiquette.* Albion books
Strawbridge, M. (2006). *Netiquette: Internet etiquette in the age of the blog.* Software Reference Ltd.
Suler, J. (2004). The online disinhibition effect. *CyberPsychology & Behavior, 7*(3), 321–326
Walther, J.B. (1996). Computer-mediated communication: Impersonal, interpersonal, and hyperpersonal Interaction. *Communication Research, 23*(1), 3–43
Wouters, C. (1995). Etiquette books and emotion management in the 20th century: Part one – the integration of social classes. *Journal of social History, 29*(1), 107–124

RAIL AND TRANSPORT

ASSESSING BEHAVIOURAL ADAPTATION TO PHYSICAL RISK

D.E. Haigney[1] & N. Gkikas[2]

[1]*Psychology Department, University of Worcester, UK*
[2]*Autonomics, Northampton, UK*

The investigation of behavioural response and task performance in relation to physical risk is fraught with ethical, legal and practical issues. In these studies, weather conditions have been suggested as a means of operationalising physical risk against the normalized risk taking activity of driving thus allowing ethical, empirical testing of behavioural adaptation. The article details a combination of data from simulator, closed road track and in-depth accident investigation studies. All three types of data supported the existence of behavioural adaptation with changes in objective physical risk, potentially mediating individual accident likelihood. Issues surrounding the collection, analysis and validity of such data are discussed.

Introduction

The design of research investigating human response to a threat of physical harm is problematic due to self-selection bias (Harrison *et al.*, 2009) and a variety of other ethical and legal considerations (Haigney & Westerman, 2001). Whilst on-road driving behaviour can be regarded as a socially normalised active process of 'threat avoidance' (Gibson & Crooks, 1938; Fuller, 1984; Matthews, 1992) with the influence of demand characteristics (Rosnow, 2002) participant consent (ibid.) and the potential statistical significance bias of journals (Sridharan & Greenland, 2009) it could be argued that reported behavioural adaptation to threat, is – however unwittingly – distorted. As the extent and nature of any behavioural adaptation pathway cannot necessarily be predicted reliably (Wilde, 1988) this raises issues regarding uncontrolled, unassessed, ungeneralisable and unpredicted physical risk exposure for all other road users who have not given consent – in which the participant, the experimenter and any supporting institution may well be culpable (Coolican, 2009).

Since the logistics and pragmatics of overcoming these issues are formidable, the range of legally unambiguous, ethical and analytically sound empirical techniques available to researchers investigating responses to physical risk through driving behaviour is restricted. It would still be possible for example, to observe the driving behaviour of participants through analogous on-road environments however, e.g. closed track (Gkikas *et al.*, 2009b; Han & Yi-Lang, 2006) computer simulations (Blana & Golias, 2002; Jamson, Westerman, Hockey & Carsten, 2004) or non-experimental manipulation (e.g. following involvement in a road traffic accident such as the 'On The Spot' (OTS) Co-operative Crash Injury Study, CCIS (CCIS, 2009).

The use of meteorological condition as means of operationalising physical threat within these techniques allows ethical, socially normalised physical risk for all road users. Objective measurement in terms of weather event intensity and duration is also possible (Edwards, 1998) against which perceived risk, affective and behavioural response may also be assessed.

These original, albeit summarised studies detail data from simulator, closed road track and in-depth accident investigation studies in which physical threat is operationalised through meteorological condition. Data are discussed in terms of their potential contribution to the debate surrounding the possibility of driver adaptive response in the face of threat.

Simulator Study

Given driver concern over conditions of low visibility in fog and darkness (Brisbane, 1992; Andrey and Knapper, 1993) and the increased objective road traffic collision (RTC) risk under such conditions *ceteris paribus* (Andrey and Knapper, 1993; Edwards, 1998) this study will concentrate specifically on the relation between driver behaviour and visibility conditions. The assessment of subjective risk across experimental conditions has been undertaken through of the 'UWIST Mood Adjective Checklist' or UMACL (Lazarus and Folkman, 1984; Matthews, 1992; Matthews *et al.*, 1990). This gives participant ratings on the dimensions of EA (energetic arousal, i.e. alertness through tiredness) and TA (tense arousal, i.e. tension through relaxation).

Forty individuals participated in the study (25 male and 15 female). Mean age in years was 22.8 (sd. 5.64), mean annual driving exposure 13,453 miles (sd. 1.11) and mean driving experience 4.63 years (sd. 4.02). Participants negotiated in the left carriageway along a simulated road via the manipulation of the steering wheel, the accelerator and brake pedals, populated with simulated 'other vehicles'. Performance variables include: speed, braking, collisions (also referred to as 'error') and left headway (distance between participant vehicle and 'other' in front of them in the left carriageway). Visibility and contrast per condition followed meteorological definitions as well as empirical chroma contrasts and cycles (Commission Internationale de l'Eclairage, 1992; Meteorological Office,1969).

In this within-subject design, participants in the study were required to complete the UMACL immediately prior to and following counterbalanced visibility conditions with a 2-minute rest period post-condition (Matthews *et al.*, 1992).

Figure 1 shows that statistically significant variation was established for mean speeds across all experimental conditions ($df = 3$; $f = 8.57$; $p < 0.000$). Post-hoc Scheffe analysis, indicated that significant differences existed between the day,

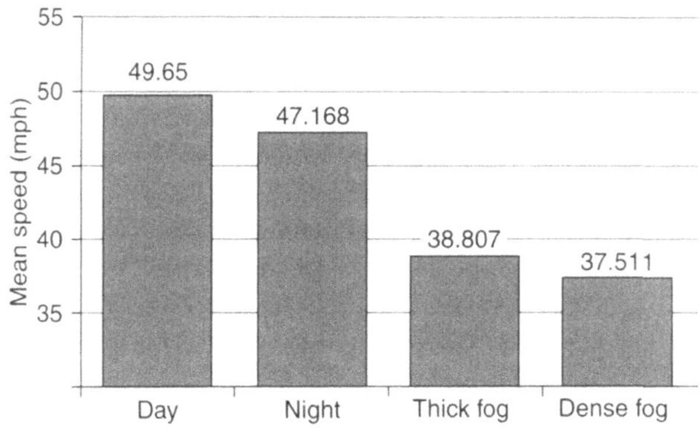

Figure 1. Mean speed across conditions.

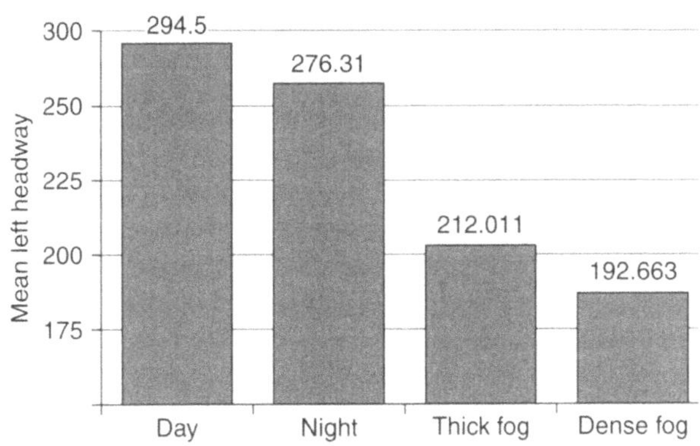

Figure 2. Mean left headway across conditions.

thick fog and dense fog experimental groups at the $p < 0.05$ level, supporting speed survey studies (Tenkink, 1988; Sumner *et al.*, 1977).

Mean braking was not statistically significant across conditions (df = 3; f = 1.125; $p < 0.933$).

The four types of collision recorded were not found to vary significantly across the experimental driving conditions at the 5% level (left carriageway: participant car and 'other' car [df = 3; f – 0.242; $p < 0.866$]; right carriageway: participant car and 'other' car [d£ = 3; f – 0.193; $p < 0.900$]; left edge of road [df – 3; f = 0.551; $p < 0.647$; right edge of road [df – 3; f = 0.799]).

Mean left headway (Figure 2) was found to vary significantly across conditions, with the greatest values recorded in the daytime condition falling through to the lowest values in dense fog. Post-hoc Scheffe analyses show that significant differences between groups ($p < 0.05$) lie between the daytime and thick fog conditions.

Table 1 shows that the same distribution of statistically significant differences across conditions were noted for both post-run EA and post-run TA mood factors, with no

Table 1. Statistically significant variation in post-drive mood dimensions across condition.

Mood dimension		Visibility condition	
		Day	Night
EA	Thick Fog	$t = -2.92$; $p < 0.006$	$t = 3.39$; $p < 0.002$
	Dense Fog	$t = -3.44$; $p < 0.001$	$t = -4.55$; $p < 0.000$
TA	Thick Fog	$t = -3.49$; $p < 0.001$	$t = -2.59$; $p < 0.013$
	Dense Fog	$t = -3.92$; $p < 0.000$	$t = -3.47$; $p < 0.001$

significant differences being determined between day and night, or between dense fog and thick fog.

Test-track study

For further evidence of weather effects on driver behaviour a closed road-track was employed. The vehicle was instrumented with force/pressure and position sensors on the control-pedals, a video camera mounted on the dashboard and a second camera in the footwell (Gkikas, Richardson, & Hill, 2009a). Forty-eight drivers took part (24 male, 24 female) whose ages ranged from 21 to 84 (mean 31.3) years, mean driving experience was 12 years (min 1, max 48), and mean mileage was 9653 miles/year (min 2000, max 30000). They all held a full driving license (UK/EU or equivalent international) and had 3 or less penalty points.

Participants were asked to drive the instrumented vehicle at their preferred distance behind a confederate vehicle, the latter being fitted with a trailer.

The lead-vehicle accelerated to 30 mph and kept a constant speed until releasing the trailer after 0.2 miles (321.86 m) on the main straight of the track. Upon release, the trailer would decelerate at an average rate of $-6.81\ m/s^2$. Participants would brake to avoid a collision. Weather conditions were recorded as "clear", "cloudy" or "rain/wet" as per the Road Accident Statistics (STATS 19) coding system used by police forces in the UK (Department for Transport, 2009).

Weather conditions had a significant effect brake pedal force ($F = 5.37$, $p = 0.01$; Figure 3). However, analysis of variance did not provide statistical evidence of weather effects on throttle-off ($F = 0.61$, $p = 0.94$).

Road-accident data

With the above in mind, the accident files of 3024 road accidents investigated by OTS (op. cit) investigators in Nottinghamshire and Thames Valley areas between

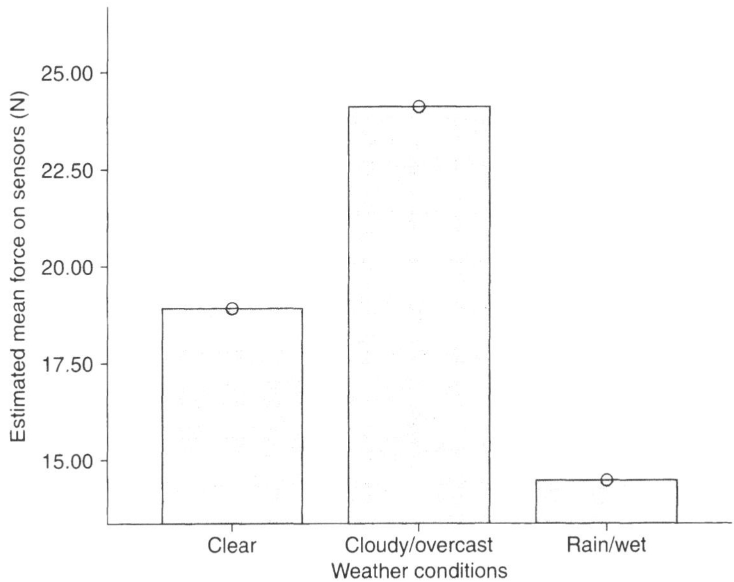

Figure 3. Brake-pedal force against weather conditions.

2000 and 2006 were examined. In comparison to conventional accident data, OTS offers much more depth and detail. OTS accident data for example, includes information on events antecedent to and concomitant with the accident for which investigators can also determine the relative influence on accident causation. OTS data has been compared against National Accident Statistics (STATS 19, op. cit) and validated as representative of the UK road accident data (Hill, Thomas, Smith, & Byard, 2006).

For the purpose of the current article, not only the weather conditions at the site of the accident but also the estimated contribution of weather related visibility and traction to the incident will be examined.

On average, 120 hours of sunshine/month (approx 33% of daytime), 15 days/month of rainfall over 1 mm (approx 50% of time) and just over 1 day/month of snowfall (approx 3% of time) were recorded in the UK region of the Midlands for the period of study .(Meteorological Office, 2008).

Evidence that different meteorological conditions were associated with accident occurrence were heavily skewed by weather event frequency. Conditions salient at the OTS accident sites were identified as "clear" in 59.9% of all cases, wet/rainy weather was present at 13.3% of accident cases and about 20% were classified as "other" or "unknown". Blinding sunlight, windy and foggy conditions were each recorded in about 1% of cases. In terms of meteorological events contributing towards accidents, slippery roads were recorded in 2.1% of accident cases, dazzling sun impaired vision in 0.3% of cases, whilst rain, sleet, snow and fog affected vision affected in 0.1% of accident cases.

The OTS evidence suggests then that while adverse meteorological conditions are experienced relatively infrequently by the UK driver, a lack of a linear association between weather event, accident occurrence or causal link was noted. Whilst no direct behavioural data were available, it would seem likely that behavioural adaption had occurred, since otherwise the shifts in objective risk through changes in visibility and traction would have been realised through the accident data available.

General Discussion

To summarise, in order to investigate behavioural response (if any) to changes in physical risk, driver behaviour – a socially normalized physical risk taking activity - was examined against variable levels of threat afforded by meteorological condition. Data suggest that behavioural adaptation influenced through affective state and perceived risk contingencies, can be realised through mean speed, headway and braking, that it occurs across weather types and that such adaptation may mediate objective accident risk under reduced visibility and traction (Wilde, 1988).

Examination of the simulator data however also suggests that consideration ought also be given to participant mood state and subjective perception of risk contingencies. Some instances of driver adaptation has previously been regarded as 'irrational' since the risk of collision rises – such as increasingly close-following in fog (Ayres *et al.*, 2001; Naatanen & Summala, 1976). Such adaption may be explicable through a discrepancy in the 'safe' and 'risky' labels applied to behaviours by experimenters and participants (Haigney *et al.*, 1997; White and Jeffrey, 1980). Participants experiencing reductions in the availability and contrast of visual cues may regard close following as a 'safety' behaviour, maximising the number of cues available to them (Haigney *et al.*, 1995).

Behavioural adaption may also be evident to some extent in the closed track study, with heavier braking indicative perhaps of the realisation of a performance benefit through greater traction relative to wet conditions. This is in line with the theoretical framework of behavioural adaptation (Fuller, 1984; Naatanen & Summala, 1976; Wilde, 1988).

The OTS accident data suggest that even with objectively decreased vehicle control under wet, foggy and icy conditions, most road accidents take place during clear weather conditions even if the meteorological condition frequency controlled for. This again could agree with behavioural adaption by potentially reducing perceived risk against target (Wilde, 1988) although a number of caveats warrant further investigation. Whilst accident data statistics provide an image of the real-world road safety, behavioural adaption has to be inferred through discrepancies between observed and expected accident occurrence and these are potentially distorted by inclusion and omission error (Jacobs & Aeron-Thomas, 2000), the possibility of greater number of vehicles on the roads in fair weather, with further confounding from time of year, time of day etc.

It is apparent in these studies that the relationship between the affective, the behavioural and their relationship with error (collision) are suggestive of adaptation but that such relationships are not necessarily linear, or on the face of it 'rational'. Consideration needs to be made of numerous, possibly associated confounding variables which may distort behavioural response overall or second-by-second, skewing the results and/or the statistical analysis – or in the case of a lack of affective data collection – our ability to interpret it. The weight with which various data types contribute towards the debate surrounding adaption ought also be discussed – from the contention that the determination of behavioural adaption with accident statistics is akin to 'proving the null hypothesis' (Wilde, 1988) and through behavioural measures akin to 'proving the existence of God'. (Joubert, 1985)

These issues merit further investigation and are the focus of an on-road driver response research programme throughout 2009–10.

References

Adams, J.G.U. (1988). Risk homeostasis and the purpose of safety regulation. *Ergonomics* 31:407–428

Allen, T. W., Lunenfeld, H., & Alexander, G.J. (1971). Driver information needs. *Highway Research Record, 366*, 102–115

Andrey, J. & Knapper, C. (1993). Weather hazards, driver attitudes and driver education. *Report to the co-ordinator of Highway Safety Research Grant Program*, Safety Planning and Policy Branch, Ministry of Transportation, Ontario

Ayres, T. J., Li, L., Schleuning, D., & Young, D. (2001). *Preferred time-headway of highway drivers*

Blana, E., & Golias, J. (2002). Differences between vehicle lateral displacement on the road and in a fixed-base simulator. *Human Factors: The Journal of the Human Factors and Ergonomics Society, 44*, 303–313(11)

Brisbane, G. (1992). Driver behaviour during periods of restricted visibility. *Proceedings 16th ARRB Conference* 16(4) 313–329, Perth Western Australia, 9–13 November.

Co-operative Crash Injury Study (2009, December 2) Improving car safety through accident research. Retrieved from http://www.ukccis.org/

Commission Internationale de l'Eclairage (1992) Technical report: *Fundamentals of the visual task of night driving*. CIE Central Bureau, Kegelgasse 27 A-1030 Vienna, Austria

Coolican, H. (2009). *Research methods and statistics in psychology* (5th ed.). London: Hodder Education

Department for Transport (2009) Accidents and casualties. Retrieved from http://www.dft.gov.uk/pgr/statistics/datatablespublications/accidents/

Edwards, J. B. (1998) The relationship between road accident severity and recorded weather. *Journal of Safety Research* 29(4) 249–263

Fuller, R. (1984). A conceptualization of driving behaviour as threat avoidance. *Ergonomics, 27*(11), 1139–1155

Gibson, J. J., & Crooks, L. E. (1938). A theoretical field-analysis of automobile-driving. *American Journal of Psychology*, *51*, 453–471

Gkikas, N., Hill, J., & Richardson, J. (2008). Getting back to basics: Using road accident investigation to identify the desirable functionality of longitudinal control systems. In D. de Waard, F. O. Flemisch, B. Lorenz, H. Oberheid & K. A. Brookhuis (Eds.), *Human factors for assistance and automation.* (pp. 203–216). Maastricht, the Netherlands: Shaker Publishing

Gkikas, N., Richardson, J., & Hill, J. (2009a). Towards a driver-centred brake assist system. *Braking 2009,* St Williams College, York, UK. 85–92

Gkikas, N., Richardson, J. H., & Hill, J. R. (2009b). Exploitable characteristics of driver braking. *Proceedings of the 21st International Technical Conference on the Enhanced Safety of Vehicles (ESV)*, Stuttgart. (Paper Number 09–0247)

Gkikas, N., Richardson, J. H., & Hill, J. R. (in press). Reset to zero and specify active safety systems according to real world needs. *Journal of Transportation Engineering*

Han, D., & Yi, K. (2006). Validation of a virtual test track for intelligent driver assistance systems. *Proceedings of AVEC '06: The 8th International Symposium on Advanced Vehicle Control,* Taipei, Taiwan. 403–408

Haigney, D. E.; Hoyes, T. W.; Glendon, A. I. and Taylor, R. (1995). Objective risk, perceived risk and the compensation process. *Presented to the Ergonomics Society*, April

Haigney, D.E.; Kennett, C. and Taylor, R. (1997). Assessing the determinants of behaviour: Utility vs. engineering. In *Contemporary Ergonomics 1997*, Edited by S.A.Robertson. Taylor and Francis: London

Harrison, G., Lau, M. & Elisabet Rutstrom, E. (2009). Risk attitudes, randomisation to treatment and self-selection into experiments. *Journal of Economic Behaviour & Organisation* 70(3), 498–507

Hill, J. R., Thomas, P. D., Smith, M., & Byard, N. (2006). The methodology of on the spot accident investigations in the UK. *Enhanced Safety in Vehicles 2001*, Amsterdam

Jacobs, G. D., & Aeron-Thomas, A. (2000). *A review of global road accident fatalities.* Retrieved March 28, 2007, from http://www.transport-links.org/transport_links/filearea/publications/1_771_Pa3568.pdf

Jamson, A. H., Westerman, S. J., Hockey, G. R. J., & Carsten, O. M. J. (2004). Speech-based E-mail and driver behavior: Effects of an in-vehicle message system interface. *Human Factors: The Journal of the Human Factors and Ergonomics Society*, *46*, 625–639(15)

Joubert, P. 1985. Comment. In L. Evans and R.C.Schwing (eds.) *Human Behaviour and Traffic Safety.* New York: Plenum Press.

Lazarus, R.S. and Folkman, S. (1984) *Stress, appraisal and coping.* New York: Springer

Lee, J. D., & Strayer, D. L. (2004). Preface to a special section on driver distraction. *Human Factors, 46*(4), 583–586

Matthews, G. (1992). Mood. In *Handbook of Human Performance.* Volume 3. ISBN 0-12-650353-2

Matthews, G.; Jones, D. M. & Chamberlain, A.G. (1990) Refining the measurement of mood: The UWIST Mood Adjective Checklist. *British Journal of Psychology* 81, 17–42

Meteorological Office (1969). *Observers handbook*. HMSO ISBN 0-11-400329-7

Meteorological Office. (2008). *Regional climate: Midlands*. Retrieved September 17, 2009, from http://www.metoffice.gov.uk/climate/uk/mi/

Michon, J. A. (1985). A critical view of driver behaviour models; what do we know, what should we do. In L. Evans, & R. C. Schwing (Eds.), *Human behaviour and traffic safety* (pp. 485–520) Plenum Press

Moore, R. L. and Cooper, L. (1972). Fog and road traffic. *Transport and Road Research Laboratory* Report LR 446

Naatanen, R., & Summala, H. (1976). *Road-user behavior and traffic accidents*. Amsterdam ; Oxford: North-Holland Publishing Co. etc.

Rosnow, R. L. (2002) The nature and role of demand characteristics in scientific inquiry. *Prevention & Treatment* 5(1)

Sridharan, L. & Greenland, P. (2009) Editorial policies and publication bias. *Archives of Internal Medicine* 169(11) 1022–1023

Stigson, H., & Hill, J. (in press). Use of car crashes resulting in fatal and serious injuries to analyse a safe road transport system model and to identify system weaknesses. *Traffic Injury Prevention*

Summala, H. (1996). Accident risk and driver behaviour. *Safety Science, 22*(1–3), 103–117

Sumner, R.; Baguley, C. and Burton, J. (1977). *Driving in fog on the M4*. Transport Research Laboratory Digest SR281. Department of Transport: Crowthorne, Berkshire

Tenkink, E. (1988). Lane Keeping and speed choice with restricted sight distances. In *Road User Behaviour*, Edited by Rothengatter and de Bruin. New York: Van Gorcum

van Winsum, W. (1998). Preferred time headway in car-following and individual differences in perceptual-motor skills. *Perceptual and Motor Skills, 87*(3 PART 1), 863–873

van Winsum, W., & Brouwer, W. (1997). Time headway in car following and operational performance during unexpected braking. *Perceptual and Motor Skills, 84*(3 PART II), 1247–1257

White, M. E. and Jeffrey, D. J. (1980). *Some aspects of motorway traffic behaviour in fog*. Transport Research Laboratory Digest LR958, Department of Transport, Crowthorne, Berkshire

Wilde, G. J. S. (1988). Risk homeostasis theory and traffic accidents - propositions, deductions and discussion of dissension in recent reactions. *Ergonomics, 31*(4), 441–468

COGNITIVE WORK ANALYSIS OF SIGNALLING PROTECTION FOR RAIL ENGINEERING

David Golightly, Brendan Ryan & Sarah Sharples

Centre for Rail Human Factors, Faculty of Engineering, University of Nottingham, Nottingham, NG7 2RD

Cognitive Work Analysis is a means to represent the complexity of cognitive work in domains such as rail engineering, in order to facilitate the introduction of new technology or new procedures. We present the application of two stages of Cognitive Work Analysis (Work Domain Analysis and Control Task Analysis) to gain an understanding of the complexities of signalling protection for short-term rail engineering work. We discuss our experiences in using these stages of Cognitive Work Analysis, and the potential benefits.

Introduction

A critical component of safe rail engineering work is to use signalling to protect track workers. There are specific challenges with shorter-term engineering and maintenance work, or brief inspections, as they often take place during the day to fit around the running train service (as opposed to longer pieces of work which usually involve a full 'possession' of the line by an engineering team). A signaller will protect track workers by disabling the capacity to set routes through the area where the work is taking place. On Entry and Exit (NX) signalling this is achieved by placing physical 'collars' on the buttons, corresponding to signals, used to set a route. On VDU-based signalling this is achieved by placing virtual 'reminders' on the appropriate points. In both cases, being unable to set a route ensures the signals surrounding the work remain at 'danger' (i.e. show 'red'). As well as visually instructing approaching trains to stop at the signals, accidentally passing a signal at danger (a 'SPAD') will trigger alarms and failsafes.

In practice, this protection involves pre-arrangement of time and location, and close coordination between the signaller and track workers to ensure that workers can operate safely within the required window of opportunity with no disruption to the current service. As such, protecting track workers is an example of a socio-technical system where decision making and action is distributed and coordinated across actors and artefacts (Walker *et al.*, 2006). It is critical to have a full understanding of the complexity of such a system, in order to understand the human factors implications of procedural or technical change (Wilson *et al.*, 2009). Without this full understanding, there is a the risk that a purely procedural review will fail to take into account the range of contextual, social and technical factors that may influence safe and effective performance. Likewise, technical change to supporting systems (e.g. the introduction of open channel communications) needs to consider the sometimes indirect implications for tasks such as track-worker protection.

It is valuable, therefore, to have a systematic approach for understanding and representing both the scope and detail of a complex work system. Cognitive Work Analysis (CWA) (Vicente, 1999) is such an approach. CWA draws on five forms of analyses, each with related representational forms. These forms enable the expression of the broad context of the work and work domain, the cognitive strategies used by agents within the domain which are in turn linked to cognitive processing modes (based on Rasmussen's 'Skills, Rules and Knowledge' processing modes approach), and analyses of social coordination and competency. There are now a wide range of examples in domains such as nuclear/process control, Air Traffic Control and military operations. Rail-related examples include train driving (Jansson *et al.*, 2006) and more, recently, describing rail control (Millen *et al.*, 2009). CWA results in a formative model of work that is used primarily for design (e.g. the design of ecological interfaces for process control (Vicente, 1999)), but for other purposes such as training needs analysis, procurement and evaluation.

While the full details of CWA are available elsewhere (e.g. Vicente, 1999; Lintern, 2009) we examine the first two stages – Work Domain Analysis (WDA) and Control Task Analysis (CTA) – to assess their value for signalling protection. We had faced some specific issues in our work on short-term protection to fully understand all of the elements of the socio-technical system. Feedback from Subject Matter Experts to our analysis of observation had shown that there were many subtle procedural and technological factors, for example whether reminders could be placed in advance of a request for protection, that might have major implications for safe and effective work. Also, the priorities of signallers (to keep a safe but running service) were not always the same as those tasked with improving track work performance (to carry out work as quickly and safely as possible). We wanted a modelling approach that would allow us to represent our current understanding of both the overall work domain, and the observable (and potentially cognitive) processes that signallers were engaging in. This could be used for future validation and discussion with SMEs. WDA and CTA appeared to be suitable candidates, and we wanted to test whether this was the case.

As well as exploring the utility of WDA and CTA, we wanted to understand whether they could be carried out quickly (days, rather than weeks or months) through self-teaching. We also wanted to capture what decisions would need to be made, and what the potential limitations of the analyses might be. While the work presented is not strict hypothesis testing, we hope to offer qualitative support for applying part of the CWA approach in the given domain, as well as some insight into potential issues that face the analyst.

Data collection and scope

Before presenting details of the analyses themselves, we briefly cover the source and scope of the data used. This came generally from a number of programmes of work to understand signaller expertise (funded as part of EPSRC RRUK B6) and a specific programme to examine short-term protection methods (funded by Network Rail). This included around 30 observations of protection arrangements within signal boxes (approximately half VDU-based, and half NX), using a structured codification and analysis approach (Balfe *et al.*, 2008). This analysis gave us sequences of actions linked to events taking place in the signal box, and more generally within the short-term protection work. This was augmented with subjective observer comments at the time, by SME discussion before, during and after the observations, and also by reviewing relevant procedural documentation (RSSB, 2007).

Work domain analysis

WDA has a track record in its own right for supporting system design and evaluation (e.g. Naikar & Sanderson, 2001). The approach offers a means to understand the scope of the work being analysed over five levels of abstraction from high-level domain purpose through to the physical artefacts used for carrying out the work. The core of WDA is the Abstraction Hierarchy, with the example for short-term protection presented in Figure 1. Lack of space prevents a full exposition of what each of the nodes means, though it is hoped that the majority are self-evident. Importantly, WDA captures the scope of the work domain, rather than expressing any particular task sequence. Therefore it is not necessary to evaluate a system in terms specific scenarios. Instead the WDA actively supports design and evaluation in terms of domain constraints, and therefore in a manner that should accommodate unanticipated events (Vicente, 1999).

An important first step of our WDA was to understand the purpose of each level, as there is some variety in labelling across different examples of the Abstraction Hierarchy in the literature. In the end, we elected to use the naming strategy of Lintern (2009) as this proved to be most useful in highlighting what should lie at each level. That said, we found it useful to add some additional terms, emphasising that the Domain Purpose was the 'rationale' for the work domain, the Domain Functions were what needed to be 'achieved' with the system, and the physical form were the 'resources' used (this is indicated on Figure 1). The notion of 'affordances' was particularly useful in establishing the means-end relationships across the hierarchy. As an example, 'Shared Displays' and CCF (Control Centre of the Future) 'affords' (or 'supports') 'wide area views of the network', which in turn facilitates the function of 'traffic regulation and route setting'.

The actual process of building the hierarchy involved iteration between the levels, rather than a sequential bottom-up or top-down approach. This would often lead to 'weeding out' nodes that did not have links to either a higher or lower level. For example, a Domain Function of 'Protection Documentation' was removed because it did not have direct links to the next level – in practice, this may be a sub-function of Implementation and Cancellation of protection, but we have not, as yet gone to this level of detail.

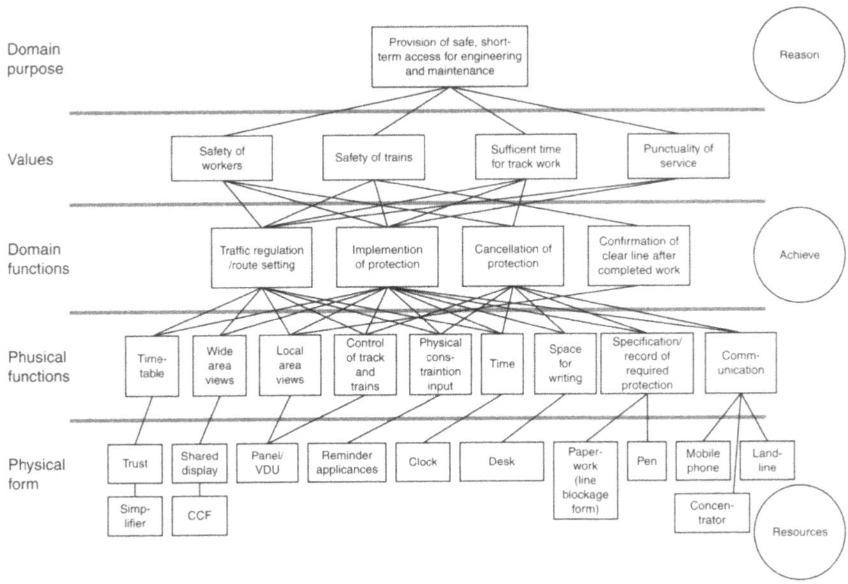

Figure 1. Abstraction Hierarchy for short-term protection arrangements.

The potential value of using this approach is that the hierarchy allows us to express not just the work itself but, importantly, some of the higher level contextual constraints of the work, and we have seen examples of this in the hierarchy that we have developed. The inclusion of Priorities and Values brings a dimension that would not be apparent just from a traditional task analysis. In practice, we observed that not all protection could be granted by the signaller, because they were often requested at a time when the surrounding traffic was so great that it would not be possible to carry out the track work without causing delay or compromising safety. However, procedural documentation only requires the signaller to check that the area where the work is to take place is clear before the protection is put in place. By giving making explicit the priority of 'punctuality of service', we can make it apparent that a future system ensures that protection arrangements are planned to make better use of windows of opportunity in the schedule. The inclusion of 'traffic regulation' within domain functions also serves to emphasise that the signaller is actively running the rest of the rail service while the protection is in place. On the other hand, the function of confirmation of a clear line, where the signaller monitors the first train through the area where the work has taken place to check track circuits, is not immediately observable. As long as there is a problem detected with the functioning of the track circuits, there is no outward indication that this function has been performed, and was not commented on by signallers. The function is only apparent by reading through procedural documentation (in this case, for T2 (RSSB, 2007)), but we can now place this within the hierarchy to make it explicit and visible for future work using this hierarchy as a reference.

While much of the hierarchy is technology-independent, the last two levels (Physical function and Physical objects) are intended to reflect the physical equipment and their configuration. We decided therefore to only to include physical objects i.e. the equipment used by the signaller, as this is the most common implementation of the analysis found in the literature. However this means we were not able to include all information. For example, we did not find a means at this level of including physical configuration information, though this was important as signal box layout had an important influence on whether the signaller had line-of-sight of their workstation while completing paperwork, and therefore was able to keep awareness of panel status. The abstraction hierarchy is useful as a placeholder, or almost an *aide memoire*, Another example is information about social organisation, and roles. We anticipate this will be expressed in the Social and Organisation analysis stage of CWA but as this is a stage we have yet to tackle, it is yet to be explicitly stated in the analysis.

Another important decision is the scope of the analysis. We made an explicit decision to only include nodes that were immediately relevant and visible to the signaller in the signaller environment. So, for example, we did not consider the role of signals, track etc., at a physical level – instead, this is represented to the signaller via their panel, which is included in the analysis. This is a practical strategy to limit our analysis to those areas we had ready access to, and were deemed to be most relevant. This is, to an extent, a judgement decision on the part of the analyst, and a constraint of attempting to do the analysis in a relatively short period of time, but may mean that critical information is missed. Validation is obviously crucial to understanding the implications of this – a point we discuss at the end of the paper.

Control task analysis

CTA is a means for representing the activity required to achieve the functions defined in the WDA. In strict practice CTA comprises two stages, Work Context Analysis and Decision Ladders. Work Context Analysis maps domain functions, as defined in the WDA, against different work situations, as differentiated by time, location or conditions (Naikar *et al.*, 2006). For example, the function of cancelling protection may involve different decisions depending on whether it is planned or unplanned, and therefore may require a unique instantiation of a Decision Ladder, defined at the next stage. For simplicity, we have omitted this first phase from our analysis as our example, ‘protection implementation’, should be consistent over situations.

In the second phase, control of a function is mapped against a Decision Ladder. The Decision Ladder is a template (shown in Figure 2) for defining potential knowledge states and intervening cognitive processes (states of knowledge are shown in ovals; cognitive processes are shown in rectangles), required to complete a Domain Function. The left leg represents the stages of attention and awareness, the right leg represents planning and action, and the top of the template covers instances that require evaluation of options against current system state and system goals (Vicente,

1999). We have populated the template with the function of protection implementation (text in italics). The signaller would be alerted to the need to arrange protection through an incoming call (Stages 1 and 2). Observation and discussion with signallers showed the importance of studying the surrounding area and upcoming traffic to ensure the protection can take place (Stages 3–7). Normally, they would accept (Stage 11), and then protection arrangements would be confirmed, the documentation completed and the reminder appliances put in place (Stages 12–16). In some circumstances the situation may be more complex. For example, if the protection request lies at the edge of the signaller's area of control, they may have to engage in conscious reflection and utilise wide area views and the timetable, to decide on which option to take (Stage 8, 9 and 10).

It is important to stress that the decision ladder is a template rather than a sequential model. It is unlikely that an expert would go through all stages of the decision ladder, and is likely to jump from one process to a state of knowledge, based on experience. Another form of alternative path is that it is not necessary to start at the bottom left, or to exit out of the decision ladder at the bottom right. There might be other reasons for the operator to start their chain of control tasks and cognition, and that chain may not end in explicit action. We can see illustrative examples of this in scenarios typically found during protection arrangements. First, it will not always be necessary for the signaller to go through explicit evaluation of the system state in order to decide on their goal state. If traffic is very light and/or the protection is required on a section of track that is used infrequently, then the signaller is likely to grant the protection without further reflection (moving from Stage 5 to Stage 11). Another variant is that it may often be a trivial task to identify the protecting arrangements from the track workers request, thus making it possible to go from the goal state (Stage 11) to formulating a procedure (Stage 14). However, there may be complications, such as the track work taking place within 200m of the signals, which requires the signaller to also ensure the preceding signal is placed at danger, which may require a more complex formulation of the task (Stages 12 and 13).

Using CTA in this way allows us to build up an understanding of a range of steps that a signaller may go through to complete the function of setting protection. The analysis also serves as a template for further investigation of different factors that might change a trajectory through the decision ladder – for example, what kind of task conditions (or levels of experience) cause the signaller to engage in more active reflection of possible goals against performance criteria (Stage 8, 9 and 10) rather than simply accepting (or declining) protection. The main issue we faced in populating the Decision Ladder comes in using observation data to tease apart what are often implicit or highly procedural aspects of the control tasks, and placing them within the framework. Because very low level activity is explicit in the template (Stage 1–3, Stage 14–16) it forces the analyst to address these stages even though, in practice, they may not be easy to distinguish. For example, knowledge of task (Stage 13) and knowledge of procedure (Stage 15) may be tightly bound together and, as with the Abstraction Hierarchy, there is something of a judgement call on the part of the analyst to separate the two. This is also an issue with the use of leaps and

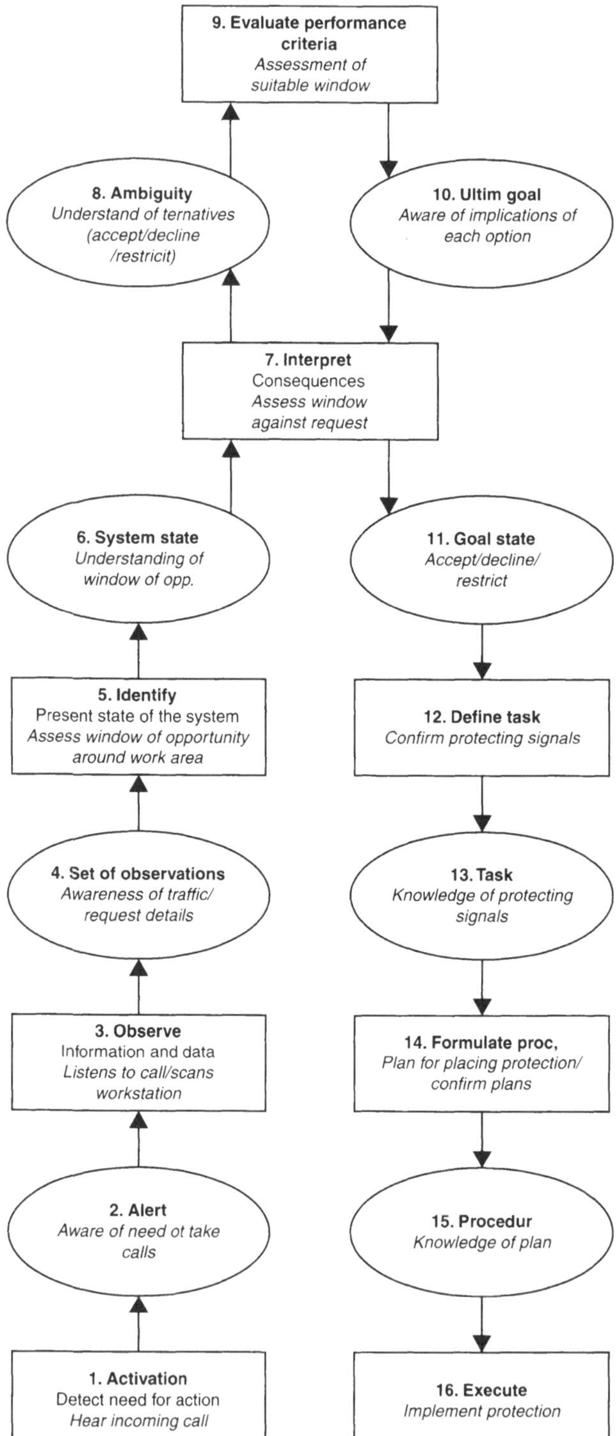

Figure 2. Decision Ladder for Implementing Protection.

shunts, as it is possible that an activity that appears as one of these shortcuts may only do so because we have been unable to tease apart the relevant steps. With that point noted, the use of the Decision Ladder in this (so far) relatively approximate manner gives us a theory-based representation of the activity of the signaller which can discuss and refine with SMEs. Ideally, with more time and resource a number of these judgement calls can be explored, verified or refuted with real data, for instance by using Critical Decision Method.

Concluding remarks

Our overall intention was to examine the value of two stages of CWA to help understand short-term protection arrangements in engineering work. One aim was to understand the general utility of the approaches, and we believe the examples of how to structure protection both in terms of the WDA (e.g. the value of specifying priorities) and CTA (e.g. structuring different paths for decision making) illustrate this. Other aims were to carry out the analyses relatively quickly, and to self-teach. The analysis presented here has taken around two weeks (not including the time for data collection) and in practice, we found CWA to be a much more approachable technique than might be first thought. In fact, applying CWA to a real task proved an essential step to understanding the analyses and their underlying principles.

As well as the specific issues raised against WDA and CTA, some general issues remain. First, validation is not a trivial task. While we have validated the interpretations of the data used in the analyses, it is an ongoing task to validate the analyses themselves. SMEs are not necessarily familiar with the CWA approach, and the risk is that by changing the analyses into a more accessible representation (e.g. a scenario) some of the power is lost. Also, because there are a number of judgement calls on part of the analyst, there is a question of whether we are better off with an approximate model, or whether this carries the risk of leading to design decisions that are erroneous. We have yet to use our analysis in 'front-line' human factors work, but our initial conclusion is that even as an approximate model, these analyses serve as a useful reference point for clarifying the constraints of rail engineering and similar socio-technical systems, and we expect that validation and use of CWA, as a secondary representational form for now, will highlight how much of the analysis is subjective interpretation, and how crucial a factor this is.

References

Balfe, N., Wilson, J.R., Sharples, S. and Clarke, T. (2008) Structured Observations of Automation Use. In Bust, P (Ed.) *Contemporary Ergonomics*, 552–557

Jansson, A., Olsson, E. and Erlandsson, M. (2006) Bridging the gap between analysis and design: improving existing driver interfaces with tools from the framework of cognitive work analysis. *Cognition, Technology and Work*, 8 (1), 41–29

Lintern, G. (2009). *The Foundations and Pragmatics of Cognitive Work Analysis: A Systematic Approach to Design of Large-Scale Information Systems.*

Available online at http://www.cognitivesystemsdesign.net/home.html [accessed 22/09/09]

Millen, L., Sharples, S. C., Golightly, D., and Balfe, N. (2009) The application of Cognitive Work Analysis in Rail. To appear in *Proceedings for the Conference in Rail Human Factors, 2009, Lille, France*

Naikar, N. and Sanderson, P. (2001). Evaluating Design Proposals for Complex Systems with Work Domain Analysis. *Human Factors*, 43 (4), 529–542

Railway Safety and Standards Board (2007) Rule Book Modules (GE/RT8000/T2 and GE/RT8000/12). Available online at http://www.rssb.co.uk/rgs/rulebook/rulebook_modules.asp [accessed 22/09/09]

Vicente, K. (1999). Cognitive Work Analysis: Toward safe, productive, and health computer-based work. Mahwah, NJ: Lawrence Erlbaum Associates

Walker, G. H., Gibson, H., Stanton, N. A., Baber, C., Salmon, P. and Green, D. (2006) Event analysis of systematic teamwork (EAST): a novel integration of ergonomics methods to analyse C4i activity. Ergonomics, 49 (12-13), 1345–1369

Wilson, J.R., Ryan, B, Schock, A., Ferreira, P., Smith, S. and Pitsopoulos, J. (2009) Understanding the safety and production risks in rail engineering planning and protection. *Ergonomics,* 52 (7), 774–790

AN OPTIMISED PROCESS FOR THE DEVELOPMENT OF A PERFORMANCE-CRITICAL CONTROL DEVICE

M.B. Edwards

Bombardier Transportation Passengers Division,
Derby, DE24 8AD

The paper presents a rigorous, industry proven, process to efficiently develop a performance-critical device to meet complex design requirements. This process was applied by the Bombardier Transportation human factors team on recent rolling stock projects during the design of a performance-critical control device, the Traction Brake Controller (TBC). Through the review of standards and human factors good practice, design requirements were identified which required a development method incorporating a range of human factors techniques. The methods used are presented, advantages and disadvantages discussed and lessons learnt summarised including the large influence of small design changes on comfort, a clear relationship between comfort and mis-use, the need for testing with a full cab mock-up to identify issues associated with integration and the importance of an iterative design process.

Introduction

The Bombardier Transportation (BT) human factors team, working with industrial designers, are experienced in applying human factors within new Rolling Stock design. This is applied primarily within the design of the cab and saloon, although other areas are included as required by the needs of the project, for example maintenance and signalling. Given the large scope of rolling stock design projects, it is often necessary for this work to be done in close collaboration with the Customer.

The process presented in this paper has been applied during two recent rolling stock projects (design and build) executed by BT, specifically during the design of the Traction Brake Controller (TBC) used to control train speed. The TBC is considered a performance-critical design as accurate control of train speed is required for safe operation of the train and for the train to perform efficiently in terms of journey time. As such, a train with a non-working TBC cannot enter service.

Both projects possessed a number of similarities which enabled the approach used to develop the TBC to be shared between each project. Similarities included the use of a sit/stand cab and identical target population user groups (current 5th percentile British female to 2049 95th percentile British male). Operational and engineering limitations also meant the location of the TBC had to be fixed on one side of the operator, mounted on the control workstation. This is in line with many existing

Workstation mounted Cab seat mounted

Figure 1. Example Traction Brake Controllers found on existing rolling stock.

rolling stock where the TBC can be found either mounted on the control workstation, or on the arm of the cab seat (see Figure 1).

Differences between each project included the design requirements for the physical cab design and the intended period of use for the TBC. The latter was determined by the signalling infrastructure. On one project the TBC was expected to have low levels of use due to automatic train control, the maximum expected period of TBC use being 1½ hours driving at constant speed with no station stops. The other project was designed to have relatively long periods of use with the train operated mainly, but not exclusively, in manual modes. The longest period of TBC use in this case was 2 hours 50 minutes with fine speed control required for 78 station stops. Whilst these differences resulted in tailored design solutions being required for each project, the same robust development method was used in each case.

The paper aims to present an optimised method for developing a performance-critical device such as a TBC. This is done through critical analysis of the method followed on two projects conducted by the BT human factors team which allows for the identification of lessons learnt and an optimised design development method.

Initial design requirements

To begin the TBC design, a number of high level TBC design requirements were identified through a review of relevant industry standards, human factors good practice and the project requirements:

a) Low levels of user discomfort from both sitting and standing operating positions
b) Low risk of injury e.g. RSI
c) The provision of a high degree of control over train speed when used in conjunction with the feedback on train speed provided by the speedometer and external visual cues
d) The incorporation of a deadman's safety device (DSD) to disable the control should the operator be incapacitated

e) The prevention of mis-use through design
f) Product strength/durability, given the safety and performance critical nature of a TBC
g) A fixed TBC location within the cab

These requirements were used as the initial starting point to develop a design that addressed each requirement.

Initial design development

The design methodology chosen was to derive lower level requirements for each of the high level requirements already defined and then design against these requirements. This was performed by reviewing subject specific human factors literature (e.g. Kroemer *et al.*, 2003), industry standards and obtaining input from Subject Matter Experts (SMEs) on relevant operational and technical considerations. This allowed a design specification to be produced with sufficient detail to allow an initial TBC to be developed.

Requirements in the design specification included the need to minimise the vertical movement and maximise the horizontal travel to ensure ease of use, minimisation of the impact on access to other cab equipment and the importance of a neutral wrist posture. Recommendations found in literature were used to generate specific requirements including handle length (95th percentile hand breadth, 95 mm – Open Ergonomics, 2000), diameter (22–38 mm – Tilley and Dreyfuss *et al.*, 1993) and operation forces (<10N – London Underground Ltd, 2007). However, given the combination of movements required to set the DSD and engage traction, these were all caveated with the need to validate them with testing. A key decision at this stage was the operation of the DSD. Various types of DSD were reviewed as part of the literature review and a twist-type DSD, requiring the TBC handle to be rotated away from the operator for the DSD to be set, chosen as optimal for preventing mis-use.

The design specification and iterative model production and evaluation were used to enable the rapid development of a design against the requirements. The modelling performed included 3D Computer Aided Design (CAD) modelling of the cab interior and user reach envelopes to allow for the TBC to be integrated into the cab design at an optimal position for the target user population. The TBC handle was then designed against the specification and a range of low fidelity handle mock-ups produced and evaluated in user trials. The production and evaluation of the models was performed iteratively to allow the design to develop. This allowed for a variety of different handles to be assessed to determine the optimal handle shape [see Figure 2].

Evaluation of the handle shapes was done in user trials performed using the low fidelity models in a wooden cab mock-up to allow for representative location with respect to the operator [see Figure 2]. Office staff were used as trial participants as they could be selected according to body size, a factor considered more important than operational experience. Participants were asked to move the TBC over its

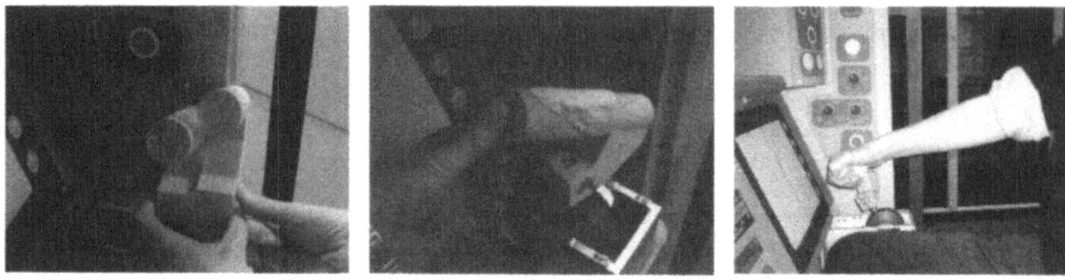
(a) (b) (c)

Figure 2. TBC handle mock-up variations in (a) grip angle and (b) a palm rest variation and (c) in use.

full range of movement in standing and seated postures. Subjective questions were asked and used to rank the prototype handles in terms of favourability. During these evaluations, an iterative approach was utilised for the design development was employed to develop the design based on the subjective responses in addition to the rankings. The iterative nature of these trials was found to be critical for the development of a suitable handle as small changes in the handle design often had a large influence on user comfort.

This process provided a TBC design suitable for manufacture and installation on a pre-series train built for test purposes. Once installed on a train, two problems were identified. These showed the limitations on the testing already performed. It was found to be possible to mis-use the TBC by jamming an item of the correct size between the handle and the cab sidewall, or by slumping against the handle, thereby bypassing the DSD and holding the TBC in a motoring position. This had obvious and serious safety implications. This factor was not highlighted in earlier design development due to the emphasis placed on comfort and reach in the user trials. The forces required to twist the handle to set the DSD, something not included in the early assessments due to the mock-up limitations, was also found to be a problem. This caused discomfort for users holding the handle for long periods of time. To address these limitations, further design development was performed.

Further design development

To address the limitations identified in the work performed, the direction of the twist required to set the DSD was changed to require an operator to twist the handle towards themselves (anticlockwise for a TBC positioned on the operator's right). This change prevents the TBC from being held in the motoring sector with the DSD set by slumping on the TBC, so addressing the problem of mis-use. However, this had an impact on the comfort of the developed handle, as a different hand movement was needed to set the DSD.

To address the problem of discomfort due to force and the changed DSD direction, low fidelity handles were again produced and evaluated iteratively to refine the existing design. The evaluations were performed in a cab mock-up using a prototype TBC unit to which the low fidelity models of the handle were attached. This ensured that operational forces required of the trial participants to set the DSD where representative of the final unit (2.6N), which met the 10N requirement defined earlier in the design process.

The evaluations were performed by human factors engineers from BT and the Customer for each project. This allowed the designs to progress quickly with less than a week between design iterations, with expert knowledge being utilised from BT's previous experience on similar projects and the Customer's knowledge of their own drivers and operations. During this development a clear relationship was found between user comfort and the ease of mis-use. This has been attributed to the inclusion of a DSD that requires the TBC to be operated in a specific fashion. Loosening the requirement for operation in a set manner increased comfort as the operator could vary their position, however this also made it simpler to mis-use.

Once the design had progressed to a stage where BT human factors engineers and the Customer's human factors Engineers were happy that an optimum compromise had been reached between comfort and mis-use, it was taken forward to user trial to provide assurance that the design was suitable and the set design requirements had been met.

Two user trials and a workshop were performed in order to provide a rigorous assessment of the proposed design. The location of the TBC within the cab and user comfort was evaluated using BT office staff sampled against the target population body size (5th female to 95th male) with discomfort assessed using 5-point rating scales. This assessment was performed in a cab mock-up to ensure correct placement of the TBC and external signalling positions were marked outside the mock-up to ensure the participant was aware of the external sightlines they must maintain. This assessment found no discomfort issues due to design of the TBC, when used correctly. Discomfort was noted when the TBC was held using a sub-optimal grip or when gripped too firmly as it was perceived by some that this would provide greater control over train speed. To ensure users are aware of the correct method of operation, training recommendations were produced. The Customer carried out scenario-based testing performed with train drivers simulating TBC use against a video of the train route for each project. This testing activity identified no issues.

A mis-use focus group was run with Customer human factors engineers, SMEs selected by the Customer and train drivers. The majority of the input in the focus group was gained from the train drivers who attempted to mis-use the TBC and bypass the DSD using a variety of methods. It was found that a pragmatic approach to mis-use was needed to identify which methods are risks as they can be easily performed and which methods provide no practical risk due to the difficulty in performing them. This resulted in minor modifications being made to the TBC handle on one project and to the adjacent cab panels on the other [see Figure 3].

Completion of the user trials and the mis-use workshop allowed design assurance to be provided for the TBC on each project. These designs have since been accepted

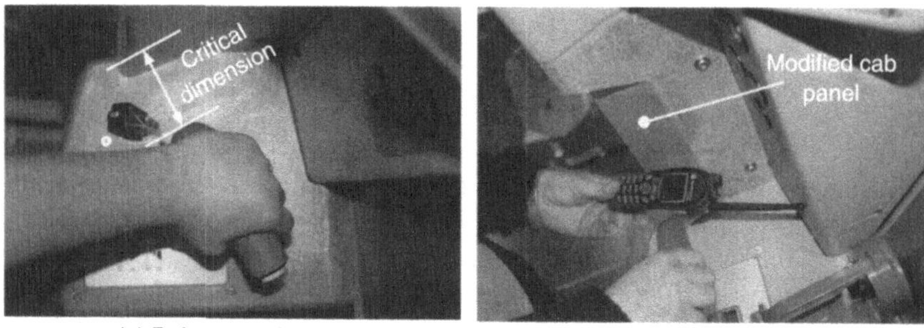

(a) Before modification (b) After modification

Figure 3. Cab sidewall modification to prevent TBC mis-use.

by the respective project Customers and are being installed in the new rolling stock builds.

Lessons learnt

The TBC design development identified a number of lessons learnt, including both the advantages and limitations of the process followed. These should be considered in the development of any comparable safety-critical product.

Determining the optimum TBC location within the cab was performed effectively by modelling the user reach capabilities and joint angles in 3D CAD. Reach to the TBC was not identified as a problem in any of the testing performed. When considering the range of possible user body size and operating postures in a sit/stand cab, careful consideration was required in order for an effective solution to be achieved.

Small changes in handle design were found to have a large influence on comfort. Recommendations found in literature provided a firm foundation for the handle design; however, iterative design was key in developing the handle to be optimal for the context of use. A limitation of the testing performed early in the process was not including a representational operational force. Force requirements had a direct impact on comfort and should be considered as integral to the design of the handle.

A clear relationship was found between comfort and ease of mis-use with changes made to increase user comfort often making misuse easier. In the early design work, mis-use was assessed by the BT personnel. This formed a limitation as the expert knowledge provided by drivers in latter stages showed a much greater likelihood of mis-use than initially considered. Mis-use should be assessed with the users and a carefully considered compromise found between comfort of use and mis-use.

Whilst much of the development work was shared between the two projects, the differing cab layouts resulted in differences in the location of the TBC in relation to the user and the clearance provided around the TBC handle. This required the TBC handle shape to be different in order to design out mis-use whilst providing optimal user comfort.

For the designs produced, the input of iterative testing in a full cab mock-up with representative handle size and operation forces was vital. Without this, many issues would have remained unnoticed until too late in the design process.

Whilst it is possible to spend a large amount of time involving high numbers of participants in trials assessing the design options, the development work employed a more pragmatic approach using quicker trials involving fewer participants. This was found to be an effective method of quickly generating and evaluating designs, particularly in the early stages. This is in line with recommendations given by Nielson and Molich (1990).

A limitation of the process is that no testing was performed with drivers using the TBC in service on either project. As such, all of the testing addressed the TBC being operated in isolation, rather than as part of the larger cab system, e.g. in combination with the speedometer and external speed cues. This limitation cannot be easily avoided given the cost to build a train and commission it for running. As a working TBC is needed for this, changes to the TBC design after the train has been commissioned are undesirable. This limitation is not considered an issue given the full range of target user body size included in all trials, the design process followed, driver input being gained where appropriate and the considerations to the larger cab system made in the final set of tests, for example using signal positions marked outside the cab mock-up.

Generic development method

The work performed allows a generic development method to be proposed for the development of safety-critical control devices, such as a TBC, that is based on industry proven methods and experience. This method considers all influential factors at appropriate points within the design process and aims to provide a design that can be assured from a human factors standpoint as being safe for use by the target user population. The proposed method has four stages and is summarised in Figure 4.

The first stage is the identification of requirements and limitations. This involves a review of project level requirements to identify how the device will be used, the target user population and relevant technical limitations. This guides a review of literature and the derivation of both high and low level design requirements. As the method of operation is a key decision for the design of the device, this should be decided through consideration of the project requirements and literature review. This stage will finish with the production of a comprehensive design specification with measurable design requirements.

The second stage takes the design specification and produces an initial design. This will include modelling of user reach capabilities to ensure optimal integration within the larger system and the design of the physical control interface, for example the TBC handle, against recommendations found in literature in the previous stage. This stage is performed with no SME or user inputs to allow the initial design to be rapidly generated against ergonomic principles.

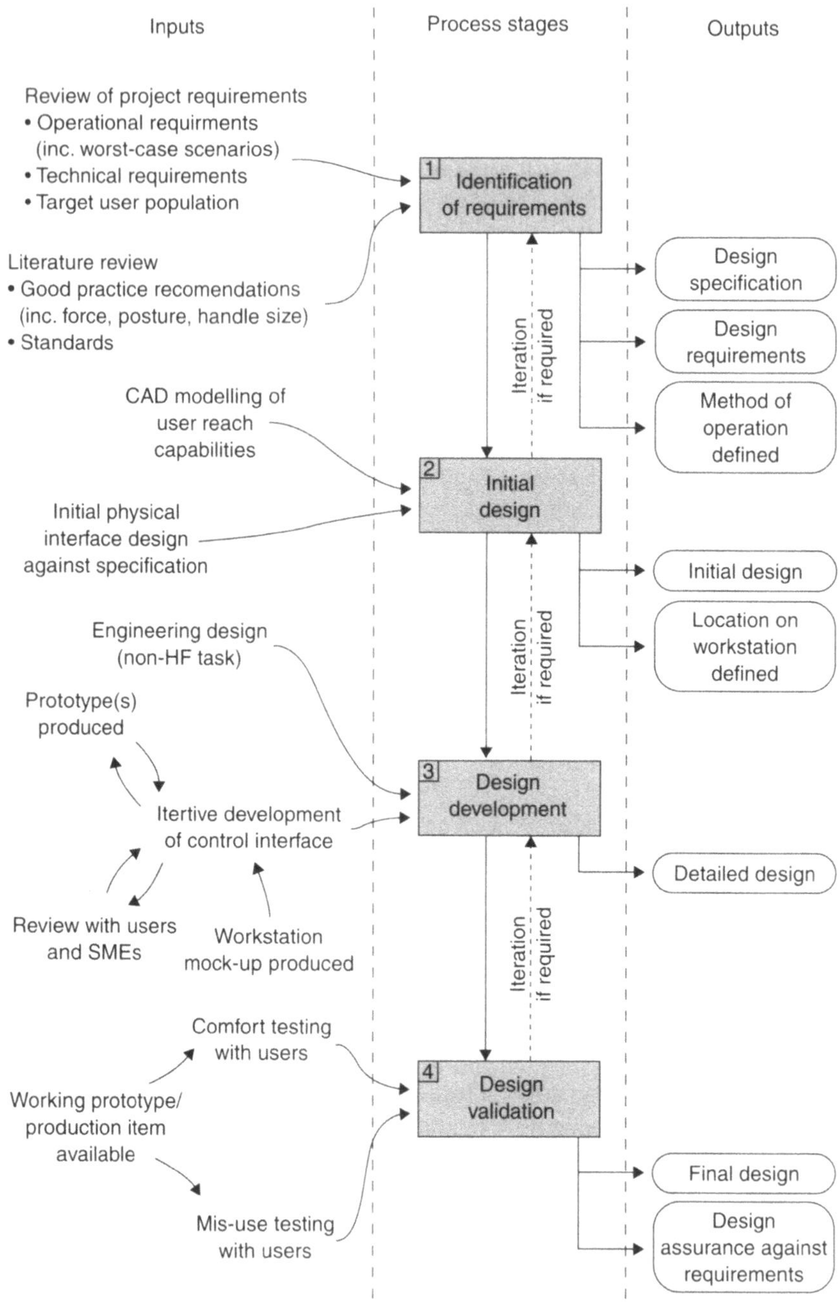

Figure 4. Generic development method flowchart.

The third stage takes the initial design and develops it by reviewing prototypes in a full workstation mock-up with SMEs and end users. Early prototypes can be of limited functionality, however, it is important to incorporate as much functionality as possibility into the models to ensure the testing can be generalised to the final device. It is recommended that assessments are made with small groups of participants. The choice of participants should be made against the purpose of the specific evaluation, for example sampling according to body size to address the full target user population. The output of the third stage is a detailed design of the proposed device from which a prototype can be manufactured that is representative of the final design, both in terms of size and operation.

Once the detailed design has been completed, the design should be validated by user testing against the identified requirements. For systems where no similar device is currently in use, this testing should also include SMEs to ensure that their specific knowledge is incorporated into the assessments. Testing should aim to provide evidence that the requirements identified in the first stage have been fulfilled. The two issues identified as requiring testing in the TBC development described in this paper are user comfort and the risk of mis-use. Satisfaction arguments could be created for all requirements from either this testing, or earlier design work completed.

The final stage provides assurance that the design fulfils all of the identified requirements, so providing assurance that the device is suitable to go into service. It is important to note that whilst each stage should be performed in sequence, it is possible for a stage to show that a previous stage has not been satisfactorily completed. In this case the previous stage should be revisited. This may also need to occur if changes occur at a higher system or project level. If a previous stage must be repeated, elements of the work completed that are still valid should be retained to minimise the time and cost implications.

Conclusions

The paper presents a rigorous, industry proven, iterative process to efficiently develop a performance-critical device to meet complex and sometimes conflicting design requirements. The process has been developed during the design of a TBC on recent rolling stock projects. Through the review of existing standards, human factors good practice and project requirements, high level design requirements were identified including comfort in sitting and standing operating postures, low risk of injury, the incorporation of a DSD and the prevention of mis-use through design.

The paper presents the techniques used to develop a TBC to meet these requirements, taking into account the conflicting nature of some requirements, operational requirements, user capabilities and the integration of the device within the cab. Techniques employed include literature review, CAD assessments and user trials with a range of prototypes. Lessons learnt from this development have been summarised and include the influence on comfort that small design changes can have, the clear relationship between comfort and mis-use and the importance of an iterative design process.

This analysis has been used to define a rigorous and efficient process for designing a performance-critical device, such as a TBC, to meet complex design requirements.

References

Kroemer, K. H. E., Kroemer, H. B., and Kroemer-Elbert. K. E., 2003, *Ergonomics, How to Design for Ease and Efficiency*, 2nd Edition, (Prentice Hall)

London Underground Ltd. 2007, *Brake System Components.* RSE/STD/006 PART 3, Issue D

Nielsen, J. and Molich, R. 1990, Heuristic Evaluation of User Interfaces, in Chew, J. C., and Whiteside, J. (eds). *Proceedings of the SIGCHI Conference on Human Factors In Computing Systems: Empowering People* (Seattle, Washington), 249–256

Open Ergonomics. 2000, *Peoplesize 2000 Professional.* Open Ergonomics Ltd, Melton Mowbray, Leicestershire

Tilley, A. R. & Henry Dreyfuss Associates. 2002, *The Measure of Man and Woman, Human Factors in Design.* John Wiley and Sons, New York

DELIVERING SMART DRIVING FEEDBACK THROUGH A HAPTIC PEDAL

S.A. Birrell, M.S. Young & A.M. Weldon

Human-Centred Design Institute, School of Engineering and Design, Brunel University, Uxbridge, Middlesex, UB8 3PH, UK

This study evaluates the effects of a vibrotactile (or haptic) accelerator pedal on car driving performance and perceived workload using a driving simulator. The stimulus was triggered when the driver exceeded a 50% throttle threshold, past which is deemed excessive in the literature. Results showed a significant decrease in mean and maximum acceleration values, also a reduction in the maximum and excess throttle values when the haptic pedal was active versus a baseline condition. In addition, a further decrease in maximum and excess throttle was observed compared to when participants were simply asked to drive economically. As well as the positive changes to driver behaviour, subjective workload decreased when driving with the haptic pedal. The haptic processing channel offers a largely untapped resource in the driving environment, and could provide information without overloading the other attentional resource pools used in driving.

Introduction

Eco-driving has become a regularly used phase in the motorised transport arena; it is used to describe a driving style which results in an increase in fuel economy. Increasing the miles per gallon of a journey not only results in a financial saving for the driver, but helps to reduce their carbon footprint and the impact of other emissions. One of the golden rules of eco-driving is to avoid excessive acceleration events, reducing the need for high engine loads and helping to regulate a smooth diving style. A reduction in acceleration rates has been shown to be beneficial to both fuel economy and emissions (El-Shawarby *et al.*, 2005; Ericsson *et al.*, 2001; Waters and Laker, 1980). However 'Smart' driving encompasses both fuel efficient and safe driving behaviours, with acceleration rates also being independently linked to the risk of accidents (AAA Foundation, 2009; af Wahlberg, 2006). In contrast, the effect of aggressive driving (characterised by frequent and excessive acceleration use) on other vehicle emissions is more marked, specifically when considering excessive acceleration levels. Research has suggested that aggressive driving styles increased hydrocarbon emissions by between 200 and 600%, and nitrous oxide by 50 to 200% (De Vlieger *et al.*, 2000; El-Shawarby *et al.*, 2005). These findings have lead to the recommendations that to facilitate eco-driving throttle use should be 'positive' but not exceed a threshold of 50% (Johansson *et al.*, 1999; van de Burgwal and Gense, 2002).

There is a plethora of economical driving advice available through websites, information leaflets and also through driver training programmes. However, whilst it has been demonstrated that such benefits can be achieved through driver education alone (e.g., Haworth and Symmons, 2001), evidence also suggested that the learnt positive effects of eco-driving may be lost in time, as drivers forget or do not feel motivated to maintain aspects of the eco-driving style (Johansson *et al.*, 2003). A longer-term, sustainable solution would be to provide feedback to the driver in the car. This however does provide its challenges as providing the driver with more information in the vehicle may increase workload and cause distraction, which ultimately affects performance.

Clearly an approach is needed which does not cause the associated negative effects of an additional task, but does facilitate the benefits of adopting an eco-driving style. One such method may be to give information to the driver via unused or underused attentional resources. The commonly cited multiple resource model for human information processing (Wickens, 2002) suggests that performance degradation is limited when complementary independent sensory resources are used to present information. Since driving is a predominantly visual task (Kramer and Rohr, 1982), using these other sensory modalities may reduce overload to the visual channel, leaving an increased capacity for visual driving tasks and theoretically improving performance. Van Erp (2001) has suggested that overload may be avoided by presenting feedback in either an auditory or haptic form, thereby not competing for the driver's visual resource pool.

A limited number of studies have investigated haptic feedback in the driving environment. Van Erp and Van Veen (2004) investigated in-vehicle navigation information presented to the driver via vibrating elements mounted in the seat, against traditional visual feedback. They found that the haptic navigation display reduced the driver's workload, and led to the fastest reaction time when combined with a visual display. Faster reaction times and reduced number of missed responses was also observed in a driving task with vibrotactile feedback provided through the steering wheel (Enriquez *et al.*, 2001). Ho *et al.* (2006) found faster braking reaction times and greater safety margins when vibrotactile feedback was given, versus no feedback, when assessing rear-end collision potential in a driving simulator. Haptic warnings are emerging within the driving domain specifically to alert drivers to safety discretions such as lane departure; this feedback is currently supplied either through the seat (as with certain Citroen models) or steering wheel (BMW 5 Series). However, as yet no products are on the market which provides haptic feedback through pedals, or relating to eco parameters.

In this paper, we describe the use of haptic feedback provided through a vibrating accelerator pedal in order to facilitate eco-driving. A vibrotactile stimulus was activated when the throttle was depressed past a 50% threshold, with participants' driving behavior responses and mental workload measured in a laboratory based driving simulator study.

Methodology

Participants

Twelve participants (eight male and four female; mean age 21.3 years, SD = 0.78 years) were recruited from Brunel University to take part in the study. All participants held full UK driving licences, had at least one year's driving experience and had normal or corrected to normal vision.

Equipment

The Brunel University Driving Simulator (BUDS) was used for this study. BUDS is a fixed-based immersive simulator with a 2006 Jaguar S-Type as the donor car. The driving simulator software is provided by STISim, which enables a real-time, anti-aliased, 3-D graphical scene of the projected virtual world. The images are projected onto three 2.5 m × 2.1 m (viewable area) screens, thus giving the forward facing scene plus the left and right peripheral scenes, giving a 150° horizontal and 45° vertical field of view. The simulator is controlled by a Logitech multimedia driving unit (G25 Racing Wheel) consisting of steering wheel, gear lever and pedal block (including clutch pedal), fitted as a UK-standard right-hand drive vehicle allowing manual transmission. The frame rate and data capture rate throughout the study were fixed at 20 Hz.

The vibrating alert on the accelerator pedal was activated when the throttle exceeded a predetermined threshold of 50% (i.e., half the pedal's travel potential, not 50% of the vehicles acceleration limits). The 50% threshold was set for this study following recommendations from two reports which analysed the effects of an eco-driving style on fuel consumption and emissions (Johansson *et al.*, 1999; van de Burgwal and Gense, 2002). The prototype haptic pedal (Figure 1) was fitted to the existing pedal block, and consisted of a mechanical arm which, on contact with the base plate of the pedal block, activated a local vibration alerting the participant that they had exceeded 50% throttle. The vibration was provided by a 3-volt motor which drove an offset weight attached to the spindle, creating a noticeable vibration. The mechanical arm on the back of the pedal held a micro-switch in place, ensuring that the switch remained activated when the throttle input exceeded the predetermined threshold.

Figure 1. Prototype haptic pedal illustration (left) and actual version (right).

Driving scenario and experimental conditions

The simulated driving scenario used for this study consisted of a six-minute run, through a mixed urban and extra-urban route. Speed limits varied from 30 to 50 mph throughout the scenario, and traffic light controlled intersections were located within the urban sections. Two 8m-wide driving lanes were present for each direction of traffic, with light traffic density used throughout, which was specifically programmed not to obstruct the participant. To increase the realism of the scenario, buildings, pedestrians and trees etc were added where appropriate.

In order to assess the influence of the haptic pedal on eco-driving parameters, three experimental conditions were used:

- Baseline – Participants were asked to drive normally and were not specifically informed of eco-driving techniques.
- Eco – Participants were asked to drive economically, both according to their own perceptions, and it was also suggested that a simple rule for driving economically is to avoid exceeding 50% throttle.
- Haptic – Participants were asked to drive economically, with the haptic pedal indicating when they had exceeded 50% throttle.

Procedure

Participants were given a full verbal and written explanation of the study before signed, informed consent was gained. Following this a two-minute practice run (using a different scenario to that described above) was given for participants to familiarise themselves with the simulator and driving setup. When comfortable with the simulator, participants completed the baseline condition first, then they were randomly assigned to complete the eco or haptic condition second, with the remaining condition completed last. The experimental aims of the study would have been compromised if the order of the conditions was fully randomised, as some participants would have been exposed to the notion of using a maximum of 50% throttle as an eco-driving technique before they completed their baseline run. Whilst this creates a potential source for error within the study, it was essential to achieve the study aims. Following each of the experimental runs the participants completed the NASA Task Load Index (TLX; Hart and Staveland, 1988) subjective workload questionnaire.

Data collection and analysis

A series of driving parameters were collected to assess the effectiveness of the haptic pedal in achieving the desired behaviours, including mean and maximum vehicle acceleration, throttle position and driving speed. The final parameter was a measure of excess throttle, calculated as a product of the magnitude of the throttle position (when over the 50% threshold) and time spent over the threshold. This measure could also be interpreted as the area under the curve of the throttle position versus time history. Subjective workload was also measured using the TLX. Although the driving scenario used for the study totalled six minutes in length, only a specific section (approximately 2 minutes) of the run was used to calculate the driving parameters. This was decided because the primary objective of the study was to evaluate participants' acceleration behaviour, therefore sections when cruising at a constant speed would have been of comparatively little interest. For this reason a section of the scenario was highlighted as appropriate which began from stationary at a set of traffic lights, the participant then accelerated to a steady speed (speed limit of 40 mph), then decelerated to a stop for a second set of traffic lights, and once more accelerated to the 40 mph speed limit. This gave the participants two acceleration and deceleration events, with a short section of steady speed.

The main effects of the experimental conditions on driving performance parameters were statistically analysed using a MANOVA, with related Bonferroni corrected pairwise comparisons evaluating potential differences between the conditions. The TLX was analysed using Friedman and Wilcoxon Signed Rank tests. Statistical testing was completed using SPSS 15.0 and significance was accepted at $p < 0.05$.

Results

The driving data collected revealed some interesting effects between the experimental conditions adopted for this study (Table 1). Statistical analysis revealed that both mean and maximum driving speed did not differ significantly ($p > 0.05$) between the conditions; the same was true for mean throttle position throughout the abridged section of the driving scenario. The acceleration parameters did change with the conditions; a significant main effect was observed with mean acceleration ($F_{(2,33)} = 12.1, p < 0.001$) and maximum acceleration ($F_{(2,33)} = 7.0, p < 0.01$). This difference was primarily made up of a significant reduction in acceleration parameters (assessed using Bonferroni corrected pairwise comparisons) from the baseline to eco and haptic conditions, but not between the eco and haptic conditions. Also seen in this study was a significant decrease ($F_{(2,33)} = 15.3, p < 0.001$) in the maximum throttle input. The pairwise difference between the haptic condition and both the baseline and eco conditions was significant. The final driving parameter was excess throttle where a significant main effect was also observed between the three conditions ($F_{(2,33)} = 3.6$, $p < 0.05$). However this difference was only significant between the haptic and baseline conditions.

As well as the driving data presented above, notable differences were also observed when considering the subjective ratings of workload during the driving scenario for

Table 1. Mean data for the driving parameters for each experimental condition.

	Speed (mph)		Acceleration (m/s^2)		Throttle Input (%)		Throttle Excess
	Mean	Max	Mean	Max	Mean	Max	
Base	27.5	48.1	0.73	2.38	61.0	99.8	6864
Eco	24.9	45.3	0.52	1.59	62.6	88.6	4584
Haptic	25.4	46.0	0.54	1.83	62.4	78.0	3311

each experimental condition. The mean TLX ratings for the three conditions were: Baseline = 40.8, Eco = 45.0, Haptic = 35.1. As well as a significant main effect between the conditions ($df = 2, x^2 = 5.9, p < 0.05$) being revealed by the Friedman test, the ratings given for the haptic condition were significantly lower than those given in the eco condition.

Discussion

Results from this study suggest that the use of a vibrating pedal to warn drivers when they exceed a 50% threshold had many positive effects on acceleration and throttle parameters associated with eco-driving, as well as on perceived workload of the driving task. An interesting finding was that average throttle position did not differ between the conditions, but maximum throttle did (Table 1). Maximum throttle decreased from nearly 100% in the baseline condition to 78% in the haptic condition, which was a further 10% lower than when participants were simply asked to drive economically. From this we could assume that a more stable and consistent throttle use pattern emerged from the haptic feedback. Since research has shown emissions data (excluding CO_2) to be extremely sensitive to large differences in throttle position (represented by instantaneous acceleration; De Vlieger *et al.*, 2000; El-Shawarby *et al.*, 2005), a more stable throttle position can be assumed to be wholly beneficial to reducing vehicle emissions.

Maximum and mean acceleration values were also seen to decrease in the eco and haptic conditions compared to the baseline (Figure 2). This was an expected outcome from the study, since throttle position is directly proportional to vehicle acceleration. Although speed consistency or stability was not measured in this current study, research suggests that a smooth driving speed is associated with decreased fuel consumption (Haworth and Symmons, 2001). This study did

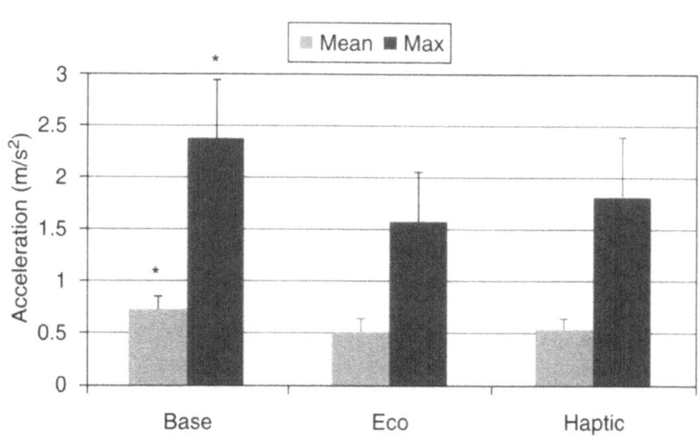

Figure 2. Mean and maximum acceleration data for each condition. Asterisks indicate significant (p < 0.05) difference to both Eco and Haptic condition. Error bars represent standard deviation.

however show that the haptic condition resulted in a slightly smaller spread of the acceleration data compared to baseline (standard deviation of 0.10 compared to 0.12 respectively). A reduction in maximum acceleration values has been shown to have positive effects on eco parameters such as fuel consumption (Ericsson *et al.*, 2001; Waters and Laker, 1980). Safety aspects of 'Smart' driving are also positively affected by a reduction in acceleration values as they have been shown to be independently to this the risk of accidents (Af Wahlberg, 2006). In addition a high number of excessive acceleration events (amongst other parameters) are a prerequisite for a driver with an 'aggressive' style, with aggressive driving being associated with 56% of all fatal crashes in the US (AAA Foundation, 2009).

The lack of difference to either mean or maximum driving speed can be taken as a positive effect for the haptic pedal. It suggests that decreases in high throttle events came at no expense to mean driving speed, which will ultimately relate to journey time. However, a decrease in driving speed has been observed when drivers are engaged in a mobile phone conversation while driving; this is considered to be a compensatory behaviour in an attempt to reduce workload, as well as to increase perceived safety margins (Haigney *et al.*, 2000). This behaviour is indicative of increased driver workload or distraction, which was not the case with this present study.

The final driving parameter was excess acceleration. This parameter decreased from the baseline condition when participants were asked to drive in an economical way, and further decreased when the haptic pedal gave additional feedback. The implications of this are that drivers spent less time with the throttle depressed beyond the 50% threshold, and what time they did spend over 50% was of a lower magnitude. Within a simulator it is not possible to measure actual fuel consumption or emissions, however given the recommendations of Johansson *et al.* (1999) and van de Burgwal and Gense (2002), that an eco-driving technique should not involve throttle positions of greater than 50%, this result can be seen as a major positive for the haptic pedal.

Participants rated their perceived workload from the driving task following each of the experimental runs. Participants rated workload to be significantly lower in the haptic compared to the eco condition. This is despite the fact that in both conditions participants were asked to drive economically, and use the 50% throttle rule as a guideline for economical driving; the only difference being with the haptic condition drivers were alerted to this threshold. A similar result was seen by Van Erp and Van Veen (2004) who found a decrease in workload when navigation advice was administered using haptic feedback. The results from the current study suggest that simply asking a driver to limit throttle use to 50% does result in beneficial changes to throttle use and acceleration values, at least in the short-term, but a resulting negative is that workload was increased. This observation might occur as it may be unreasonable to expect anyone other than highly skilled drivers to be able to pinpoint a set threshold on the accelerator. Even if they could, such a task is likely to increase the attentional demands on the driver, which in turn may increase risk if the driver becomes overloaded, as observed with this current study.

This study was exploratory study and raises interesting issues which need to be addressed in future work. Such studies could include comparisons with a visual and/or auditory display, to investigate if vibrotactile alerts still offer the advantages proposed in this study. A replication of this study should also be conducted to investigate responses of participants from differing age ranges and driving experience to haptic feedback. A limitation to this, and all simulator studies, is that no force feedback or vibrational feedback is available from the car or road itself, this may change driver's behaviours in the real world or mask any benefits observed in the simulator. Therefore, given the safety critical nature of pedals in cars, further on the road testing needs to be conducted. Later versions of the haptic pedal may utilise a variable threshold, rather than fixed at 50%, to allow greater throttle use at lower speeds or in lower gears.

Conclusion

The use of a vibrotactile stimulus on the accelerator pedal which is triggered when the throttle is depressed beyond a predetermined threshold of 50% resulted in positive changes to driving parameters compared to a baseline condition. A significant decrease in mean and maximum acceleration values, as well reductions in the maximum throttle position were observed, along with a decrease in excess throttle use with the haptic pedal. In addition, a further decrease in maximum and excess throttle was observed compared to when participants were simply asked to drive economically. As well as the positive changes to driver behaviour, a decrease in subjective workload was observed when driving with the haptic pedal. It could be argued that simply informing drivers of eco-driving rules would allow them to adjust their driving style accordingly. However, a haptic pedal solution has several advantages. Firstly, as has been pointed out already, such education-based interventions tend to be short-lived. The most significant advantages of a haptic pedal is that by providing the driver with instant, naturalistic feedback on when those thresholds have been reached, it not only relieves them of additional workload, but actually maintains a level of automaticity on the task. There is considerable evidence that haptic interfaces impose significantly fewer demands than visual or auditory displays – and, indeed, that haptic information can to some extent be automatically processed (Gustafson-Pearce, 2007; Sklar and Sarter, 1999; Van Erp and Van Veen, 2004). Moreover, since driving is a predominantly visual task (Kramer and Rohr, 1982), the haptic processing channel offers a largely untapped resource in the driving environment, and could be used to provide information without overloading the other attentional resource pools used in driving (cf. Wickens, 2002).

References

AAA Foundation. 2009, *Aggressive Driving: Research Update*. AAA Foundation Research Report, pp 1–12

Af Wahlberg, A. 2006, Speed Choice versus Celeration Behaviour as Traffic Accident Predictor. *Journal of Safety Research*, 37, 43–51

De Vlieger, I., De Keukeleere, D. and Kretzschmar, J. 2000, Environmental effects of driving behaviour and congestion related to passenger cars. *Atmospheric Environment*, 34, 4649–4655

El-Shawarby, I., Ahn, K. and Rakha, H. 2005, Comparative field evaluation of vehicle cruise speed and acceleration level impacts on hot stabilized emissions. *Transportation Research Part D*, 10, 13–30

Enriquez, M., Afonin, O, Yager, B. and MacLean, K. 2001, *A Pneumatic Tactile Notification System for the Driving Environment*. Workshop on Perceptive User Interfaces (PUI '01), Orlando, FL, November 15–16, 2001

Ericcson, E. 2001, Independent driving pattern factors and their influence on fuel-use and exhaust emission factor. *Transportation Research Part D*, 6, 325–345

Gustafson-Pearce, O. 2007, Comparison between audio and tactile systems for delivering simple navigational information to visually impaired pedestrians. *British Journal of Visual Impairment*, 25, 255–265

Haigney, D., Taylor, R. and Westerman, S. 2000, Concurrent mobile (cellular) phone use and driving performance: task demand characteristics and compensatory processes. *Transportation Research Part F*, 3, 113–121

Haworth, N. and Symmons, M. 2001, *The relationship between fuel economy and safety outcomes*. Victoria, Australia: Monash University, 1–57

Ho, C., Reed, N. and Spence, C. 2006. Assessing the effectiveness of intuitive vibrotactile warning signals in preventing front-to-rear-end collisions in a driving simulator. *Accident Analysis and Prevention*, 38, 988–996

Johansson, H., Farnlund, J. and Ergstrom, C. 1999, *Impact of EcoDriving on emissions and fuel consumption: A pre-study*. Borlange, Sweden: Swedish National Road Administration Report No: 1999: 165E, 1–41

Johansson, H., Gustafsson, P., Henke, M. and Rosengren, M. 2003, *Impact of EcoDriving on emissions*. International Scientific Symposium on Transport and Air Pollution, France

Kramer, U. and Rohr, G. 1982, A Model of Driver Behaviour. *Ergonomics*, 25, 891–907

Sklar, A.E. and Sarter, N.B. 1999, Good vibrations: tactile feedback in support of attention allocation and human-automation co-ordination in the event-driven domains. *Human factors*, 4, 543–552

Van de Burgwal, H. and Gense, N. 2002, *Application of 'New Style Driving' and effects on passenger car fuel and emissions*. TNO Report 02.OR.VM.004.1/HVDB, 1–59

Van Erp, J. 2001, Tactile navigation display. In S. Brewster and R. Murray-Smith (Eds.), *Haptic human computer interaction. Lecture notes in computer science*, (Berlin: Springer Verlag), 65–173

Van Erp, J. and Van Veen, H. 2004, Vibrotactile in-vehicle navigation systems. *Transportation Research Part F*, 7, 247–256

Waters, M. and Laker, I. 1980, *Research on fuel conservation for cars*. Crowthorne: Transport and Road Research Laboratory, pp. 1–38

Wickens, C.D. 2002, Multiple resources and performance prediction. *Theoretical Issues in Ergonomic Science*, 3, 159–177

WHOLE-BODY VIBRATION & TRUNK ROTATION: A SURVEY OF EXPERT OPINION

L.J. Morgan & N.J. Mansfield

Loughborough University, LE11 3TU, UK

Off-road occupational driving is often associated with low back pain, the causal factors being associated with the drivers' exposure to whole body vibration (WBV) and trunk rotation. The aim of the present study was to evaluate the level of agreement between academic experts in the field of ergonomics and human response to whole-body vibration, and agricultural operators, on the effects of exposure to WBV and trunk rotation. A total of 83 individuals responded to the paper based questionnaire, on risk levels from single and combined exposures, discomfort development, exposure duration limits and tasks within agriculture.

A general consensus was seen among experts in judging a risk progression with increasing exposure. Differences in responses occurred between the groups in areas regarding acceptable exposure durations, areas of discomfort under exposure and recommendations for cab developments.

Introduction

Within the many facets of modern ergonomics, there is interest in how individuals are affected by their work; often this includes investigation of their working postures. Unfortunately working in undesirable postures is still often a requirement and is reflected in the health of the operator. Poor and/or long term seated postures are known to contribute to the development of symptoms in the low back (Pheasant and Haslegrave, 2006). Within driving environments the posture can often be constrained for long periods of time (Frymoyer *et al.*, 1983), in addition adverse postures maintained over extended periods increase risks of development of musculoskeletal disorders (Kumar, 2001; Marras, *et al.*, 1993). Trunk rotation (twist) and has been found to be significantly related to low back pain (Frymoyer, *et al.*, 1998). In addition to the posture of the driver, there may be other factors in a work environment that cause stress upon the body. In a driving environment, one of these additional stressors is WBV, which is widely accepted as a risk factor for low back pain (LBP) amongst exposed populations (Wikstrom, *et al.*, 1994; Lings & Leboeuf-Yde, 2000). Cervical and lumbar spine deformations have been seen in individuals exposed to vibration for many years (e.g. Rosegger and Rosegger,1960). Increased lifetime vibration dose increases the risk of developing a low back condition (Schwarze, *et al.*, 1998; Tiemessen, *et al.*, 2008; Bovenzi, 2009), with sufferers often experiencing not only discomfort but pain and disability.

It has been suggested that a person will react to the vibrating environment better if they are not under additional postural stress, and that combinations of vibration and posture increase the probable stress leading to the development of low back pain (Donati, 2002). The posture of the operator may make the muscles and circulatory system less tolerant to the stresses of vibration (Magnusson and Pope, 1998). The lumbar vertebral discs exhibit increased fracture risk under both vibration and constrained postural stress (Wilder and Pope, 1996). Physical loading factors (e.g. manual handling), postural stress and WBV are all factors in the complex nature of LBP development in occupational drivers (Bovenzi, *et al.*, 2006), and some researchers believe that interaction effects of posture, vibration and/or manual materials handling are the main contributors for precipitation of LBP (Okunribido, *et al.*, 2008). The exposure-response relationship between trunk rotation, WBV and associated health risk has proven difficult to identify. ISO 2631 requires that operator posture is taken into consideration but there are no criteria by which judgments can be made.

Within industrialised agriculture, few tasks are carried out without agricultural wheeled tractors, which typically have rear mounted linkages and power take-off where machinery, implements and trailers are attached, controls for these are banked to the right of the operator. Depending on the task, operators may be required to monitor rear-mounted implements, operate in-cab controls, and maintain vehicle control. This combination of factors dictates the operating posture, resulting in periods of trunk rotation (Bottoms and Barber, 1978; Mehta and Tiwari, 2000). Previous studies within agricultural driving populations have attempted to adjust for these other factors and have developed odds ratios demonstrating the link between vibration dose and symptom risk; ranging from 2.0 (Boshuizen, *et al.*, 1990) to 2.4 (Bovenzi and Betta, 1994). Vibration exposures within agriculture are also known to exceed the guidance for acceptable daily exposures (Sorainen *et al.*, 1998). The three-year averaged prevalence for LBP in agriculture is 1410 reported per 100,000 employed (HSE, 2007/8), which is the largest prevalence of all industries in the UK; for this reason agriculture is used to provide a context for the tasks and exposures, although it is suggested results and any recommendations could be transferred to other off road driving professions.

Methods

This investigation was designed as a survey of current opinion of the relative risks of exposures seen in agriculture as described above. Target groups for attaining current opinion were experts in WBV, postural ergonomics and operators within agriculture. Experts were recruited through conference attendance (UK conference on Human Response to Vibration, 2009; Ergonomics Society Annual Conference, 2008), transport ergonomics special interest group (Ergonomics Society), ergonomists at Loughborough University, agricultural shows (Royal Welsh Show, 2009) and farming unions (National Farmers Union; Farmers Union of Wales). The survey took the form of a questionnaire to elicit direct responses on particular topics with open ended comment boxes for additional information which respondents felt important. The postural ergonomics expert (PE) and WBV expert (VE) response groups completed the same questionnaire, which was split into five main categories. A demographics section contained questions on the area and length of expertise of the respondent. The second section required respondents to rate the risk of common vibration and posture combinations seen within off-road driving industries. Experts marked an outline of the body where they considered discomfort would be experienced in common combinations of posture and vibration. This data was then coded by body segment, as defined by Kuorinka *et al.* (1987) to facilitate comparison between response groups. Those with experience in assessing WBV were asked whether they considered that the posture of the operator should be reflected as part of the assessment of WBV exposure. Previous research has suggested possible additions to agricultural tractors to avoid the stresses put upon the body (e.g. large mirrors, Sjofløt, 1980; swivelling seat, Bottoms and Barber, 1978). Respondents were asked to consider how effective they would perceive such additions to be. An operator expert (OE) group completed a largely similar questionnaire, however questions about risk assessment were removed, and supplementary questions on working durations and least preferred tasks were added.

Results

13 WBV experts (VE) completed the questionnaire with 46% having >10 years experience, the main focus of these experts was within measurement and assessment of vibration. 37 postural ergonomics experts (PE) completed the questionnaire, with >50% having >10 years experience; these experts were mostly working within anatomy, anthropometry and physiology in human activities. 33 operators (OE) completed the operator questionnaire with >90% of respondents having >10 years experience. Of the respondents, 67% spent >10 hours driving per day (5–10 hours, 21%; 2–5 hours, 12%) and of these 82% took two or fewer breaks from driving in the day.

Subjects were asked to evaluate the risk for LBP posed by exposure to WBV, trunk rotation and a combined exposure. The consensus was that a risk was posed by the exposures, which increased when exposures are combined. In judging the risk of LBP from WBV, the same proportion (66%) of OE (N = 33) and PE (N = 37) thought there was a causal relationship, a greater number of VE (N = 13) respondents (78%) judged LBP risk from WBV exposure. The same proportion (69%, 68% and 67%) of the OE, PE and VE respondent groups respectively considered a risk is posed by trunk rotation. 86%, 63% and 78% of OE, PE and VE respondents thought that the risk for LBP increased when exposures were combined.

PE and VE groups were asked to assess the relative risk of nine predetermined exposure combinations, with trunk rotation and WBV as variables. Tables 1 and 2 display

Table 1. Level of risk associated with exposure to vibration (V) and trunk rotation (T), expressed as percentage of posture expert responses (median highlighted; N = 38).

Combination	Not at all	Slightly	Moderately	Very	Extremely
1	43%	**57%**	0%	0%	0%
2	5%	**62%**	29%	5%	0%
3	0%	23%	**50%**	18%	9%
4	0%	**50%**	45%	0%	5%
5	0%	5%	**65%**	25%	5%
6	0%	0%	26%	**53%**	21%
7	0%	0%	30%	**60%**	10%
8	0%	0%	5%	**60%**	35%
9	0%	0%	0%	16%	**84%**

Table 2. Level of risk associated with exposure to vibration (V) and trunk rotation (T), as percentage of vibration expert responses (median highlighted; N = 13).

Combination	Not at all	Slightly	Moderately	Very	Extremely
1	**67%**	33%	0%	0%	0%
2	22%	**44%**	33%	0%	0%
3	0%	11%	**44%**	**44%**	0%
4	22%	**56%**	22%	0%	0%
5	11%	22%	**67%**	0%	0%
6	0%	11%	22%	**44%**	22%
7	11%	11%	**56%**	22%	0%
8	0%	0%	33%	**56%**	11%
9	0%	0%	22%	11%	**67%**

the PE and VE group responses respectively, giving the percentage of responses for each rating. The exposures were as follows:

(1) no twist, low vib. (smooth driving ~0.25 m/s^2)
(2) no twist, medium vib. (rough driving ~0.5 m/s^2)
(3) no twist, high vib. (off road driving ~1 m/s^2),
(4) medium twist (70°), low vibration
(5) medium twist, medium vibration
(6) medium twist, high vibration
(7) high twist (170°), low vibration
(8) high twist, medium vibration
(9) high twist, high vibration

Table 3. Suggested exposure time limit for exposures across the three groups, median responses, hh:mm (25th %ile, 75th %ile).

	No twist, no vib.	Med. twist, no vib.	High twist no vib.	No twist, with vib.	Med. twist, with vib.	High twist, with vib.
VE (N = 13)	06:30 (2.8, 8.0)	01:45 (0.9, 2.0)	00:53 (0.4, 1.0)	03:00 (1.5, 3.8)	01:00 (0.7, 1.0)	00:30 (0.4, 0.5)
PE (N = 20)	04:00 (2.8, 8.0)	01:00 (0.5, 1.4)	00:15 (0.2, 0.5)	02:00 (2.0, 3.0)	00:30 (0.3, 1.1)	00:10 (0.0, 0.4)
OE (N = 18)	08:00 (8.0, 12.0)	07:00 (5.3, 9.5)	04:00 (4.0, 8.0)	06:00 (4.5, 10.0)	05:00 (4.0, 6.8)	03:00 (2.0, 4.0)

A measure often employed within risk control is the limiting of exposure durations, appropriate to the exposure magnitude. The three respondent groups were asked to specify time limits for exposures involving increasing levels of rotation and WBV. A large proportion of the OE group (45%) disputed the requirement for formalised exposure time limits, common stated answers included 'until the job is completed' 'if I can do it at my age, anyone can' and 'increase the number of breaks taken'.

Those with experience in WBV risk assessment (VE, N = 9; OE, N = 5) were asked to consider at what level of the evaluation operators' posture should be considered. All PE (100%) and the majority of VE (89%) agreed that there is a need for acknowledgement of additional risk at risk assessment level. A lower proportion (PE, 80%, VE, 67%) thought there should be an inclusion within standards, and fewer (PE, 80%, VE, 56%) concurred that a consideration of combined exposures should be recognised within legislation. All experts were in agreement that exposure duration should be reduced when vibration exposure is combined with operator twist. For high vibration and twist the mean duration limits given were 3.5 hrs (OE, n = 27, s.d. 1.98), 0.6 hrs (PE, n = 11, s.d. 1.08) 0.5 hrs (VE, n = 7, s.d. 0.18).

Experts marked on body outlines where they felt operator discomfort would be felt under vibration and posture stress. The results were categorised by body segment and are shown below in Figures 1, 2 and 3. All groups associated low back discomfort with WBV. OE only associated trunk rotation with neck discomfort, unlike PE and VE groups. When considering combined exposures, all experts judged an increase in discomfort.

OE (N = 32) provided details on which tasks caused them particular discomfort. Travelling over rough ground was the most common response, with some indicating they hit their heads on the cab roof. Hedge-brushing and ploughing were also indicated, due to the high level of twist required along with the requirement for control activation. Shock vibration was considered particularly uncomfortable, especially when pile driving with a twisted trunk.

The suggestions made by Sjofløt (1980) and Bottoms and Barber (1978) were considered by the experts for their value in a typical agricultural driving task with twist and vibration components. 42% OE (n = 21), 75% VE (n = 8) did not recommend

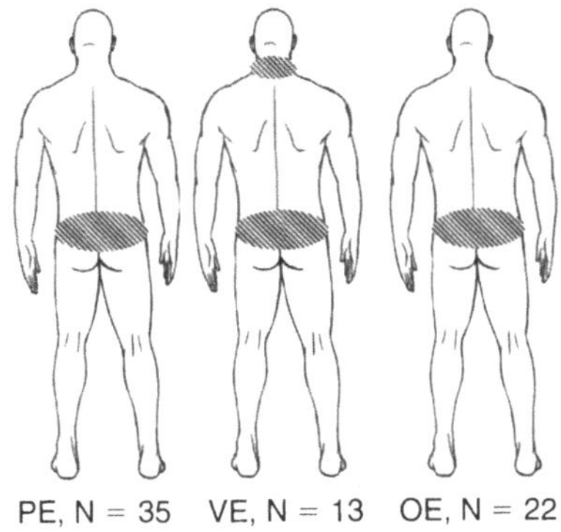

Figure 1. Areas where >40% judged to be uncomfortable when exposed to whole-body vibration.

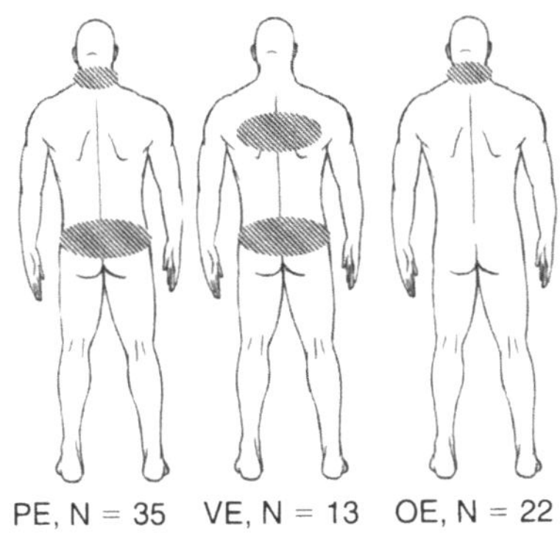

Figure 2. Areas where >40% judged to be uncomfortable when exposed to trunk rotation.

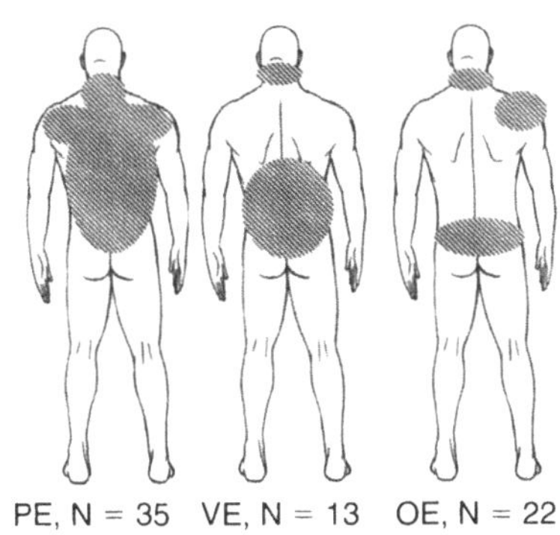

Figure 3. Areas where >40% judged to be uncomfortable when exposed to *both* whole body vibration and trunk rotation.

a full backrest seat, in contrast to only 12% PE (n = 14). 100% of OE, 50% of VE and 76% of PE would recommend a swivel seat. 38% of OE would recommend the addition of enlarged mirrors in comparison to 100% of both VE and PE. 100% of PE and VE would also recommend the addition of CCTV, only 23% of OE would agree with this addition for the task in question.

Discussion

Overall, the three expert response groups provided similar patterns of response to the questionnaire mostly being in agreement that there is an increased risk posed by the combined exposure to WBV and twist. This is in agreement with the conclusions in Bovenzi and Betta (1994).

When controlling the risks posed by a hazard, reduction of the causal factor at source is often the preferred solution. However, when this is not an available option, reduction of exposure duration is utilised. For WBV exposure, the physical agents (vibration) directive provides guidance on current limit values. All groups reduced their recommended exposure duration with increasing twist and vibration levels. The greatest disparity was between the operator and academic experts, perhaps unsurprisingly. Only 55% of operators provided a duration recommendation, the remainder affirming that despite any associated risk, they would continue with the task until completion. This highlights the issue of risk prevention in agriculture and therefore the suggestion is made that risk reduction strategies employ any possible engineering solutions firstly, in an attempt to avoid the issue of non-compliance.

Body maps are often used to evaluate self-reported and expert opinion of exposures (Robb and Mansfield, 2007; Kuorinka *et al.*, 1987). In this study, the body map is utilised for comparison between operator first-hand experience of the exposures and academic experts experience in judging risk scenarios. In general there was a consensus between the 3 response groups on where symptoms would display. The low back area was represented in all but one occasion (trunk rotation, OE). This is in agreement with the multifactorial nature of LBP development (Frymoyer *et al.*, 1983). The prevalence of the right shoulder in the OE responses may be indicative of the issues caused by control placement in the cab (banked to the right of the operator) or the direction of twist (in general to the right). This issue of control placement is also raised by Donati (2002) in his review of cab ergonomics. It would be interesting to note if the increasing trend to mount controls on the armrest and move the seat reduces this discomfort. In the OE response group, the thigh backs were represented in 36% of responses for a combined exposure; this was not seen in either the PE or VE responses. This discomfort may be due to the anthropometry of the operator population investigated (Gyi and Porter, 1999) or from badly fitting, or improperly adjusted seats (Magnusson and Pope, 1998). Comments from the questionnaire suggest that leg pain stems from the requirement for pedal force to the front of the operator, when the seat pan is rotated (OE, n = 3). This may suggest a requirement for training interventions for operators within agriculture of possible redesign of the seat.

The operator group were asked to consider which tasks caused the greatest discomfort, those requiring twist with a force requirement and high levels of vibration, particularly shock vibration were well represented. Control placement, armrest length, insufficient head room and requirement for foot pedal force were all issues that operators raised as causing particular discomfort. Many of these could be solved with relatively simple ergonomics interventions. Donati (2002) arrived at similar conclusions in 2002, so it seems more advice or maybe pressure may be required to compel the industry to make the necessary design interventions.

In this study three groups with knowledge and expertise in the exposures commonly experienced by agricultural tractor drivers were chosen as representative of the key stakeholders in the area. However, it is conceded that it may have been beneficial to include a response group from a medical and biomechanics backgrounds to allow for completeness. It is possible that collection bias may have affected the results from this survey, with the low numbers of vibration experts responding. However, there was little spread in the conclusions in each of the response groups, suggesting the sample was representative of the expert opinion.

Conclusion

Results show the consensus that exposing operators to WBV and trunk rotation simultaneously increases the risk for low back pain. Operators reported discomfort in the right shoulder and thigh backs when exposed to both risk factors. This was not highlighted in the academic responses. Those with experience in risk assessment were in agreement that the issue of postural stress should be considered when assessing the risk from vibration exposure. Interventions aimed at reducing the requirement for twist and improving seat discomfort should therefore be prioritised as a risk reduction strategy for off road drivers

References

Boshuizen, H. C., Bongers, P. M., Hulshof, C. T. J. 1990, Self-reported back pain in tractor drivers exposed to whole-body vibration, *International Archives of Occupational Environmental Health* 62: 109–115

Bottoms, D. J., Barber, T. S. 1978, A swivelling seat to improve tractor drivers' posture. *Applied Ergonomics*, 9, 2, 77–84

Bovenzi, M., Betta, A. 1994, Low-back disorders in agricultural tractor drivers exposed to whole-body vibration and postural stress, *Applied Ergonomics*, 25, 231–241

Bovenzi, M., Rui, F., Negro, C., D'Agostin, F., Angotzi, G., Bianchi, S., Bramanti, L., Festa, G., Gatti, S., Pinto, I., Rondina, L., and Stacchini, N. 2006, An epidemiological study of low back pain in professional drivers, *Journal of Sound and Vibration*, 298, 514–539

Bovenzi, M. 2009, Metrics of whole-body vibration and exposure–response relationship for low back pain in professional drivers: a prospective cohort study, International Archives of Occupational and Environmental Health, 82, 7, 893–917

Donati, P., 2002, Survey of technical preventative measures to reduce whole-body vibration effects when designing mobile machinery. *Journal of Sound and Vibration*, 253, 1, 169–183

European Parliament and the Council of the European Union (2002) Directive 2002/44/EC on the minimum health and safety requirements regarding the exposure of workers to the risks arising from physical agents THE EUROPEAN VIBRATION DIRECTIVE 479 (vibration). Official Journal of the European Communities, OJ L177, 6.7.2002, 13–9

Frymoyer, J. W., Pope, M. H., Clements, J. H., Wilder, D. G., MacPherson, B. and Ashikaga, T., 1983, Risk factors in low-back pain. An epidemiological survey, *Journal of Bone & Joint Surgery.*, 65, 213–218

Frymoyer, J. W., Pope, M. H., Costanza, M. C., Rosen, J. C., Goggin, J. E. and Wilder, D. G. 1998, Epidemiologic studies of low back pain, *Spine*, 215, 4, 965–976

Gyi, D. E., Porter, J. M., 1999. Interface pressure and the prediction of car seat discomfort. *Applied Ergonomics* 30, 99–107

HSE statistics, 3 year average back pain prevalence 2005/6 – 2007-8 http://www.hse.gov.uk/statistics/lfs/0708/backind2_3yr.htm, date accessed 30/11/09

Kumar, S. (2001) Theories of musculoskeletal injury causation. Ergonomics 44, 1, 17–47

Kuorinka, I., Jonsson, B., Kilbom, A., Vinterberg, H., Biering-Sørensen, F., Andersson, G. and Jørgensen, K., 1987, Standardised Nordic questionnaires for the analysis of musculoskeletal symptoms, 18, 3, 233–237

Lings, S., Leboeuf-Yde, C., 2000, Whole-body vibration and low back pain: a systematic, critical review of the epidemiological literature 1992–1999. *International Archives of Occupational and Environmental Health* 73, 5, 290–297

Magnusson, M. L. and Pope, M. H. 1998, A review of the biomechanics and epidemiology of working postures (It isn't always vibration which is to blame), *Journal of Sound and Vibration* 215, 4 965–976

Marras, W. S., Lavender, S., Leurgans, S. E., Sudhakar, R., Allread, W. G., Fathallah, F. A., Ferguson, S. A. 1993, The role of three-dymensional trunk motion in occupationally-related low back disorders: the effects of workplace factors, trunk position and trunk motion, *Spine*, 18, 5, 529–677

Mehta, C. R., Tewari, V. K. 2000, Seating discomfort for tractor operators: a critical review, *International Journal of Industrial Ergonomics*, 25, 661–674

Okuniribido, O. O., Magnusson, M., and Pope, M. H., 2008, The role of whole body vibration, posture and manual materials handling as risk factors for low back pain in occupational drivers, *Ergonomics*, 51, 3, 308–329

Pheasant and Haslegrave, 2006, *Bodyspace: Antropometry, ergonomics, and the design of work*, 3rd Ed, CRC Press, London

Robb, M. J., Mansfield, N. J. 2007, Self-reported musculoskeletal problems amongst professional truck drivers, *Ergonomics*, 50, 6, 814–827

Rosegger, R., Rosegger. S., 1960. Health effects of tractor driving, *J Agric Eng Res*, 5, 241–272

Schwarze, S., Notbohm, G., Dupuis, H., Hartung, E., 1998. Does-response relationships between whole-body vibration and lumbar disk disease – a field study on 388 drivers of different vehicles, *Journal of Sound and Vibration* 215, 4, 613–628

Sjøflot, L. 1980, Means of improving a tractor drivers working posture, *Ergonomics*, 23, 751–761

Tiemessen, I. J. H., Hulshof, C. T. J., Frings-Dresen MHW. 2008, Low back pain in drivers exposed to whole body vibration: analysis of a dose–response pattern, *Occupational and Environmental Medicine*, 65, 667–675

Wikstrom, B. O., Kjellberg, A., Landstrom, U. 1994, Health Effects of long-term occupational exposure to whole-body vibration: A review. *International Journal of Industrial Ergonomics* 14, 273–292

Wilder, D. G., Pope, M. H. 1996, Epidemiological and aetiological aspects of low back pain in vibration environments: an up-date. *Clinical Biomechanics*, 11, 61–73

SLIPS, TRIPS AND FALLS

SYSTEMIC STAIR STEP GEOMETRY DEFECTS, INCREASED INJURIES, AND PUBLIC HEALTH PLUS REGULATORY RESPONSES

Daniel A. Johnson[1] & Jake Pauls[2]

[1]*Daniel A. Johnson, Inc., Olympia, Washington, USA*
[2]*Jake Pauls Consulting Services, Silver Spring, Maryland, USA*

Steep stairs, caused by risers that are too great and or by goings (runs) that are too small, contribute to an increased risk of missteps, falls and injuries. Traditionally, building codes have allowed such conditions in homes and this appears to be a major factor in relatively high injury occurrences related to home stairs. Risk of missteps and falls is even more potently influenced by variations in step geometry. Systemic, top-of-flight dimensional uniformity defects currently appear pervasive in new stair construction in the US and Canada. This might be the major factor in an apparent doubling, over the last decade, of home stair-related injury risk, relative to other settings, in the US. Designers, builders, regulatory officials, consumers and others appear unaware of the defects, even though they are prohibited by building codes and easily identified by using simple tests.

Introduction

Over thousands of years stair step geometry has been a subject of fascination, speculation, rule making, analysis, controversy and, more rarely, formal ergonomics study. This paper provides only a scant introduction to this history. While there are multiple reasons—such as comfort, energy expenditure, and safety—for choosing one geometry over another, the focus here is on fall-related injury prevention. Two potent patterns of systemic stair step geometry defects have been identified recently by US-based ergonomists. The first systemic defect has two components that, in combination, increase the steepness of a stairway. The first is a riser that is too great, and the second is an undersized going. "Going", "run" and "tread depth" are interchangeable terms used in the UK, Canada and the USA, respectively, for the horizontal dimension of the tread *measured nosing to nosing*. (The "nosing" is the tread's leading edge). An undersized going is generally a more potent defect than an oversized rise in contributing to missteps, departures from normal gait, which can lead to falls. The second systemic defect, non-uniformity of step dimensions at the top of stair flights and, to a lesser extent, at the base of stair flights, has only recently been recognized by ergonomists as warranting special attention. Non-uniformity has long been identified as a danger although, traditionally, there might have been more emphasis on rise uniformity than going uniformity. So stair design and construction regulations have traditionally addressed the need for step dimensions to be uniform. However, there have been serious deficiencies in the ways rules are stated and then enforced, and these are now the subject of relatively intense examination.

Using stairs as a learned behaviour

Archea *et al.* (1979) state that, as a result of climbing stairs, we build an internalized image and response to the structure, the stairs. Successfully using the stairs occurs when there is a match between the internal image and the actual structure. In the first few steps users will test to see if the actual stair conditions and the initial perceptions are matched. If their internalized images and expectations are confirmed in the first several steps, users will probably conclude that their image and responses are correct and will proceed to use the stairs without calling forth unusual behaviours needed to successfully negotiate the steps. If, from the first few steps, users notice that the internalized images and expectations are at variance from the actual geometry of the steps, appropriate modifications in the behaviour will be made. If there is a subsequent change in step geometry it will not be expected and can lead to a misstep and perhaps a fall.

An important question for future research is how few steps need a person experience to establish an expectation for a particular stair descent? A related question is how does information from the first step down a stair flight affect the foot placement on the second step, typically done by the opposite foot? Can we process the information from, say, the right foot taking the first step and put that information to work almost immediately in taking the second step with the left foot? Or is the information from the right foot more completely applied in the movement of that foot from the first stair tread to the same foot's descent to third stair tread? These questions are being increasingly considered by ergonomists (like the authors) and will be prominent in research priority and methods discussions planned during and immediately after the Ergonomics Society Annual Conference 2010.

Steep stairways cause falls

The height of the riser and the size of the going are each critical in determining the safety of a stairway. Probably everyone recognizes that the steeper the stairway the more difficult the descent. Templer (1974) provided a good history and Pauls (2002) reviewed the history with greater attention to how US building codes and standards responded in recent decades. Experimental studies confirmed the relationship between stair steepness and falls. Subjects were tested as they ascended and descended stairs at three different slopes (25, 35 and 45 degrees), with three different riser-to-tread ratios. In descent, risers with the fewest missteps ranged from 117 mm (4.6") to 183 mm (7.2). Risers greater than 183 mm (7.2") had a greater number of missteps. And "treads that were 312 mm (12.3") or more had the fewest missteps at all speeds; treads that were 269 mm (10.6") to 292 mm (11.5") had more missteps, followed by treads that were 254 mm (10.0") to 262 mm (10.3"). All treads that were 229 mm (9.0") or less performed uniformly poorly regardless of riser height" (Templer, 1974). Later Templer *et al.* (1985) videotaped workers using 31 flights of stairs where there had been high frequencies of severe stair-related injuries. They identified 98 stair users who experienced "incidents" (falls, slips, trips, missteps, and moments of temporary instability). Examination determined that incident rates were associated with stairs with risers greater than 180 mm (7") and having tread depths less than 280 mm (11").

The most recent and best research in this area, conducted in the UK, has been presented and published, mostly within the context of Ergonomics Society conferences (Roys, 2001; Roys and Wright, 2003, 2005, 2008; Wright and Roys, 2005, 2008). For example, Wright and Roys (2005) also found that the foot pivoting and/or slipping off of a nosing could be attributed to risers that are too high or goings that are too short. They found that preferred goings were longer than those usually found on stairs. Goings greater than 300 mm (11.8") were optimal. *Further, higher risers resulted in more of the shoe being placed further forward on the tread regardless of going length.* This indicates that a foot pivoting/slipping off of a tread could be due to a riser that was too high. Optimal rise height was 190 mm (7.5") and less.

The probability of a shoe slipping over the nosing, based on going length and exposure, has been calculated (Roys and Wright, 2003). For instance, a flight with 14 steps, and having goings of 225 mm (8.9") and being used by five people every day, can be expected to result in a slip of the foot over the nosing once every three years. If there were 2,000 uses each day a slip could be expected every three days.

Later research by Wright and Roys (2008) utilized a maiback survey where respondents measured the stair in their home and reported whether or not the stair had been the site of an "accident" in the prior two years. Their conclusion: "Across the range of stair geometries found in a random sample of English dwellings there is a very significant influence on accident rate of going size, with larger goings reducing the likelihood of a stair accident being recorded." Table 1, derived from Wright and Roys (2008) with respect to estimated relative risk of falls, also shows

Table 1. Risk of falls on home stairs with various going dimensions Risk estimates derived from Wright and Roys (2008) Figure 4.

Going dimension	Relative risk of falls	Used for home stairs by
255 mm (10.0")	0.03	ICC Codes in USA
245 mm (9.6")	0.07	
235 mm (9.3")	0.11	
225 mm (8.9")	0.12	NAHB in USA
215 mm (8.5")	0.13	
205 mm (8.1")	0.14	NBCC in Canada
195 mm (7.7")	0.21	

ICC—International Code Council; NAHB—National Association of Home Builders; NBCC—*National Building Code of Canada* the minimum going dimensions permitted by widely used US and Canadian model building codes as well as by the National Association of Home Builders.

The NAHB is politically powerful enough to dictate the minimum going dimension requirement—230 mm (9")—actually used in some state and local codes that have been adopted in place of the minimum going required in the model. ICC's *International Residential Code* stipulates the minimum going at 250 mm (10"). The *National Building Code of Canada*—with a minimum going of 210 mm (8.25") for homes—has what appears to be the least demanding going requirement of English-language model building codes.

Variations between risers or between goings cause falls

Templer (1992, 151–152) reports that Velz and Hemphill (1953) found that 15 percent of stairs examined had variations, but on steps where people fell, 75 percent had variations. And Templer also reports that Miller and Esmay (1958) found tread depth irregularities on 47.6 percent of the stairs where falls had occurred as compared to only 20.7 percent of the sample group of stairs where falls had not been reported. The average irregularity was 15 mm (0.6"). Jones (1963) found more variations in the rise heights of stairs where falls had occurred (78.8%) as compared to stairs where no fall had been recorded (28.4%).

Unpublished studies by one of the authors suggest that dimensional non-uniformities can result in greatly increased risk—*by as much as two orders of magnitude size*—in the rate of missteps on stairs. Both were documented in the course of field observations, the first as part of the large documentation of spectator behaviour at the 1976 Olympic Games in Montreal (in which both John Archea and John Templer participated on the ten-person study team). The second was in 1987 at the Molson Indy race in Toronto. In the Olympics case, an unfinished stair was being used for crowd descent but the top steps were non-uniform in both rise and going dimensions. Misstep rates averaged one in 20. Study team recommendations to close the stair were not heeded by Olympic authorities but spectators were verbally warned by Olympics Games staff of stair dangers.

Later, under comparable crowd conditions, when the stair construction was complete and steps uniform, misstep rates were reduced by a factor of 50 to approximately one per 1000. (This was still not a good rate but there were other stairway defects.) In the 1987 observations a stair used exclusively for crowd ascent had an odd rise dimension in mid flight; the misstep rate here was one in 16. In this case the stair was rebuilt during the event, simply by replacing an odd-thickness wooden tread with one comparable to other treads in the flight. Within an hour of the pre-condition, post-condition observations revealed no missteps in about a couple of thousand uses, a ratio again of about two orders of magnitude improvement over the defective condition. Of course both these cases posed unusual ethical challenges for the observers; mitigation or elimination of the defects took precedence over meticulous scientific documentation; thus only order of magnitude estimates could be made.

More recently, Cohen *et al.* (2009) analyzed 80 actual stairway falls that were the subject of forensic investigations. They report that the most significant factor associated with falls was the variability between risers, and, secondly, the variability between goings. They found that in 60 percent of the cases there were riser variations in excess of 9.5 mm (0.375"). In 34 percent of the cases there were variations between goings in excess of 9.5 mm (0.375").

Variations in stairway geometry are often both unexpected and difficult for the user to perceive. A person must accurately place the ball of the foot near to, but not beyond, the nosing of the tread. Any variation in riser height or going increases the chance that the ball of the foot will be placed too far beyond the nosing so that when weight is transferred to that foot it will pivot and/or slide over the edge—an overstepping misstep, possibly resulting in a fall.

While the ball of the foot, in descent, is usually placed near to or on the nosings, there is some variability within subjects as to actual foot placement (Roys and Wright, 2005). When those variabilities are combined with excessive and unexpected variability in stair construction the chance of a misstep increases.

Top-of-flight defect

This defect, found increasingly often in forensics work performed over the last 30 years by first one of the authors, and then the other, has been found to exist in many structures, but most commonly in homes, especially newer homes. The first going below the top landing is greater than subsequent goings because the floor nosing does not project beyond the riser as do all other nosings in the flight. One reason this might be occurring more often now is the increased use of (off site) manufactured stair flights, with projecting nosings, being connected to a floor structure, the top landing of which does not have a projecting nosing. The top-of-flight defect conveys erroneous information to the user who undoubtedly expects goings on subsequent treads to have similar dimensions to that of the first going.

One of the authors, collaborating with a professional home inspector, has learned that up to 90 percent of new homes could have this defect which was featured in a paper by Pauls and Harbuck (2008). Instances have been recorded where the second and third goings were as much as 50 mm (2") shorter than the top going. A foot placed on the second or third tread is much more likely, under this condition, to be placed too far forward and will pivot and/or slip over the nosing. Since this defect is found at the top of the flight, the consequences of a fall are likely to be more severe than they would be for a fall from a lower level. This defect was found, by both of the authors (in their forensics practices), to be associated with falls that resulted in serious and even fatal injuries (Johnson, 2009).

Although the top-of-flight defect violates building codes, the violations were not detected prior to the buildings being occupied. This may be because inspectors did not receive specific information on the problem or how to easily detect it. Pauls and Harbuck (2008) describe a simple detection technique illustrated in Figure 1. By

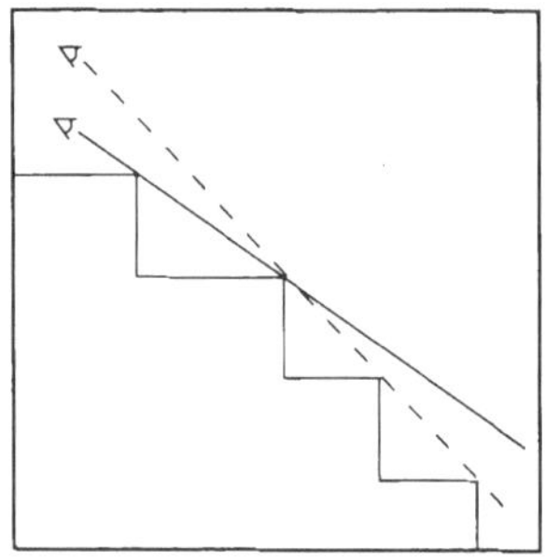

Figure 1. The "crouch and sight" technique can detect a top-of-flightdefect.

sighting down the stairs the top-of-flight defect can be detected with a high degree of reliability.

Accuracy of this "crouch and sight" test is based on there not being a simultaneous defect of riser height that is exactly proportional to the increased tread going dimension. Thus the prudent inspector will only use this test to confirm a stair is *not* uniform, not to confirm it is uniform. A reliably high degree of confidence that a stair is indeed consistent in dimensions requires additional measurements, beginning with measurements of the nosing-to-nosing distances of at least the top few steps. Also, the sightline test is of no value at all for bottom-of-flight riser height non-uniformities, occurrences that are also relatively common.

Generally, while the "crouch and sight" test is extremely useful, the prudent inspector will utilize the proper tools to accurately measure the nosing-to-nosing dimensions. This will require use of a spirit or digital level, the latter becoming the tool of choice for inspectors, such as the two authors, who have developed methods based on two measurements per step. The first is the nosing-to-nosing angle (relative to horizontal) and the second is the nosing-to-nosing distance—*being very careful to measure to the same points on each nosing, a difficult task with beveled, rounded or thick-carpet-covered nosings*. The rise and the going are then calculated using the sine and cosine functions respectively (Johnson, 2006; Pauls, 1998).

Public health and regulatory responses (or lack thereof)

Thus far, although the public health ramifications of defective stairs come from data collected by major US Government agencies (e.g., the US Consumer Product Safety Commission) there has been no official response to public release of graphs such as Figure 2 which indicate a rapidly growing problem with home stairs. Such stairs have systemic defects of all the kinds discussed thus far in this paper. Currently, it is only possible to speculate if the long term, five times ratio is mainly due to the systemically lower standard for step geometry in homes, relative to all other settings and if the newly reached, ten times ratio is due to recent increases in the systemic top-of-flight occurrence of non-uniformity.

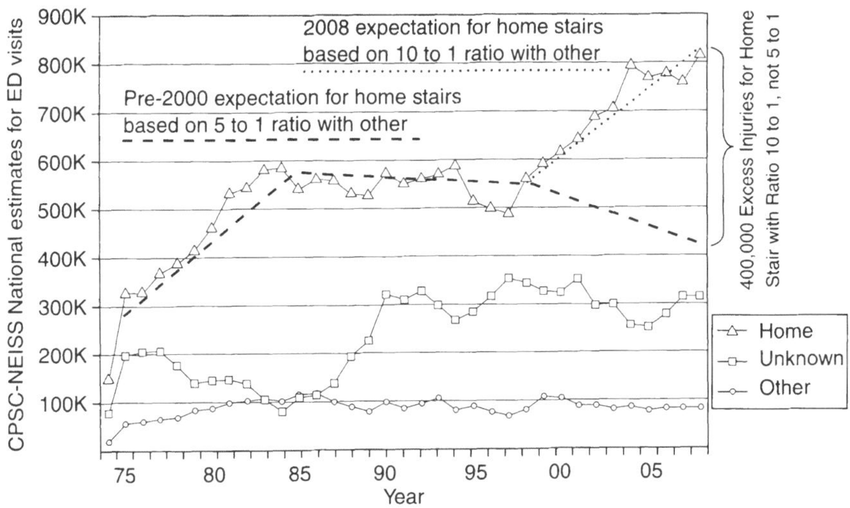

Figure 2. Trends in hospital-treated stair-related injuries in the USA based on data accessible at www.cpsc.gov/library/neiss.html

Both trends should be seen as unacceptable. So far, however, neither the public health community nor the building regulatory community has acknowledged, let alone discussed, the need for strategies plus tactics to mitigate or, better still, prevent the huge injury toll. The toll for medical care alone in the US for stair-related injuries is about one million dollars per hour. The societal cost is about ten times that amount.

Conclusions

Currently the average annual increase in US home stair-related injuries is about 5 percent, or five times greater than US population growth. At the same time stair-related injuries in all other (non-home) settings are decreasing at about 1 percent annually. Bringing these epidemiology insights to the attention of public health authorities in the USA has had only mixed success. Trying to persuade public health and regulatory authorities to respond to the rapidly growing injury toll has been frustrating. Moreover, US and Canadian organizations responsible for model code development, adoption and enforcement seem to be heavily influenced by homebuilder management. Traditional defects in home stair step geometry—of both systemic types—appear largely to blame. There are some hopeful signs. One is the proposed BS 5395-1:2009 which increases minimum going dimensions for both home and other stairs. Another, in Ontario, Canada, is a proposed mainstreaming of the widely used 180 mm maximum rise with 280 mm minimum going (the so-called "7–11" step geometry in inch units) for use in new homes and elsewhere.

What is needed is the systematic recording of falls and injuries as they occur, especially in homes, and determining if and how those falls are related to the design characteristics of the stairs. Further, additional information is needed to determine if the stairs on which falls have occurred met the standards specified in local codes (in which case the standards may be implicated) or whether the stairs did not meet standards (in which case the building design and inspection process may be implicated).

References

Archea, J., Collins, B., and Stahl, F. 1979, Guidelines for stair safety, *NBS Building Science Series 120,* (NTIS, Washington, DC) 11–13

BS 5395-1:2009, 2009, British Standard, *Stairs – Part 1: Code of practice for the design of stair with straight flights and winders*

Cohen, J., LaRue, C.A., and Cohen, H.H. 2009, An ergonomics analysis of 80 cases, *Professional Safety, 54*, 27–41

Johnson, D.A. 2006, An improved method for measuring stairways, *Proceedings of IEA2006, 16th World Congress on Ergonomics*, Maastricht

Johnson, D.A. 2009, Improper construction results in dangerous stairs: Large top runs produce fall hazard, *Proceedings of the 53rd Annual Meeting of the Human Factors and Ergonomics Society,* (Human Factors and Ergonomics Society, Santa Monica, CA)

Jones, R.A. 1963, New safety developments in home construction and equipment. *National Safety Council Transactions 6* (National Safety Council, Itasca, IL)

Miller, J.A. and Esmay, M.L. 1958, Winter, Stairway falls, Home *Safety Review*, 23–25 As reported in Templer, J. 1974, *Stair shape and human movement*, (Dissertation Information Service, UMI, Ann Arbor, MI)

Mowat, W. and Mowat, A. 1900, *A Treatise on Stairbuilding and Handrailing,* Republished with additions, (Stobart Davies Ltd., London, 1989), 15

Pauls, J. 1998, Techniques for evaluating three key environmental factors in stairway-related falls, *Proceedings of the Human Factors and Ergonomics Society*, (Human Factors and Ergonomics Society, Santa Monica, CA), 1630

Pauls, J. 2002, Proposal and detailed justification for Life Safety Code adoption of the 7-11 minimum step geometry for dwelling unit stairs, in Report on Proposals for NFPA 101, National Fire Protection Association, Quincy, MA, NFPA 101, F2002-ROP, 228–241. Downloaded 9/17/2009: www.nfpa.org/itemDetail.asp?categoryID=817&item ID=23683&URL=Codes & Standards/Code development process/Proposals (ROP) and Comments (ROC)

Pauls, J. and Harbuck, S.C. 2008, *Ergonomics-based Methods of Inspecting, Assessing and Documenting Environmental Sites of Injurious Falls,* presented at American Society of Safety Engineers, Downloaded 10/24/2008: www.asse.org/education/pdc08/show-session.php?id=704

Roys, M. 2001, Serious stair injuries can be prevented by improved stair design, *Applied Ergonomics*, 32, 135–139

Roys, M. 2006, Stairs and Steps, in R. Haslem and D. Stubbs (eds.) *Understanding and Preventing Falls,* (Taylor and Francis, London), 51–68

Roys, M. and Wright, M. 2003, *Proprietary nosing for non-domestic stairs,* information paper 15/03, (BRE Centre for Human Interaction, Garston, Watford, UK)

Roys, M. and Wright, M. 2005, Minor variations in gait and their effect on stair safety, in Bust, P.D. and McCabe, P.T. (eds.) *Contemporary Ergonomics 2005,* (Taylor and Francis, London), 427–431

Roys, M. and Wright, M. 2008, The minimum cost of non-fatal fall accidents to the NHS in England, in Bust, P.D. (ed.) *Contemporary Ergonomics 2008*, (Taylor and Francis, London), 681–686

Templer, J. 1974, *Stair shape and human movement*, (U.M.I. Dissertation Information Service, Ann Arbor, MI)

Templer, J. Archea, J. and Cohen, H. 1985, Study of factors associated with risk of work-related stairway falls, *Journal of Safety Research,* 6, 183–196

Templer, J. 1992, *The staircase: studies of hazards, falls and safer design*, Vol. 2, (MIT Press, Cambridge, MA)

Velz, C.J. and Hemphill, F.M. 1953, *Investigation and application of home injury survey data in the development of preventive procedures.* (University of Michigan, School of Public Health: Ann Arbor) as reported in Templer, J. 1992, *The staircase: studies of hazards, falls and safer design*, (MIT Press, Cambridge, MA) 151–152

Wright, M. and Roys, M. 2005, Effect of changing stair dimensions on safety, in Bust, P.D. and McCabe, P.T. (eds.) *Contemporary Ergonomics 2005,* (Taylor and Francis, London), 469–474

Wright, M. and Roys, M. 2008, Accidents on English dwelling stairs are directly related to going size, in Bust, P.D. (ed.) *Contemporary Ergonomics 2008*, (Taylor and Francis, London), 632–637

REDUCING HOSPITAL INPATIENT FALLS

Mike D. Williams[1]**, Bernadette George**[2] **& Ros Wade**[2]

[1] *University of Exeter Business School*
[2] *Royal Devon & Exeter NHS Foundation Trust*

Hospital inpatient falls are common. The literature shows that there is no single intervention that makes a significant difference. This paper reports on a case study of a large National Health Service hospital where an improvement methodology (Plan, Do, Study, Act) was used to introduce a checklist approach, combined with intentional visiting (intentional rounding) of high risk patients every hour by nurses. Initial success was limited but once a formal training programme was implemented a significant reduction in falls within the hospital occurred.

Introduction

Hospital inpatients have about a one in ten chance of suffering some form of adverse event (de Vries *et al.*, 2008). An 'adverse event' is defined in this paper as 'any unintended event caused at least partly by healthcare and which resulted in harm' (Sari *et al.*, 2007). A frequent cause of adverse events in hospital is patient falls. Falls in hospital are more routinely reported as incidents then medication errors or adverse drug reactions (National Audit Office 2005). For a small number of patients, a fall causes significant harm or even death (Healey *et al.*, 2008). A 'fall' is defined as 'all situations in which a patient suddenly and involuntarily came to rest upon the ground or surface lower than their original station' (Oliver *et al.*, 1997).

A review of the research literature indicates limited impact of a range of interventions in reducing the number of falls (Oliver *et al.*, 2007). Few studies have considered how and when hospital staff interact with patients as part of the daily ward activity in hospitals to reduce the risk of falls. This paper presents results from a case study of an NHS hospital where two fatalities had occurred from falling. Nurse leaders used an improvement method, Plan, Do, Study, Act (PDSA) to introduce a more rigorous risk assessment tool, the use of a checklist for high risk patients combined with a process on 'intentional rounding' (visiting) patients every hour when they are awake.

Literature

The size of the problem of patient falls in hospital is significant. The reported rate ranges from 3 to 14 per 1000 bed days, and are associated with a range of injuries and an increased length of stay in hospital (Oliver *et al.*, 2007; Healey *et al.*, 2008). In England from September 2005 to August 2006, some 200,000 falls were reported to the National Patient Safety Agency. Analysis of the reported incidents of falls shows that 65% of patients resulted in no harm, 31% in 'low harm', 3.6% in 'moderate harm' and 0.6% in severe harm with 26 reported deaths (Healey *et al.*, 2008). However, it is well recognised that there is serious under reporting of incidents in hospital (Olsen *et al.*, 2008), so the scale of the problem is likely to be greater.

A systematic review and meta-analysis of studies to prevent falls and fractures suggested that there 'is reasonable evidence that using a multifaceted intervention for hospital inpatients may have a modest effect on falls but not on fractures' (Oliver *et al.*, 2007). Interventions included risk assessment, medication review, care planning, and education. The literature supports the view that patient falls is a complex issue and that there appears to be no simple solution (National Institute for Clinical Excellence 2004).

Given the relatively common occurrence of inpatient falls on wards for elderly patients there is the danger of staff seeing them as part of the process of healthcare. From a conceptual perspective staff can 'normalise' to a level of adverse events believing that there is little or nothing that can be done about the situation. Vaughan (1996), shows that normalisation of deviance is key issue for safety. Her study of the Challenger Space Shuttle disaster illustrated that the system within which people work can produce a culture where, through small incremental steps, new situations are seen as 'normal'. Outsiders looking at the situation are more likely to see the situation as deviant, not acceptable and therefore potentially dangerous.

Waring *et al.* (2007), point to types of behaviour by health care workers that illustrate 'taken for granted assumptions about clinical risk…' Such behaviours 'normalize risk' and as such mean that risk of harm to patients is not addressed. It has been suggested that medical staff regard error as 'inevitable' which can lead to errors being seen as 'normal' (Waring *et al.*, 2005). These errors may not be reported as incidents and therefore do not provide the basis for improvements in safety. It has been suggested that in the hospital environment, staff can regard a patient fall as an inevitable part of the rehabilitation process (Grenier-Sennelier *et al.*, 2002). Such a view may then influence the reporting of such incidents (Health Care Commission 2009).

Quality improvement

There is a vast body of management literature relating to quality improvement in a wide variety of settings (Juran, 1989; Deming, 1994). Aspects of this work have been developed as a methodology for improvement in the processes and reliability in healthcare (Resar, 2006). The methodology is based on making a small change that is tested and adapted before it is implemented more widely. The bedrock is the cycle of improvement achieved through the cycle of 'plan, do, study, act (PDSA) (Langley *et al.*, 2006).

Research question

Can hospital inpatient falls be reduced by the nurses use of a checklist combined with the intentional rounding of patients assessed as being at high risk of falling?

Method

A case study approach was used as being the most appropriate to gather the contextual detail needed (Yin, 2003) and where there is complexity in the subject matter (Stuart *et al.*, 2002). A criticism of case studies is that of being context specific and therefore transferability of findings to other situations can be problematic. However, when 'they are used to contextualize key points of an explanation, and the explanation, and not the cases, carries the argument, case studies can be very powerful (Nightingale *et al.*, 2003).

A mixed method approach using the simultaneous collection of qualitative and quantitative data was used (Creswell, 2003; Bergman, 2008). The underlying philosophy is that to obtain a better understanding of the phenomenon, a range of epistemological perspectives were used. Multiple sources of data collection were utilized. This assists in achieving triangulation in data collection, the interpretation and subsequent theory development.

A 760 bed NHS hospital was chosen as the case study where an improvement methodology was being used and significant change was beginning to occur in the level of falls prevention. The case studied the medical wards where a new approach to falls was being implemented set within the context of the wider hospital. Initial implementation of the new ways of working started on those wards specialised in care of the elderly (48 beds), stroke (30 beds) and neurology (24 beds) where the risk of falls was highest.

The hospital reported 1631 in-patient ward falls in 2008. The five wards that started the change had 568 (34%) of that total. On admission patients were assessed for the risk of falling. A scoring system of over 30 indicates that the patient is at very high risk of falling. The hospital's Falls Management Policy in place at the beginning of 2008, required that certain actions must be taken by staff for any patient with a risk assessment score greater than 13.

Twenty eight staff (doctors, nurses and managers) were interviewed as part of a wider study using an interview protocol. A further six semi structured interviews were conducted with nurses working within the five wards focusing on the issue of patient falls. Meetings about patient falls were observed, papers and reports analysed and data on falls plotted using statistical process control charts (Wheeler, 2003). The interviews were transcribed and then coded using NVivo 8 to develop themes which were triangulated with the observational data and content analysis of meeting papers and reports.

Findings

The analysis of the interviews showed that during the first part of 2008 there were different histories within the wards of staff attitudes towards patient falls. The histories were related to the leadership of the ward areas. On two of the wards (A & B) the attitude was described as falls "just being one of those things that happens in hospital". Risk assessments often were not done and when they were, little if any action was then taken. There had been a gap in the senior nursing leadership on those wards and for other nurses the situation was described as rather "chaotic". The other three wards (C – E) had more stable senior nursing leadership at ward level. The attitude to falls was that they were "not what you want to happen but an inevitable part of getting patients mobile". Risk assessments were undertaken and actions put in place especially when mobilising patients.

The hospital's Falls Management Policy did not appear to be well known by staff and therefore poorly implemented and supervised. There was little if any awareness of the national initiatives to reduce harm from patient falls (Department of Health 2001; National Institute for Clinical Excellence 2004). This situation was set within the context of the hospital achieving the highest rating for its services from the Healthcare Commission and with a Hospital Standardised Mortality Rate (HSMR) below the expected rate.

The Lead Nurse with a colleague developed a tool for staff to use to improve their care of patients at high risk of falling. The tool combined two patient safety techniques; a 'checklist' approach (Frank, 2006) and 'intentional rounding' (Owensboro Medical Health System, 2008, Meade, 2006). The checklist was a simple list of questions and actions (Figure 1) for staff to undertake with any patient who had been assessed as having a high risk of fallings (score >30). The 'intentional rounding' (IR) is the requirement to go round the ward and speak to those high risk patients every hour using the checklist. A simple checklist form was devised for staff to use which then became part of the nursing record for that patient. Using the PDSA methodology IR was first tested by one nurse on one patient on one ward on one shift. Amendments to the checklist form were made before the tool was then disseminated to other nurses and then the whole ward team. A training pack was developed which helped the tool to be implemented in other ward areas within a short period of time.

In wards where a number of patients were assessed to be high risk they were cohorted in a six bedded bay and a nurse allocated to that area all the time. This made the process of IR much quicker and focused staff resource towards the higher risk patients. The use of bed rails was also questioned and staff were reminded of the need to carefully risk assess and document their use. The checking of the bed rail status was then incorporated into the intentional rounding checklist. For many patients bed rails can increase the risk of harm from falling so should be used selectively. Monthly audits were undertaken of the number of bed rails raised.

There was some opposition to the introduction of IR. The concern was whether staff had the time to dedicate to IR given the many other priorities on a busy ward. The principle of PDSA was again employed and nurses realised that in some ways it helped them in prioritising and managing their workload. PDSA was also used to introduce IR to Care Assistants who provide much of the practical hands on care for patients. Registered nurses could then delegate IR to them and focus their attention on those tasks requiring their skills, such as dispensing medications.

1. Continence: Do you need to go to the toilet?				
2. Pain: Do you have any pain?				
3. Orientation: Fully Alert (FA); Mildly Confused (MC) Severely Confused (SC) Asleep (A)				
4. Position / Comfort: Are you comfortable?				
5. Drink / Mouthcare: Would you like a drink?				
6. Call bell within reach: If you need me press this button.				
7. Bed rails down				
8. Bed to floor				
9. Is there anything else I can do because I have the time?				

Figure 1. Intentional Rounding (IR) checklist.

The Lead Nurse set up weekly then monthly meeting (Falls Meeting) to review progress with wards that implemented IR. As more wards took on the tool they were invited to the meeting. The reliability of the tool being used was tested informally by senior nurses visiting the wards and looking at the records of high risk patients.

Initial results from the first two wards to use IR (Wards A&B) were encouraging. The staff attitude towards patient falls changed, the number of falls reduced and there was a determination to learn from those incidents that did occur. The hospital incident form was supplemented by asking for further information as nurses suspected from studying the results that patients with risk scores of between 20–30, and who had some form of cognitive impairment, could also benefit from IR. This change was implemented.

The IR tool was spread rapidly from wards A&B to C-E and then within two months to a further four wards and ten wards by six months. Using the PDSA methodology what became evident during this rapid spread was that the training of staff was diluted and results were not sustained. A more formal training programme was then developed and started in March 2009 for all the nurses and physiotherapists on the ten medical wards. The number of staff attending was recorded and monitored by the Falls Meeting. Talks were given by the Lead Nurse to other areas of the hospital to raise awareness of the potential level of harm and the new practice for high risk patients.

Statistical analysis of the number of falls before and after the 30 March 2009 was conducted. An independent t test was undertaken to compare the mean number of falls before and after the 30 March 2009. The results showed that the Levene's test for equality of variance was significant ($p > .05$) and that the single tailed t test was also significant ($p > .05$). We can therefore conclude that on average, the number of falls from the April 2008 to the 30 March 2009 was greater ($M = 52$, $SE = 0.89$)

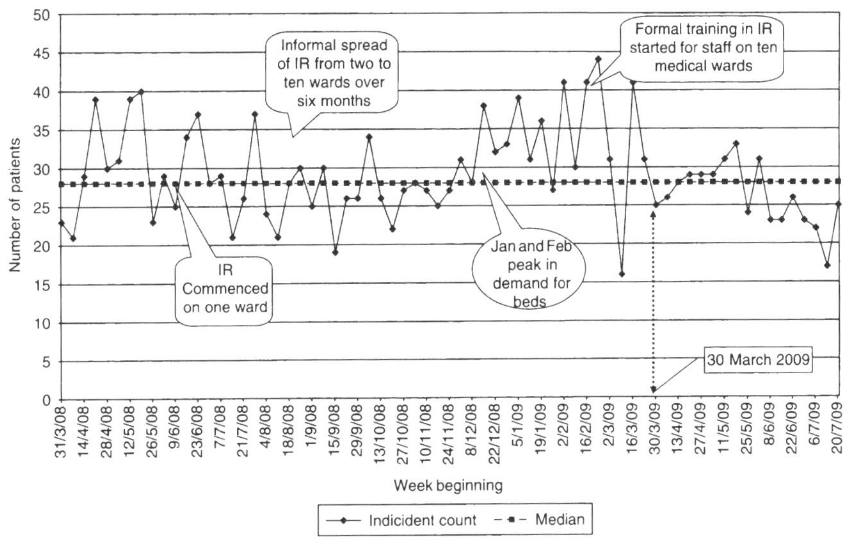

Figure 2. Weekly number of patient falls within the case study hospital April 08 – July 09.

than the number after the 30 March 2009 (m = 17, SE = 0.98). The difference was significant t(44.055) = 2.879, $p > .05$. This represents a medium sized effect r = .4.

Discussion

The nurse leaders put in place new goals (any fall is unacceptable) and institutional practice (staff using a checklist approach every hour to speak to patients at high risk of falling). It is accepted that to achieve *no* falls for patients in hospital is not a feasible goal. However, the unquestioning acceptance (normalisation) of the likelihood of falls in hospital results in a failure to implement systematic methods to reduce the risk for certain patients. The adoption of a reliable method to reduce risk required considerable leadership, education, training and persistent communication. However, even with the level of leadership given, the initial improvements were not sustained to the level desired. Using the PDSA methodology lessons were drawn as to why the initial improvement was not sustained as the new practice was spread. A number of changes to the training and auditing of results have since been put in place, which have contributed to a period of reductions in the number of patient falls from the end of March 2009.

The findings support the theory of practical drift from a designed way of working (Snook, 2000) as set out in the 'Falls Management Policy'. Staff drifted from the 'Policy' due to the competing priorities and finding practical ways to achieve their workload. In doing so they traded effectiveness for thoroughness (Hollnagel, 2004) and normalised their position as acceptable (Vaughan, 1996; Weick *et al.*, 2003). The danger of using the normal practice of a root cause analysis approach to investigate patient falls that result in significant harm, is that such a method often fails to look at the broader systemic issues such as practical drift and trade offs which are endemic amongst staff due to the competing pressures (Dekker, 2006). Senior managers can be left with the impression that the problem is isolated to particular wards rather than being a system wide issue that relates to more than one safety policy area.

Whilst PDSA is a powerful tool to achieve change to the cultural norms that have been established incrementally over many years, it takes a considerable effort by leaders and training to achieve sustained change in practice. Regular meetings with staff were necessary plus constant feedback and learning when patient falls did occur. Those staff in this study who had been directly involved or felt responsible for a patient falling with a serious subsequent injury, engaged with the change process at a much deeper level supporting the theory that 'emotions can act as drivers or motivators, of subjects' engagement with discourses' (Garrety *et al.*, 2003).

Conclusions

Improving patient safety in the midst of many competing demands requires considerable effort. Despite national initiatives and a revised Falls Management Policy, staff in the case study hospital, which was externally assessed as 'excellent', largely accepted the cultural norm that patient falls were part of being a patient in hospital. Theory relating to change in clinical practice to improve safety must take account of the strong cultural bias for staff to accept what is regarded as 'normal' and not challenge the status quo. Ways to achieve a change in culture and practice include engaging with staff about the emotional stories of individual patients who suffer harm through a lack of proactive action. Providing staff with a method (IR with a checklist) to reduce falls, backed by systematic training, appears to help in reducing the number of falls. The development of the method and training was achieved through the iterative use of the PDSA cycle of improvement.

Whilst there are limitations to this single case study there are aspects that have relevance to the wider practitioner interest in improving patient safety through cultural and practice-based change. From a theoretical perspective it seeks to build on the literature by developing the understanding of the drivers that reversed the normalisation of deviance and how safety was improved for patients in hospital.

References

Bergman M. M. Ed. (2008), *Advances in Mixed Methods Research*. (Sage, London)

Creswell J. W. (2003), *Research Design*. (Sage, London)

Dekker S. (2006), *The Field Guide to Understanding Human Error*, (Ashgate, Aldershot)

de Vries E. N., Ramrattan M. A., *et al.* (2008), The incidence and nature of in-hospital adverse events: a systematic review. *Qual Saf Health Care*, 17(3); 216–223

Deming W. E. (1994), *The New Economics*. (MIT Press, Cambridge, Massachusetts)

Department of Health (2001). *The National Service Framework for Older People*. (Stationary Office, London)

Frank C. (2006), Evidence based checklists for objective structured clinical examinations. *BMJ*, 333; 417

Garrety K., Badman R., Morrigan V., Rifkin W., and Zanko M. (2003), The use of personality typing in organisational change : Discourse, emotions and reflexive subject. *Human Relations,* 56; 211–235

Grenier-Sennelier C., Lombard I., *et al.* (2002), Designing adverse event prevention programs using quality management methods: the case of falls in hospital. *Int J Qual Health Care* 14(5); 419–426

Healey F., Scobie S., *et al.* (2008), Falls in English and Welsh hospitals: a national observational study based on retrospective analysis of 12 months of patient safety incident reports. *Qual Saf Health Care,* 17(6); 424–430

Health Care Commission (2009), *Safely does It, implementing safer care for patients*. (Commission for Healthcare Audit and Inspection, London)

Hollnagel E. (2004), *Barriers and Accident Prevention*, (Ashgate, Aldershot)

Juran J. M. (1989), *Juran on Leadership for Quality*. (New York, Free Press)

Langley J., Nolan K., *et al.* (2006), *The Improvement Guide*. (San Fransisco, Jossey Bass)

Meade, C. M., Bursell, A. L. & Ketelsen, L. (2006) Effects of Nursing Rounds: on Patients' Call Light Use, Satisfaction, and Safety. *The American Journal of Nursing,* 106, 58–70

National Audit Office (2005), *A Safer Place for Patients: learning to improve patient safety*. (NAO, London)

National Institute for Clinical Excellence (2004), *Falls: the assessment and prevention of falls in older people*. (NICE, London)

Nightingale P., Brady T., *et al.* (2003), Capacity utlization revsited: software, control and the growth of large technical systems. *Industrial and Corporate Change*, 12(3); 477–517

Oliver D., Britton M., *et al.* (1997), Development and evaluation of evidence based risk assessment tool (STRATIFY) to predict which elderly inpatients will fall: case-control and cohort studies. *BMJ*, 315(7115); 1049–1053

Oliver D., Connelly J.B., *et al.* (2007), Strategies to prevent falls and fractures in hospitals and care homes and effect of cognitive impairment: systematic review and meta-analyses. *BMJ*, 334(7584); 82–88

Olsen S., Scobie S., *et al.* (2008), What can we learn about patient safety from information sources within an acute hospital: a step on the ladder of integrated risk management? *Qual Saf Health Care*, 17(3); 209–215

Owensboro Medical Health System. (2008), *Nursing Notes*. Retrieved 4 April, 2009, from http://omhs.org/docs/newsletter/September%202008.pdf

Resar R. K. (2006), Making Noncatastrophic Health Care Processes Reliable: Learning to Walk before Running in Creating High-Reliability Organisations. *Health Services Research*, 41(4); 1677–1689

Sari A. B-A., Sheldon T. A., *et al.* (2007), Extent, nature and consequences of adverse events: results of a retrospective case note review in a large NHS hospital. *Qual Saf Health Care*, 16(6); 434–439

Snook, S. A. (2000) *Friendly Fire,* (Princeton University Press, Princeton)

Stuart I., McCutcheon D., *et al.* (2002), Effective case research in operations management – a process perspective. *Journal of Operations Management*, 20(5); 419–33

Vaughan D. (1996), *The Challenger Launch Decision.* (UCP, Chicago)

Waring J. J. (2005), Beyond blame: cultural barriers to medical incident reporting. *Social Science & Medicine*, 60; 1927–1935

Waring J. J., Harrison S., *et al.* (2007), A culture of safety or coping? Ritualistic behaviours in the operating theatre. *Journal of Health Research and Policy*, 12(S1); 3–9

Weick K. and Sutcliffe K. (2003), Hospitals as Cultures of Entrapment: A Re-Analysis of the Bristol Royal Infirmary. *California Management Review*, Vol. 48 No. 2: 73–84

Yin R. K. (2003), *Case Study Research, Design and Methods.* (London, Sage)

IMPROVEMENTS TO THE EN SLIP RESISTANCE TEST FOR FOOTWEAR

Graeme Hunwin, Steve Thorpe & Kevin Hallas

Health & Safety Laboratory, an agency of the Health & Safety Executive, Harpur Hill, Buxton, Derbyshire, SK17 9JN

The Health and Safety Laboratory (HSL) use a ramp test method for assessing the slip resistance of footwear. The test involves an operator walking over an inclinable surface until the angle is reached at which a slip repeatedly occurs, indicating the Coefficient of Friction (CoF). The current International and European Standard test method for the measurement of the slip resistance of footwear, BS EN ISO 13287: 2007, is a mechanical test, which measures the CoF by attempting to recreate the action of a slipping shoe. The two tests show limited agreement, with the mechanical test often recording a higher CoF. This paper reports work to modify the parameters of the mechanical test, in an attempt to establish a relationship between the two test methods. The improved agreement between the test methods is reported, and ideas for further modification are outlined.

Introduction

HSL currently use a ramp test method to assess the slip resistance performance of both floor surfaces and occupational footwear. Ramp tests replicate the conditions of a slip, using a human subject to generate the Coefficient of Friction by walking over a surface. Confidence in the results that the HSL Ramp Test generates has been established through extensive testing and research carried out within the laboratory, and through the correlation of the results it presents with slip accident data collected from the workplace. The ramp test has successfully identified footwear with good slip resistance properties. This has been supported by a workplace trial, where the same footwear made a significant contribution towards a reduction in the incidence of slip accidents (Thorpe *et al.*, 2002). Likewise, footwear that has performed poorly in a workplace environment has been shown to have poor slip resistance when subsequently tested using the HSL Ramp Test.

The current International and European Standard test method for the measurement of the slip resistance of footwear, BS EN ISO 13287: 2007, is a mechanical test, which measures the Coefficient of Friction (CoF) by attempting to recreate the action of a slipping shoe. The test conditions and slip resistance requirements according to this test, with regard to safety, protective and occupational footwear, are defined within an amendment to the series of standards relating to personal protective equipment, BS EN ISO 20344–20347: 2004.

HSL considers the pass thresholds for slip resistance specified within these standards to be too low (Hallas *et al.*, 2009). It is estimated that 35% of slip accidents in the workplace occur in places where safety footwear is worn. Based upon work by the Building Research Station (Harper, Warlow and Clarke, 1961), and later follow-up work (Pye, 1994), the United Kingdom Slip Resistance Group (UKSRG) suggest that a minimum Coefficient of Friction of 0.36 or above, between the shoe and floor surface, presents a low slip risk for a pedestrian (UKSRG, 2005). It is estimated that 1 in 20 people would be at risk of slipping when the CoF is 0.24. Under the current requirements of the safety, protective and occupational footwear standards, footwear can receive a slip resistant marking if it achieves a CoF as low as 0.18, a threshold at odds with biomechanical data.

Ongoing testing at HSL suggests that the current test does not recreate the lubrication effect of a pedestrian slip, particularly when operated under test conditions including less viscous, water-based contaminants. Stick-slip (or sticktion) is a phenomenon caused by the repeated formation and breaking of adhesive contact between an elastomer and a smooth surface, that can lead to unrealistically high measurements of slip resistance in the wet (Di Pilla, 2003). It has been suggested that strong stick-slip effects made the combination of sodium lauryl sulphate solution with a steel surface, an unsuitable test condition for the standards (Sebald, 2009). Stick-slip effects, with respect to slip tests, are due to the delay between the application of the vertical force and the horizontal force or, as with the mechanical friction test, the movement of the test surface. The adhesive contact becomes stronger, the longer the contact time (De Pilla, 2003) as the interfacial fluid is displaced (Strandberg 1985). The slip studies upon which the parameters of mechanical friction test are based suggest that a forward slip starts approximately 0.1 seconds after heel contact (Perkins, 1978, Perkins and Wilson, 1983). This delay is reproduced within the parameters of the mechanical test as the static contact time. BS EN ISO 13287:2007 allows for a static contact time of up to 1 second, ten times the delay observed in the development studies. Studies also suggest that during the measurement of slip resistance, the slip velocity should be at least $0.5 ms^{-1}$, possibly as high as $1 ms^{-1}$ (Gronqvist, 1999). At present, the slip velocity defined within the standard is only $0.3 \pm 0.03 ms^{-1}$.

This paper describes work to study the effect of modifying various test parameters of the mechanical friction test on the measured CoF, with a view to establishing a set of test parameters, which present results that show closer agreement with the HSL Ramp Test under the same test conditions.

Test Apparatus and Method

The STM 603 Slip Resistance Tester

The STM 603 Slip Resistance Tester is a computer-controlled friction test, which can operate to the current European and International standard BS EN ISO 13287:2007 Personal Protective Equipment – Footwear – Test Method for Slip Resistance, or to a user defined test method, within the limitations of the machine and software.

Figure 1. An illustration of the heel test mode (left), the forepart test mode (centre) and the flat test mode (right). Diagrams adapted from BS EN ISO 13287:2007.

Before each test the footwear is washed and rinsed with water and ethanol solution, then prepared by conditioning with P400 abrasive paper. The footwear is fitted onto a shoemaking last (or an artificial foot) and attached to the cage plate of the friction rig. A test floor surface is located upon a carriage beneath the footwear. When the test operates, a pneumatic cylinder lowers the cage plate, and thus the footwear, onto the contaminated floor surface below and applies a predetermined vertical force to it. After a short, static delay time, a variable speed motor draws the carriage and floor surface beneath the footwear. Two transducers measure the horizontal force developed between the footwear and floor surface, and the ratio of the horizontal and vertical forces is displayed as the Coefficient of Friction. The software allows a maximum of ten measurements of the CoF to be made within a single series of tests, from which the CoF is taken as a mean from the last 5 measurements.

Three different test modes (as shown in Figure 1) for footwear are possible, including the foward heel test, a backward forepart test, or a forward flat test. All measurements made throughout this study were made in the 7° forward, heel slip mode.

Test parameters

The software allows for the modification of the following test parameters:

Vertical Force – the vertical force applied to the footwear, perpendicular to the test surface. This has a value of 500!25N for footwear sizes of 40 paris points and above, or 400!20N for footwear sizes below 40 paris points, in BS EN ISO 13287:2007.

Slip Velocity – the velocity at which the floor surface is drawn along under the footwear. The slip velocity is defined at 0.3!0.03ms^{-1} in BS EN ISO 13287:2007.

Snapshot – the snapshot defines time interval over which the measurement of the horizontal force is made. The snapshot is between 0.3 – 0.6 seconds after the start of sliding movement in BS EN ISO 13287:2007.

Static Contact Time – the time interval between the footwear making contact with the test surface and the start of the movement of the test surface. BS EN ISO 13287:2007 defines a maximum static contact time of 1.0s. The static contact time is defined within the software.

The HSL Ramp Test

The HSL ramp consists of an adjustable platform, upon which the floor surface to be tested is positioned. A fall arrest facility is located over the platform to prevent injury to the operator during a test. Water is applied to the test surface at a flow rate of approximately 6 litres per minute via spray bar. Other contaminants can also be used with the ramp to replicate specific workplace conditions, including glycerol, motor oil, sodium lauryl sulphate, and dry contaminants.

The HSL Ramp Test method (HSL SOP-12) requires the trained operator to carry out a series of carefully controlled walks over the floor surface. The walking speed is maintained at 144 steps per minute by a metronome. If the operator completes the walk without a slip or any interference to the operator's gait occurring, the angle of inclination of the ramp platform is increased by approximately 1°. The whole process is repeated until an inclination is reached where a slip occurs, and the angle is recorded. The ramp platform is returned to an angle of inclination a few degrees below the slip angle and the process is repeated. This continues until twelve slip angles have been obtained. The highest and lowest angles are disregarded, and a mean slip angle is calculated from the remaining ten values. The mean slip angle of a second operator is generated in the same way and, if the mean slip angles of the two operators agree to within a tolerance of two degrees, the average slip angle of the two operators (the critical angle) is used to calculate the Coefficient of Friction. If the mean slip angles of both operators fail to agree within the required tolerance, a third operator will generate a mean slip angle.

The UKSRG use a standard rubber compound, slider 96, to assess the slip resistance of floor surfaces. HSL use slider 96 soled footwear to assess the slip resistance of floor surfaces. Slider 96 simulates a shoe soling material of moderate performance, which has low resistance to abrasion and, therefore, can be reconditioned to provide a reproducible surface between tests. Footwear is prepared before each set of walks using an orbital sander and P400 grit abrasive paper.

Results and discussion

Initial CoF measurements and parameter modifications

Initial measurements of the Coefficient of Friction (CoF) were made on a selection of 10 examples of occupational footwear according to the parameters of BS EN ISO 13287:2007, using the mechanical friction test on a steel plate surface with water contamination. The footwear had previously been tested according to the HSL Ramp Test method (Loo-Morrey and Houlihan, 2007), under the same test conditions. The results are displayed in Figure 2.

The results indicate that the mechanical friction test appears to overestimate the measured CoF, when operated to the parameters of BS EN ISO 13287:2007, under this test condition. This is particularly the case at values of the ramp CoF below 0.36.

Initial measurements of the Coefficient of Friction were also made according to the HSL Ramp Test method, and according to the parameters outlined within BS EN ISO 13287:2007 using the mechanical friction test, with slider 96 footwear. Tests were carried out on a smooth, steel surface, with water as the contaminant. The CoF according to the HSL Ramp Test was 0.03, whereas the CoF according

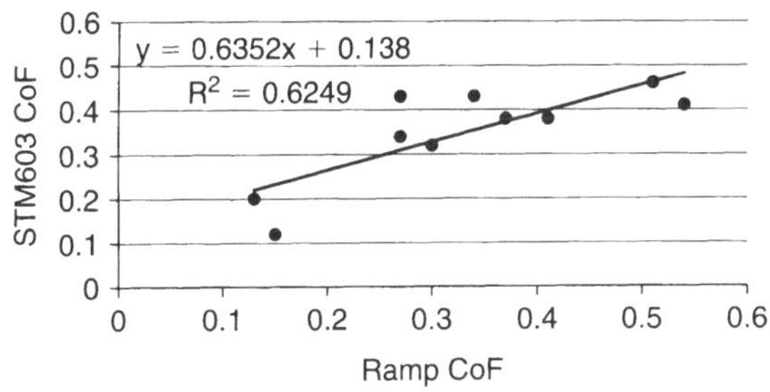

Figure 2. Measurements of the CoF made according the test parameters of BS EN ISO 13287: 2007, for 10 samples of occupational footwear plotted against the results for the HSL Ramp test, on a steel surface with water contamination.

Table 1. The variation in CoF with the mechanical friction test, with increasing slip velocity, for slider 96 footwear, on the steel surface with water contamination.

Velocity (ms^{-1})	0.1	0.2	0.3	0.4	0.5
CoF	0.27	0.27	0.27	0.21	0.17

to the mechanical test to BS EN ISO 13287:2007 was 0.27. The results indicate that the mechanical friction test is overestimating the measured CoF for the steel and water test condition, with slider 96 footwear. Further tests were carried under the same test conditions to study the effect of modifying the parameters of the mechanical test on the CoF.

In order to investigate the effect of the slip velocity upon the measured CoF, tests were also carried out at 0.1 ms^{-1}, 0.2 ms^{-1}, 0.4 ms^{-1} and 0.5 ms^{-1} (0.3 ms^{-1} is the slip velocity defined within BS EN 13287:2007). At velocities of 0.4 ms^{-1} and 0.5 ms^{-1} the snapshot interval was altered to start after 0.15 s and end at 0.35 s. The results are shown in Table 1.

Modifications to the slip velocity of the mechanical friction test, indicates that there is no change to the measured CoF at velocities up to 0.3 ms^{-1}. As the velocity increased beyond 0.3 ms^{-1}, the CoF started to fall significantly, although still remained higher than the value measured upon the ramp under the same test conditions. A slip velocity of 0.5 ms^{-1} was therefore applied for subsequent tests.

Measurements of the CoF were then carried out with the mechanical friction test, at a slip velocity of 0.5 ms^{-1}, with vertical forces of 400N, 300N, 200N and 100N, under the same test conditions. The results are displayed in Table 2. The CoF measurement at 500N is included from the previous tests at modified slip velocities.

The effect of reducing the vertical force at a velocity of 0.5 ms^{-1} also resulted in a decrease in the CoF, bringing the results of the mechanical test closer to those of the ramp test under the same test conditions. The final CoF at 150N is much closer to the result generated using the ramp test.

Table 2. The variation in CoF with the mechanical friction test, with decreasing vertical force at a slip velocity of 0.5 ms^{-1}, with slider 96 footwear, on the steel surface with water contamination.

Vertical Force (N)	500	400	300	200	150
CoF	0.17	0.16	0.13	0.11	0.10

Table 3. The CoF data for the ten samples of footwear on the steel surface with water contamination, and quarry tile surface with glycerol contamination.

	Steel and Water		Quarry Tile and Glycerol	
	HSL	Mechanical	HSL	Mechanical
Footwear	Ramp CoF	Test CoF	Ramp CoF	Test CoF
Sample 1	0.54	0.47	0.40	0.24
Sample 2	0.51	0.35	0.34	0.23
Sample 3	0.41	0.33	0.37	0.28
Sample 4	0.37	0.29	0.32	0.21
Sample 5	0.34	0.22	0.26	0.13
Sample 6	0.30	0.23	0.24	0.11
Sample 7	0.27	0.25	0.17	0.15
Sample 8	0.27	0.13	0.20	0.09
Sample 9	0.15	0.13	0.10	0.07
Sample 10	0.13	0.10	0.11	0.05

Attempts were made to alter the static contact time, but the modified parameters affected the operation of the machine in such a way that any measurement was not possible.

Testing occupational footwear

The modified test parameters from were then used in a series of tests to evaluate the effect on the larger sample of footwear. Measurements of the CoF were made using the mechanical friction test with a snapshot interval of 150–350 ms, a vertical force of 150N, a slip velocity of 0.5 ms^{-1}, a static contact time of 0.5s, and a slip distance of 0.2m. Two different test conditions were used; a steel plate surface contaminated with potable water, and a quarry tile surface contaminated with glycerol at a concentration of 90%. All tests were carried out in the heel test mode. The results are shown in Table 3, along with ramp test results under the same test conditions.

The results for both test conditions indicate that the mechanical test is underestimating the measured CoF compared to the HSL Ramp Test, when operated according to the modified parameters. Changes to the vertical force and the slip velocity seem to have a more significant effect on the measured CoF for some footwear samples more than others. For example, the ramp test indicates that samples 7 and 8 are very similar to each other in terms of CoF, under both test conditions. But for sample 8, the CoF

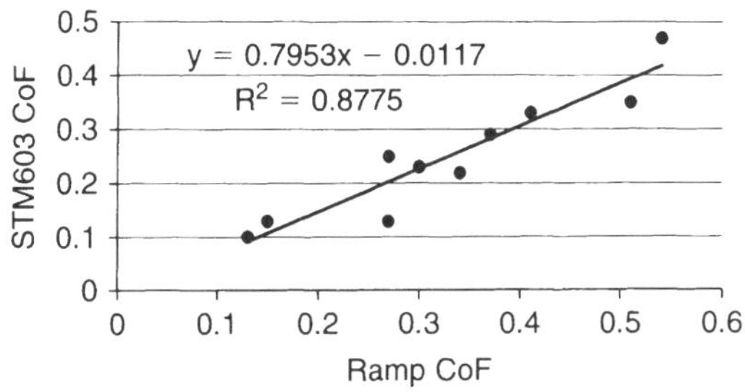

Figure 3. Measurements of the CoF for the mechanical friction test for 10 samples of occupational footwear plotted against the results for the HSL Ramp Test, on a steel surface with water contamination.

is much lower under both test conditions, according to the mechanical test. Differences in the hardness and cleating pattern of the shoe sole may cause a different response to changes in the vertical force, in terms of the measured slip resistance.

Figure 3 displays the data from Table 3, with the R^2 value shown. The plot indicates a linear relationship exists between the results from the ramp test and the mechanical test, for the limited number of samples tested. The mechanical test operated to the modified parameters (Figure 3) has a higher R^2 value, than the test operated to the parameters of BS EN ISO 13287:2007 (Figure 2), for the steel and water test condition. This suggests a closer correlation between the ramp and the modified mechanical test, compared to the results generated according to the parameters of BS EN ISO 13287:2007, for this test condition. Although the gradient of the trend line through the data should be closer to 1 before the mechanical friction test can be considered to show a strong correlation with the ramp test. A larger sample of footwear should be tested, in order to establish the relationship between the two tests with greater confidence.

It is not possible to measure the slip resistance of footwear under all possible conditions. The test conditions used within this study have been selected to be representative of common workplace contaminants and floor surfaces, which are typical of slip accident conditions. Water or water-based contaminants, are considered to be amongst the most common workplace contaminants (Carpenter *et al.*, 2006). The steel surface is representative of smooth floor surfaces with very low surface roughness, which can become very slippery when contaminated with water or water-based contamination. Quarry tile floor surfaces are commonly used in kitchens and food preparation areas, where the floors can be subject to more viscous contaminants, such as oils or fats. Glycerol is representative of viscous contaminants found in such work areas.

Conclusions

Improved correlation between the results generated according to the HSL Ramp Test method and the mechanical friction test is possible, through modification of the parameters of the mechanical friction test. The initial results generated according to the modified parameters of the mechanical friction test indicate that the test is underestimating the measured friction, when compared to the results generated according to the HSL Ramp Test method, under the same test conditions. Separating the contributions of slip velocity and vertical force should make it possible to address both of these points.

The ramp test is itself a biomechanical test and is known to work from empirical data (Thorpe, 2002). Aligning the results of the mechanical test to the ramp test should give a mechanical test that recreates the biomechanics of a slip event. If the parameters of the mechanical friction test can be modified to generate results that present a correlation with those generated according to the HSL Ramp Test, the modified test parameters and conditions may feed into the revision of BS EN ISO 13287:2007. Developing a test method that generates results that show close agreement with results generated according to the HSL Ramp Test, will offer a flexible test that presents a reliable measure of slip resistance. If this can be achieved the mechanical test has a number of potential advantages.

The limitations of any test of slip resistance are acknowledged, but establishing a relationship between the data such slip resistance test methods as those used in this study generate, has to increase confidence in the reliability and validity of the data as an indication of slip potential. Further validation is planned using footwear from actual slip accidents, where HSL have characterised the floor surface, and the contaminant is known.

References

BS EN ISO 13287:2007, *Personal Protective Equipment. Footwear. Test Method for Slip Resistance*, (British Standards Institute, London)

Carpenter J., Lazarus D. and Perkins C., (2006), *Safer Surfaces to Walk On, Reducing the Risk of Slipping*, (CIRIA, London)

Di Pilla S., 2003, *Slip and Fall Prevention: A Practical Handbook,* (Lewis Publishers, United States)

Gronqvist R., 1999, Chapter 19: *Slips and Falls. Biomechanics in Ergonomics*, edited by Kumar S., pp. 351–375, (Taylor & Francis, London)

Hallas K., Thorpe S., Liddle M., Thomas M., (2009), *The Role of Safety Footwear in Slipping Accidents,* Proceeding of the 17th IEA World Congress on Ergonomics, August 9–14, 2009, Beijing, China

Harper F. C., Warlow W. J., and Clarke B. L., (1961), *The Forces Applied by the Foot in Walking: 1. Walking on a Level Surface,* National Building Studies Research Paper 32, (Her Majesty's Stationery Office, London)

HSL SOP-12, (2009), Standard Operating Procedure – Operation of the Ramp Friction Test, Pedestrian Safety Team, Health and Safety Laboratory

Loo-Morrey M., Houlihan R., (2007), *Further Slip-Resistance Testing of Footwear for use at Work*, http://www.hse.gov.uk/research/hsl_pdf/2007/hsl0733.pdf. Last accessed 18/08/2009

Perkins P. J., (1978), *Measurement of Slip Between the Shoe and Ground During Walking, Walkway Surfaces: Measurement of Slip Resistance,* ASTM STP 649, edited by Anderson C., Senne J., American Society for Testing and Materials, pp. 71–87

Perkins P., and Wilson M. P., (1983), *Slip Resistance Testing of Shoes – New Developments*, Ergonomics, Vol. 26, No. 1, pp. 73–82

Pye P. W., (1994), *A Brief Review of the Historical Contributribution Made by BRE to Slip Research,* Slipping Towards Safer Flooring, Proceedings of Seminar at Rapra Technology, Shawbury, Shropshire, 29th September 1994

Sebald J., (2009) *System Oriented Concept for Testing and Assessment of the Slip Resistance of Safety, Protective and Occupational Footwear*, (Pro BUSINESS, Berlin)

Strandberg L., (1985), *The Effect of Conditions Underfoot on Falling and Overexertion Accidents*, Ergonomics, Vol. 28, No. 1, pp. 131–147

Thorpe S., Karnon J. and Lemon, P., (2002) *Industrial Slip and Fall Accidents. Can Footwear Make a Difference?* The Proceedings of the XVI Annual International Occupational Ergonomics and Safety Conference

The United Kingdom Slip Resistance Group, (2005), *The Assessment of Floor Surface Slipperiness: The UK Slip Resistance Group Guidelines Issue 3 2005,* (The UK Slip Resistance Group (UKSRG))

OPEN PAPERS

APPLIED

HOW ERGONOMICS MAKES THINGS BETTER: LESSONS FROM AND FOR PUBLIC ENGAGEMENT

Mark S. Young[1], Fergus J. Bisset[1], Laura Grant[2], Bella Williams[2], Reg Sell[3] & Roger A. Haslam[4]

[1]*Human-Centred Design Institute, Brunel University, UK*
[2]*Laura Grant Associates, London, UK*
[3]*The Ergonomics Society, Loughborough, UK*
[4]*Loughborough University, UK*

'Ergonomics: Real Design' was an exhibition at the Design Museum in London to celebrate the 60th anniversary of The Ergonomics Society. The main aim of the exhibition was to enhance the appreciation of ergonomics in society amongst researchers and the general public, and as a public engagement project a key objective was to facilitate two-way dialogue between ergonomists and non-ergonomists. This paper presents some of the background work to the project, in particular the formative evaluation work which was directed at determining perceptions of ergonomics from both within and outside the discipline. The output was then used to guide the development of the exhibition itself – both in terms of the content and the structure of the message. The paper concludes by extracting key lessons from the exhibition project for future public engagement with ergonomics.

Introduction

In 1949, ten scientists of differing backgrounds got together and formed an interdisciplinary research group to enable the study of human work – a group which ultimately became the Ergonomics Society (Waterson & Sell, 2006). Sixty years on, the discipline and the Society have both grown probably beyond the imagination of that landmark meeting in Oxford. This paper is not about a history of ergonomics, but instead details the development of an exhibition to mark the 60th anniversary of the Society, with the intention of providing insights about public engagement with ergonomics, and how we can better relate to those outside the field.

'Ergonomics: Real Design' was on show at the Design Museum in London, UK, from 18th November 2009 to 14th March 2010. The project was primarily supported by an EPSRC public engagement grant, and – in keeping with the sponsors' remit – its aim was to enhance the appreciation of ergonomics in society amongst researchers and the general public. This meant a two-way dialogue between ergonomists and non-ergonomists – not simply an awareness-raising exercise for 'the uninitiated'.

Public engagement

Historically, the science communication movement has been something of a one-way street. From the Royal Society's publication of the Bodmer report in 1985, to the (then) Office of Science and Technology's Wolfendale report ten years later (both cited by Stilgoe & Sykes, 2009), the focus was on communicating science *to* the public, with little regard for conversations *with* the public. Whilst such reports did act as a catalyst for several worthy science communication initiatives (popular science books, science festivals etc.), the didactic attitude did little to foster trust towards science in an increasingly consumer-oriented public. The prejudices against scientists as being disconnected from the real world, confined in their ivory towers conducting research and making policy decisions that affected us all, came to the fore later in the 1990s as issues of genetic modification, cloning, nanotechnology and nuclear power hit the headlines. Thus, in 2000, the 'Science and Society' report from the House of Lords Select Committee (again cited by Stilgoe & Sykes, 2009) advocated dialogue and engagement with the public in order to not only communicate research, but also to learn from each other about the important issues of concern.

The EPSRC's Partnerships for Public Engagement funding scheme reflects this agenda. Its objectives are consistent with the two-way dialogue – to stimulate the public's interest in contemporary research, to inspire future generations of researchers or engineers, and to encourage public debate about the role of research in society. Most importantly, the EPSRC explicitly state that it is interested in projects that not only increase the public's awareness of science and engineering, but also include some level of dialogue between the public and scientists or engineers.

Why engage with ergonomics?

It is the contention of this paper, as it was of the exhibition project, that ergonomics is better placed for, and stands to benefit more from, public engagement than perhaps many other scientific fields. As an applied, interdisciplinary science, it has direct relevance to the public on a number of levels, as consumers, as workers, and as travellers. Indeed, it has been colloquially referred to as 'the science of everyday life'. Yet, for all of its potential benefits, there seems to be a lack of appreciation of its role in the lives of the everyday public. Thus one half of the engagement agenda might be to foster more informed consumers – in a consumer-driven market, demand for more usable products would have a direct effect on manufacturers. Similar effects may be forthcoming at an organisational level, to make for safer, more productive and more satisfying work environments.

The knock-on benefit for ergonomics would, of course, be that there is then more demand for ergonomics in these areas. But, as a user-centred field, we should be less selfish than this and seek to understand what the users' requirements of ergonomics itself are – a kind of 'meta-ergonomics'. The general public are, after all, both users in the traditional sense as well as consumers of our science. Knowing what they prioritise could direct the future of the field – but first we need to know where we are starting from.

What is ergonomics?

To an ergonomics audience, this may seem like a redundant section of the paper. However, whilst we may all feel that we know what ergonomics is, the explanations and scope of the discipline can vary widely, making communication of the concept difficult. The situation is further complicated by the fact that we have two (or more) apparently competing terms for the subject, with major national societies (including, most recently, the UK) incorporating 'human factors' and 'ergonomics' into their names. Dempsey *et al.* (2000) explicitly propound the importance of a single name for the discipline, and suggest that 'ergonomics' holds sway.

We interviewed eight Registered Members of the Ergonomics Society for this project, to see how experts from within the field defined and communicated their work. They admitted that they found it hard to communicate their work, giving examples of case studies rather than any definition as such, and even avoiding mention of the word 'ergonomics' as it tended to complicate matters. In order to engage with a wider audience, we need to be able to describe the discipline and explain its relevance in a concise, non-technical manner.

The official definition of ergonomics, as set out by the International Ergonomics Association, is 'the scientific discipline concerned with the understanding of interactions among humans and other elements of a system, and the profession that applies theory, principles, data and methods design in order to optimize human well-being and overall system performance' (www.iea.cc). Whilst this may well be an accurate and comprehensive definition of the discipline, it does not easily lend itself to a plain English translation for a non-specialist audience.

At the other end of the spectrum, the original project title – 'making things better' – attempted to capture the essence of ergonomics in a soundbite, as appropriate for the funding proposal. It was felt that this phrase worked on different levels ('making' being a reference to design as much as the overall optimisation process), and was suitably generic to cover a range of applications. However, its weakness is that it would always need qualification. The ultimate exhibition title – 'Ergonomics: Real Design' (which itself went through several iterations, including 'Real World Design' and a suggestion for 'evidence-based design') – emphasises the practical nature of the subject, its direct relevance to the key audience of the Design Museum, and the synergy with the public engagement mission. Again, though, this title lacks a certain explanatory power.

In the inaugural paper of *Theoretical Issues in Ergonomics* Science, Dempsey *et al.* (2000) conducted an informal content analysis of textbook definitions of ergonomics and/or human factors, with the aim of deriving an inclusive and accessible definition of the field. They came up with "…the design and engineering of human-machine systems for the purpose of enhancing human performance" (p. 6). Whilst Dempsey *et al.* (2000) did not explicitly mention public engagement, it is clear that their ultimate goal was in line with that of the present exhibition project – to "[advance] ergonomics to a higher level of societal recognition" (p. 6).

Figure 1. Word cloud for contents listings of Contemporary Ergonomics 2008–2009.

Part of the problem in defining ergonomics lies in the broad, interdisciplinary scope of the field – although this is one of its biggest strengths as a science, it presents a challenge in communicating ergonomics, since it is many things to many people (cf. Salas, 2008). Although this issue is not unique to ergonomics, other fields (e.g., physics, engineering, architecture) are historically blessed with a much more familiar 'core', such that there is wider appreciation of the underlying discipline. With its relatively new nomenclature, ergonomics needs to identify its core and send a consistent message to the outside world. The ergonomists we interviewed underlined this view, finding it difficult to define an archetype example of ergonomics.

To some extent, though, it could be argued that ergonomics has already made some headway in defining its niche, such is the apparently popular (mis?)conception that it is 'all about chairs'. Many ergonomists take umbrage with this assumption, variously rebuking it as dated or unfairly restrictive in scope. Whilst acknowledging that many ergonomists undoubtedly *do* ply a successful trade with chairs, background research for the exhibition project illustrates the much wider variety in the field, and the significant shift in more contemporary research topics.

We looked at the contents listings of the Contemporary Ergonomics book series for the years 2008–2009 and performed a keyword analysis of the conference programmes to form a 'word cloud', a method commonly used for summarising the content of websites (e.g., Kaser & Lemire, 2007). The word cloud was created by copying the contents lists into TextStat (http://neon.niederlandistik.fu-berlin.de/textstat/), an open source concordance analysis software tool (Diniz, 2005; Wiechmann & Fuhs, 2006), which analyses the frequency of words in a body of text, and produces a list of the words with a numerical indication of their respective frequencies. The list was manually refined to remove superfluous terms unrelated to the subject, and the final list was then imported into the online open source word cloud generator 'Wordle' (http://www.wordle.net). Wordle processes the list such that the font size of each word in the cloud is representative of its frequency (see Figure 1).

Although much of the cloud is unreadable at the published level of reproduction, in many ways this is a positive side-effect in that only the significant keywords stand out. The most immediate observation is the overriding dominance of 'ergonomics' as a definitional keyword, appearing 63 times, in comparison to 'human-factors' (hyphenated in the analysis so that it was treated as a single entity, separate from 'human' or 'factors') with a frequency of 19. It was a conscious decision to leave these terms in the analysis, and whilst we appreciate the bias of being at the

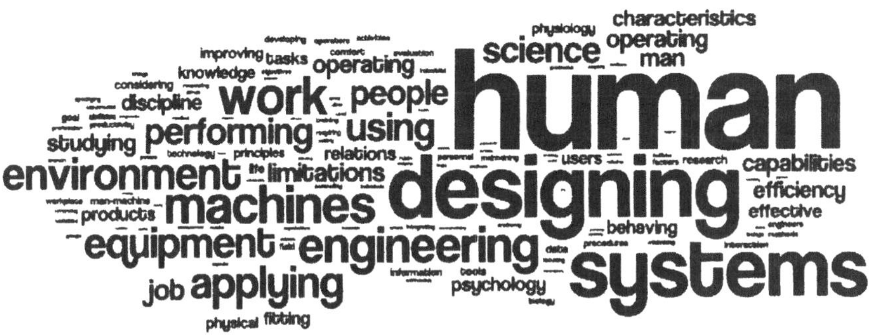

Figure 2. Word cloud for content words from definitions of human factors / ergonomics as identified by Dempsey *et al.* (2000).

Ergonomics Society conference, this analysis appears to support the proposition of Dempsey *et al.* (2000) to adopt 'ergonomics' as a unifying descriptor for the science. As for the application domains, 'design' emerges as the most popular (36), with 'health' and 'safety' both prevalent (29 and 34 respectively), followed by 'systems' (20). The industrial sectors of rail (20) and construction (15) stand out, with transport (8) further down the list.

Where the analysis of Contemporary Ergonomics gives us some idea of current trends in research, it does not necessarily represent the consensus of definitions found in the literature. A similar process was carried out on the content words from definitions of human factors / ergonomics identified by Dempsey *et al.* (2000), the results of which are in Figure 2. Here we can see the obvious primacy of 'human', but after that 'designing' is most common, adding weight to the argument that ergonomics is a design discipline and ratifying the partnership with the Design Museum for the exhibition. Another interesting observation in this analysis is the slight prevalence of 'engineering' over 'science'.

As intriguing as these analyses are, they only serve to reinforce the point that ergonomics is an interdisciplinary, multi-domain science – meaning we still have the challenge of how to communicate this nebulous topic to a wider audience. Nevertheless, we do now have a firm foundation on which to base the development of the exhibition.

Development of the exhibition

The discussion in this paper so far has focused on what ergonomics is, and what ergonomists think ergonomics is. However, this only covers one side of the equation. We (as authors in this paper, and collectively as a discipline) have made many assumptions so far about how others outside the discipline perceive ergonomics. One of the key objectives of the project was to overcome popular stereotypes about ergonomics, but there is actually no firm evidence about what those stereotypes are. The formative evaluation phase of the project entailed finding out what the target audience thought of ergonomics. Such information can then be used to inform the design of the exhibition, as well as acting as a baseline for further summative evaluation following the exhibition run (which, unfortunately, occurred after this paper went to press.

Formative evaluation

Nine informal interviews were conducted with 16 visitors (nine male and across a representative spread of ages) to the Design Museum, on-site at the Design Museum's café. The aim of the interviews was to explore visitors' perceptions of ergonomics to better understand the target audience for the exhibition.

Many of the interviewees had some kind of design experience, especially in architecture, although a significant proportion claimed to have no particular experience. Nevertheless, on the whole there did not appear to be a greater level of awareness of understanding about ergonomics among the more design-literate visitors.

Most people had an idea about what ergonomics was and attempted an explanation. As anticipated, a large number of people cited 'chairs' as an example, though some also mentioned comfort or health and/or safety. In fact, most interviewees felt that ergonomics was about the physical fit of an object or environment; few acknowledged the role of cognitive ergonomics. It was, however, reassuring to see the link made between ergonomics and usability or user-centred design, although a few interviewees expressed misconceptions in this regard:

"*I mean, not that it's an ergonomic thing, but the usability is so fantastic.*"

Similarly, some interviewees felt that ergonomics was less important to them as they were prepared to adapt to poor design in different situations.

In discussing examples of good or bad ergonomics, there were positive and negative examples given for transport (personal and public), environments and spaces (including the Design Museum itself!), technology (i.e., computers), furniture, and tools or equipment. Surprisingly, clothing was also raised as a perceived domain of ergonomics, which caused some internal debate on the project. Whilst there is a valid body of ergonomics research in areas such as military clothing and personal protective equipment, the interviewees perceptions were about everyday clothing such as ladies' shoes and the relationship between clothes and identity. Whether or not this is, or should be, an area of research for ergonomics is a stimulating question for ergonomists; such dialogue captures the ethos of public engagement.

When asked 'where is ergonomics most important', most people related to familiar objects or environments that they used on a daily basis. Again, though, one respondent prioritised clothing as 'the first thing that we're ergonomically connected to'. Interestingly, some interviewees gave an indication of where they had picked up their ideas about ergonomics, with one citing advertising as their source of information, while another stated that it was difficult to find information on ergonomics. There are further lessons in here for ergonomics engagement regarding reliable sources of information, which we will return to later.

The interviews served their purpose in informing the design of the exhibition since they largely confirmed the perceived stereotype that ergonomics is primarily concerned with physical factors, with gaps in understanding identified for more cognitive elements of ergonomics. Moreover, it was clear that people related ergonomics to familiar circumstances they encountered on a daily basis. Most promisingly, though, all were interested in the topic and expressed some level of awareness, indicating that the exhibition would be well-received by Design Museum visitors. These findings helped us to develop a story for the exhibition and pitch it at a level such that it would first attract visitors by appealing to their sense of familiarity, then stimulate further exploration around the wider themes and applications of ergonomics.

The exhibits

From the background work on defining ergonomics and the formative evaluation process, there emerged different options for taking the exhibition story forward. The first was directly driven by the interview results, and was formulated as a 'day in the life' model illustrated with themed real world contexts and associated products. Thus there would be the familiar chairs, consumer products, transport, workplace factors, and even potentially an analysis of the ergonomics of the Design Museum. Whilst this had advantages of accommodating the audience and encouraging reflection on the ergonomic issues within their everyday environment, the counterargument was that it did little for exploring those areas of ergonomics that were less familiar, and so perhaps did not satisfy the aim of 'enhancing the appreciation of ergonomics'.

An academic structure model focused instead on the processes and methods of ergonomics, highlighting through example the user-centred design process. Again, there were recognisable advantages in this approach, but it perhaps swung too far away from the needs of the audience.

The remaining option was to structure the exhibition around classifications of ergonomics, with the two most notable contributions being the models of Pheasant (Pheasant & Haslegrave, 2003) and Wickens (Wickens *et al.*, 2003). Pheasant classifies ergonomics according to anthropometry, physiology, psychology, design, engineering and management, whilst Wickens advocates safety, efficiency, usability, accessibility, comfort and aesthetics. Both offer a useful notation for understanding the breadth of ergonomics as well as the depth of application.

In practical terms, the exhibition development followed a mixed top-down and bottom-up approach, being led as much by pertinent examples that were made available to us by willing colleagues, as it was by the theoretically driven structures described above. Nevertheless, as the exhibition evolved a hybrid of these structures emerged, with elements of each to try and exploit their respective advantages for the public engagement mission. Everyday examples of ergonomics (such as mobile phones, remote controls and tape measures) reflected the 'day in the life' model, but were contrasted with more complex, safety-critical systems such as the design of the CERN control room. Even in these unfamiliar environments, though, the exhibits were designed to highlight parallels with everyday life (such as the design of workstations in the CERN example). Each exhibit was overlaid with notation to describe the particular area of ergonomics that it primarily represented, be it physical, cognitive, organisational, or sub-categories thereof. Naturally most examples did not treat these themes in isolation since it is an interactive system we are dealing with, but the primary theme was singled out for communication purposes. Finally, and through much discussion with our partners at the Design Museum, it was felt important to communicate the ergonomic process as the unique contribution that the science can make to design. The use of structured methods and data is probably one of the most pertinent messages that we as ergonomists can put across to the design community – as opposed to the familiar argument that "it's all just common sense".

At the time of writing, the exhibit list is yet to be finalised, and so a detailed description of the exhibition and its summative evaluation is beyond scope for this paper. To close, then, we discuss the lessons learned so far from this project for public engagement with ergonomics.

Lessons learned

The work conducted thus far for the exhibition project has served to confirm a number of assumptions about ergonomics. Firstly, the popular stereotype of ergonomics being predominantly concerned with physical aspects of design was held up in our interviews with museum visitors. On the positive side, though, most respondents seemed to be at least familiar with ergonomics and some even related it to aspects of usability or user-centred design – although there were also some clear misconceptions in this regard. Secondly, the view from within the discipline was that it is difficult to communicate the essence of the subject due to its broad application and interdisciplinary nature.

One of the key objectives of the public engagement project was to feed these issues back to the ergonomics community and foster a two-way dialogue with the public, an objective to which this paper is aimed. In terms of the lessons learned, then, clearly there is work to be done in broadening perceptions of ergonomics beyond the typical 'chairs' angle, but the fact that people do recognise the role of ergonomics in their everyday life at least provides a hook to get them engaged. On our part, as ergonomists we need to get better at communicating our message and even agree on a common brand or 'core' that will capture the discipline for the outside world – no small challenge in itself. During the course of this project, we have noticed that it is a small minority of the ergonomics community that is consistently willing to get involved in public engagement – and surprisingly, most of this is from industry rather than academia. We hope that one outcome of this project would be to encourage more ergonomists to take up the mantle.

Finally, future developments of this work are planned to feed lessons from ergonomics into the public engagement community, based on the human-centred design process we have adopted for the exhibition. Since ergonomics is to some extent about effective communication with end-users, it seems ironic that we struggle to achieve this at a meta level for the users of ergonomics. One would expect that a user-centred approach to public engagement, as we have tried to exemplify in the exhibition, would serve both communities well.

Acknowledgements

'Making Things Better: an ergonomics exhibition at the Design Museum' was a project funded by the Engineering and Physical Sciences Research Council under its Partnerships for Public Engagement scheme, and the Office of the Rail Regulator. The partnership comprised Brunel University, the Design Museum, Loughborough University, and Laura Grant Associates.

References

Dempsey, P. G., Wogalter, M. S. and Hancock, P. A. 2000, What's in a name? Using terms from definitions to examine the fundamental foundation of human factors and ergonomics science, Theoretical Issues in Ergonomics Science, 1, 3–10

Diniz, L. 2005, Comparative review: Textstat 2.5, Antconc 3.0, and Compleat Lexical Tutor 4.0, *Language Learning & Technology*, 9, 22–27

Salas, E. 2008, At the turn of the 21st century: Reflections on our science, Human Factors, 50, 351–353

Stilgoe, J. and Sykes, K. 2009, A little more conversation. In J. Stilgoe (ed.) *The Road Ahead: Public Dialogue on Science and Technology*. Available at http://www.sciencewise-erc.org.uk (accessed 18th September 2009)

Waterson, P. and Sell, R. 2006, Recurrent themes and developments in the history of the Ergonomics Society, Ergonomics, 49, 742–799

Wiechmann, D. and Fuhs, S. 2006, Concordancing software, *Corpus Linguistics and Linguistic Theory*, 2, 107–127

DO WE REALLY NEED ANOTHER – IST?

Guy Osmond

Osmond Group Ltd

Managers in today's work environment are surrounded by a variety of professionals proffering their services. Health and safety specialists, ergonomists, physiotherapists, occupational therapists and many others (whose titles don't end in – ist) claim to enhance the output of an organisation.

How can ergonomists distinguish themselves and demonstrate that they can really add value to a business? Ergonomics impinges on every aspect but is widely perceived in commerce as an academic subject with little direct bearing on day-to-day activities.

Ergonomics is inextricably connected to many "hot topics" such as environmental impact, return on investment, lean management, absenteeism, performance and productivity yet the word hardly ever features in discussions of these matters.

Introducing ideas and practices from Health & Safety and Human Factors practitioners around the world, this presentation explores ways of making the business community perceive ergonomics as a pragmatic and essential discipline.

Topics covered will include Stealth Ergonomics, Ergonomics ROI, Presenteeism and Environmental Ergonomics.

ALARM REVIEW AND RATIONALISATION IN A HIGH HAZARD FACILITY

Richard Simcock

Ergonomic Systems Ltd

This paper is concerned with a project to rationalise the alarm management arrangements and improve non-alarm indicators within an existing control room located in a high hazard plant. Shortcomings with the current design relate to alarm prioritisation, layout and the number of hardwired alarms and indicators. A multi-discipline approach has been adopted calling on engineering, operational, safety case and human factors personnel to improve the arrangements therein. The processes followed to date include setting alarm principles, a large alarm re-prioritisaton exercise, redistribution of alarms across hardwired and software systems, and the development of alarm and indicator layouts to support operational and safety case requirements. This has been a multi-stage process and the paper reports on the practicalities and lessons learnt from prioritising and rationalising alarms within an existing facility, where the potential for opposition between operator preferences, safety case requirements and possible system improvements is high and needs to be carefully managed. Key principles incorporated into the design proposals have included the integration of hardwired and software alarms, application of rigorous alarm prioritisation principles, setting clear limits on priority between software and hardwired systems, and the incorporation of user requirements whilst avoiding preferential engineering.

DEFENCE

AN INTEGRATED MULTI-DISCIPLINARY RESEARCH FRAMEWORK

K. Tara Smith

HFE Solutions Ltd

It has become commonplace within Government circles to discuss human factors in terms of a Human Factors Integration Plan (HFIP), which identifies the human factors activities and their relationship to the other aspects of a development programme. However, most of these programmes focus on the work that happens after the initial requirements for a system have been established.

Research studies as part of an HFIP are all too often driven by the Early Human Factors Analysis Process, looking at the critical issues, and are often not integrated at an early enough stage to form a coherent understanding of the problem domain – or are not integrated at all. This is often due to the domain being too big to analyse in one coherent manner.

This presentation outlines a framework for investigating capability at the early applied research phase. It looks at the relationship with other aspects of research as opposed to the default system engineering and integrated logistics support, which is so often the main part of a HFIP.

It presents this framework as a case study of research work carried out under the Network Enabled Capability for Close Combat (NEC4CC) programme, which integrated human factors research with operational analysis and technological capability mapping. This culminated in a coherent understanding of what combinations of introduced changes, both technological and procedural, are likely to deliver operational performance benefits.

DESIGN

CONSIDERATION OF USERS IN THE DESIGN OF ASSISTIVE TECHNOLOGY AND TELECARE PRODUCTS

Andree Woodcock

Coventry University

Involvement of end users in the design of assistive technology and telecare products is essential if such products are going to meet the needs of older and disabled people and their carers. Previous work has indicated that designers in other areas of product development find it hard to understand and design for end users who are not like themselves. For assistive technology and telecare products, the risks associated with inappropriate design are high as such products will be relied upon in critical situations.

In order to understand the type and extent of user involvement in the design of assistive technologies, the managing directors of eight SMEs in the UK were interviewed. The results indicate that SMEs in this domain appreciate the need to enter into a meaningful dialogue with end users throughout an iterative design process. Indeed, many of the companies rely on, or have developed out of the experience of clinicians, or sales teams working as, or with, service providers. In some circumstances this may mean that they are not directly involved with end users, or may understand the needs of the service provider more than the direct , primary user. Evidence was found of different types if user studies being undertaken – such as focus groups, usability trials and longterm field trials.

Barriers to greater user involvement included cost; difficulty in accessing users; collating, prioritising and using information from distributed trials and finding end users (from elderly and disabled communities) who are able to provide the right level of feedback needed. The results will be used to inform the development of systems to advise SMEs in their evaluation strategies.

ENVIRONMENTAL

THERMAL ENVIRONMENTS FOR RAIL PASSENGERS

Lisa Kelly

Loughborough University

This presentation describes both laboratory and field studies into the thermal requirements for environments in railway carriages. Rail passengers arrive at a station platform in a number of thermal states ranging from hot through comfortable to cold. They then get on the train and take a seat in the rail carriage. A journey is then undertaken where the environment is usually relatively constant and then the passenger gets off the train having completed the journey. The thermal environments experienced during that whole journey, can include changes of environment from hot to cold for example, and a period of steady state, but maybe the sun comes in through the window or there is a draught. This paper describes research and how it has been brought together to form an holistic model for predicting the rail passenger's thermal experience. When combined with models for other environmental components an ergonomics guideline for railway carriage environments can be proposed.

CLIMATE INDICES FOR OUTDOOR WEATHER CONDITIONS

Ju Youn Kwon

Loughborough University

People die in heat waves and cold snaps due to a lack of advice and preparation. There is a need to provide an early warning that extreme conditions are coming so that appropriate advice can be given and preparation made. This paper describes research into the relationship between outdoor weather conditions and human responses. The objective is to identify a valid thermal index that can be used in weather forecasting to provide a warning of potential threats to comfort of health. 168 people were exposed to outdoor conditions in the compound of the Loughborough University weather station. The experiments took place over all seasons of a year to provide a wide range of conditions for analysis. Physiological and subjective measures allowed existing thermal indices to be evaluated for this application. The results are described and a proposal made.

ERGONOMICS OF THE PHYSICAL ENVIRONMENT – INTERNATIONAL STANDARDS

Ken Parsons

Loughborough University

Professor Bernard Metz was a highly respected Ergonomist with a truly international perspective. He had great vision, a fundamental understanding of Ergonomics and the gravitas, personal warmth, skills and understanding to ensure that the vision was achieved. He leaves a legacy in Ergonomics second to none. In the early 1970s the IEA approached the ISO and ISO TC 159 'Ergonomics' was created. Professor Metz made a significant contribution to that initiative and his subsequent contribution in getting the work started and ensuring that Ergonomics standards were produced was outstanding. There are over forty International Standards concerned with the Ergonomics of the Physical Environment and these have been produced by groups of Ergonomists from across the world under a formal structure of committees, working groups and democratic voting procedures. This paper describes those international standards and demonstrates the considerable amount already achieved and the strategy for developing future standards. This is a world wide activity and a cornerstone of the internationally recognised achievement of ergonomists. Many Ergonomists have been involved and a new generation of standards makers is emerging. Bernard Metz was always a gentleman, friendly and encouraging to all with a profound understanding of different cultures and perspectives. His work in international standards as well as his many other contributions in Ergonomics is a tribute to him.

HAZARDS

A HUMAN FACTORS ROADMAP FOR THE MANAGEMENT OF MAJOR ACCIDENT HAZARDS

Alan J. Jackson

HSE

The EEC Seveso II directive is concerned with the prevention of major accidents which might result from certain industrial activities and with the limitation of their consequences for man and the environment. It is implemented within each member state through local regulation. For the UK this is the COMAH regulations. Both the directive and our regulations require sites which fall under the relevant definitions to demonstrate that they have taken all measures necessary to prevent and/or mitigate the consequences of major accidents. Where humans have a safety related role in this prevention and mitigation then all measures will include human factors. However, many operators find it difficult to link their major accident scenarios to the assured performance of safety related personnel.

This presentation describes a framework that is intended to guide operators through a practical approach for linking their major accident hazards (MAH) to the assured performance of humans engaged on the safety related tasks associated with those hazards. The framework is presented as a human factors journey with key milestones. For each of the milestones there is a link to human factors topics which may be investigated by Seveso inspectors. Most of these topics are described in more detail in the UK Human Factors Inspectors Toolkit published on the HSE web site.

The framework was developed as part of the inspection work done by human factors specialist inspectors in the UK Health and Safety Executive and the UK contribution to the development of guidance for Seveso inspectors in EEC member states. This human factors roadmap for the management of major accident hazards is becoming established as a practical tool for operators who need to get started with human factors and those who have already done some work but need to add some structure to their approach.

A CONSULTANT RECOUNTS HIS EXPERIENCE OF BEING IN A FIRE

Andy Brazier

A B Risk Ltd

The author was recently on a coach that caught fire. Looking back he can see there is a lot to learn from behaviour of himself and the other involved. In particular how the initial signs that something was wrong were dealt with, how people reacted to the first signs of smoke and how the coach was evacuated. Actually all the behaviours observed were already known about based on accidents of the past, but it may be the case that until we are personally involved we never really know what it will be like. This has major repercussions on how we learn from accidents, which may restrict our ability to manage risks.

The author hopes that presenting his experiences from this event at the conference will prompt people working in health and safety, especially where major hazards are involved, to remember that the way people behaviour in emergencies may not always be logical, but is quite predictable.

MEDICAL

HUMAN FACTORS AND ERGONOMICS OF PATIENT AND CLIENT HANDLING INCIDENTS

Ruth Boulton

HSE

Anecdotal evidence suggests that although the use of patient/client handling devices may help to reduce the risks of staff musculoskeletal disorders, their use may be increasingly becoming a significant cause of injury to patient/clients. This (open) paper investigates the scale of the problem and the underlying ergonomic factors involved in patient/client handling incidents.

To explore this subject, a two stage approach was taken, consisting of a quantitative methodology to assess the scale and nature of the problem, followed by a qualitative approach to consider the questions of why there is a problem and what could be done to address this issue.

Data on incidents reported to the Health and Safety Executive (HSE) between April 2001 and December 2007 was examined to identify the key ergonomics influences. Discussions with stakeholders and examination of equipment gave more insight into the background to the problems and indicated some potential solutions.

Data analysis showed that there were over 160 incidents in which patients/ clients were injured whilst being moved. Over 90% of them involved a hoist and of these over 60% involved the injured person slipping from the hoist. The outcome of the incidents varied from minor injuries to fatalities and appeared to be dependent upon what the injured person made contact with when they fell.

Three key topics emerged from the study, all of which were closely inter-related. Failures to undertake suitable and sufficient risk assessments (principally concerning the patient's functional ability) often led to the use of inappropriate equipment. Issues relating to the quantity, quality and safe use of equipment were identified. Finally, there were training issues relating to the safe use of equipment.

Many of these incidents appeared to be foreseeable and with the adoption of better practices of patient/client handling probably avoidable. The study indicated a need for a simple information document to advise about the potential risks when moving and handling patients/clients, to give guidance on best practice and to highlight sources of advice and further information and guidance.

DIGNITY IN DISTRESS: MANAGING AND MOVING OBESE PEOPLE BY WORKING IN PARTNERSHIP

Liz Leigh

Manual Handling Adviser/Ergonomist
Southend University Hospital NHS Foundation Trust
Bariatric Working Group/National Back Exchange Essex Group

As the population gets heavier, there is a need for hospitals and emergency services to establish plans to work together to cope with obese patients, while maintaining their dignity and privacy. This paper will endeavour to explain the processes involved in improving patient safety and establishing a safe solution to the difficulties arising from the moving and handling of bariatric individuals from community care to acute care in the event of a hospital appointment or emergency admission, and back to community care on discharge. It will also explain how the group convinced Chief Executives from the county of the importance to sign up to work together and share equipment with other emergency services to provide more integrated care.

Bariatric: for the purposes of this paper a Bariatric person is defined as someone whose weight is over 19 stone – 120 Kg or has a Body Mass Index of greater than 30.

Introduction

In the United Kingdom the national audit office reported the prevalence of obesity in England had tripled over the last 20 years and continues to rise. Most adults in England are overweight, and one in five is obese. It was estimated that obesity accounted for 18 million days of sickness absence and 30,000 premature deaths in 1998. Treating obesity costs the NHS at least half a billion pounds per year (Lambert 2004).

Based on current trends 50%–60% of adults and 25% of children will be obese by 2012 (DOH Healthy Weight, Healthy Lives A cross-government strategy for England).

Obesity and managing bariatric patients safely is a growing problem well documented by researchers; Rush (2005) said that the number of obese patients had more than doubled in the last decade, and Palmer (2004) suggests handling obese patients can be more complex.

Within the Acute care setting systems were in place for looking after these patients, however the problems experienced by members of the local National Back Exchange Essex group were the difficulties transporting these people. Some patients had been brought into the hospital in the past on the floor of an ambulance because they were over the safe working load for the equipment the ambulance service had. This was very distressing for the patient who was already in crisis and is now being subjected to an undignified transfer. Discharging patients was also problematic when attempting to send them home with dignity and ensuring they had an adequate package of care.

Addressing the problem

To address these issues a bariatric working party was set up with representatives from Acute care, Fire, Ambulance (this included the private sector) and Community care services in an attempt to work in partnership with outside agencies. The ultimate aim of the group was to work in partnership, sharing experience, skills, knowledge and equipment through effective communication.

The group began by establishing a core group of multidisciplinary representatives. Invitations were extended to other disciplines to attend meetings on an ad hoc basis. A care pathway was drawn up (Fig 1 below) and dependant on the section to be discussed the appropriate person was invited to attend (see chart below). This worked very well and a good working relationship has been established

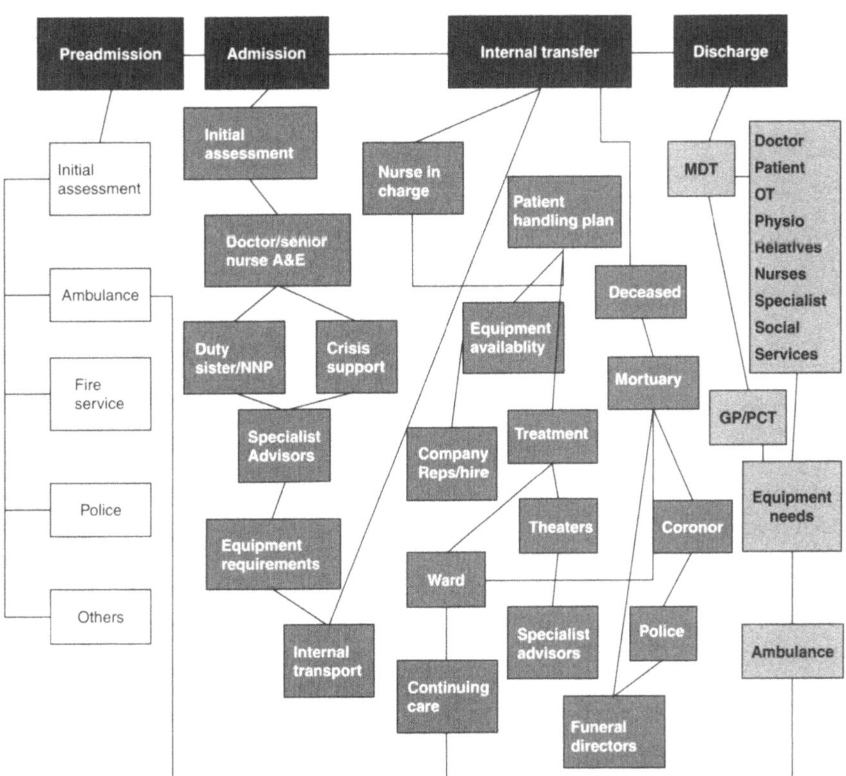

Figure 1. The care pathways of a Bariatric Individual.

throughout. It has improved patient privacy, dignity and safety as well as establishing the sharing of information and ultimately equipment.

First practical mock up exercise

The group wanted to work proactively so they organised a practical exercise to establish how each discipline worked and what equipment they used. A suitable venue was found to enable all disciplines to participate in the emergency evacuation of a bariatric patient. Manufacturers of equipment were invited to bring some of their products along to enable members of the group to have the opportunity to see what is available to them.

Scenarios were developed to enable to participants to practice taking a patient from an upstairs room out into the corridor, down the stairs and eventually outside into a waiting vehicle.

A lot was learned by all and the group had a better understanding of the difficulties encountered by colleagues when faced with heavy patients. Some of the difficulties were due to the unavailability or lack of access to appropriate equipment. This was due to a number of reasons:

- Some groups were unaware of the existence of some of the equipment
- Lack of space on vehicles to store some of the items
- Cost of supplying items that may only be used occasionally
- Competency to use equipment they are unfamiliar with

Following the success of the first exercise the group have staged a number of practical exercises to allow participants to problem solve. The aim was to allow them to be proactive when they are faced with bariatric patients and some of the difficulties this can entail.

Second practical mock up exercise

The second exercise was focused on confined spaces such as stairwells and bathrooms. The group were very grateful to the volunteer who spent the day participating in various scenarios. This allowed the group to problem solve and use a lot of different equipment to evacuate the patient from the house. A number of diverse methods were successfully tried and tested and an excellent time was had by all. The lessons learned by some of the group were the amount of equipment that is available and how it can be used. The problem some of the members continued to experience was accessing equipment due to cost and the amount of times some items may be required.

Sharing equipment

Following this exercise a member of the group experienced a problem with a deceased patient and called a colleague for help. A deceased patient had been brought into their hospital and staff were having difficulty moving them for a post-mortem which was to take place a couple of days later. They had some equipment but it was not appropriate for the needs of the situation. The hoist they had would not allow the person to be prone and the problem facing them was could they share equipment? The decision they made was yes they could. The colleague had the same manufacturers hoist with extra fittings to allow the person to be hoisted prone. Arrangements were made to transport the missing sections to the hospital to enable them to use the equipment in the appropriate manner. This was a great example of sharing equipment which should be extended to others in the county. The difficulty was getting agreement.

Third practical mock up exercise

A third exercise was arranged by the group with the emphasis this time on the fire service. The aim was to remove a patient from a first floor flat. Due to health and safety reasons the volunteer on this occasion was a bariatric manikin. The manikin mimics the movements of a bariatric person. The fire service removed a window to allow the patient to be evacuated on a cherry picker with the assistance of the ambulance service. This gave everyone the opportunities to problem solve together as the access to the building was difficult and some of the equipment was items that are only used occasionally.

Moving obesity forward

The group had successfully been working together in partnership for three years with the support of their organisations. It was now time for them to invite senior people from those organisations to take things to the next level. An awareness day was organised and invitations sent to Chief Executives and Board members of each disciplines organisation. The aim was to encourage the attendees to sign up to sharing information, equipment and expertise to improve the management of the bariatric person in crisis. At present there is a lot of reluctance to sharing equipment for a number of reasons. These range from legal issues such as who is responsible for the equipment to who has the expertise to use it.

Attendance at the event was very good, despite there being a couple of Trusts who did not acknowledge receipt of the invitation or attend. A number of presentations were given to the audience in an effort to given them a better understanding of the difficulties the disciplines experienced when the patient is bariatric. The first presentation was 'scene setting' from the ambulance service and examples were shown of the sort of problems the group had identified when attending an emergency call. This was followed by a video clip of a patient's perspective of being handled in this sort of situation. The patient told the audience how they had felt about being handled by an ambulance crew and her worries about possible injury to staff that had been forced to carry her at the time. They also revealed their feelings on how they felt when attending a clinic and the embarrassment it caused. A representative from the Health and Safety Executive went on to tell the audience 'that they had a duty of care to this patient group and they would be expected to provide the appropriate equipment etc. They also said that the client group was foreseeable and there was a need to plan, communicate and train staff for those occasions these patients required health care'. The final presentation informed the audience of the need to take things up a level. The group had done a lot of work over the past three years but they were unable to take this forward without input from the right people. There were lots of benefits to signing up to working in partnership and sharing equipment and information. The group wanted to be proactive rather than reactive. Resources needed to be cost effective. Hiring equipment can be very costly to an organisation and there is the added risk it will not be available when it is needed. This could create a risk to staff and patients if someone is injured when equipment is not there.

The reputation of an organisation is fundamental. Taking proactive steps to manage bariatric patients can enhance their quality of care. This will help ensure a positive outcome and encourage them to recommend the service provider as the provider of choice. It could make cost savings to some organisations. Essex County Fire and Rescue Service have the highest number of calls to bariatric people in distress or at risk in the UK. If equipment were to be shared this may reduce the number of occasions the fire service needed to respond to provide the manpower to assist with patient type. The ambulance services need to provide more re-enforced specialist vehicles for transporting this patient group. The private sector currently provides the bariatric ambulances in the county and this service goes a long way in improving the dignity and care of these patients. There are an insufficient number of vehicles for the number of patients and more are needed. This would reduce the length of stay in hospital for some patients who remain for longer periods when appropriate transport is not always available when they are ready for discharge.

The day also allowed attendees to view some of the equipment the group had used during their practical exercises. This was to persuade those with responsibility for funding how important some of the items could be in their organisations areas of work. The purchase of equipment could be either for total use by the organisation or with a view to share.

Findings

The group had some very positive feedback from attendees. Some were very keen to sign up to the prospect of sharing resources. The group finally received the commitment from the audience to plan the next level. The plan now for the group is to work out:

- How we share equipment successfully
- Who purchases which items of equipment
- Where we store equipment to give access to various disciplines
- Who is responsible for monitoring the use of the equipment

- Who and how do we train for competency
- Which other organisations do we need to include

Discussion

The mock up exercises were beneficial to the group as they identified areas that could be improved. As a result of working in partnership there have been a number of improvements. We have achieved enhanced communication exchange and information is given to the appropriate group of people much earlier when dealing with a particularly difficult transfer. The private sector of the ambulance service have invested money and commissioned more vehicles to supply a service to those caring for this group of individuals. One in particular commissioned a support vehicle which carries a variety of equipment to assist in the transfer of a bariatric patient. The vehicle can take extra crew members as well as equipment to support the ambulance crew. This has not only improved the working conditions but also the morale of staff and gives the patient the dignity they deserve. Where local ambulance service do not have a bariatric ambulance they have sub contracted the service of transporting bariatric patients to the private sector.

Community staff contacts the acute trust more frequently to discuss the needs of patients and equipment when discharge planning is proceeding. Occasionally staff from the acute Trust have assisted in assessing the patients home in an effort to discharge the patient earlier as this has been more cost effective.

The fire service now benefit by not receiving as many calls to assist in the manual handling and transporting bariatric patients in the area.

The project has been a learning curve for all those involved. The group are still enthusiastic and are contributing to the improvement of the service. There is still a lot of work to do if they are to complete the project. It is comforting to know they have the support of their organisations to take things forward. In the meantime the group are organising another practical mock up exercise. This time it will focus on the ambulance service and the issues they face with evacuating bariatric patients involved in road traffic accidents.

References

Rush, A. (2005) Use of specialised equipment to mobilise Bariatric patients. *International Journal of Therapeutic Rehabilitation.* Jun: 12(6): 269

Lambert, C. (2004) The way we eat now. Harvard Magazine (May – June 2004)

Palmer, R. (2004) Moving and Handling Bariatric patients safely: a case study. *International Journal of Therapeutic Rehabilitation.* Jan: 11(1): 31

Health Services Administration. (2006) Avoid Risk from obese patients with specific policies. *Healthcare Risk Management.* Jun 28(6): 65

METHODS AND TOOLS

EVIDENCE-BASED ERGONOMICS

Joanne O. Crawford

Institute of Occupational Medicine

Evidence-based medicine is seen as the gold standard in supporting and integrating clinical expertise and systematic evidence. The uses of systematic reviewing techniques are already being used by professionals in ergonomics to summarise evidence and identify good practice in a variety of working environments. Internationally the Cochrane Collaboration Occupational Health Field hold information on musculoskeletal health and mental health. In the UK, NICE have also expanded their role to include workplace public health interventions. The aim of this presentation is to summarise the different methods used in the review process. One of the difficulties in reviewing ergonomics research is that review methods normally used to assess randomised control trials are not always relevant because more research is either cross-sectional in design, observational or based on consensus. However, other more pragmatic methods can be used to assess and synthesise evidence and examples of this will be shown. The use of systematic reviewing techniques in ergonomics should be a positive step in increasing the evidence-base for decision making and guidance occupational environments and interventions. Importantly, using systematic reviewing identifies both the evidence that exists and just as significantly, data gaps and where future research needs to be focussed. Furthermore the use of economic analysis techniques, if applicable, will be able to identify the costs and benefits of interventions. In conclusion, the authors will suggest that the use of an evidence-based approach will help develop the scientific basis of both ergonomics research and practice.

A VALID OBSERVATIONAL DATA COLLECTION SYSTEM FOR AIRPORT SECURITY CHECKPOINTS

L. Hay[1], A. Zar[1], A. McClumpha[1] & C.G. Drury[2]

[1]*Booz & Co. UK*
[2]*Applied Ergonomics Group Inc., USA*

Performance measurement for airport security systems is a difficult trade-off between cost and validity. For X-ray screening the Threat Image Projection System (TIPS) provides a performance measurement tool that is at least face-valid, but X-ray is only one checkpoint task. This project developed and validated an observational measure for non-X-ray checkpoint tasks such as hand search of baggage or passenger pat-down.

Introduction

Measuring the performance of Security Officers (SOs) on airport checkpoint tasks other than X-ray screening of carry-on items has traditionally meant using Covert Test trials to probe the security of the overall aviation security system. Covert Test is expensive and difficult to schedule on a regular basis, for example to provide performance feedback to individual SO or to aggregate into overall performance metrics for a checkpoint or an airport. For X-ray screening the Threat Image Projection System (TIPS) data collection system can provide such measures and it is at least face-valid. The challenge of this project was to produce a cost-effective system with characteristics similar to TIPS for evaluating the non-X-ray tasks at checkpoints. There have been proposals in a number of countries for "key performance indicators", but no validated system has been developed. Stolzer et al (2008) showed a system for aviation safety. The basis for the system developed here was a detailed analysis of the tasks involved to determine where the opportunities for error existed, followed by developing an observation tool that concentrated observers onto these error opportunities.

Developing the Observation System

The observation system consisted of observation forms for three of the five major checkpoint tasks: People Search, Bag Search and Electronic Trace Detection (ETD) Search. Their aim was to measure how well Security Officers performed each task. Each form focused on those actions taken by the SO that directly impacted probability of detection of security threats. For all checkpoint tasks a detailed Hierarchical Task Analysis (HTA) was performed, using SO interviews and observations. For each task at the lowest level, the possible errors were listed. These HTAs and error types were edited by personnel at a security company for sequence and content. A group of these security personnel then rated the impact of each error on the failure to detect a threat object, using a scale from 0 meaning no effect on detection to 100 meaning this failure would lead to missing a threat if present. The non-zero errors were then formed into observation questions and observation forms designed for three different checkpoint tasks. The initial forms were tested with users and modified to be better suited to user needs. Each form produces an overall measure of task performance impact: The Total Impact Score or TIS, which is the number of observed errors multiplied by their impact.

Evaluating the Observation System

Despite this care and detail in developing the observation forms, a direct test was still required of whether the outputs recorded by observers agreed with the objective reality of how a SO performs each task, i.e. a validation experiment. Of course having a valid measure is no use unless that measure can be consistently observed, i.e. have high reliability. For an aviation security measure relying on observation we defined this as the observer-to-observer reliability. A third aspect of any measure is its sensitivity. Not only should the observation agree with ground truth, it must be sufficiently sensitive to differentiate between good and poor SO performance on a task. The validation experiment was designed and conducted to provide quantitative evidence for reliability, validity and sensitivity.

In designing the validation experiment we used 12 experienced aviation security observers specially trained in using the developed forms, and had them view video scenes of SOs performing the three tasks. Video was used as it ensured that each observer saw exactly the same SO/passenger interaction, removing one source of variability that would have been present using live passengers at a checkpoint. In fact, to ensure that observers were not time-constrained, we allowed them to pause and rewind each video scene until they were satisfied that they had seen all errors. For each task, twelve video scenes were scripted and produced at a checkpoint using SOs. These video scenes comprised three levels of ground truth severity: Perfect Search, Minor Errors and Major Errors, with four different examples of each level. A doubly-balanced experimental design was used to control the order in which each observer viewed each video scene and the order of the three tasks. Each observer recorded his/her evaluations on the appropriate task form, and the whole data set (10,800 observations) was coded for statistical analysis.

The analysis measured reliability by calculating the correlations between all pairs of Total Impact Scores across the 12 observers. The correlations were uniformly high, averaging about +0.9 for all three tasks. Out of the 792 correlation coefficients calculated, only 6 (4.5%) were not statistically significant. This indicates a remarkable level of agreement between observers and thus very high reliability. To analyze the validity, correlations between the mean TIS scores for each scene and the ground

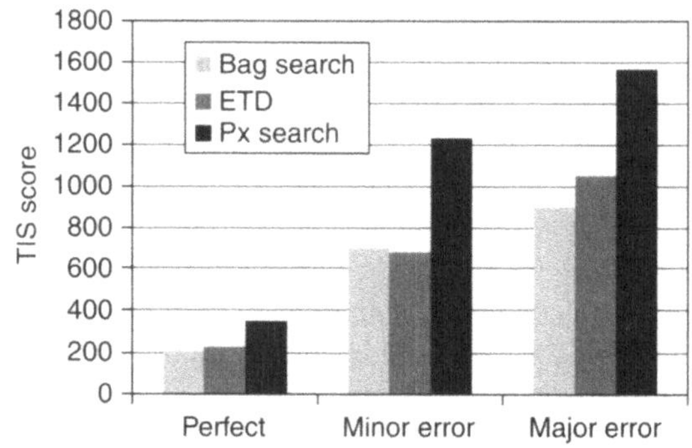

Figure 1. Sensitivity of observation system for detecting errors at each severity level.

truth TIS score were calculated. All were highly significant. The actual validity values ranged from about 0.5 to 0.8, again remarkably high for an observation system. Sensitivity was measured by whether ANOVAs could detect differences between perfect search, minor errors and major errors. For each task, the difference in severity TIS scores was highly statistically significant, with $F(2, 99) > 100$ ($p < 0.001$) for each task. Perfect searches scored lower than minor error searches which in turn scored lower than major error searches, as requited to demonstrate sensitivity. The mean values for each task are shown in Figure 1. Note that the absolute values are higher than expected, meaning that the observers detected more errors in each scenario than had been scripted.

Conclusions and Future of the Observation System

The conclusions from the validation study are that the observation system was proven reliable, valid and sensitive. This is important information as other proposed systems for observation of performance indicators have never been tested formally for these measures. However, there are still some issues to be resolved before large scale deployment of the observation system:

1. Allowing unlimited time to view video clips may not be realistic in an operational setting. The effect of observation time on observer detection performance would be useful to measure. Choosing an appropriate observation time may also bring the absolute values of the TIS into line with the "ground truth" values in the scenarios.
2. Not all checkpoint tasks were covered in the work so far, which concentrated on those seen by management as most error-prone. The methodology is replicable for these other tasks, and indeed for any new tasks added at checkpoints. Other countries can also use the methodology, modifying the HTAs to reflect their tasks.

3. The observation forms themselves can be improved. Although an initial trial gave some improvements, the more extensive validation study yielded more places for improvement.

Reference

Stolzer, A. J., Halford, C. D. and Goglia, J. J. (2008) *Safety Management Systems in Aviation* (Ashgate Studies in Human Factors for Flight Operations). Ashgate

OCCUPATIONAL HEALTH AND SAFETY

TASK PERFORMANCE DURING STANDING EXPOSURE TO WHOLE-BODY VIBRATION

William Baker & Neil Mansfield

Standing people experience whole-body vibration in many environments, performing a wide range of tasks (Mansfield, 2004). This paper compares two studies, designed to investigate the performance of specific types of manual control tasks, namely discrete and continuous tasks.

Method (Overview)

In both studies, participants were exposed to horizontal random vibration stimuli (frequency band-limited up to 4Hz), and were required to perform the necessary task in two standing postures (fore-and-aft and lateral feet positions). Objective measures of performance and subjective ratings of workload were recorded in each study.

Results

Study 1: sixteen participants performed a timed pegboard task, representing discrete manual control. Performance was based on the time taken to complete the task and showed a progressive increase (impaired performance) with increases in vibration magnitude (Figure 1 (a)). Subjective ratings showed similar trends to performance data. Impairments due to dual axis exposure were well predicted using r.s.s. summation calculations based on single axis components. The results indicated that best performance for those standing in moving environments could be achieved if the adopted posture is in the same direction of movement, for example, a lateral posture with the most severe vibration in the y-axis.

Study 2: twenty-one participants performed a lane-change task using a driving simulator (continuous manual control). Objective measures of performance were based on deviation from the desired position on the lane. Vibration exposure showed no significant decrements to objective performance, however subjective ratings showed that the workload experienced by participants increased with corresponding increases in vibration magnitude (Figure 1 (b)). For both objective and subjective measures, no postural effects were found and the effects of vibration direction were variable. This study identified that participants were able to maintain continuous control performance, at the expense of increased subjective workload.

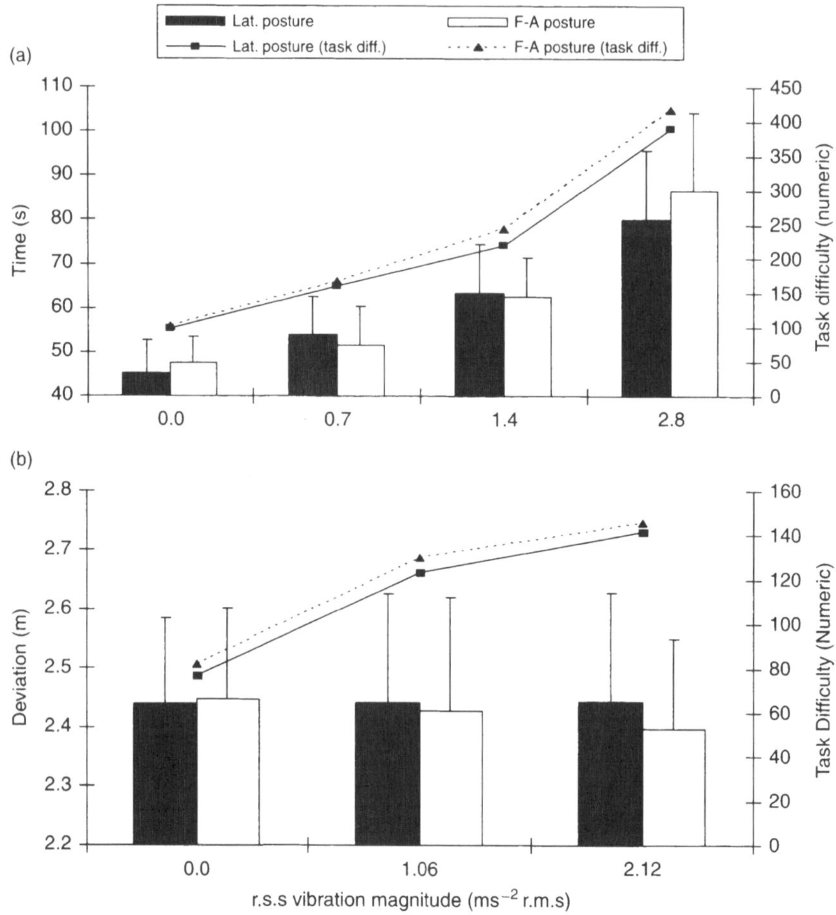

Figure 1. Task performance and subjective ratings of task difficulty during dual-axis vibration exposure for (a) discrete pegboard task and (b) continuous lane change task.

Discussion (Comparison)

Ratings of task difficulty for the discrete pegboard task were significantly higher than those for the lane change task. Additionally, the difficulty ratings for the lane change task showed a slight trend towards a 'plateau', with little further increase in task difficulty ratings. These results suggest that despite the increased workload (due to vibration exposure) experienced for both tasks, individuals were able to adapt and maintain performance in the less difficult (lane change) task. For the more difficult (pegboard task), exposure to vibration could have exceeded the participant's capacity for adaption and therefore resulted in degraded performance.

Conclusion

Exposure to vibration resulted in progressive decrements to objective performance for the discrete pegboard task, whereas for the continuous lane change task, objective performance was largely unaffected. Regardless of type of task, subjective workload (task difficulty) increased with vibration magnitude and in contrast to previous work (Buffington *et al.*, 2006), body posture showed no significant influence on objective performance.

References

Buffington, C. W., MacMurdo, S. D. and Ryan, C. M. 2006, Body Position Affects Manual Dexterity. *Anaesthesia Analgous*, 102: 1879–1883

Mansfield, N. J. 2004, *Human Response to Vibration*. London: CRC Press

ORGANISATIONAL

A PRELIMINARY STUDY ON TRAINING INTERVAL FOR COMPUTERIZED EMERGENCY OPERATING PROCEDURES

Zhizhong Li

Tsinghua University

The increasing complexity and the use of computerized displays in the control system of nuclear power plants have brought new issues to the training of operators, especially for emergency situations which are known of high stress. This study focuses on the time interval of training and its influence on the operation performance (operation time and error rate) of computerized emergency operating procedures. Three levels of time interval (one-week, two-week, and four-week) were examined. The results indicate significant influence of time interval on operation performance. In general, frequent training results in shorter operation time but not necessarily lower error rate. Learning and forgetting effects can be seen in the retraining process.

WORKSHOP: HF PRACTITIONERS – IS IT TIME TO LOOK AT WHO WE NEED ON THE GROUND?

Richard Simcock

Ergonomic Systems Limited

This workshop follows on from a piece I wrote for The Ergonomist regarding the potential for embedding human factors into the fabric, culture and day-to-day operations of organisations. I am sure we are all aware of that even now in 2009 the question of 'how to effectively integrate ergonomics…' still arises on a reasonably regular basis. I think it is time to take a different view of how we could go about doing this.

Behaviour Modification, Behavioural Safety, Human Performance Improvements processes and teams are more and more prevalent, particularly within high-risk organisations. Often, such initiatives focus largely on improving the behaviour of front line personnel in order to reduce errors (which is not ideal!) but they can have a lot of clout with senior managers. Concerns in this area have been raised previously by industry regulators. As human factors professionals I believe we need to drive the processes for making sure sound principles are embedded into organisations and that our discipline is not misunderstood.

I have worked with operators and team leaders on the ground on various assessments and these people are a real asset in highlighting and addressing human factors shortcomings. A set of 'eyes and ears on the ground' if you will.

I believe human factors and ergonomics professionals need an open and honest review of our training and capability arrangements. Current models need to be challenged, and if appropriate, updated. Maybe it is time for us to look beyond just having degree qualified human factors professionals, what about human factors people on the ground, embedded into organisations? The intention is not to supplant professional ergonomists, but assist them, for example with supporting personnel embedded within organisations who could have some level of responsibility for policing and implementing human factors requirements.

Using a simple comparison, NEBOSH recognises the need for different levels of occupational health and safety roles. Certificates are a recognised qualification for people (often on the ground) with some responsibility for safety. These people don't define or set up the systems, but they monitor and police them. The need for safety professionals is still recognised, and this is what the diploma is in place for (much like we have registered ergonomists). These different roles have helped occupational safety become part of the fabric of organisations. If we expect this of human factors, I think we have to at least consider such possibilities.

I would like the workshop to openly discuss current arrangements, the potential need for change, possible alternative and supporting arrangements, along with potential benefits, shortcomings and areas to be managed.

PSYCHOLOGY AND COGNITIVE

BEHAVIOURAL MODIFICATION PROGRAMMES, THE NEED FOR A FULLY INTEGRATED APPROACH

Farah Waheed

AMEC

Annually, the Health and Safety Executive (HSE) reports on the accident figures and fatalities across different sectors and industries in Britain. Despite being one of the safest in Europe, the British construction industry consistently reports high levels of fatalities, work place accidents, occupational health illnesses and near misses when compared with other sectors and industries (e.g., manufacturing). Although the situation is improving and we are witnessing a decline in incidents and accidents over recent years, the legacy and reputation of the construction sector in practising poor safety related work practices and behaviours still remains.

Studies (e.g., BOMEL Ltd and Loughborough University) have cited a number of factors leading to such a high rate of accidents and fatalities. Most of these factors are related to worker behaviour and poor supervision, neatly summed up by this press release by the HSE in 2003 "The biggest single cause of construction accidents is simply getting to the workface".

It is unsurprising therefore that promoting and encouraging the practice of safe behaviour at work has become an increasing priority and gained in prevalence in many industries over recent years. The various strategies and interventions include behavioural-based safety, safety observation and behavioural safety management systems which, despite the different titles share the same objective and purpose of achieving behaviour modification to promote safety at work.

This open presentation describes an attempt to introduce the first phase of a behaviour modification programme to a bio-fuel plant undergoing construction. The findings of the intervention compliments the literature (e.g., Anderson 2004) which highlight a number of factors influencing the success of behavioural modification programmes and illustrates that they are highly susceptible and influenced by compounding individual, job and organisational factors.

THE FORENSIC ERGONOMICS OF DISTRACTION ERRORS: A COMPUTER SIMULATION

Derek J. Smith

UWIC

Reason (1990) recognised that the applied science of cognitive ergonomics needs a better understanding of how the human "information-handling device" can be capable of operating correctly most of the time, only to fail unpredictably from time to time (Reason, 1990). Unfortunately, models of cognition still struggle with the phenomenological and real-time cybernetic aspects of nervous system activity. With Reason's theoretical challenge in mind, this paper will introduce Konrad, a prototype artificial intelligence testbed based upon a powerful metamodel of the human mind. Konrad's data model identifies more than 250 functionally discrete memory types and 21 major information processing mechanisms, and records the respective involvement of each of these resources in simulated cognitive tasks to print-out. Konrad's performance on a simulated kitchen chopping task, in which higher cognition was distracted from the task at hand by a verbal interruption, will be presented, and attention drawn to the detailed mechanisms by which one motor schema is discontinued in favour of another. Results show how oversensitivity or undersensitivity of the interrupt mechanism are both capable of impairing the overall performance. Possible improvements to the conventional models of hierarchical control are discussed.

DIGITAL INCLUSION: IS IT TIME TO START TAKING AN EXCLUSION APPROACH TO INTERFACE DESIGN?

M.D. Bradley[1], Y. Barnard[2] & A.D. Lloyd[3]

[1]*School of Engineering and Information Sciences, Middlesex University, London, UK, N14 4YZ*
[2]*Institute for Transport Studies, University of Leeds, Leeds, UK, LS2 9JT*
[3]*University of Edinburgh Business School, The University of Edinburgh, Edinburgh, UK, EH8 9JY*

The digitally excluded (those who do not engage in the digital economy) form a substantial proportion of the UK population. It is suggested that some of the factors that cause exclusion can be designed out through understanding the prior knowledge, cognitive requirements and mental models that the interface requires for the use of the application. This paper proposes an 'exclusion calculator' approach for assessment of digital interfaces, as well as for development of guidelines for interaction designers.

Introduction

In the UK, 21% of adults and 64% of those over the age 65, have never accessed the internet (ONS, 2009). An estimated 17 million people over the age of 15 are considered digitally excluded, and 70% of non-internet users fall into the socio-economic grouping C2DE; manual workers, state pensioners, casual & lowest grade workers and welfare state beneficiaries (Milner, 2009). These 'non-users' can be classified either as 'Rejectors'; those that do not wish to engage with the internet, and others as 'Excluded'; those who do not have the skills, time or the resources to use it – the split between these groups for non-users is roughly 50:50 (Milner, 2009).

Selwyn (2003) warns against assuming that the use of information technology is always good for everyone. For some non-users the use of the internet would not fit in with their way of living. However, before people can decide whether or not to become a 'Rejector' they have to experience computer use. This poses the dilemma that some non-users may profit from being introduced to information technology and decide to make it part of their lives, while others may experience how it works but will fail to see how this could be useful for their personal goals.

There are many aspects (social, psychological, financial, as well as technical) to introducing non-users to internet use, but in this paper we focus on the interaction with computer interfaces. When a novice user has taken the step to get behind a computer, for example in a community centre or in the house of a friend, and he or she tries it out, the novice may come to the conclusion that it is too difficult and impossible to learn. A novice user who is insecure about this, will need an interface that boosts confidence and invites the user into further exploration.

This paper proposes that some of the skills that are currently required for internet use are due to the software interfaces which dominate the Personal Computer (PC) user experience. These interfaces have been developed and honed over many years for PC users (regardless of skill level as a PC user), rather than having been optimised for lowering the barrier to entry for non-users. Interfaces designed specifically for elderly users, such as the 'Eldy' interface developed by Vegan Solutions, and incorporated into the SimplicITy computer also marketed to older users and those who find computers too complicated, (Eldy, 2009 and SimplicITy Computers, 2009) tend to reduce the range of options and visual clutter, as well as facilitate mouse use through use of larger cursor and buttons to click onto. However, detailed strategies and guidance for user interface designers to design for this group are not readily available.

Guidelines for designing to minimise exclusion, exist in detailed form e.g. through the inclusive design toolkit (Clarkson *et al.*, 2007), however information for designing to minimise digital exclusion, is not specifically available. Within the toolkit are exclusion calculators, which allow designers to assess the demands that their product requires of the users, and enables them to estimate the percentage of population that will be excluded. This currently uses data from a national survey of disability from a follow-up to the 1996/97 Family Resources Survey, completed for the Department of Social Security. Whilst the exclusion calculator does address cognitive issues, it requires the user of the calculator to relate the cognitive load of their interface to tasks that were used in the original survey, such as 'Remember a message and pass it on correctly' or 'Do something without forgetting what the task was whilst in the middle of it', and sums the number of tasks (out of a possible 13) to provide an overall score. This is insufficiently detailed to provide the ability to compare or assess detailed interaction elements.

The concept of designing for the user's mental models rather than for the designer's mental models, is one which has been around for some time (Norman, 1988), however details of how to achieve this for regular users is not made explicit for designers. In the context of designing for the 'non-user' this task becomes both more difficult and more important.

The view of digital technology from the digitally excluded

One current internet browser designed for the PC presents the user with 19 icons and 8 visible menu items to choose from, excluding any webpage content. Other applications can offer significantly more – one popular word processing application offers 86 visible icons to the user on opening and a popular web based email 47 visible icons. Since expert users are able to ignore visual noise to find and select the options that they need to achieve their tasks and goals, this level of complexity does not hinder their performance unduly. Novice users are less able to do so and are therefore hindered in their tasks by the availability of too many options early on in their exposure to the interface.

One solution to this is to develop a different interface for novice users and should they wish to, they can migrate at a later date to a more sophisticated interface when they are ready. This solution has been dismissed (Raskin, 2000), but implicitly this is the approach used in the Eldy interface which once mastered, gives users some of the skills they need to migrate to using a fully featured conventional interface.

Basic PC use requires the user to have mental models regarding mouse interaction with the on-screen cursor, that a circle with a vertical line intersecting one part of it indicates an 'on' and 'off' switch, a long button press can have different functionality to a short press (long buttons presses can be necessary to activate a computer, as a short stab at the power button often will not suffice), that there is a difference between a button that may require only one left mouse button press, and an icon that might require two rapid left button mouse presses with no movement of the mouse to activate. The list of current interface elements is extremely long but not, as yet, quite endless. This prior knowledge that many experienced users have is now almost second nature – most of us no longer need to consciously process the difference between a 'double click' icon and a single click icon due to the sophisticated mental models we have developed over extended periods of using the software and which are now automated below the conscious layer. So has the barrier to entry for the digitally excluded to make use of digital technology interfaces been increased needlessly to meet the perceived sophisticated needs of expert users?

Some mental models held by older users can be in direct conflict with mental models required to interact with technology. For example, the belief that playing with something will 'break it', stifles the urge for some to explore the response of the system to varied inputs which would enable them to develop an accurate mental model of an interface, and hence lengthens the learning process. Some mental models of the system will need to be learned (e.g. that emails need email addresses in a particular format, and are sent electronically rather than by printing and posting, or that a space in a typed web address will render it useless), but it is argued that complex mental models for interactions should only be required during graduation from a novice to an expert user.

Conversion of non-users to using digital technology, usually is prompted by a specific functionality or facility that it can offer. Figure 1 shows a simplification of how a novice user might perceive the journey of learning how to use a piece of technology, taking into account the perceived difficulty of learning and then the opportunity of how much easier the particular task might be once they have acquired the necessary skills. It is proposed that it is both possible and desirable to lower both the actual height and the perceived height of the 'Max learning pain' part of the figure, to encourage and facilitate the use of digital technologies, through interface design which takes into account the needs of novice and digitally excluded potential users.

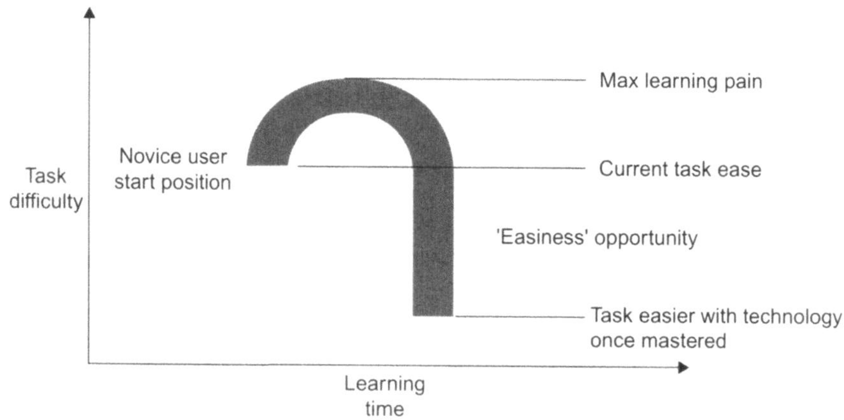

Figure 1. The Novice Digital Technology Learning Hump – Task Difficulty vs. Learning Time.

Digital interface design elements, 'noviceability' and prior learning

Langdon *et al.*, (2009) showed that novice users' ability to adapt and learn different interface interactions was strongly dependent on age and cognitive capability, particularly for interactions requiring working memory, perception and reasoning. They also argue that product interfaces should be designed to be similar to those of products made in the past to maximise inclusivity, and also that interfaces should permit the discovery and chaining of action rules. Clearly with some software interfaces, the functionality and references may not be applicable to products that the users will have experienced before, and in such situations the interaction elements should be explicit in their function, operation and feedback for novice use.

Capturing and communicating excluded user's needs

If we can identify through the use of a task analysis technique in combination with a cognitive model the interface elements that will prove difficult to learn to use and are prone to errors in use for excluded users, then it will be possible to identify the tasks within the interfaces that are problematic for this group. This can then be used to provide guidance to interaction designers. Using the exclusion calculator approach, we can test interfaces with non-users and generate data to show what proportion of the digital novices are likely to be excluded by different parts of the user interface. This information can be used in an inclusivity audit, or used as an evaluation part of a design iteration, to improve the usability of an interface.

Conclusions

An exclusion approach to designing digital interfaces (i.e. an exclusion estimation for particular types of interaction element) could provide rationale for designers to use (i.e. 'if I use this interaction element I get only 10% exclusion vs. 30% if I use this one'). The use of a cognitive modelling tool could also provide a predictive tool to test new interfaces, thereby facilitating early evaluation of concept interaction elements for the excluded potential users.

Acknowledgements

This project is funded through the EPSRC via the 'Building Relationships with the Invisible in the Digital (Global) Economy' (BRIDGE) collaborative sandpit project.

References

Clarkson, P.J., Coleman, R., Hosking, I. and Waller S. 2007, *Inclusive Design Toolkit,* (Engineering Design Centre) and at: www.inclusivedesigntoolkit.com (accessed December 2009)

Eldy, 2009, available at: www.eldy.eu (accessed December 2009)

Langdon, P.M., Lewis, T. and Clarkson, P.J. (2009), *Prior experience in the use of domestic product interfaces.* Universal Access in the Information Society, (Springer)

Milner, H. (ed.). 2009, *Freshminds Research Report for UK Online Centres: Does the internet improve lives?* Available at: http://www.ukonlinecentres.com/images/stories/downloads/does_the_internet_improve_lives. pdf

Norman, D.A. 1988, *The Design of Everyday Things*, (Doubleday/Currency) Office of National Statistics, 2009, *Internet Access Households and Individuals,* (ONS)

Raskin, J. 2000, *The Humane Interface,* (ACM Press)

Selwyn, N. 2003. Apart from technology: understanding people's non-use of information and communication technologies in everyday life. Technology in Society 25, 99–116

SimplicITy Computers, 2009, available at http://www.discount-age.co.uk/simplicity_computers, (accessed December 2009)

PROF. TOM REILLY MEMORIAL

ERGONOMICS AND FOOTBALL

Barry Drust

The pioneering work that was completed by Professor Thomas Reilly in the 1970's provided a framework for our current understanding of the physiological demands of football match-play. This early research not only developed methodological approaches to record and analyse the activity profiles of players (approaches that still form the basis of similar research completed today) but also highlighted the energy demands that are associated with English professional soccer. Other innovative research projects further contributed to the development of the knowledge base in the area by examining specific factors that influence football performance (e.g. playing position, style and level of play etc). Tom's role in the development of laboratory-based football-specific simulations based around these "real world" investigations enabled the energetic costs of football-specific activities (e.g. backwards and sideways movements), technical actions and football-specific intermittent activity patterns to be evaluated. This data deepened our understanding of the physiology of intermittent exercise as well as provided a suitable platform to investigate how footballers can be better prepared to meet the demands of the sport. Significant contributions to the literature base associated with the design, planning and implementation of football-specific training programmes has also been a legacy left by Tom's research activities. This research activity not only relates to the elite male but also the recreational footballer and the female player.

TOM REILLY AND RESEARCH ON THE BODY CLOCK

Greg Atkinson

Tom's research work on the influence of the human body clock on aspects of sport, leisure and ergonomics began in the 1970s. Tom completed the first well-controlled studies on the combined effects of sleep deprivation and physical activity, including a case-study in which he completed 100 hours of continuous exercise. As well as being relevant to sports competitors and occupations in the military, the results of these studies provided valuable information about the interaction between circadian variation and time spent awake on human performance. In 1980, Tom worked with George Brooks at the University of California, Berkeley, where he gained valuable experience through a series of seminal studies on circadian variation in the physiological responses to exercise. The results of these studies showed how the thermoregulatory responses to human physical work are altered during the night. Over the years, Tom published over 900 peer-reviewed manuscripts, including over 200 papers relevant to chronobiology. His review papers on circadian rhythms and exercise continue to be highly cited. Tom became the person who elite sports competitors, including Olympic gold medalists, turned to for ergonomics-based advice on jet lag alleviation, having co-authored two reviews on this topic in The Lancet. He also applied his knowledge about circadian rhythm disruption to occupational contexts and gained a major grant from the Health and Safety Executive (HSE) to study the impact of physical activity on human circadian rhythms and their disturbance during shiftwork. Another grant from the HSE funded important research work on the effects of the menstrual cycle and pregnancy on lifting performance in an occupational context. He also published much research work on the influences of Ramadan on human biological rhythms. In 1998, Tom was awarded the highest academic degree of D.Sc. for all these contributions to chronobiology.

ANTHROPOMETRIC APPLICATIONS TO ERGONOMICS

Laura Sutton

Given that ergonomics is the application of scientific information concerning humans to the design of objects, systems and environment for human use, it is of great importance that we are able to quantify with accuracy various aspects of the human body. Tom recognized this need and, through his work in the field of kinanthropometry, made significant contributions to the understanding of the influence of the size, shape, proportion and composition of the human body on human movement and physical activity, and conversely, the influence of physical activity on the composition of the body.

Tom's earliest work in which body composition is addressed as an ergonomic factor formed part of his PhD, completed in 1975. He considered the physical demands and effects on the body of the occupational stress associated with professional soccer, focusing on factors such as body composition and energy requirements.

Subsequent methodological work provided much-needed guidelines for anthropometric assessment in a variety of distinct populations. In the 90s, in his role as advisor to the British Olympic Association, Tom provided advice for assessing body composition that, upon first impression, appeared so simple that it seemed odd that his recommendations were not already in practice! To this day, data are still being produced supporting these guidelines, and what is probably one of Tom's shortest manuscripts is, in my opinion, one of his most pertinent.

Tom's research outputs included detailed descriptive work and attempts to identify appropriate field-based measurement tools for many different athletic groups, including, but by no means limited to, track and field athletes, games players and disabled athletes. Where there did not appear to be a valid and reliable method for the estimation of body composition, Tom produced his own.

In addition to sports ergonomics, Tom investigated and provided advice to workers in physically-demanding occupations, for example, sports officials, nurses, the fire service and the military. He was interested in identifying valid techniques for the assessment of body composition, and investigating both whole-body and site-specific adaptations to occupational demands. His articles and books now provide important reference information and guidelines for kinanthropometry, applicable to a diverse audience.

RAIL AND TRANSPORT

DEVELOPING AND PROMOTING ALARMS AND ALERTS GUIDANCE FOR THE RAIL INDUSTRY

D. Pennie[1], P. Leach[1], B. Cash[1], J. Berman[1], R. Wicksman[2], D. Fletcher[2] & J. Edworthy[3]

[1] *Greenstreet Berman Ltd*
[2] *RSSB*
[3] *University of Plymouth*

RSSB contracted Greenstreet Berman Ltd (GSB) to develop guidance, in a form that could be readily used by anyone concerned with the design of alarms and alerts, specifically within the rail industry.

The guidance uses web-based technology and is freely available from the RSSB website. There are two parts to the guidance. The first part is an Alarms and Alerts Guidance and Evaluation Tool. This provides a set of guidelines covering good practice principles in the design of auditory and visual alarms and alerts. It also provides a facility for users to conduct an evaluation of their own alarms and alerts system against these principles. The evaluation provides feedback indicating areas of design which do not currently comply with good practice and could benefit from improvements.

The second part provides a 'Sound Library' of different types of playable auditory alarms and alerts. The tool as a whole does not seek to promote the use of the specific alarms and alerts provided in the Sound Library. Instead the application presents information to help support decisions about the selection of appropriate auditory alarms and alerts for the rail environment.

Work to develop this guidance and application has been a collaborative effort with support from alarm and alert specialists from the UoP. This paper reports on the: background to this guidance: how it was developed and current plans for its future promotion amongst relevant rail industry groups.

Introduction

Background

RSSB is owned by major railway stakeholders and facilitates their cooperation to continuously improve the level of safety in the rail industry, drive out unnecessary cost and improve business performance. To support this, RSSB builds industry-wide consensus and facilitates the resolution of difficult cross-industry issues. It provides knowledge, analysis, a substantial level of technical expertise, powerful information and risk management tools. This delivers a unique mix to the rail industry across a whole range of subject areas.

RSSB is home to a broad range of technical expertise, including in the field of Human Factors (HF). A large number of traffic and operational management issues have provided the opportunity to apply and promote the HF discipline in partnership with stakeholders. It has also featured in a number of projects managed as part of the Research and Development (R&D) programme that RSSB coordinates on the industry's behalf, and which is funded primarily by the Department for Transport.

One research issue identified by RSSB and its stakeholders was that there were no regulatory standards or official guidance available to the rail industry on the design and fitting of alarms and alerts. There was, therefore, a requirement to provide the GB rail industry with specific guidelines for designing and implementing alarms and alerts. In 2005, work commenced to develop guidance alarms and alerts with the aim of improving the performance of human operators and to reduce risks due to inappropriate alarm handling in the train cab environment. Developing guidance, however, is ineffectual if it does not reach or is not read by its intended audience.

This paper reports on the work to develop this guidance and current plans for its future promotion amongst relevant rail industry groups.

The aim

The aim of the project was to deliver practical and useful guidance on alarm and alert design for the GB rail industry. The guidance was aimed specifically at helping: manufacturers; designers; operators; safety and HF specialists, to assess the acceptability of current alarm and alert systems and provide advice to help address any concerns relating to their design.

Method

The work to develop the guidance changed from the original brief as more was understood about the topic area and the requirements of the user group. Essentially, however, it was developed in three key phases. These phases are described below and then covered in more detail under relevant headings:

1. Derive guidance and good practice principles as a preparation for Phase 2;
2. Develop an electronic tool to provide more easily accessible information with greater interactivity to illustrate design principles;
3. Develop a 'Sound Library' application to further enhance the tool.

Derive guidance and good practice principles

Phase 1 of the work was undertaken by the UoP to derive good practice guidance by assimilating a wide range of current thinking on the design and implementation of alarms and alerts.

Sources of this information included:

- The academic research literature;
- Existing industry standards and guidelines;
- Observations of current alarm and alert usage in train cabs in GB.

A review of the academic and applied research literature (from the last twenty years) was conducted using relevant search terms to search the following databases: Web of Science; 'PsychLit'; 'Medline'; 'PsychINFO' and 'Google Scholar'. The aim of this review was to summarise information about the design and implementation of different warning systems and determine the factors that make warnings more or less effective. In addition, a range of US and GB industry standards and guidelines were reviewed from other sectors such as nuclear and aviation. Finally, observation work was conducted, to establish the extent and variation of alarms and alerts currently used in train cabs in GB. Observations of different train types were made at the Southern Trains simulator facility in Croydon. In addition six different real cab environments were also observed: Class 221 Tilting Super Voyager; Class 82 DVT (Driving Van Trailer); Class 87 Electric Locomotive; Class 175 Passenger Train; Class 153 DMU (Diesel Multiple Unit) and a Class 47 Diesel Freight Locomotive. During each observation session, the alarms and alerts were recorded in conjunction with an event log of alarm and alert activations. Semi-structured interviews were also conducted to assess driver evaluation of the alarm and alert environment. These subjective data were supplemented by 488 questionnaires which were posted to drivers from six train operating companies across GB.

Two case studies were used to evaluate the paper-based version of the guidance. First it was used to assess an existing train cab environment and then to review the alarms and alerts provided as part of the Global System for Mobile Communications on Railways (GSM-R) radio system. The results from the case studies were used to revise the content of the guidance and also led to the proposition to develop an interactive web based electronic tool.

Develop an interactive web-based electronic tool – Alarms & alert guidance & evaluation tool

It was acknowledged that leaving the presentation of the report as a paper based document derived from academic sources, made the information less accessible to the intended audience of manufacturers, designers and engineers. Therefore, the second phase was to develop an interactive electronic tool.

A requirements capture exercise was facilitated by GSB. A number of plausible scenarios were developed; then discussions were held between HF specialists, RSSB and stakeholders, such as the Association of Train Operating Companies (ATOC). These scenarios were then used to explore how different user groups might set about designing an in–cab alarm system.

This method of using scenarios was chosen because it is intuitive and relatively simple approach that is easier for non-HF specialists to apply.

In total 16, different user scenarios were then developed for four different user groups. These were used to identify:

- The information each user might require;
- The types of question each user might have and how best to answer these questions;
- The best way to present information.

One of the scenarios considered a train manufacturer and what support might be required in helping them to produce a specification for the alarms and alerts design of a new system. Exploring this scenario helped to identify different requirements, for example, the tool should:

- Provide general guidance on good practice for alarms and alerts design;
- Provide more specific guidance on different types of alarms (visual, auditory);
- Help designers develop an alarm strategy (e.g.: establish levels of alarm priority).

This scenario was also used to identify the following types of general questions a train manufacturer might want answers to:

- Why is considering HF important to a new system specification?
- When should I use this application?
- What are the key areas I should consider when designing alarms and alerts of a new system?

The findings from the case studies, meetings with stakeholders and the requirements capture exercise were used to develop a technical specification document. This outlined:

- User requirements (e.g. provide high level guidance on alarm strategy and prioritisation and more specific guidance on alarm and alert design);
- System requirements (e.g. users should be able to print information);
- The proposed structure for the tool (e.g. no more than three levels of screen hierarchy);
- Presentation (e.g. guidance should have a navigation menu to make specific items of interest easy to find);
- Navigation (e.g. provide back button and hyperlinks);
- The layout of web pages (e.g. navigate through the tool by using a menu on the left hand side of the page);
- Technical details (e.g. screen resolution (1024 × 768), meaning that the whole application will be completely visible on any computer).

This work led to the development of the Alarms and Alerts Guidance and Evaluation Tool.

Develop a sound library

During its development the Alarms and Alerts Guidance and Evaluation Tool underwent user testing to evaluate both the usability of the software and the usefulness of the information provided within it.

Users participating in the testing (particularly those responsible for decisions associated with the design and procurement of rail alarms and alerts) indicated that in addition to guidance on the design of alarms and alerts, it would be useful to have a library of playable sounds. This would provide further support to decision making and make it easier to select appropriate alarm/alert combinations. To enhance the main tool it was proposed that such a 'Sound Library' be developed.

The work to research and develop the sounds was again conducted by alarm and alert specialists at UoP. First a scoping document was produced that outlined eight different categories of sound types, with six to eight examples in each category. The categories are based on the different categories of sounds which are relevant to alarm design. In addition, the scoping document also outlined a number of sound combinations to help demonstrate different sound design principles. When these categories were agreed work was carried out to generate the sounds.

No examples of actual guidance content are included in this paper. The main tool and the sound library are freely available to view on the RSSB website. They can be accessed by using the RSSB website search function. Descriptions of the types of information provided in the main tool and the sound library are provided in the next section.

Description of the alarms and alert guidance and evaluation tool and sound library

Content of the alarms and alert guidance and evaluation tool and sound library

The main tool provides a set of guidelines detailing good practice principles in the design of alarms and alerts, for example, designing so as to reduce operator error rates in detecting, comprehending and responding to them. The guidance also considers the factors that might contribute to error such as, the overall number of alarms and alerts, their possible confusability with each other and with those of other information systems. The tool also provides an evaluation facility to assess current alarm and alert systems against a set of good practice principles. This facility provides feedback indicating areas of design which do not comply with good practice and could benefit from improvements. The evaluation facility links to relevant pages within the main guidance section providing feedback and suggestions on how to enhance or improve current designs.

Both the guidance and evaluation cover the following areas:

- General issues (modality, false alarms, alarm/alert philosophy, standardisation, characteristics of alarm/alert users);
- Auditory issues (number, rate, prioritisation, confusability and loudness);
- Speech issues (usage, number, detection, intelligibility, speech forms, voice attributes and prioritisation);
- Visual issues (number, visual field, prioritisation, colour, flash rates, legibility, luminance, glare and reflectance and language).

In addition the guidance provides general information on areas such as who should use this tool, what does this tool provide and why should I evaluate alarms and alerts.

The Sound Library provides a number of different types of playable auditory alarms and alerts. It also provides playable examples of different combinations of alarms to help to illustrate alarm design principles. These sounds are supported with a commentary note that provides: a brief description of the sound; its advantages and disadvantages and under what circumstance it might be effective. The information is presented in two sections:

- Sound Design Principles – 16 alarm and alert combinations illustrate good and bad alarm and alert design, these have been grouped into four areas: priority and detection; urgency; noise; confusability;
- A Sound Library of 54 alarms and alerts – topics include speech, 'earcons' progressively more complex sounds based on a simple sound) traditional alarms, abstract alarms, auditory icons, Patterson alarms (abstract alarms based on short bursts of simple melodies), GSM-R sounds and European Rail Traffic Management sounds.

Together with the main tool, the sound library provides advice to improve understanding of good design principle and information to help determine the most appropriate types of sounds to be used within a current system or new design.

Screen layout and presentation of the alarms and alert guidance and evaluation tool and sound library

The main tool was originally designed to launch directly from a compact disc onto a personal computer. During its development, however, it was decided that the tool should be launched from the RSSB website because this makes the information more widely available and easier to maintain and update. One disadvantage, however, of launching from a website is that an application such as the sound library which uses a lot of a memory (an unavoidable aspect of its integration within the programming language) can take some time for its content to download, especially when first opened. In order to minimise delays in operability the applications are launched from two separate web pages. To compensate for this, the two applications have been

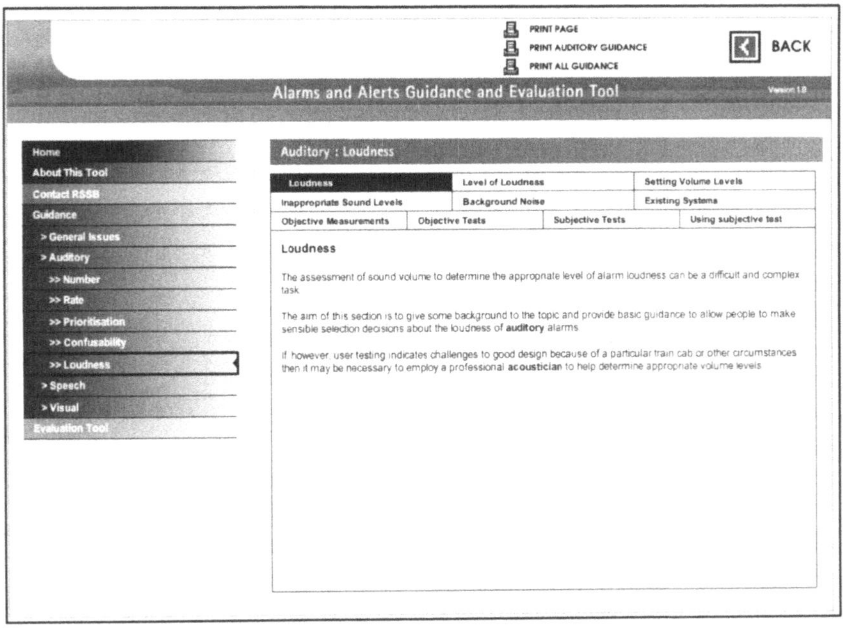

Figure 1. Screen shot of the Alarms and Alert Guidance and Evaluation Tool.

designed to sit side-by-side on the RSSB website. They are linked with clickable hyperlinks so that a reader using one application can open up the other and view it at the same time.

The content of the applications has been written so that each topic covered, e.g. 'modality', is divided into a series of sub-sections: visual; auditory; speech; tactile and combination. Each sub-section is presented as a single page. The discipline of presenting all the information relating to a topic or sub-topic in this way ensures that the content is very concise. This was an important requirement identified during user testing. Using single pages also means that important information is less likely to be missed because the reader does not need to scroll through a long web page. Additional hyperlinks (text highlighted in blue) are used so that the reader can easily open related sections of the guidance. Certain topic areas, particularly auditory alarms and alerts, can be technically complex. The tool allows the reader to click on words highlighted in red for an explanation of technical terms, such as: 'decibels' or 'fundamental frequency'. The layout of the information has been designed to be consistent with the RSSB website. For example, it uses the same colour conventions and the same menu layout and fonts.

The content of the sound library is presented in a similar format to the Alarms & Alert Guidance & Evaluation Tool. Information is grouped by topic and related sub-sections are presented on a single page.

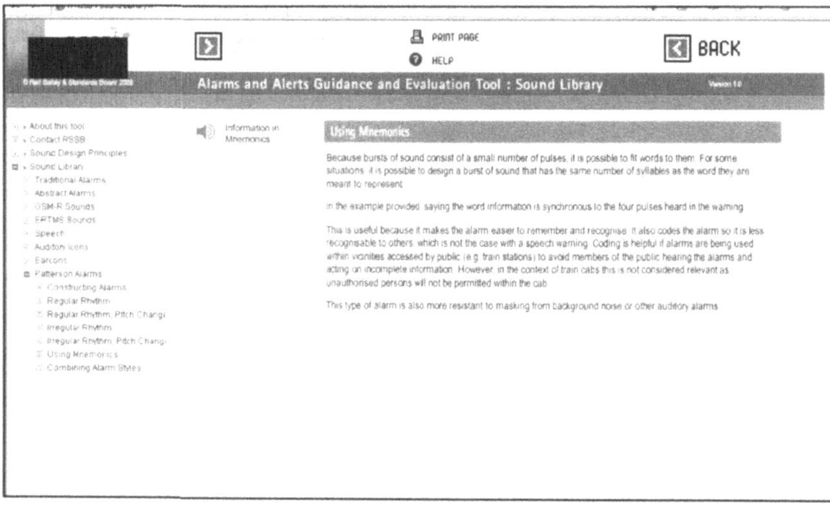

Figure 2. Screen shot of the sound library.

The key difference between the two applications is the design of the menu. This was due to the different functions of each application. The sound library required a very simple design to provide more memory space for hosting the individual sound files. By contrast, the Alarms and Alert Guidance and Evaluation Tool had a more complex menu allowing the evaluation part of the tool to link to related areas of guidance.

Promotion of the alarms and alert guidance and evaluation and sound library

It is important that the tool and sound Library are promoted so that they reach their intended audience and bring benefits to the rail industry. A number of potential options are currently under consideration to help promote the applications more widely.

Case studies

The applications can be used for a number of relevant rail industry case studies. These may involve organisations that are either planning to design new alarms and alerts or that may wish to assess and revise their existing systems. The case studies will help to raise the profile of the applications. If the rail industry parties involved in the case studies find that the applications have been of benefit to the design process then they are likely to promote them through partner companies and other industry contacts. The results of the case studies can be published as reports (available on the RSSB website). These would illustrate how the applications are used and showcase the benefits they can bring.

Demonstrations

Short demonstrations (e.g. 30 minutes) can be provided to interested parties to show how the applications are used to improve the design process. A good way of conducting a demonstration is to use a relevant scenario, for example actual alarms re-design, and then use this scenario to show how the applications are used to help identify and solve design problems. They could be held at different industry events and meetings, for example: NFRIP (National Fleet Reliability Improvement Programme). Different stakeholder groups might include, for example: ATOC; individual train/freight operating companies; rolling stock owners; infrastructure maintenance and renewal companies; London Underground.

Workshops

Workshops would help to enhance understanding of the applications' contents, how to apply them and what benefits they could bring. Workshops are designed to allow participants to learn by directly applying the applications to real-life scenarios. Workshops would typically be a half to a full day session.

Conclusion

The objective of RSSB is to achieve continuous improvement in the health and safety performance of the railways in GB. The aim is to reduce risk to passengers, employees and the wider public. Developing guidance that seeks to enhance performance is only effective if it is read by the intended audience and actively used by industry.

There is greater realisation today of the importance of presenting Human Factors guidance in a way that can be directly applied by an organisation to address its particular design issues. The Alarms and Alerts Guidance and Evaluation Tool was created to fulfil this aim, for example, by presenting information in digestible bite sized chunks, by providing practical advice in plain language, and by providing direct feedback to the organisation about the concept and design of its own alarms and alerts system.

Under the direction of RSSB, GSB in conjunction with UoP have tried to achieve this aim by developing electronic guidance launched from the RSSB's website. It is anticipated that evidence that this aim has been achieved, will come from the wider use of this guidance, by the rail community, to evaluate and enhance the design of alarm and alerts.

References

Professor J. Edworthy, Dr. E. Hellier, Professor J. Noyes, Dr. K. Aldrich, A. Naweed, W. Gall, R. Metcalfe. 2008 *Good Practice Guide for the design of alarms and alerts RSSB* (Engineering good practice guides).

RSSB Research and Development Programme 2008 *Human factors good practice guide to managing alarms and alerts* RSSB (Research Brief).

HIGHLIGHTING THE PRESENCE OF EMERGENCY LEVEL CROSSING PHONES IN EMERGENCY SITUATIONS

Anna Scott

Human Engineering Ltd

In 2008, RSSB commissioned a research project as part of the rail industry's R&D programme it manages on its behalf, and funded by Department for Transport, to establish how best to highlight the presence of emergency phones and associated signs at level crossings. The research followed two collisions where crowded trains hit stationary vehicles causing significant loss of life. In one of the instances the decision to use mobile communications via 999 prolonged the railway response time where use of the emergency level crossing phone could have potentially averted the incident.

Human Engineering was tasked with identifying not only the public's reasons for use or non use of existing level crossing phones and any methods for encouraging use, but also to explore alternative, efficient methods of communicating level crossing emergencies to the relevant authorities. Key questions addressed by the project included.

- What are the communication links between members of the public reporting an incident and the information being received by the railway? Can these be made more efficient?
- How much do the public know about emergency phones at level crossings and can/would they use them?
- How can we best present level crossing phones and associated signs so that the public will use them?
- How can we optimise the visibility and use of the phone/signs to different user groups?
- What are the cost/benefit associations with each of the options?

The project reviewed information from primary and secondary sources ranging from naïve potential users to subject matter experts and representatives from the organisations involved in receiving and routing a 999 emergency phone call. Methods utilised by the study included a review of standards and research, a communications link analysis, a survey with members of the public, subject matter interviews and workshops.

Ideas for improvement were sub-divided into 5 areas

- Phone design – improving the physical appearance of the telephone system in order to increase visibility and significance to a potential user
- Phone operation – contrasting different phone operating procedures to establish the most intuitive and practical means of operation
- Design of user instructions – facilitating use of the telephone system by different user groups
- Design of associated signs – directing attention and provide information to potential users
- Improvement of the 999 communication flow – expediting the relay and improving the clarity of information passed from the user to the railway.

Consideration of proposed solutions is still underway at the time of writing, with final design and procedural recommendations to be made at the project's conclusion in September 2009.

DEVELOPMENTS IN THE TRAIN PROTECTION AND WARNING SYSTEM ON THE UK RAILWAYS

Anita Scott

RSSB

The train protection and warning system (TPWS) is the primary train protection system used on UK railways. TPWS reduces the consequences from trains passing signals at danger (SPADs). A train approaching a red signal too quickly or passing a red signal will be braked by TPWS. Following the incidents at Southall in 1997 and Ladbroke Grove in 1999, TPWS was rolled out across the UK railway with fitment largely complete by 2002.

Although TPWS was very effective in reducing SPAD risk, the system also had limitations. The system allows the driver to clear a brake demand following a SPAD by removing power from the driving desk and continue forwards, potentially into conflict with another train. In addition, the driver machine interface is poorly designed, not discriminating between different causes of brake demand.

This paper will discuss some of the initiatives that the UK rail industry has undertaken to address these issues. The initiatives include research to re-design the TPWS driver machine interface, the development of a training DVD and monitoring of TPWS brake demands using software by train operators. The future of TPWS in relation to the introduction of ERTMS will also be discussed.

REPETITIVE STRAIN INJURY

REPETITIVE STRESS INJURY PREVENTION PROGRAM (RSIPP)

Trish Roy

Chevron Australia Pty

Chevron Australia works hard to make its workplace incident and injury free. It pays close attention to discomfort, because it recognises that physical discomfort can be the first sign of repetitive stress risk. Every effort is made to quickly intervene through a systematic ergonomic response program and see improvements in health and safety.

Chevron Australia's program goes well beyond a quick response mechanism to a member of its workforce reporting discomfort. The RSIP program aligns individual behaviours with risk monitoring and education to encourage safe individual performance. This approach is expected to reduce the risk of developing computer-related repetitive stress injuries (RSIs).

The key elements of the RSIP Program include risk assessment and training using computer based assessment, ergonomic evaluations by specialised workstation evaluators normally from the field of Occupational Therapy or Physiotherapy, prevention measures such as ergonomic equipment, break reminder software, risk self-assessment software, speech recognition software, sit/stand workstations, laptop guidelines and early reporting and rapid response to discomfort.

The RSI Prevention Team completes over 100 workstation evaluations and 20 rapid response interventions per month in addition to the preventative RSI activities. This program has reduced the total number of RSI related injuries and has increased the awareness of risk associated with using computer workstations.

POSTERS

ASSESSMENT OF WORKLOAD AND OCCUPATIONAL HEALTH HAZARDS OF HOSPITALITY INDUSTRY WORKER

Vinay Deepa[1] & Moharana Gayatri[2]

[1]*Department of Family Resource Management, College of Home Science*
[2]*G. B. Pant University of Agriculture & Technology; Pantnagar – 263 145 (Uttarakhand)*

The study was conducted for ergonomic assessment of Hotel workers (n = 78) working in Three Star Hotel. The mean heart rate was found to be 126.68, 123.45 and 137.12 (beat/min) respectively in Kitchen, Housekeeping and Laundry department. The average energy expenditure rate of the workers of different departments i.e. Kitchen, Housekeeping and laundry was 12.59, 12.33 and 13.27 (kj/min) respectively. The results of the study implicated that the laundry workers were doing very heavy work followed by the kitchen and house keeping workers because the energy expenditure, total cardiac cost of work and physiological cost of work among laundry workers were very high as compared to the other department of the hotel. Study indicated that the workers of these three areas of the hotel were performing very heavy tasks which affects their health and in turn leads to occupational health hazards.

Introduction

The hospitality industry is a part of larger enterprise known as the travel and tourism industry. Hospitality is defined as "The friendly reception and treatment of strangers". Hospitality means entertaining guest with courtesy and warmth and is also an industry made up of business that provide lodging, food and other services to travelers. People of hotel industry are engaged in different jobs like housekeeping, cooking, catering, laundering and managing the front office etc. These are very tedious jobs at their work place. In the fast-paced environment of hotels and restaurants, a common attitude is that accidents are inevitable and a part of doing business. But injuries mean losses i.e. loss of money, time and productivity. In different sections of hotel like kitchen, laundry and housekeeping workers perform varied types of job which are physically very demanding and may lead to occupational health hazards such as back pain or pain in upper extremities like upper limbs, neck and shoulders. The other aches arising from manual handling injuries and most common type of occupational health hazards are lifting, carrying heavy items or pushing and pulling the object from one place to other. Besides these, forceful or repetitive activities and poor postures can be linked to upper limb injuries. These ultimately affect the health of the workers and leads to occupational health hazards.

Therefore, the study was carried out to find out the physiological parameters of the workers working in the areas like laundering, housekeeping and production of foods and beverages. The study also aims to find out occupational factors influencing their health status. Ergonomic cost of selected drudgery prone activities was calculated by measuring the physiological parameters of selected activities and developed training module was used for successful management of musculoskeletal disorders.

Methodology

The ergonomic evaluation of various hotel activities was conducted with hotel workers of a Three Star Hotel situated at Udham Singh Nagar District of Uttarakhand State, India. Out of six departments in hotel, four departments were chosen purposively for the study which included front office, housekeeping, kitchen and laundering. Out of these departments seventy eight respondents i.e. thirty eight respondents from kitchen, twenty six respondents from housekeeping, six from the front office and eight of them from the laundering department were selected through purposive random sampling. Energy expenditure and grip fatigue for the various hotel activities were measured. The standardized Nordic questionnaire validated was used to measure the prevalence of musculoskeletal disorders among the hotel workers.

Physiological Cost of Work

Circulatory stress was evaluated from the cardiac cost of work and cardiac cost of recovery. The cardiac cost of recovery is the total number of heartbeats above the resting level occurring between the end of the work and return to the resting state. Formula given by (Singh *et al.*, 2007) was used to calculate the total cardiac cost of work (TCCW) and physiological cost of work (PCW).

Visual Analogue Scales (VAS)

Postural discomfort questionnaire based on Visual Analogue Scales (VAS) was used to measure subjective qualitative judgments of different staffs of laundering, front office, house keeping and kitchen workers regarding postures and discomfort at various times throughout the day.

Results

Results shows that both female and male worker were working in housekeeping and front office department whereas, kitchen and laundry section were male exclusive data of physiological parameters of laundering, house keeping and kitchen work

Table 1. Physiological workload of different activities performed by hotel workers.

Sl. No.	Physiological parameters	Kitchen	House keeping	Laundry
1.	Average working heart rate (beats/min)	127.68	123.45	137.12
2.	Average energy expenditure (kj/min)	12.59	12.33	13.27
3.	Peak heart rate (beats/min)	134.00	131.00	146.00
4.	Peak energy expenditure (kj/min)	12.76	12.93	14.40
5.	Average TCCW (beats)	1120.56	1093.37	1277.51
6.	Average physiological cost of work (beats/min)	48.15	47.70	50.73

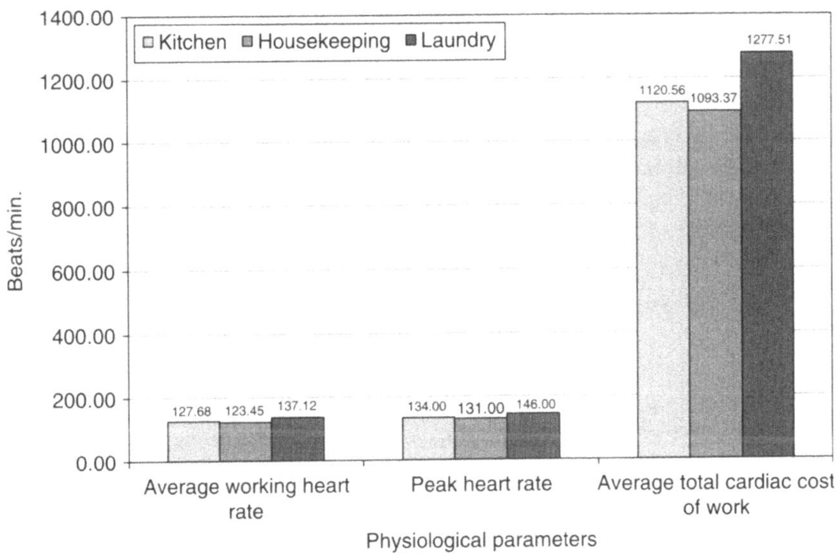

Figure 1. Average working, peak heart rate (beats/min) and Average TCCW (beats) of workers of different departments while performing the activities.

was collected and analyzed. It was found that the average working heart rate and peak heart rate was found maximum for laundering activities followed by kitchen and house keeping activities. (Table 1)

Similar trend was obtained for the average energy expenditure also. The average total cardiac cost of work was found 1277.50 beats for laundering activities followed by kitchen activity (1120.56 beats) and house keeping activity (1093.37 beats). Similar trend was also observed for the average physiological cost of work i.e. 50.73 beats/min for laundry, 48.15 beats/min for kitchen and 47.70 beats/min for the house keeping activities.

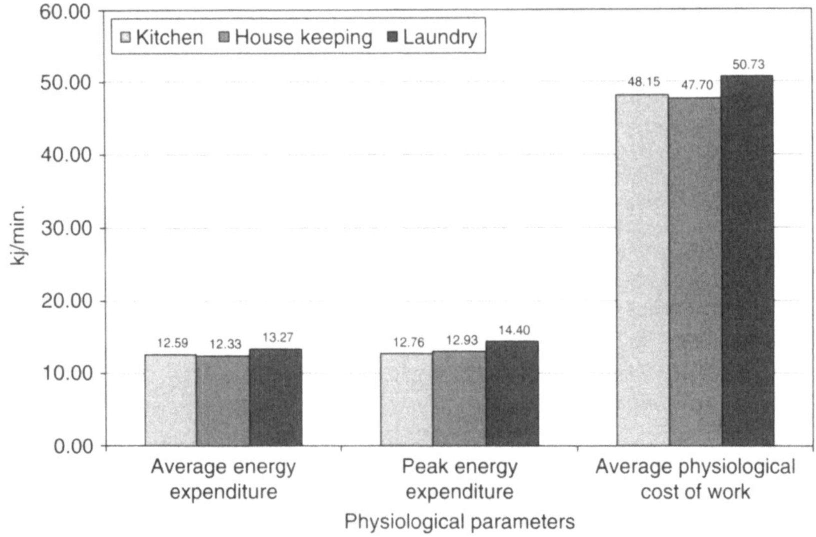

Figure 2. Average, peak energy expenditure (kj/min) and Average PCW (beats/min) of workers of different departments while performing the activities.

Table 2. Percentage change in grip strength of the subjects, before, during and after activities.

	Rest		During work		After work		% change in grip strength	
Different area	Left	Right	Left	Right	Left	Right	Left	Right
Laundry	34.38	33.38	32.87	31.75	33.62	31.12	4.39	4.88
Kitchen	39.29	38.26	36.42	35.82	37.92	37.05	7.03	6.32
House Keeping	38.57	39.42	36.23	36.28	37.31	38.04	6.06	7.53

Results also shows that all the physiological parameters i.e. the average working heart rate, peak heart rate and average energy expenditure was found higher for laundering activities as compared to kitchen and house keeping activities. Sharma *et al.* (1998) studied the physiological work load of the women performing the task of fetching water from the near water resources and found the similar trend of energy expenditure as laundry workers.

The Table 2 depicts the percentage change in grip strength of the subjects, before, during and after activities performed by the workers working in laundering, kitchen and house keeping department. The findings shows that the percentage change in grip strength of the respondents of laundry was found less as compared to the activities performed by the workers of house keeping and kitchen. The percentage change in grip strength of kitchen workers was more as compared to the respondents of house keeping department this shows that grip fatigue among the workers of kitchen department was more as compare to workers of other departments.

Occupational Health Hazards

The Visual Analogue Scales (VAS scale) was administered to the four departments and the activities of whole day were assessed. The awkward posture adopted by the workers at work place was the main reason of pain and discomfort in body parts. Kitchen workers reported severe pain and discomfort in legs whereas discomfort in buttock, thighs, and mid back was comparatively low. Among the housekeeping workers highest discomfort was reported for shoulder followed by upper back, lower arm and leg. However the pain was highest in legs, followed by shoulder, upper back, mid back and neck among laundry workers. The front office workers reported highest discomfort in neck followed by shoulder, upper back, lower back and mid back.

It was evident from the above findings that the risks associated with the work place or work station includes extreme conditions, working climate and adapting unnatural postures while working. All these factors may be responsible for generating musculo-skeletal problems. It was assumed to be the ultimate factor which lower the work output of the workers. Therefore a self developed training module have been developed for giving knowledge to the workers to enhance their capacity and to increase the output by correcting different posture adopted while performing different activities.

A training schedule was developed in the form of flap book to train the worker and management regarding the MSDs. The training module was based on introduction on ergonomics, principal of ergonomics, job risk factors, posture for performing task, use of body muscles and preventive measures.

In conclusion the present study explores the physical stress and strain in the working postures of the workers working in the kitchen, house keeping and laundering in the hotel. All the workers reported that they had to work continuously for a longer period in different postures which produces pain in different body parts. This study revealed that the workers of different departments of the hotel were engaged very heavy task with awkward postures and repetitive motion. Their physiological cost of work was very high as they had to spent more energy while performing the tasks. The environmental factors like temperature, humidity and noise affect their working activities and also affect the physical and mental health of the worker since the worker are the real stars behind the stars. Therefore, to abide the chances of occupational health hazards immediate attention of ergonomists is required. Developed training module was found very effective in successful management of musculoskeletal disorder of workers.

References

Sharma, D. and Sharma, S. 1998–99. Ergonomic study on farm women's drudgery, Annual Report, All India Coordinated Research Project, Department of Family Resource Management, College of Home Science, G.B.P.U.A.&T.; Pantnagar

Singh, S. P.; Mathur P. and Rathore. M. 2007. Weeders for drudgery reduction of women farm workers in India. *Journal of Agricultural Engineering* 44(3): 33–38

Preventing injuries to hotel and restaurant workers: a focus report. workers compensation board. British Columbia. www.labour.gov.bc.ca/wab cited on 18th October 2007 at 5 P.M.

THE INFLUENCE OF THICKNESS OF SUBCUTANEOUS LAYER ON THE EMG SIGNAL AS A FUNCTION OF FORCE LEVEL AND TYPE OF MUSCLE

P. Bartuzi, T. Tokarski & D. Roman-Liu

Central Institute for Labour Protection-National Research Institute, Warsaw, Poland

The aim of this study was to evaluate the influence of subcutaneous layer on electromyography (EMG) parameters taking level of force and type of muscle into consideration. The subjects were classified on the basis of body mass index (*BMI*) and the amount of fatty tissue (*FT*). The EMG signal was measured from 3 muscles on 5 levels of load. The EMG signal parameters determined for the time domain and for the frequency domain were analysed. The results indicate, that both the amount of fatty tissue and the muscle contraction level influence the EMG signal parameters. For the frequency domain parameters this influence is ambiguous, which may be a result from the proportion of the muscle fibres of the type I and type II. More unambiguous results can be obtained by analysis based on more detailed parameters of the power spectrum of the EMG signal.

Introduction

Disorders of the musculoskeletal system can be caused by excessive muscle load and fatigue (Visser and van Dieën, 2006). One of the commonly used methods to access the load and fatigue from the muscles is a surface electromyography (EMG) (Piscione and Gamet, 2006).

The parameters of the EMG signal depend not only on muscle force and fatigue, but also on subcutaneous layer (Kuiken *et al.*, 2003; Farina and Rainoldi, 1999; Farina and Mesin, 2005; Nordander *et al.*, 2003). Studies indicate fatty tissue, located between the electrode and the muscle, as a cause of inaccuracy in the results of the EMG signal measurement obtained with usage of surface electromyography (Farina and Mesin, 2005; Nordander *et al.*, 2003).

Farina and Rainoldi (1999) show that the subcutaneous tissue layers produce an attenuation of the potential distribution present at the muscle surface. Kuiken *et al.* (2003) suggest that with increasing fatty tissue layer decrease the values of amplitude EMG signal and increases the cross-talk in muscles adjacent to activated muscle.

The aim of this study was to evaluate the influence of subcutaneous layer on global EMG parameters in a function of level of force and type of muscle. The influence of fatty tissue on EMG characteristic expressed quantitatively will allow to improve the accuracy of the EMG signal measurement with usage of surface electromyography.

Methods

The studies were performed in 30 men in age between 20 and 27 years, divided into two equal groups: with BMI ranging from 18 to 23 and level of fatty tissue (*FT*) below 13% (*LF* group) as well as BMI ranging from 27 to 35 and level of *FT* above 18% (*HF* group).

During measurements participants activated *rectus femoris (RF), biceps brachii (BB)* and *rectus abdominis (RA)* muscles. The tests involved maintaining the constant force with constant length of muscle at the five levels: 5%, 15%, 30%, 50% and 70% MVC.

Amount of *FT* was measured with Futrex 6100 (Futrex, USA). For measurements and registration of the raw EMG signal the **Bagnoli-16** device (Delsys, USA) was used. The registered EMG signal was sampled at the frequency of 4 kHz.

Results

The *RMS* (root mean square) amplitude and the *MPF* (mean power frequency) determined on the basis of the power spectrum of the EMG signal were analysed. In order to eliminate influence of individual factors the *RMS* was expressed in % MVC (maximal voluntary contraction). As the example the mean values and standard deviations for *RMS* from *BB* muscle and *RA* muscle were presented (Figure 1).

Table 1 presents the effect of the amount of fatty tissue (*HF* and *LF* group) and the force on the values of the analysed parameters (*RMS* and *MPF*).

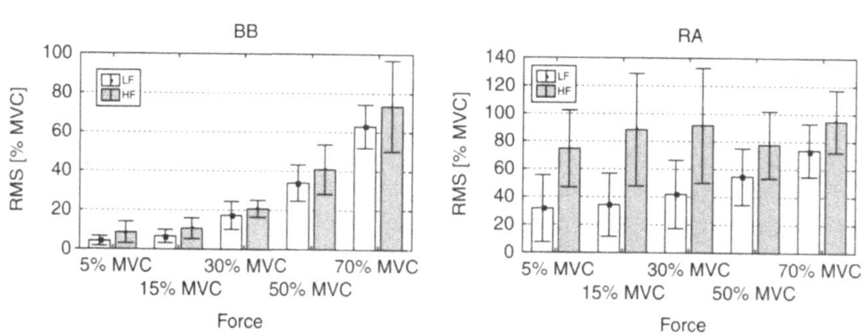

Figure 1. Mean values and standard deviations for *RMS* from *BB* muscle and *RA* muscle for *LF* and *HF* group during muscle tension at the 5 levels of load (5, 15, 30, 50 and 70% MVC).

Table 1. The effect of the amount of fatty tissue (*HF* and *LF* group) and the force on the values of the analysed parameters, obtained by ANOVA application.

		Fat layer		Force		Fat layer * Force	
Parameter	Muscle	F	*p*	F	*p*	F	*p*
RMS	*BB*	**10.86**	***0.0012***	**168.57**	***0.0001***	0.52	*0.7208*
	RA	**63.78**	***0.0001***	**4.61**	***0.0016***	2.04	*0.0923*
	RF	**7.84**	***0.0058***	**48.91**	***0.0001***	1.08	*0.3691*
MPF	*BB*	**19.17**	***0.0001***	0.78	*0.5399*	1.92	*0.111*
	RA	**30.51**	***0.0001***	**13.79**	***0.0001***	0.3	*0.8771*
	RF	0.77	*0.3815*	**3.57**	***0.0082***	0.18	*0.9463*

As shown on Figure 1 in muscle *RA* the *RMS* in *HF* group is similar in all forces, whereas in *LF* group the *RMS* increases with force, which indicate the influence of fatty tissue on values of EMG signal amplitude in this muscle.

Discussion and conclusion

The results of the analysis indicate, that both the amount of fatty tissue and the muscle contraction level influence the EMG signal parameters in all analysed muscles. The fat layer and load influence values of *RMS* parameter in muscles *BB*, *RF* and *RA*. However, for *MPF* parameter the influence of the fatty tissue and contraction level is ambiguous. Values of *MPF* parameter depend on fat layer in *BB* and *RA* muscles, while dependence on level of load in muscles *RA* and *RF* is visible. This can suggest that there is an influence of thickness of subcutaneous layer on EMG power spectrum. However, it can not be reflected by standard parameter like *MPF*.

More unambiguous results can be obtained by analysis based on more detailed parameters based on the power spectrum of the EMG signal. The results of this study suggest, that if the research group isn't homogeneous, the amount of fatty tissue should be taken into consideration during research with usage of EMG.

Acknowledgments

This study was supported by a grant Nr N N404 167134 from the Ministry of Science and Higher Education, Poland, Agreement nr 1671/B/P01/2008/34.

References

Farina, D. and Mesin, L. 2005, Sensitivity of surface EMG-based conduction velocity estimates to local tissue in-homogeneities-influence of the number of channels and inter-channel distance. *J Neurosci Methods*, 142(1), 83–9

Farina, D. and Rainoldi, A. 1999, Compensation of the effect of sub-cutaneous tissue layers on surface EMG: a simulation study. *Med Eng Phys*, 21(6–7), 487–97

Kuiken, T.A., Lowery, M.M. and Stoykov, N.S. 2003, The effect of subcutaneous fat on myoelectric signal amplitude and cross-talk. *Prosthet Orthot Int*, 27(1), 48–54

Nordander, C., Willner, J., Hansson, G.A., Larsson, B., Unge, J., Granquist, L. and Skerfving, S. 2003, Influence of the subcutaneous fat layer, as measured by ultrasound, skinfold calipers and BMI, on the EMG amplitude. *Eur J Appl Physiol*, 89(6), 514–9

Piscione, J. and Gamet, D. 2006, Effect of mechanical compression due to load carrying on shoulder muscle fatigue during sustained isometric arm abduction: an electromyographic study. *Eur J Appl Physiol*, 97, 573–81

Visser, B. and van Dieën, J. 2006, Pathophysiology of upper extremity muscle disorders. *J Electromyogr Kinesiol*, 16(1), 1–16

THE PHYSIOLOGICAL DEMANDS OF HURLING MATCH-PLAY

K. Collins[1], D.A. Doran[1] & T.P. Reilly[2]

[1]*Institute of Technology Tallaght, Ireland*
[2]*Liverpool John Moores University, UK*

Despite the participant and spectator appeal, research into the practice of hurling has lagged considerably behind that of other field games. The physiological demand of hurling has been hypothesised to resemble Gaelic football. The aim of the present study was to determine the work rate and physiological stress imposed during elite level competitive match-play. The results indicate that during match-play elite hurling players were subjected to a high physical load during the game. The aerobic load during the game is high, with episodes of high anaerobic activity. Some, corollary between current data and that reported for Gaelic football is apparent. Based on these findings, specific training regimens should be implemented that replicate the demands of math-play to help improve the fitness perpetration of elite players.

Introduction

Hurling is the national game of Ireland and one of the world's most dynamic field games (O' Donoghue *et al.*, 2004). Notwithstanding, the popularity of hurling in Ireland, research into physiological demands of hurling has lagged behind Gaelic football and other field games (Reilly and Collins, 2008). Few attempts to directly measure the work rate and physiological responses of hurling match-play have been made (Reilly and Doran, 2003; O' Donoghue *et al.*, 2004). Inferences as to the physiological demands and training requirements of hurling have been extrapolated from other field games particularly Gaelic football based on the obvious similarities that exist with regard to field dimensions and match duration (Reilly, 2000). Whilst these observations on work rate and physiological responses for Gaelic football may provide some analogy (Reilly and Doran 2001), they have been derived mainly from training sessions and non-competitive match-play; elements which may provide little insight into hurling and in particular elite level hurling match-play (Reilly and Collins, 2008).

In essence hurling constitutes a form of intermittent exercise in which the timing of high-intensity efforts are acyclical following the ebb and flow of the game (O' Donoghue *et al.*, 2004). During match-play a wide range of offensive and defensive skills are executed at high speed, play shifts rapidly from end-to-end due to the large distances the ball can be delivered (~100 m +), rapid accelerations and decelerations, changes of direction, unorthodox movement patterns make hurling match-play unlike other field games (Reilly and Collins, 2008). These patterns of play are likely contributors' to high levels of physiological strain and energy expenditure (Collins and Reilly, 2008). The demands of elite level inter-county hurling would be expected to be higher, than that reported for club level players by O' Donoghue *et al.*, (2004). Confirmatory motion analysis and physiological response to substantiate this does not currently exist (Reilly and Doran, 2001; Reilly and Collins, 2008). The purpose of the present study was to determine the work rate and physiological stress imposed during elite level competitive match-play. Understanding this will help inform suitable preparatory practices that reflect the demands of the game.

Methods

Work Rate Analysis

Time-motion analysis was conducted using commercially available software (Sports Code Pro, version 6.5, Sportstec, Australia) on eight elite level competitive matches, these included seven All-Ireland hurling finals, plus the game in which heart rate was recorded. The same researcher analysed the selected footage of games in their entirety. Each match was observed using the software and coded using a pre-defined structure, modified from Gilmore (2007).

Operational Definitions

- Puck out: The keeper or chosen player pucks (strikes) ball back into play after a wide or a score.
- Sideline Cut: The ball is struck back into play from the ground after exiting over the sideline boundary of play
- Free and penalties: A free or penalty is awarded when a player is fouled depending on the area they were fouled in.
- A 65 is a free strike 65m from the end line which is a warded when a player in the defence puts the ball wide at there own net.

The activity data were expressed by its duration in seconds. Time was defined as that elapsed since the beginning of each half of the match. The time of actual play and aggregate stoppage time was documented. The classification of stoppage was also recorded. Play was judged to start immediately the ball was pucked out, a side-line or free was taken, However if a free was taken and no purposive action followed due to a score or a wide, then this period of time was included in the previous recovery period. To ensure acceptable reliability an intra-observer test was calculated (Hughes *et al.*, 2002), reflecting an error percentages of less than 5%. The author (over 100 hours experience on the software system) viewed randomly selected 20-minute segments over a four-week period under the same conditions.

Physiological Responses to Match-play

Quantification of physiological strain during elite level competitive hurling was obtained on fifteen elite male adult hurlers (26 ± 3 years) during a competitive match. Heart rate was measured using the Polar Team System heart rate monitors (Polar Ltd, Kempele Finland). For each player the relative intensity of match-play was expressed as a percentage of age-predicted maximum heart rate utilising the equation 220-age. This value was then used to estimate the duration of time spent in aerobic and anaerobic work zones (Billows *et al.*, 2004). Heart rate was recorded at 5-s intervals throughout the game.

Statistics

The means and standard deviations of the designated variables were calculated. A paired t test was used to assess the significance of differences between the first and second halves with significance accepted at $P < 0.05$.

Results

The duration of a hurling game was 4530 ± 46 s, with a playing time of 1846 ± 81 s, representing 41% of the duration of the game. During the games there were 100 ± 5 activity periods with the ball in play for an average of 19 s. The ball was out of play for an average of 28 s. There was no significant difference between the activity profiles of the first and second half. The stoppages during a game were largely as a result of scores or attempted scores, reflecting 35 and 23% of the number of stoppages respectively. Foul play only contributed 27% of the total of stoppages during a game. The remainders of stoppages were the result of 65's (2%) and sidelines (13%). The findings are similar to the game in which heart rate responses were recorded.

The mean maximum heart rate (HR_{max}) was 194 ± 3 beats $\cdot$ min^{-1}. The heart rate range during the game was 100–197 beats $\cdot$ min^{-1}and the mean heart rate for the game was 83% of HR_{max}. The mean heart rate for the first and second half of the game was 163 ± 14 beats $\cdot$ min^{-1} and 160 ± 15 beats $\cdot$ min^{-1}. This corresponds to 84% and 82% of age related HR_{max} respectively. No significant difference was observed between the first and second halves. Fifty nine percent of total match time was spent working below 85 % HR_{max}. It is widely accepted that in adults the energy required to sustain work at heart rates below 85% of HRmax can be generated aerobically and above 85% the energy is generated anaerobically (Billows, Reilly and George, 2004). The relationship of aerobic and anaerobic work represents a 4:3 ratio.

Discussion

The purpose of this investigation was to determine the work rate patterns and physiological stress imposed during elite competitive hurling match-play. Traditionally, researchers using time-motion analysis attempt to gauge the work-load imposed during match-play by examining the temporal sequencing of either work to rest or high intensity to low intensity activities game related activities. The combination of both time-motion analysis and surrogates of physiological strain i.e. heart rate present a global index of the imposed physical demand (Reilly and Doran, 2001). Present data suggest that contested match-play occurred for 41% of the of the game duration, whereby there where ~100 activity periods with the ball in and out of play on average for 19 s and 28 s respectively. Concomitant with these periods of play, heart rates where in excess of 80% of HR_{max} indicating a sub-maximal but intense physiological load across the game. Super imposed on this sub-maximal activity were episodes of high intensity activity as reflected in the exercise intensity ratio analysis. Such findings imply a predominance of aerobic metabolism during match-play with superimposed periods of high-intensity activity where the phosphagen and glycolytic pathways are likely to be heavily recruited as previously inferred (Reilly, 2000). These findings clearly differentiate both the work rate and intensity of the elite game from that of the club level game as reported by O' Donoghue *et al.*, (2004). In the current model, elite play is characterized by more frequent bouts of activity, for a more sustained period of contact followed by shorter periods of recovery between episodes, with a higher associated physiological strain.

Some, corollary between current data and that reported for Gaelic football is apparent (Doggart *et al.*, 1993), supporting the belief that the physiological strain of the games may be comparable (Reilly, 2000; Reilly and Doran 2001; Reilly and Keane, 2001). Similarities also exist in the number of stoppages during match-play; however, the stoppages in hurling are largely attributable to scores and attempted scores, whereas stoppages in Gaelic football are largely attributed to foul play (Doggart *et al.*, 1992).

In conclusion, elite hurling match-play can be characterized as intermittent exercise with numerous brief bouts of high-intensity activity interspersed by low intensity recovery periods. Surrogates of physiological strain i.e. heart rate during match-play is relatively high. These data reflect the importance of sprint performance and the ability to recover between high intensity exercise periods. In addition, specific training regimens should be implemented that replicate the demands of match-play to improve the fitness preparation of elite players. The Yo-Yo Intermittent Recovery Test may provide a simple and valid way to obtain important information of an individual's capacity to perform repeated intense exercise and to examine changes in performance (Krustrup *et al.*, 2003). The test has not previously been utilized in Gaelic games research (Collins and Reilly, 2008). Further research is required to identify the individual movement patterns of the game, particularly to identify movement characteristics specific to positions. The information garnered could potentially support the design of physiological tests that specifically reflect the game of hurling and potentiate the development of suitable training regimen. The utilization of small-sided games following the ergonomic model of training as described by Reilly *et al.*, (2005) could be a potential focus.

References

Billow, D., Reilly, T. and George, K. 2004, Physiological demands of match-play on elite adolescent footballers. *Journal of Sports Sciences*, 22, 524–525

Doggart, L., Keane, S., Reilly, T. and Stanhope, J. 1993, A Task Analysis of Gaelic Football. In T. Reilly, J. Clarys and A. Stibbe (ed) *Science and Football II*, (E and FN Spon, London), 186–189

Hughes, M., Cooper, S.M. and Neville, A. 2002, Analysis procedures for nonparametric data from performance analysis. *International Journal of Performance Analysis in Sport*, 2, 6–20

Krustrup, P., Mohr, M., Amstrup, T., Rysgaard, T., Johansen, J., Steenberg, A., Pedersen, P.K. and Bangsbo, J. 2003, The yo-yo intermittent recovery test: physiological response, reliability, and validity. *Medicine and Science in Sports and Exercise*, 4, 697–705

O' Donoghue, P.G., Donnelly, O., Hughes, L. and McManus, S. 2004, Time-motion analysis of Gaelic games. *Journal of Sports Sciences*, 22, 255–256

Reilly, T. 2000, Endurance aspects of soccer and other field games. In R.J. Shephard and P.-O. Astrand (ed) *Endurance in Sport*, (Blackwell Science, London), 900–930

Reilly, T. 2005, An ergonomics model of the soccer training process. *Journal of Sports Sciences*, 23, 561–572

Reilly, T. and Collins, K. 2008, Science and Gaelic sports: Gaelic football and hurling. *European Journal of Sport Science*, 8, 231–240

Reilly, T. and Doran, D. 2001, Science and Gaelic football: a review. *Journal of Sports Sciences*, 19, 181–193

Reilly, T. and Keane, S. 2001, Estimation of physiological strain on Gaelic football players during match-play. In W. Spinks, A. Murphy and T. Reilly (ed) *Science and Football IV*, (E and FN Spon, London), 157–159

THE PHYSIOLOGICAL VALIDATION OF A LABORATORY-BASED SOCCER-SPECIFIC TRAINING SIMULATION ON A MOTORIZED TREADMILL

Tae-Seok Jeong, Tom P. Reilly, James P. Morton & Barry Drust

Research Institute for Sport and Exercises Sciences, Tom Reilly Building, Liverpool John Moores University, Liverpool, L3 3AF, UK

In order to investigate physiological adaptations to soccer training under controlled conditions, a lab-based training-specific simulation is necessary. Soccer-specific simulations available in the literature can not be used to simulate the physiological loads associated with soccer training.

The aims of this study were to devise and validate a laboratory-based soccer-specific training simulation (LSSTS) that replicates demands of professional players (n = 20) during a training session.

Work-rates observed in training were used to devise the LSSTS on a motorized treadmill. All major movements (walking, jogging, high speed running and sprinting) and static pauses were included. The durations of static pauses, walking, jogging, high speed running and sprinting were 14 s, 12 s, 6 s, 5 s and 3 s. The speeds of each movement were determined from the data obtained during match-play (Mohr *et al.*, 2003). The validity of the devised protocol was evaluated by comparing the physiological responses of the LSST with those recorded in the actual training session. Ten healthy active males (mean age, 27 ± 7 years) participated in this study. The physiological measurements such as heart rate (HR), % of maximal HR (HRmax) and ratings of perceived exertion (RPE) were included. Heart rate was recorded every 5 s using a short-range telemetry system (Polar Team System®, Kempele, Finland). RPE was also determined using a modified 10-point rating Borg scale after the completion of exercise. The time distribution in the 6 specific HR zones (50% ~ 100% of HRmax) was also compared between both conditions. Student T-test was used for analysis of independent samples.

The LSSTS replicated both the work-rate profiles and the durations of discrete activities observed in the actual training session. The mean HR and % of HRmax associated with the simulation (136 ± 10 beat · min-1, 71 ± 5% of HRmax) were similar to those obtained in training (137 ± 8 beat · min-1, 72 ± 3% of HRmax). Time spent in high intensity HR zones between 80% and 100% of HRmax during the simulation (24%) was also similar to that seen in training (22.4%). The mean RPE associated with the simulation

(6 ± 1) was significantly lower than that of the training session (7 ± 1) ($p < 0.05$).

These results suggests that the LSSTS successfully re-created a soccer-specific training session in the laboratory, and that such simulations could be used as a model in laboratory-based investigations to study the physiological responses to soccer-specific training exercise.

EFFECTS OF INTERACTION BETWEEN PHYSICAL AND MENTAL WORKLOAD ON HUMAN PERFORMANCE

A.M. Basahel, M.S. Young & M. Ajovalasit

Brunel University, Uxbridge, UK

Workload is an important factor that affects human performance in an operating system. This research study investigates the interaction of physical workload, mental workload and arousal level on performance and working memory. Fifteen participants (ages 25–35) performed a physical task (pedalling on a bicycle-ergometer) under three different levels of resistance load (low, mid and high) concurrently with a mental task (arithmetic) at three levels of demand (simple, intermediate and difficult). Participants also completed two subjective assessments following the task, Borg-CR10 and NASA-TLX to evaluate the perceived physical and mental workload respectively. The hypothesis of the study is that optimum performance occurs at intermediate levels of workload, whereas poor performance is observed at the low and high levels of physical and mental demand due to underload and overload.

Introduction

For decades, scientists have been studying the impact of task workload on operator performance. Moreover, workload is a significant factor which impacts human performance in an operating system, especially in complex systems and demanding tasks (Hwang *et al.*, 2007; Sluiter, 2006). Balancing task demands and attentional resources is essential in order to achieve acceptable performance (DiDomenico and Nussbaum, 2005).

Workload may be divided into two types: mental demands and physical demands (Hwang *et al.*, 2008). Physical workload may be defined as tasks which require that the operators' muscles work, with the participation of the musculoskeletal, cardiorespiratory and nervous systems (Sluiter, 2006). In contrast, mental workload is a multidimensional concept and it can be identified by operator characteristics (e.g., experiences, attention and skills) and task features such as task load and procedures (Young and Stanton, 2004). Many tasks in the workplace impose a physical workload (PWL), which in turn places loads on mental tasks and cognitive resources (Perry *et al.*, 2008). The human response to both mental and physical workload, as well as their interactions, leads to an increase in arousal level, which can be measured by a rise in both heart rate (HR) , blood pressure (BP) and rate pressure product (RPP) (Fredericks *et al.*, 2005).

Table 1. Nine conditions of PWL and MWL interaction in this experiment and their associated hypotheses.

	Mental Workload (MWL)		
Physical Workload (PWL)	Simple arithmetic task	Medium arithmetic load	Complex arithmetic task
Low PWL 20% VO_2 max)	Performance will decrease under this condition (underload).	×	Performance may be acceptable in this condition
Medium load (50% VO_2 max)	×	Acceptable performance	×
High load (80% VO_2 max)	High PWL may compensate for mental underload for acceptable performance	×	Performance will decrease under this conditions (overload).

This study will investigate the impact of combinations of physical and mental workload on individuals' performance. The purpose of this experimental study is to identify the optimum level of human performance in a multi-task demand (physical and mental) scenario. Moreover, there may also be an interaction between these factors such that higher levels of physical workload could compensate for mental underload effects, and vice versa.

Methods

The current experiment was designed to investigate the interaction effects of physical workload (PWL) and mental workload (MWL) under different conditions and arousal levels on working memory performance tasks as illustrated in Table 1. Participants were asked to conduct a mental arithmetic task concurrently with a physical task, with the dependent variable being performance on the arithmetic task. The study also sought to examine whether an interaction with high levels of physical workload can compensate for the performance shrinkage caused by mental underload.

The experimental hypotheses predict that performance will decrease under conditions of both high level of physical demand (i.e. cycling at 80% of VO2 max) and the complex mental arithmetic task, due to overload and high arousal levels. In addition, performance deteriorations will also occur under conditions of low physical workload (i.e. cycling at 20% of VO2 max) and low mental task demands (simple arithmetic task) due to underload and low arousal levels. Conversely, acceptable performance on the arithmetic task is expected at the intermediate level of physical and mental workload. Moreover, the study also seeks to determine if high/low combinations of physical and mental workload interact, such that (for instance) higher levels of physical workload can compensate for decrements in performance due to mental underload.

Materials and Participants

Fifteen healthy participants aged 25-35cycled on a bicycle ergometer to create the three different conditions of physical workload, while Mental-Math Software was used to perform the arithmetic problems at three levels of difficulty. Blood pressure (BP), heart rate (HR) and heart rate variability (HRV) were measured electronically before and during the tests. Additionally, this experiment recorded the rate pressure product (RPP) and brain blood flow in order to identify the physiological arousal level of the participants due to the workload interaction. Also, the NASA-TLX (Hart and Staveland, 1988) and Borg-CR10 (DiDomenico and Nussbaum, 2008) were completed after each condition to evaluate subjective mental and physical workload respectively.

Future Work and Conclusion

A similar study is planned to investigate the effects of the interaction between physical and mental workload using a spatial figures task instead of mental arithmetic, to investigate the potential interactions with different attentional resource pools (Wickens, 2002). Furthermore, whereas the current study aimed to separate pure physical and mental workload in a laboratory setting, future work is planned to investigate the overlaps between physical and mental workload in a real domain. In conclusion, task workload measurement is a necessary tool in the process of operating system design and diagnosis. This research study examined the effects of the combination of physical and mental demands on human performance in dual occupations. Although the results were not available at the time of writing, it is anticipated that the findings may be used to influence the design of work systems in applied domains such as factory production lines.

References

Didomenico, A. and Nussbaum, M.A. 2005, Interactive effects of mental and postural demands on subjective assessment of mental workload and postural stability, *Safety Science*, 43, 485–495

Didomenico A. and Nussbaum M. A. 2008, Interactive effects of physical and mental workload on subjective workload assessment. *International Journal of Industrial Ergonomics*, 38, 977–983

Fredericks, T. K., Choi, S.D., Hart, J., Butt, S.E. and Mital A. 2005, An investigation of myocardial aerobic capacity as a measure of both physical and cognitive workloads, *International Journal of Industrial Ergonomics*, 35, 1097–1107

Hart, S. G. and Staveland, L. E. 1988, Development of NASA-TLX (Task Load Index): results of empirical and theoretical research. In P. A. Hancock and N. Meshkati (Eds.), *Human Mental Workload*, (North-Holland: Amsterdam), 138–183

Hwang, S.L., Yau, Y.J., Lin, Y.T., Chen, J.H., Huang, T.H., Yenn, T.C. and Hsu, C.C. 2008, Predicting work performance in nuclear power plants, *Safety Science*, 46, 1115–1124

Perry, C.M., Sheik-Nainar, M.A., Segall, N., Ma R. and Kaber, D.B. 2008, Effects of physical workload on cognitive task performance and situation awareness, *Theoretical Issues in Ergonomics Science*, 9, 95–113

Sluiter, J.K. 2006, High-demand jobs:age-related diversity in work ability? *Applied Ergonomics*, 37, 429–440

Wickens, C. D. 2002, Multiple resources and performance prediction. *Theoretical Issues in Ergonomics Science*, 3, 159–177

Young, M.S. and Stanton, N.A. 2004, Mental Workload. In N. A. Stanton , A. Hedge, K. Brookhuis and E. Salas (eds.). *Handbook of Human Factors and Ergonomics Methods*, (Taylor and Francis Group, London), 39.1–39.7

CREATING A LEARNING ORGANISATION IN WITH OPEN PAPERS

EXPLORING 'SYSTEM RESILIENCE' IN THE CONTEXT OF A HOSPITAL BED CRISIS

Mike D. Williams

University of Exeter Business School

The safety of healthcare does not compare well with many other industries. Accident theory developed from high reliability organisations and normal accident theory provide some but limited insight into how to make the healthcare systems safer. The concept of 'resilience' is developed to take a 'system' perspective in a case study of a large NHS hospital in England during a period of high demand for beds. The ability of the hospital to anticipate, mitigate and recover from peaks in demand was studied as a means to explore 'system resilience' and its relation to patient safety.

Introduction

The safety record for patients admitted into hospital in developed countries indicates that around one in ten suffer some form of adverse event (Kohn *et al.*, 2000, de Vries et al., 2008, Department of Health, 2006). It is estimated that around fifty percent of events are avoidable and not all result in lasting harm. However, such a safety record does not stand up to comparison with other industries such as commercial airlines or nuclear power, which have been described as 'ultra-safe' (Amalberti *et al.*, 2005).

This paper presents the results from a case study of a large National Health Service (NHS) hospital in England. The hospital experienced a situation where the demand for inpatient services was greater than the capacity of the staff and physical bed spaces. The aim of the study was to explore the concept of 'system resilience' as a means to build on the patient safety literature from a systems perspective. 'System resilience' is defined as '. . . a work system's ability to buffer, adapt to, absorb and prevent adverse patient outcomes in the face of disruption' (Woods, 2006). 'System' is defined as a 'set of elements which together form a whole' (Checkland, 1981). A system consists of: 'elements, interconnections, and a function or purpose' (Meadows, 2008). In a hospital there are the elements of the staff, beds and other equipment that interrelate for the purpose of treating patients. A systems view looks at how patients 'flow' into and through the hospital, the 'stocks' (where patients are held in wards) and the 'feedback loops' between the interacting parts of the whole (Sterman, 2000).

Literature

The safety of any organisational undertaking is important. However, the safety imperative competes with the need to be productive (Reason, 1997). There is considerable literature examining how complex and potentially hazardous industries appear safer than might be expected, which is known as high reliability theory (HRT) (La Porte, 1996, Roberts, 1990, Schulman, 1996, Weick and Sutcliffe, 2001). The view that HRT can be applied to healthcare has been challenged due to the unreliability of basic health treatment processes (Resar, 2006) and the structural barriers that exists in most healthcare services in developed countries (Amalberti et al., 2005). Another view on how accidents occur in complex organisations suggests that failure is an inevitable consequence of the interactions of tightly coupled parts, which is known as 'normal accident theory' (NAT) (Perrow, 1984). This paper takes a NAT position but accepts that there are ways to organise to reduce the risk of safety errors (Perrow, 1999).

Tight coupling is where parts of the system are closely connected. A problem in one area can rapidly spread. Where the systems are complex with many feedback loops and time delays combined with tight coupling of the parts, then the risks of an accident are much higher due to the unforeseen interactions of the parts (Perrow, 1984, 1999).

In health care there is considerable pressure to maximize efficiency. This means that anything regarded as excess capacity or slack has often been removed. Without slack the system becomes more tightly coupled and the risk of error increases. The ability to absorb peaks in demand is decreased as slack (buffer) capacity (empty medical beds) is used up. Medical patients are commonly accommodated on surgical wards (known as 'medical outliers') with the consequent impact that the elective surgical service becomes more tightly coupled to the whole hospital system. From a conceptual perspective we could argue that in such situations the 'vulnerability' of the system to things going wrong increases. Vulnerability to risk has been described as being the other side of the resilience coin, whilst others see them as separate concepts (Manyena, 2006, Fikret, 2007).

Cook and Rasmussen (2005), use Rasmussen's safe working envelope model (Rasmussen, 1997) to examine safety problems for hospitals that are overcrowded They suggest that it is normal for healthcare systems to work at the limit of their capacity and for Hirschhorn's 'law of stretched systems' to apply: 'every system always operates at its capacity. As soon as there is some improvement, some new technology, we stretch it…' (Cook and Rasmussen, 2005). When that capacity limit is reached, failure or restructuring has to occur (Woods and Wreathrall, 2008). When systems are under resource and or performance pressures the benefits of change are taken in increased productivity or efficiency rather than improving any capacity for safety. Studies have shown an association between hospitals that have high occupancy rates and an increase in mortality and adverse events (Richardson, 2006, Sprivulis et al., 2006, Weissman, 2007).

Lessons drawn from industries that do not have high levels of standardization and control, such as water management systems, have focused on the concept of resilience rather than just the avoidance of risk (Schulman, 2004). Rasmussen (1997) argues that to understand safety in complex socio-technical systems, analysis has to be undertaken at multiple levels. Resilience in this context is about anticipating, mitigating and preparing to recover from unsafe events (Sheridan, 2008). Developments in resilience theory have broadened the approach to help leaders understand safety in the light of competing priorities at a system level (Cook and Rasmussen, 2005). However, resilience is a very broad concept with different meanings being applied (Smith and Fischbacher, 2009). Sheridan (2008) concludes that resilience, as a concept, has yet to be operationalised in a quantifiable manner. Woods (2006) sets out some broad characteristics related to the safe working envelope. These are the ability to absorb and or buffer, the flexibility of response to disruption, the margin or how close to the edge of the safety envelope the system operates and the tolerance to any disturbance. Meadows (2008), proposes that resilience arises from the ability of the feedback loops to balance each other to ensure the system survives and persists in a changing situation.

Method

A case study approach of an NHS hospital was used during the period October to February. The case study approach was chosen because it is suitable to investigate a phenomenon within their context (Yin, 2003) and where there is complexity in the subject matter (Stuart et al., 2002). It is also used where the boundaries between the phenomenon are unclear, where multiple sources of evidence are needed to converge, and where there is a theoretical background to guide the data collection and analysis (Yin, 2003). Case studies do suffer from the weakness of being context specific. However, Flyvbjerg (2006), argues that case studies provide the depth of knowledge needed in social situations. Yin (2003) states that findings can be generalised to a theoretical level.

Within the case study the unit of analysis is the whole hospital as a system. An embedded method (Yin, 2003), combined with a stratified purposeful approach (Miles and Huberman, 1994) was used to examine three within case units of analysis. One of the reasons behind selecting three embedded areas is to cover both the formal and informal actions that occur, to understand the macro context and then the managers and individual team members (micro) response (Yin, 2003). The units were:

Organisational: Trust Board and hospital wide operational processes
Sub unit: Division of Medicine
Team: Ward / Consultant Team / Department members

A 'system resilience' approach seeks to explore the dynamic nature of how a hospital anticipates, mitigates and responds to perturbations. The theoretical safe working envelope (Williams, 2008) was used to guide data collection. During the periods of high demand on the bed capacity with large numbers of medical outliers, semi-structured interviews of twenty eight doctors, nurses, managers and directors were conducted, recorded and transcribed. Non-participative observation of staff actions and meetings were undertaken. The semi-structured interview examined how the envelope boundaries of finance, staffing, targets and patient safety were monitored and managed. Thematic analysis of the concepts was undertaken. NVivo 8 was used to analyse the interview data to identify phenomena about the competing priorities and the actions by staff to prevent potential boundary failure (Miles and Huberman, 1994).

The Case Study Hospital

The study was conducted in a large NHS Foundation Trust Hospital with over 700 beds. The hospital provides both emergency and non-emergency services. There are a number of waiting time targets that require considerable management attention (DoH 2008a):

- A maximum of waiting time of one month from diagnosis to treatment for all cancers.
- Waiting time in Accident and Emergency Department (ED) – maximum of 4 hours for 98% of patients
- Waiting time from GP Referral To Treatment at hospital (RTT) – maximum of 18 weeks with a push to reduce to 15 then 13 weeks.

The hospital has consistently achieved all the waiting time targets including the 18 week RTT target eleven months ahead of schedule (DoH 2008b).

Findings

There were two points during the study where the hospital system 'transitioned' from a managed situation into a tightly coupled position and with a surge in medical patients having to be treated on surgical wards (medical outliers) and substantial management time spent making sense of what was happening and managing the flow of patients. The first was due to an outbreak of the Norovirus, an airborne sickness virus. The second transition was due to a period of increased emergency admissions for two weeks post Christmas. The observed points of 'transition' provided a window on the system to use the conceptual envelope (Williams, 2008) to study the response to the situation.

The hospital anticipated an outbreak of Norovirus during the winter. It has policies and processes in place to contain the outbreak. Wards can only be reopened after a thorough cleaning and despite pressure on the bed capacity the infection control standards were not compromised. So whilst the patient safety boundary had been breached (patients had been infected) there were robust processes to mitigate and recover the situation, which included closing the hospital to visitors for a short period. The staff had learnt from previous outbreaks that a full scale closure to visitors helps reduce the duration of the outbreak.

At the same time the staff workload boundary became threatened due to staff sickness. In anticipating the pressure on staff, a period of over recruitment had taken place to cover for potential vacancies ahead of time. Normal practice was to only recruit once a vacancy had occurred thus leaving a gap. Action had been taken to reduce the vulnerability of the system to staff shortages during the winter months. Senior nurses monitored the staffing situation daily and undertook informal risk assessments which informed how to deploy the staff available. The financial boundary was closely monitored and in a strong position and therefore, money was made available to hire agency staff when that was needed to cover for short term sickness.

The reduction in bed capacity, in the first transition, was up to ten wards (30%) that were not able to accept new admissions or transfers. The waiting time target boundary was measured and monitored very tightly at the middle and senior management level. The combination of the ED target for emergency patients with the elective RTT and cancer targets meant that there was little margin for the system to work with. Cancelling large numbers of elective patients would ease the pressure on the beds but would lead to a breach in the target failure boundary. During the Norovirus elective activity was maintained to avoid breaching the RTT target boundary.

> "So the throughput is phenomenal yet they haven't got any beds for these patients and yet they are having to start theatre lists without having you know a bed for a patient. So it's each day trying to work how we can best make, it's safe for the patients to go to theatre so the staff feel that they are looking after a full ward plus also a corridor of patients." (Surgical Nurse)

This decision not to cancel patients made the system more vulnerable to peaks in demand as it created a situation of tight coupling due to overcrowding. The nature of the ED target is that there is no buffer that can be put in place to absorb surges in demand. The ED is largely dependent on the rest of the hospital system to achieve the four hour target by providing empty beds when required. Emergency patients who have seen a GP prior to coming to the hospital go directly to the Emergency Medical Unit (EMU). During the weekdays EMU would run out of empty beds by midday. The GP patients were then diverted to the ED, creating a reinforcing feedback loop. To generate empty beds for the medical patients in ED and EMU, patients from medical wards were moved to surgical wards. The structure of the EMU was too small to absorb peaks in demand. This created a feedback loop into ED and the wider hospital. "At this time of year we are moving into the 55/60 every 24 hours, that's a large number coming through, we are a 31 bedded unit." (Nurse) The response of staff to the situation was, whilst not liking having medical outliers, was to accept the need to move patients, even in the middle of the night, to help avoid breaches in ED. "I don't know if you could demonstrate in any trial that they did less well than the people on a general medical ward yet it's a pain having to go out and look after them and it creates more work. . ." (Doctor). Due to the pace of transfers patients were sometimes moved without their clinical notes to a new team of doctors and nurses. This obviously had potentially significant implications for patient safety.

During the second transition event, learning from the Norovirus incident about how to increase bed capacity was used. This meant that additional inpatient bed capacity was made available more quickly through staffing day case beds overnight and putting up beds in a pre-admission area. However, it was not without problems: "…it wasn't an operational area it was pre assessment area so even now we are still you know we haven't got the right stores in place and the pharmacy's not quite right because of course it's not an established in-patient area." (Senior Nurse) The staffing of these areas proved difficult and meant moving staff from other medical wards thus depleting their levels of expertise. Again there was some flexibility with money to hire staff when they were available. However, apart from measuring infection rates, there was no measure of patient safety other than retrospective incident rates and complaints.

Given the rate of emergency admissions during transition two, managers agreed to cancel a number of elective patients. However, due to the number of admissions and the need to find empty beds, in the late afternoon of Friday 3rd January previous infection control advice was overridden and medical patients transferred onto the gynaecology ward, creating more medical outliers. The decision was to move six patients and contain them in a bay. However, the practical reality during the weekend of needing to find empty beds, meant that over twenty medical patients were placed on the gynaecology ward. This increased the risk of spreading infections from medical patients into the maternity unit, which shared the same doctors as the gynaecology ward.

Discussion

Using the concept of a safe working envelop has given us some insight into how a healthcare organisation monitors and manages the potential boundaries of failure. The findings confirm the literature (Reason, 1997, Smith, 2005) that finance and workload targets are easy to measure and become the focus for management attention. Staffing levels were constantly monitored and staff moved to cover gaps. Safety is less easy to define and measure until something goes wrong. Senior managers can assume that since nothing has gone wrong it is possible to make the structure (e.g. flex number of beds) and processes (admit patients before a bed is available) more efficient and therefore eat into the buffer capacity which may have historically maintained safety (Reason, 1997, Dekker, 2006), thus making the system less resilient to disruption.

The study shows that the buffer capacity and flexibility required for a resilient response was limited to the financial boundary (spent money on extra staff) and to some extent on the RTT target (cancelled some non urgent patients). The admission process became brittle (little flexible capacity; had to overflow) leading to tight coupling of medicine with ED, surgery and obstetrics (Woods, 2006; Cook and Rasmussen, 2005).

What was unknown to hospital staff was the impact on patient safety due to the increase in the number of medical outliers and the less than ideal ways of managing the admission process for elective cases. There was no measurement of the adverse event rate. There are no studies in the literature about the safety consequences for medical outliers. There appears to be an inbuilt assumption that they are safe. However, human factors research indicates that, staff working under increased pressure, or in unfamiliar situations, are more likely to make errors (Reason, 1990). We can also speculate that where there are large numbers of patient transfers, there will be more handovers of information between staff on wards and less continuity of care for the patients. Poor handover creates discontinuity leading to potential harm (Jeffcott et al, 2009).

Sterman (2000) argues that the behaviour of the system is governed by its structure. The structure consists of feedback loops, stocks and flows and the dynamic interactions between the decision making people and the institutional structure of the system. What the case study revealed was decision makers putting in place a balancing feedback loop to reduce the spread of the airborne Norovirus by dramatically closing wards and restricting access to only necessary staff. The reinforcing feedback loop generated by the interacting waiting time targets was much harder to balance in the short-term by either reducing demand or the average length of stay as the patient safety consequences were not as evident as they were with the Norovirus.

Resilience as a concept is not easy to quantify in measurable terms. From a systems perspective there are a number of observations that can be made. The nature of socio-technical systems is that they are complex with multiple interactions which do not easily lend themselves to quantification. Stretching systems to maximise their efficiency may increase their vulnerability at times of disruption (Reason, 1997, Dekker, 2006). The focus on what can be easily measured has to be balanced with an awareness of the system dynamics gained through mindfulness of what might go wrong (Weick & Sutcliffe, 2001). Safety is dynamic and cannot be assumed.

Conclusion

Hospitals face increasing complexity as the interactions between the parts of the whole escalate. With the ease of measuring finance and targets they receive greater attention from decision makers. The pressure to live within the budget has meant over the years, the drive for efficiency has reduced much of the buffer of empty beds and staff flexibility.

The implications of a highly efficient organisation from a patient safety perspective are little understood. What this case study has shown is that the hospital was vulnerable to peaks in demand which caused a transition from a manageable situation to one that became so tightly coupled that at points in the day the system became frozen. Whilst there is no direct evidence of harm resulting from high numbers of medical outliers, there was a drift towards greater vulnerability due to practical necessity (Snook, 2000).

A single case study has weaknesses and cannot lead to generalisation although it can provide ideas for theory development. Further case study work is planned during the winter of another hospital to further develop theory.

References

Amalberti, R., Auroy, Y., Berwick, D. & Barach, P. (2005) Five System Barriers to Achieving Ultrasafe Health Care. *Ann Intern Med*, 142, 756–764.

Checkland, P. (1981) *Systems Thinking, Systems Practice*, (Wiley, Chichester).

Cook, R. & Rasmussen, J. (2005) "Going solid": a model of system dynamics and consequences for patient safety. *Qual Saf Health Care*, 14, 130–134.

De Vries, E. N., Ramrattan, M. A., Smorenburg, S. M., et al. (2008) The incidence and nature of in-hospital adverse events: a systematic review. *Qual Saf Health Care*, 17, 216–223.

Dekker, S. (2006) *The Field Guide to Understanding Human Error*, (Ashgate, Aldershot).

Department of Health (2006) Safety First. (Stationary Office, London).

Department of Health (2008a) *Developing the NHS Performance Regime*. available at http://www.dh.gov.uk/en/Publicationsandstatistics/Publications/Publications PolicyAndGuidance/DH_085215 (accessed 25 November 2008.)

Department of Health (2008b) *18 Weeks monthly RTT collection* available at http://www.performance.doh.gov.uk/rtt/downloads/monthly/RTT_Admitted_Provider_Aug08.xls (accessed 24 November 2008.)

Fikret, B. (2007) Understanding uncertainty and reducing vulnerability: lessons from resilience thinking. *Natural Hazards*, 41, 283–295.

Flyvbjerg, B. (2006) Five Misunderstandings About Case-Study Research. *Qualitative Inquiry*, 12, 219–245.

Kohn, L. T., Corrigan, J. M. & Donaldson, M. S. (2000) *To Err is Human*, (Washington DC, National Academy Press).

La Porte, T. R. (1996) High Reliability Organizations: Unlikely, Demanding and At Risk. *Journal of Contingencies & Crisis Management*, 4, 60.

Manyena, S. B. (2006) The concept of resilience revisited. *Distasters*, 30, 433–450.

Meadows, D. H. (2008) *Thinking in Systems*, Chelsea Green Publishing, White River Junction, Vermont.

Miles, M. B. & Huberman, A. M. (1994) *Qualitative Data Analysis*, Thousand Oaks, Sage.

Perrow, C. (1984) *Normal Accidents*, (Basic Books, New York).

Perrow, C. (1999) Organizing to Reduce the Vulnerabilities of Complexity. *Journal of Contingencies and Crisis Management*, 7, 150–155.

Rasmussen, J. (1997) Risk management in a dynamic society: a modelling problem. *Safety Science*, 27, 183–214.

Reason, J. (1990) *Human Error*,(CUP, Cambridge).

Reason, J. (1997) *Managing the Risks of Organizational Accidents*, (Aldershot, Ashgate).

Resar, R. K. (2006) Making Noncatastrophic Health Care Processes Reliable: Learning to Walk before Running in Creating High-Reliability Organisations. *Health Services Research*, 41, 1677–1689.

Richardson, D. B. (2006) Increase in patient mortality at 10 days associated with emergency department overcrowding. *Medical Journal of Australia*, 184, 213–216.

Roberts, K. H. (1990) Some Characteristics of One Type of High Reliability Organization. *Organization Science*, 1, 160–176.

Schulman, P. R. (1996) Heroes, Organizations and High Reliability. *Journal of Contingencies & Crisis Management*, 4, 72–82.

Schulman, P. R. (2004) General attributes of safe organisations. *Qual Saf Health Care*, 13ii, ii39–ii44.

Sheridan, T. B. (2008) Risk, Human Error, and System Resilience: Fundamental Ideas. *Human Factors*, 50, 418–426.

Smith, D. & Fischbacher, M. (2009) The changing nature of risk and risk management: The challenge of borders, uncertainty and resilience. *Risk Management*, 11, 1–12.

Smith, P. (2005) Performance Measurement in Health Care: History, Challenges and Prospects. *Public Money & Managment*, August, 213–220.

Snook, S. A. (2000) *Friendly Fire*, Princeton University Press, Princeton.

Sprivulis, P., Da Silva, J.-A., Jacobs, I. G., et al. (2006) The association between hospital overcrowding and mortality among patients admitted via Western Australian emergency departments. *Medical Journal of Australia*, 184, 208–212.

Sterman, J. D. (2000) *Business Dynamics: Systems Thinking and Modeling for a Complex World*, Irwin McGraw-Hill, Boston.

Stuart, I., Mccutcheon, D., Handfield, R., Mclachlin, R. & Samson, D. (2002) Effective case research in operations management – a process perspective. *Journal of Operations Management*, 20, 419–433.

Weick, K. E. & Sutcliffe, K. M. (2001) *Managing the Unexpected*, Jossey Bass, San Francisco.

Weissman, J. (2007) Hospital Workload and Adverse Events. *Medical Care*, 45, 448–455.

Williams, M. D. (2008) Designing System Resilience in NHS Hospitals. *Ergonomics Society Improving Patient Safety Conference 16–18 July.* Cambridge.

Woods D. (2006) Essential Characteristics of Resilience. in: Hollnegal E, Woods D, Leveson N. (eds) *Resilience Engineering: concepts and precepts*. Aldershot: Ashgate.

Woods, D. D. & Wreathrall, J. (2008) Stress-strain Plots as a Basis for Assessing System Resilience. In Hollnagel, E., Nemeth, C. & Dekker, S. (Eds.) *Remaining Sensitive to the Possibility of Failure.* Ashgate, Aldershot.

Yin, R. K. (2003) *Case Study Research, Design and Methods*, London, Sage.

Presentations without abstracts

Correlation between the fingertip forces exerted during typing and MSD
N. Ratzon, Tel Aviv University

Examining the effectiveness of a structured intervention programme among hospital nurses
N. Ratzon, Tel Aviv University

Enhanced safe riding: Preventing accidents in motorcycle mail delivery work
C.H. Brace, ScottWilson Ltd

Intervention to reduce falls on steps
S. Thorpe, HSL

Introduction: the man and his legacy
T. Cable

Thermoregulation and occupational contexts
W. Gregson

Organisational integrity: Challenging the principle of ALARP
I. Hamilton, Technical Director Human Engineering Ltd

Creating a learning organisation in the healthcare context
M. Lyons, National Patient Safety Agency

Developing process safety performance indicators for major hazards industries
P. Dawson, Principal Specialist Inspector HSE

Organisational learning: Who, when and how?
L. Carver, BAE Systems

Improving the learning culture
R. Scaife, The Kiel Centre

Learning from audits
R. Miles, Team Leader Offshore Safety Division, HSE

Establishing communities of practice
T. Hawley, Associate Director Arup

Human factors investigation in the rail industry
E. Lowe, Senior Ergonomist Network Rail

WORKSHOP: The Assessment of Repetitive Tasks (ART) tool
J. Ferreira, HSL

Author index